Topics in Fluorescence Spectroscopy

Volume 3
Biochemical Applications

Topics in Fluorescence Spectroscopy

Edited by JOSEPH R. LAKOWICZ

Volume 1: Techniques
Volume 2: Principles
Volume 3: Biochemical Applications

Topics in Fluorescence Spectroscopy

Volume 3
Biochemical Applications

Edited by

JOSEPH R. LAKOWICZ

Center for Fluorescence Spectroscopy
Department of Biological Chemistry
University of Maryland School of Medicine
Baltimore, Maryland

SPRINGER SCIENCE+BUSINESS MEDIA, LLC

Library of Congress Cataloging in Publication Data

(Revised for volume 3)

Topics in fluorescence spectroscopy.

Includes bibliographical references and index.
Contents: v. 1. Techniques—v. 2. Principles—v. 3. Biochemical applications.
1. Fluorescence spectroscopy. I. Lakowicz, Joseph R.

QD96.F56T66	1991	543′.0858	91-32671

10 9 8 7 6 5 4 3

ISBN 978-1-4757-8772-6 ISBN 978-0-306-47059-2 (eBook)
DOI 10.1007/978-0-306-47059-2

Originally published by Plenum Press, New York in 1992
Softcover reprint of the hardcover 1st edition 1992

Contributors

S. Arnold • *M*icro*p*article *P*hoto*p*hysics *L*aboratory (MP^{3}L), Department of Physics, Polytechnic University, Brooklyn, New York 11201

Daniel Axelrod • Department of Physics and Biophysics Research Division, University of Michigan, Ann Arbor, Michigan 48109

A. P. Demchenko • A. V. Palladin Institute of Biochemistry of the Academy of Sciences, Kiev 252030, Ukraine

L. M. Folan • *M*icro*p*article *P*hoto*p*hysics *L*aboratory (MP^{3}L), Department of Physics, Polytechnic University, Brooklyn, New York 11201

Bryant S. Fujimoto • Department of Chemistry, University of Washington, Seattle, Washington 98195

Robert M. Fulbright • Department of Physics and Biophysics Research Division, University of Michigan, Ann Arbor, Michigan 48109

Edward H. Hellen • Department of Physics and Biophysics Research Division, University of Michigan, Ann Arbor, Michigan 48109

William R. Laws • Department of Biochemistry, Mount Sinai School of Medicine, New York, New York 10029

Thomas M. Li • Development Department, Syva, Palo Alto, California 94304

Richard F. Parrish • Development Department, Syva, Palo Alto, California 94304

J. B. Alexander Ross • Department of Biochemistry, Mount Sinai School of Medicine, New York, New York 10029

Kenneth W. Rousslang • Department of Chemistry, University of Puget Sound, Tacoma, Washington 98416

J. Michael Schurr • Department of Chemistry, University of Washington, Seattle, Washington 98195

Lu Song • Department of Chemistry, University of Washington, Seattle, Washington 98195

Christopher D. Stubbs • Department of Pathology and Cell Biology, Thomas Jefferson University, Philadelphia, Pennsylvania 19107

Jane M. Vanderkooi • Department of Biochemistry and Biophysics, School of Medicine, University of Pennsylvania, Philadelphia, Pennsylvania 19104

Brian Wesley Williams • Department of Chemistry, Bucknell University, Lewisburg, Pennsylvania 17837

Pengguang Wu • Department of Chemistry, University of Washington, Seattle, Washington 98195

Herman R. Wyssbrod • Department of Chemistry, University of Louisville, Louisville, Kentucky 40292

Preface

Fluorescence spectroscopy and its applications to the physical and life sciences have evolved rapidly during the past decade. The increased interest in fluorescence appears to be due to advances in time resolution, methods of data analysis and improved instrumentation. With these advances, it is now practical to perform time-resolved measurements with enough resolution to compare the results with the structural and dynamic features of macromolecules, to probe the structures of proteins, membranes, and nucleic acids, and to acquire two-dimensional microscopic images of chemical or protein distributions in cell cultures. Advances in laser and detector technology have also resulted in renewed interest in fluorescence for clinical and analytical chemistry.

Because of these numerous developments and the rapid appearance of new methods, it has become difficult to remain current on the science of fluorescence and its many applications. Consequently, I have asked the experts in particular areas of fluorescence to summarize their knowledge and the current state of the art. This has resulted in the initial three volumes of *Topics in Fluorescence Spectroscopy*, which is intended to be an ongoing series which summarizes, in one location, the vast literature on fluorescence spectroscopy.

These first three volumes are designed to serve as an advanced text. These volumes describe the more recent techniques and technologies (Volume 1), the principles governing fluorescence and the experimental observables (Volume 2), and applications in biochemistry and biophysics (Volume 3).

Additional volumes will be published as warranted by further advances in this field. I welcome your suggestions for future topics or volumes, offers to contribute chapters on specific topics, or comments on the present volumes.

Finally, I thank all the authors for their patience with the delays incurred in release of the first three volumes.

Joseph R. Lakowicz

Baltimore, Maryland

Contents

1

Tyrosine Fluorescence and Phosphorescence from Proteins and Polypeptides

J. B. Alexander Ross, William R. Laws, Kenneth W. Rousslang, and Herman R. Wyssbrod

1.1. Historical Perspective and Background

The fluorescence and phosphorescence of proteins and polypeptides is the sum of the contributions from the three aromatic amino acids tryptophan, tyrosine, and phenylalanine. The work on protein and polypeptide luminescence prior to 1971 has been reviewed in detail by Longworth.[1] Another fine account of the early work, emphasizing tryptophan and tyrosine, is the monograph by Konev.[2] An excellent review on tyrosine fluorescence in proteins and model peptides, for the period up to 1975, is given by Cowgill.[3] In 1984, Creed[4] reviewed the photophysics and photochemistry of tyrosine and its simple derivatives, including a thorough coverage of steady-state fluorescence and a brief discussion of triplet-state properties, but did not include any work on proteins or polypeptides.

The first quantitative studies of the excited-state properties of the three aromatic amino acids were carried out in the 1950s. The low-temperature phosphorescence of the aromatic amino acids was initially observed by Debye and Edwards[5] in 1952, and phosphorescence emission spectra were reported by Steele and Szent-Gyorgyi[6] in 1957. In 1953, Weber[7] postulated that the fluorescence of the aromatic amino acids should occur in the near-ultraviolet region of the electromagnetic spectrum. In 1956, independently and almost simultaneously, Duggan and Udenfriend[8] and Shore and Pardee[9] reported the results of their investigations of protein fluorescence. At the same time,

J. B. Alexander Ross and William R. Laws • Department of Biochemistry, Mount Sinai School of Medicine, New York, New York 10029. *Kenneth W. Rousslang* • Department of Chemistry, University of Puget Sound, Tacoma, Washington 98416. *Herman R. Wyssbrod* • Department of Chemistry, University of Louisville, Louisville, Kentucky 40292.

Topics in Fluorescence Spectroscopy, Volume 3: Biochemical Applications, edited by Joseph R. Lakowicz. Plenum Press, New York, 1992.

Konev[10] and Vladimirov[11] were carrying out similar work in the Soviet Union. In 1957, Teale and Weber[12] reported the first careful, thorough investigation of the fluorescence excitation and emission spectra of the aromatic amino acids.

While this chapter emphasizes the more recent work on the fluorescence and phosphorescence of tyrosine in proteins and polypeptides, salient earlier results will also be discussed to provide necessary background information. With the continuing improvements in instrumentation and data analysis, the ability to investigate the excited states of tyrosine has considerably improved. As a result, a more detailed understanding of the photophysics of tyrosine in proteins is now becoming possible. We will first review the absorption properties of tyrosine. This will be followed by a description of the first excited singlet state and the lowest excited triplet state. Next, quenching mechanisms in polypeptides and proteins will be discussed, followed by examples from the literature using tyrosine emission as a biological probe. We conclude with a discussion of the potential formation of tyrosinate in proteins.

1.2. The Absorption Properties of Tyrosine

The aromatic amino acids each have two major absorption bands in the wavelength region between 200 and 300 nm (see reviews by Beaven and Holiday[13] and Wetlaufer[14]). The lower energy band occurs near 280 nm for tryptophan, 277 nm for tyrosine, and 258 nm for phenylalanine, and the extinction coefficients at these wavelengths are in the ratio 27:7:1.[14] As a result of the spectral distributions and relative extinction coefficients of the aromatic amino acids, tryptophan generally dominates the absorption, fluorescence, and phosphorescence spectra of proteins that also contain either of the other two aromatic amino acids.

A theoretical interpretation of the ultraviolet absorption transitions of the tyrosine phenol ring has been made by Hooker and Schellman.[15] Figure 1.1 summarizes, in Platt's notation,[16] the orientations of the electronic transitions. Based on the analysis of Hooker and Schellman,[15] the lowest energy singlet transition of tyrosine is due to the ^1L_b band, which has a maximum near 277 nm, and the much stronger ^1L_a band is near 223 nm. Since the absorption transitions to the ^1L_a and ^1L_b states are $\pi^* \leftarrow \pi$, hydrogen bonding is expected to lead to a shift in the absorption spectrum to lower energies (red shift).[17,18]

Chignell and Gratzer[19] have investigated the relative contributions of hydrogen bonding and solvation to the absorption shifts of both the ^1L_a and ^1L_b bands of *p*-cresol, a tyrosine model. Their results show that hydrogen bonding shifts the absorption bands to the red as well as increases their extinction. Moreover, the degree of these perturbations depends upon the

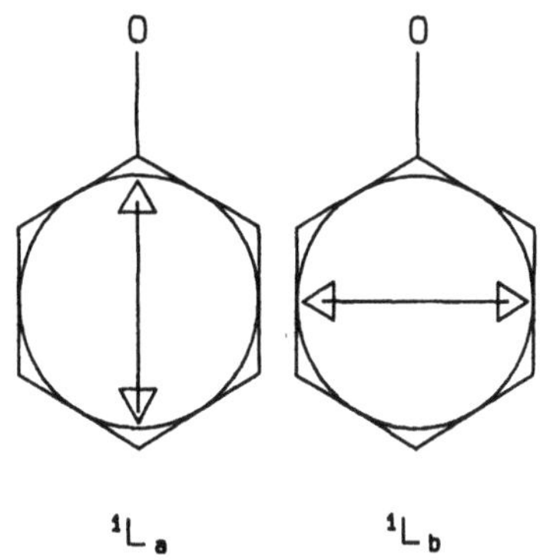

Figure 1.1. Singlet–singlet electronic absorption transition moment orientations of the phenol chromophore in tyrosine.

mutual strength of the complex between the phenol chromophore and the polar solvent. Nagakura and Gouterman[20] have shown that the red shifts of hydrogen-bonded aromatic alcohols correlate roughly with the strength of the solvent as a hydrogen bond acceptor. Chignell and Gratzer[19] also pointed out that the phenolic hydroxyl group can act as both a proton acceptor and donor in a hydrogen bond. In low-dielectric polar and nonpolar solvents, the degree of charge transfer in the hydrogen bond depends on solvent polarity and polarizability.[21] In water, aromatic alcohols are hydrogen bonded; if stronger proton acceptors are present, other hydrogen bonds are formed that cause a further shift in the absorption spectrum.[22]

Ionization of the phenol hydroxyl group in tyrosine shifts the 277-nm absorption peak to 294 nm and the 223-nm peak to 240 nm. The molar extinction coefficient for the peak of the lower energy band increases from about 1350 M^{-1} cm^{-1} to about 2350 M^{-1} cm^{-1} and for the higher energy band from about 8200 M^{-1} cm^{-1} to about 11,000 M^{-1} cm^{-1}.[13,14] In addition, the lower energy absorption band of tyrosine shows vibrational structure that is lost upon ionization of the phenol side chain.

1.3. The Excited Singlet and Triplet States of Tyrosine and Tyrosinate

Fluorescence and phosphorescence originate from the lowest lying singlet state and triplet state, respectively. The important interactions and relationships of the singlet and triplet states of tyrosine are shown in Figure 1.2, an energy level diagram. As shown in Figure 1.3, the fluorescence emission spectrum of tyrosine in an aqueous environment is a single, unstructured band with a maximum near 303 nm,[11] while the phosphorescence spectrum has weakly resolved vibronic structure with a maximum intensity near 395 nm.[23,24] Ionization of the phenol hydroxyl group, to form tyrosinate, causes a large red shift of the fluorescence spectrum to near 340 nm[25]; the phosphorescence spectrum is shifted less to the red, to near 408 nm,

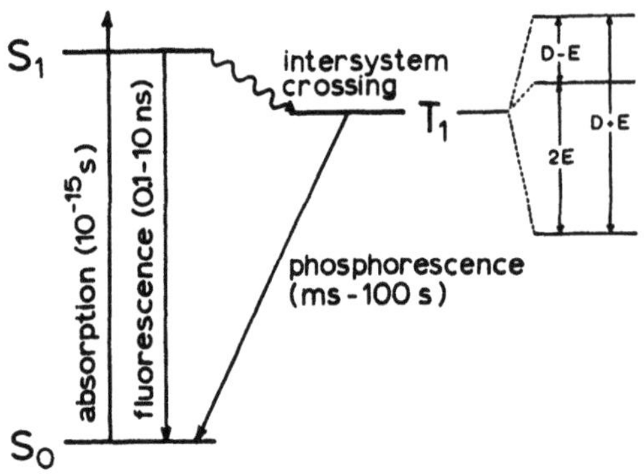

Figure 1.2. Jablonski energy level diagram showing the singlet state and the triplet state with its zero-field splittings for a planar aromatic chromophore.

becomes a slightly broader band, and essentially has no resolved vibrational structure.[23, 24]

The tendency of phenolic hydroxyl groups to ionize depends upon whether the aromatic system is in the singlet ground state, the first excited singlet state, or the first excited triplet state. In the singlet ground state, the phenolic hydroxyl group of tyrosine has a pK_a near 10, while in the first excited singlet state the pK_a^* has been calculated to be between 4 and 5.[25] These values can be compared with those for the well-studied molecule

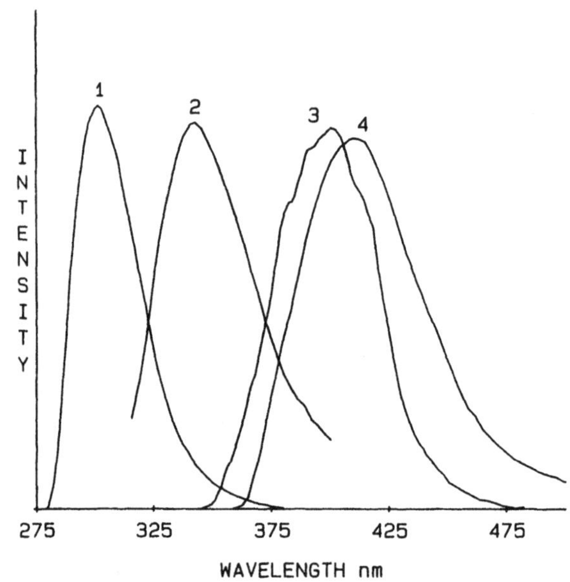

Figure 1.3. Tyrosine and tyrosinate fluorescence and phosphorescence emission spectra with 280-nm excitation. Curve 1: tyrosine fluorescence in water, pH 7. Curve 2: tyrosinate fluorescence in 0.01 *M* NaOH.[25] Curve 3: tyrosine phosphorescence in 50% (v/v) ethylene glycol/water.[24] Curve 4: tyrosinate phosphorescence in 50% (v/v) ethylene glycol/0.01 *M* NAOH.[24].

2-naphthol, which has a pK_a of 9.5 and a pK_a^* of 2.8.[26] This difference between the ionization potential of the first singlet excited state of aromatic alcohols and that of the singlet ground state can lead to ionization during the lifetime of the excited state, that is, excited-state proton transfer, if a suitable acceptor molecule is present. The pK_a^* of the first excited triplet state of aromatic hydrocarbons with ionizable groups, however, is essentially the same as that of the singlet ground state,[27] and excited-state proton transfer is not expected to occur. Consequently, it is conceivable that tyrosine emission in proteins may be quenched by excited-state proton transfer to, for example, aspartate or glutamate side chains; this would affect the fluorescence and phosphorescence intensities through nonradiative depletion of the first excited singlet state. Since the photochemistry occurs in the singlet manifold and does not involve spin–orbit coupling, its effect would be reflected in the decay kinetics for tyrosine fluorescence but not for triplet-state phosphorescence. However, as discussed in Sections 1.4.4 and 1.5.2, excited-state proton transfer to form tyrosinate is unlikely, even in the presence of a strong proton acceptor such as a carboxylate side chain in a protein.

1.3.1. The Zero-Field Splittings of the Triplet State

In the singlet state, the total spin of the electrons is zero while in the triplet state the total spin is one. Whereas a singlet state is diamagnetic and has only one level, the triplet state is paramagnetic with three distinct "sublevels," as shown in the energy level diagram of Figure 1.2. In planar, aromatic hydrocarbons that contain heteroatoms or have side-chain substitutions, the triplet sublevels are nondegenerate at zero external magnetic field, with the zero-field splitting (zfs) arising from the magnetic dipole–dipole interaction of the unpaired electron spins.[28] While the electronic transitions for the fluorescence and phosphorescence of tyrosine involve energies in the near-ultraviolet region of the electromagnetic spectrum (see Figure 1.3), the energies for the zero-field transitions within the triplet state are in the microwave region.

The three triplet sublevels are defined in terms of two independent parameters, D and E,[28, 29] which relate the energies of the zfs (see Figure 1.2). D is related to the inverse cube of the distance between the two triplet-state electrons, and E is related in a very complex manner to distances within the plane of the aromatic ring. Both D and E are affected by the symmetry of the aromatic system. E is zero for molecules with a C_3 or higher principal axis of symmetry. While the magnitudes of D and E reflect spin–spin interactions that occur in the aromatic ring plane, D is additionally influenced by interactions that occur out of the ring plane. For tyrosine, the values of D and E are greater than zero, and D is greater than E; the maximum zfs is referred to as

$D+E$, and the other two smaller zfs are $D-E$ and $2E$. The energies of the zfs are related by:

$$[D-E]+[2E]=[D+E] \tag{1.1}$$

The triplet-state splittings of tyrosine were first observed by electron paramagnetic resonance (EPR) more than two decades ago.[30–32] The initial characterization of the splittings was limited to a measurement of D^*, a root-mean-square zfs defined by

$$D^*=(D^2+3E^2)^{1/2} \tag{1.2}$$

D^* was determined by measuring the $\Delta m_s = \pm 2$ EPR transition, where m_s is the spin quantum number. Several years later, D and E were obtained directly by observation of the $\Delta m_s = \pm 1$ EPR signals.[33] In 1973, Zuclich *et al.*[34] redetermined the triplet-state splitting parameters of tyrosine using optically detected magnetic resonance (ODMR) spectroscopy.

Whereas EPR requires an external magnetic field, ODMR can be carried out at zero magnetic field.[35] In ODMR, the zfs can be observed directly from the change in the phosphorescence intensity that occurs when the spin populations of two of the triplet sublevels are perturbed by microwave frequencies corresponding to their zfs. At very low temperatures (i.e., below 4 K), spin–lattice relaxation (SLR) between the triplet sublevels becomes sufficiently slow that the spin population of each of the three levels is determined largely by its intersystem crossing rate from the first excited singlet state and its decay rate to the singlet ground state. At higher temperatures, SLR tends to equalize the spin populations. The SLR rates near 1.3 K have been measured for both tyrosine[36] and tyrosinate.[37] The ODMR and EPR values for the zfs of tyrosine and tyrosinate are compared in Table 1.1. The

Table 1.1. Zero-Field Splitting Parameters for Tyrosine and Tyrosinate[a]

	ODMR			EPR		
Compound	D	E	D^*	D	E	D^*
Tyrosine[b]	0.139	0.053	0.166			
Tyrosine[c]	0.130	0.056	(0.162)[d]			
Tyrosine[e]				0.130	0.056	0.162
Tyrosinate[b]	0.118	0.056	0.153			
Tyrosinate[e]				0.108	0.054	0.143

[a] Values given in cm^{-1}
[b] From Ref. 24.
[c] From Ref. 34.
[d] Our calculation.
[e] From Ref. 33.

major effect on the zfs due to phenol ionization is a decrease in the value of D; E is essentially unaffected.

1.3.2. Excited-State Decay Kinetics

Spontaneous emission from an excited state may be described kinetically, in many cases, as a first-order rate process. For example, simple planar aromatic compounds that do not interact in any way with the surrounding matrix will generally exhibit a single-exponential fluorescence decay. If SLR is efficient in the triplet state, the total phosphorescence decay from the triplet sublevels will be described by a single exponential. If SLR is slow but only one sublevel is radiative, the decay will also be a single exponential. Simple decay kinetics, however, are the exception for the fluorescence of the aromatic amino acids in proteins and peptides. Complex decay kinetics can result from either ground-state or excited-state interactions and reactions since most molecules do interact with their surrounding matrix. For example, if the pH of an aqueous solvent is near the ground-state pK_a of an ionizable group of the chromophore, two different ground-state chemical species will be present. The relative concentrations of each ground-state species will be determined by the pK_a of the solute molecule and the pH of the solution. As shown in Figure 1.4, Laws *et al.*[38] demonstrated that the fluorescence decay of 3-(*p*-hydroxyphenyl)propionic acid (PPA) upon titration of the carboxyl group is well described by two pH-independent lifetimes with relative weights (amplitudes) that vary in accord with the Henderson–Hasselbach relationship:

$$\log([A^-]/[HA]) = \mathrm{pH} - \mathrm{p}K_a \tag{1.3}$$

This behavior is what would be expected for a two-state, ground-state ionization. The pK_a obtained from the pH at which the amplitudes associated with the two decay constants are equal compares well with that of alkyl carboxyl groups of similar compounds.[39, 40]

In the case of phenols, both excited-state and ground-state ionization must be considered if complex decay behavior is observed. Gauduchon and Wahl[41] and Laws *et al.*[38] have examined the fluorescence decay of phenol and straight-chain phenol derivatives in water at varying pH. In agreement with Rayner *et al.*,[25] Laws *et al.*[38] found that excited-state proton transfer to water is too slow to affect the decay kinetics. This conclusion was based on three observations. First, the fluorescence decays of phenol and several straight-chain phenol analogues are single-exponential in water unless the pH is near the pK_a of an ionizable group, as in PPA (Figure 1.4). Second, tyrosine fluorescence has a constant quantum yield as a function of pH

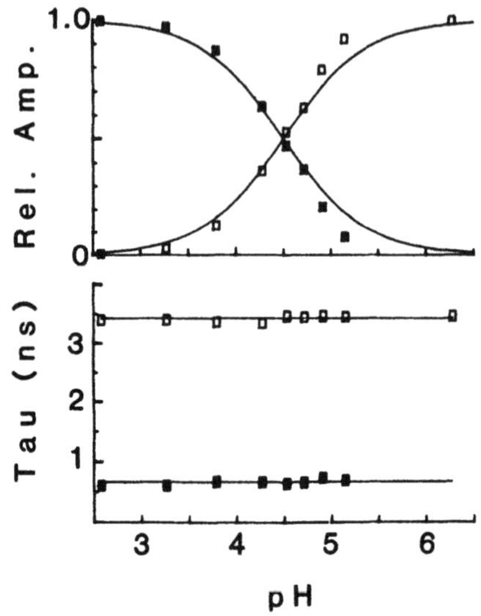

Figure 1.4. Fluorescence decay parameters of 3-(*p*-hydroxyphenyl)propionic acid as a function of pH. The filled squares indicate the association between the shorter lifetime and its amplitude; the open squares that between the longer lifetime and its amplitude. The solid lines running through the relative amplitudes express the theoretical Henderson–Hasselbach relationship for a pK_a of 4.5. The solid lines connecting the lifetimes represent their average value. (Reprinted from Ref. 38 with the permission of the American Chemical Society.)

through the region of the pK_a^*; that is, the quenching that would be expected due to excited-state proton transfer is not observed. Third, the complex kinetic behavior of the time dependence of tyrosine fluorescence as a function of pH, which initially might be ascribed to either excited-state proton transfer or titration of an ionizable group, is also seen in *O*-methyltyrosine, which does not have this proton to exchange, and in analogues without an ionizable group.[38] Consequently, there is no experimental evidence for excited-state proton transfer by tyrosine to water.

The pK_a^* of tyrosine explains the absence of measurable excited-state proton transfer in water. The pK_a^* is the negative logarithm of the ratio of the deprotonation and the bimolecular reprotonation rates. Since reprotonation is diffusion-controlled, this rate will be the same for tyrosine and 2-naphthol. The difference of nearly two in their respective pK_a^* values means that the excited-state deprotonation rate of tyrosine is nearly two orders of magnitude slower than that of 2-naphthol.[26] This means that the rate of excited-state proton transfer by tyrosine to water is on the order of $10^5\ s^{-1}$. With a fluorescence lifetime near 3 ns for tyrosine, the combined rates for radiative and nonradiative processes approach $10^9\ s^{-1}$. Thus, the proton transfer reaction is too slow to compete effectively with the other deactivation pathways.

The fluorescence decay parameters of tyrosine and several tyrosine analogues at neutral pH are listed in Table 1.2. Tyrosine zwitterion and analogues with an ionized α-carboxyl group exhibit monoexponential decay kinetics. Conversion of the α-carboxyl group to the corresponding amide results in a fluorescence intensity decay that requires at least a double exponential to fit the data. While not shown in Table 1.2, protonation of the carboxyl group also results in complex decay kinetics.[(38)]

Gauduchon and Wahl[(41)] suggested that the complex kinetics could be explained in terms of the rotamer populations resulting from rotation about the $C^{\alpha}-C^{\beta}$ bond, as diagramed in Figure 1.5. They proposed that the shorter, subnanosecond time constant, observed for analogues with an amide group, was due to quenching of the phenol ring in rotamer III by contact with the carbonyl group and that the longer time constant was the average of the decays of rotamers I and II. This differential quenching of rotamers had support based on the suggestions by Cowgill[(42)] that the peptide carbonyl, or an amide group, is responsible for quenching tyrosine fluorescence in proteins and by Tournon *et al.*[(43)] that the carbonyl of protonated carboxyl groups could quench aromatics efficiently by a charge transfer mechanism.

This rotamer model for the fluorescence decay of an aromatic amino acid also predicts that the amplitudes of the kinetic components should correspond to the ground-state rotamer populations, provided that interconversion

Table 1.2. Fluorescence Decay Parameters for Tyrosine and Tyrosine Analogues at Neutral pH[a]

Compound	pH	α_1[a]	τ_1 (ns)	α_2[a]	τ_2 (ns)
Tyrosine[b]	6.14	1.0	3.76		
Tyrosine[c]	5.2	1.0	3.38		
Tyrosine[d]	6.0	1.0	3.27		
O-Methyltyrosine[b]	6.20	1.0	5.06		
N-Acetyltyrosine[b]	5.24	1.0	3.60		
N-Acetyltyrosine[c]	6.88	1.0	3.2		
Glycyltyrosine[b]	6.46	1.0	1.65		
Glycyltyrosine[c]	5.5	0.95	1.25	0.05	2.1
Tyrosinamide[b]	6.38	0.32	0.54	0.68	1.91
Tyrosinamide[c]	5.5	0.41	0.4	0.59	1.2
N-Acetyltyrosinamide[b]	6.09	0.14	0.93	0.86	2.22
N-Acetyltyrosinamide[d]	7.0	0.35	0.11	0.65	1.66
Tyrosylglycine[b]	6.44	0.40	0.79	0.60	2.42
Tyrosylglycine[c]	5.5	0.46	0.50	0.54	1.6

[a] The αs (amplitudes) have been normalized to a sum of 1.
[b] From Ref. 38; 5°C, 284-nm excitation, 302-nm emission.
[c] From Ref. 41; 20°C, 270-nm excitation, 315-nm emission.
[d] From Ref. 171; 20°C, 286-nm excitation, 300-nm emission.

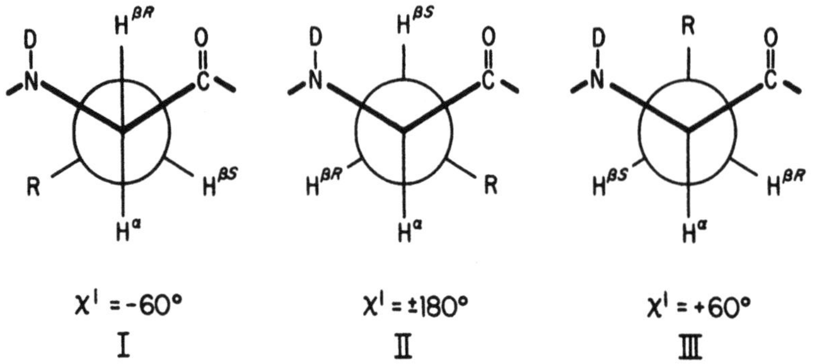

Figure 1.5. Newman projections about the $C^{\alpha}-C^{\beta}$ bond, showing the possible positions of the phenol group (R) in relation to the C^{α} substituents. The rotamer assignments I, II, and III and the corresponding values of the torsion angle χ_1 are indicated for tyrosine (and its derivatives) of the L configuration.

between rotamers is slow compared to the lifetime of the excited state and that the rotamers have equivalent extinction coefficients. Based on the NMR data available in the literature for free tyrosine and phenylalanine, Gauduchon and Wahl[41] noted that rotamer III is not favored. However, in their data the relative amplitude of the shorter lived component was much larger than would be expected on the basis of the NMR-determined populations reported in the literature for rotamer III. They therefore concluded that the rotational rate about the $C^{\alpha}-C^{\beta}$ bond is similar to the deactivation rate of the excited state. Consequently, an excited-state reaction is implied, with the amplitudes being kinetically derived and not corresponding to the ground-state rotamer populations. Furthermore, in those cases in which single-exponential decays were observed, they suggested that the rate of exchange between rotamer conformers is faster than the deactivation rate, yielding an averaged environment.

Laws *et al.*[38] have elaborated on this rotamer model for the fluorescence decay of tyrosine and tyrosine analogues. This refinement made use of two advances in technology. First, instead of using a deuterium flashlamp operating at 10 kHz, like Gauduchon and Wahl,[41] Laws *et al.* used synchrotron radiation for excitation. Synchrotrons have the advantage of having more intensity, a broad continuum of excitation energies, and higher repetition rates[44]; these features obviously make data collection much easier. Also, the synchrotron used for their studies has a narrower pulse width than a typical flashlamp, permitting better resolution of the decay constants. Second, global data analysis methods had become available to discriminate between various kinetic models. The principle of global data analysis, which has been developed and applied by Brand and co-workers to both the phase-modulation[45] and pulsed[46] methods of time-resolved fluorometry, involves the simultaneous analysis of several data sets, collected as a function of an independent variable such as wavelength, pH, or temperature. The analysis assumes that one or more parameters are common to all the data sets based on a kinetic model employing the independent variable. With a simultaneous analysis of multiple data sets, the parameters that are common to all the data sets are greatly overdetermined, there is a reduction in the total number of variables to be iterated, and inconsistencies in a particular model are easier to detect.

Since the slow-exchange rotamer model predicts that the fluorescence decay should be described by the sum of three exponentials, requiring the determination of six parameters, and since many other decay mechanisms could be adequately fit by six parameters, Laws *et al.* also chose to restrict the analysis of their tyrosine decay data.[38] If the ground-state rotamer populations are known, and if rotamer interconversion is slow compared to the lifetime of the excited state, then the normalized amplitudes should equal the rotamer populations. Consequently, two amplitudes can be made dependent on the third during the analysis of the decay data by linking them through proton NMR-determined rotamer populations for each compound. Thus, by using this linked-function approach,[47] not only is the number of iterated parameters reduced, from six to four in the present example, but the parameter search space is also highly restricted. Consequently, a complex model can be given a more critical test using just a single decay curve through the incorporation of independent information in the data analysis.

By the use of global and linked-function analyses, Laws *et al.*[38] were able to show that the rotamer model can explain the complex fluorescence decays seen for the tyrosine analogues examined. Although the decays often could be statistically fit for fewer than three exponentials, a unique solution for three exponentials could be found, possessing equal statistical parameters, provided that the rotamer populations were linked to the amplitudes. Furthermore, in all but one compound studied, the unique solution resulted in the amplitude associated with the shortest lifetime being correlated with the population of rotamer II. From these results, it appears that rotamer interconversion about the $C^{\alpha}-C^{\beta}$ bond in tyrosine is slower than the lifetime of the excited state. For those tyrosine compounds exhibiting a single-exponential decay, they were unable to establish whether (1) the slow-exchange rotamer model is the accurate description but the three rotamers have similar, unresolvable fluorescence lifetimes; or (2) rotamer interconversion is fast, averaging the emission. This rotamer model has been used to explain acrylamide quenching of tyrosinamide.[48] According to this analysis, the different environments for the three rotamers that result in distinguishable fluorescence lifetimes can also affect the kinetics of collisional quenching.

The phosphorescence decays of phenol, tyrosine, and related compounds, which had been examined extensively during the 1960s, have been reviewed by

Becker.[49] The lifetimes were all reported as being single-exponential. Rousslang and his collaborators have recently reexamined a number of these compounds at pH 3 and 5.[50] In general, the phosphorescence decays are biexponential, but are dominated by a longer lived component of about 3 s which comprises 98% or more of the decay.

1.4. Quenching Mechanisms of Tyrosine Emission in Polypeptides and Proteins

Quenching of tyrosine residues in polypeptides and proteins can occur by a number of different mechanisms. The local environment of a tyrosine residue in a protein or peptide, including its position (distance) and orientation relative to nearby quenching side chains, will govern which mechanisms are important. Using quenching mechanisms as a classification scheme, Cowgill[3] has separated tyrosine residues in proteins into the eight types listed in Table 1.3. These eight types can be grouped into four broader categories as defined by interactions with a specific moiety or quenching by a specific mechanism. These categories include quenching mechanisms involving (1) the peptide bond and specifically the carbonyl group; (2) resonance energy transfer, which can readily occur from tyrosine to tryptophan or tyrosinate; (3) disulfide bridges or sulfhydryl groups; and (4) amino acid side chains, which can act as proton donors or acceptors or as partners in hydrogen bond formation.

1.4.1. The Peptide Bond

Cowgill pointed out that there are essentially two distinct quenching processes of tyrosine fluorescence resulting from association with the peptide bond.[3] Tyrosines affected by these mechanisms are classified in Table 1.3 as

Table 1.3. Tyrosine Quenching Interactions[a]

Type	Interaction/quencher
I	Exposed and not quenched
II	Exposed and quenched by a hydrated peptide carbonyl group
III	Quenched by a disulfide group
IV	Quenched by tyrosinate
V	Buried and hydrogen bonded to a peptide carbonyl group
VI	Quenched by collisions with ionized carboxyls
VII	Quenched by types V and VI acting as energy sinks
VIII	Quenched by tryptophan

[a] After Cowgill.[3]

types II and V. The first mechanism operates in an aqueous environment, while the second is more effective in a nonaqueous environment. The aqueous mechanism was demonstrated from comparison of the enhancement of the fluorescence yield resulting from the transfer from water to dioxane of the model compound *p*-cresol versus the tripeptide glycyltyrosylglycinamide; whereas *p*-cresol fluorescence was enhanced by a factor of two, that of the tripeptide was enhanced 12-fold. This fluorescence enhancement is not unique to dioxane. The quenching process in water is intramolecular, requires a hydrated carbonyl group, has spatial requirements,[3] and probably occurs via a charge transfer mechanism[43] which does not involve a dissociable proton.

Supporting evidence for a charge transfer mechanism for the quenching of phenols by the carbonyl group is seen in the titration behavior of the model compound PPA (Figure 1.4). Since the protonated carboxyl group should be a better electron acceptor than the ionized carboxylate, PPA at a pH below the pK_a of the carboxyl group would be expected to have a shorter fluorescence lifetime than at a pH above the pK_a. As shown by Laws *et al.*,[38] the shorter fluorescence lifetime was associated with the protonated carboxyl group while the longer lifetime was associated with the ionized carboxylate. In addition, the results of Laws *et al.*,[38] supporting the rotamer model for tyrosine analogues, showed that the rotamer in which the phenol ring can come in closest contact with the carbonyl group had the shortest fluorescence lifetime.

The nonaqueous mechanism involves hydrogen bond formation between the phenolic hydroxyl and an amide carbonyl group in a nonpolar environment. This ground-state, hydrogen-bonded complex between phenols and amide carbonyls has been shown by Cowgill[3,42] to be nonfluorescent. Furthermore, it was shown that the association constant for this complex in a nonpolar, non-hydrogen-bond-forming solvent, such as hexane, is much larger than in hydrogen-bond-forming solvents, such as an alcohol. Moreover, anisole, which lacks the titratable hydroxyl proton, is not quenched by amide carbonyls in nonpolar solvents. Thus, in contrast to the aqueous process, the nonaqueous process depends upon a dissociable proton.

1.4.2. Singlet–Singlet and Triplet–Triplet Resonance Energy Transfer

Resonance energy transfer between the aromatic amino acids proceeds by very weak coupling between the donor and acceptor.[51,52] Very weak coupling implies that the interaction between the donor and acceptor wave functions is small enough so as not to perturb measurably the individual molecular spectra. This transfer process, which is distinct from the trivial process of absorption of an emitted photon, involves radiationless deexcitation of an excited-state donor molecule with concomitant excitation of a ground-state acceptor molecule. Resonance energy transfer depends upon three factors: the distance between the donor and acceptor moieties, their relative orientation, and the overlap of the donor emission and acceptor absorption spectra. If the energy transfer is between singlet donors and acceptors, both the coulomb interaction and the exchange interaction resulting from the overlap of the wave functions can be important. In general, singlet–singlet energy transfer is essentially due to dipole–dipole coupling when donor–acceptor distances are greater than 10 Å, and the transfer rate constant, k_t, varies as the inverse sixth power of the distance, R, separating the two centers of the donor and acceptor dipoles. Accordingly,

$$(R_0/R)^6 = k_t \tau_D \tag{1.4}$$

where R_0 is the characteristic distance at which the excitation energy of the donor is transferred with 50% efficiency, and τ_D is the donor excited-state lifetime, which includes all radiative and nonradiative deexcitation processes in the absence of energy transfer. R_0 can be expressed in terms of the product of the donor–acceptor orientation factor, κ^2, the donor quantum yield, ϕ_D, the inverse fourth power of the refractive index, n, and the spectral overlap integral, J (cm^3 mmol^{-1}), by the equation[51, 52]

$$R_0 = 9.8 \times 10^3 (\kappa^2 \phi_D n^{-4} J)^{1/6} \text{ (Å)} \tag{1.5}$$

With the exception of the orientation factor, all the parameters in this equation may be obtained within reasonable error by direct experimental measurement or by estimation. The problem of setting reasonable values for κ^2, which may vary from 0 to 4 for orientations in which the dipole moments are orthogonal or parallel, respectively, is nontrivial. A value of $\frac{2}{3}$, which is an unweighted average over all orientations, is often used. Dale *et al.*[53] have examined this problem in great detail and have shown that a κ^2 value of $\frac{2}{3}$ is never justified for energy transfer in macromolecules because it is impossible for the donors and acceptors to achieve a truly isotropic distribution. They do provide an experimental approach, using polarized emission spectroscopy, to estimate the relative freedom of motion for the donor and acceptor that allows reasonable limits to be set for κ^2.

At distances less than ~10 Å between donors and acceptors, triplet–triplet energy transfer becomes important. Triplet–triplet energy transfer between excited-state triplet donors (D*) and ground-state singlet acceptors (A) proceeds according to the reaction

$$\text{D*(triplet)} + \text{A(singlet)} \rightarrow \text{D(singlet)} + \text{A*(triplet)}. \tag{1.6}$$

As a consequence of the quantum-mechanical selection rules for resonance energy transfer involving the spin wave functions of the donor and acceptor, the coulomb interaction vanishes and the transfer rate is determined only by the exchange interaction. The triplet–triplet energy transfer probability is difficult to assess because the distance dependence is a function of the donor and acceptor wave function overlaps, and these are obtained only by calculation. While closed-shell calculations made of ground states can be correlated reasonably well with experimentally obtained physical parameters, open-shell calculations for excited states are not so well correlated.

1.4.2.1. Singlet–Singlet Energy Transfer in Peptides

Eisinger *et al.*[54] examined intramolecular singlet–singlet energy transfer between all combinations of donor and acceptor pairs of tyrosine, tyrosinate, phenylalanine, and tryptophan. With tyrosine as the donor, the most probable acceptor is another tyrosine, tyrosinate, or tryptophan. With tyrosinate as a donor, the most probable acceptor is tyrosinate, although in principle a tryptophan residue buried in the interior of a protein could be an acceptor, since its absorption could then have a significant overlap with the fluorescence emission band of tyrosinate. Because the absorption and fluorescence spectra of the aromatic amino acids are temperature-dependent, the density of isoenergetic states for the donor and acceptor residues is temperature-dependent, affecting the probability of energy transfer. For example, according to Eisinger *et al.*,[54] the spectral overlap between tryptophan absorption and tyrosinate fluorescence (in a 1:1 mixture of ethylene glycol and water) is sufficiently low at 300 K that singlet–singlet energy transfer is predominantly due to the exchange interaction. At 80 K, however, as a result of an increase in the spectral overlap, the dipole–dipole interaction becomes the dominant energy transfer mechanism.

Singlet–singlet energy transfer with tyrosine as a donor or an acceptor has been applied to solution conformation studies of several polypeptides. In a landmark paper on energy transfer, Eisinger[55] compared intraresidue distances in adrenocorticotropin, calculated using a random coil model of the hormone structure, to distances obtained from resonance energy transfer measurements. Based on the distances calculated from the efficiency of tyrosine to tryptophan resonance energy transfer, the random coil model was not adequate to describe the conformation of the hormone.

Eisinger[55] also noted that it is difficult to obtain accurate data with phenylalanine as the donor and either tryptophan or tyrosine as the acceptor. The source of this problem is the weak $S_1 \leftarrow S_0$ absorption of phenylalanine compared to that of tyrosine or tryptophan, which leads to considerable experimental uncertainty in measuring the sensitized acceptor emission. This error may account for the finding of Kupryszewska *et al.*[56] that the sensitization of the acceptor fluorescence was less than the quenching of the donor fluorescence in their study of phenylalanine-to-tyrosine energy transfer in leucine- and methionine-enkephalin. Kupryszewska *et al.*,[56] however, did note that sensitization of the acceptor fluorescence yielded results which were closer in accord with theoretical distance estimates. In general, donor quenching is not an accurate way to estimate energy transfer since other quenching interactions can be introduced by the presence of the acceptor.

1.4.2.2. Triplet–Triplet Energy Transfer in Peptides

From consideration of the relative energy levels, triplet–triplet energy transfer can, in principle, occur from tyrosine to tyrosine, from tyrosinate to tyrosinate, from tyrosine to tyrosinate, as well as from either tyrosine or tyrosinate to tryptophan. With the exception of tyrosinate to tryptophan triplet energy transfer, the sensitization of acceptor phosphorescence could be explained either by singlet–singlet energy transfer followed by intersystem crossing of the acceptor or by a direct triplet exchange. Direct excitation of the donor triplet state from its singlet ground state provides a direct way to distinguish between these two processes.

The most direct demonstration of triplet–triplet energy transfer between the aromatic amino acids is the ODMR study by Rousslang and Kwiram on the tryptophanyl-tyrosinate dipeptide.[57] Since the first excited singlet state of tyrosinate is at lower energy than that of tryptophan, it is possible to excite tyrosinate preferentially. The phosphorescence of this dipeptide, however, is characteristic of tryptophan, which is consistent with the observation that the triplet state of tyrosinate is at higher energy than that of tryptophan, making tryptophan the expected triplet acceptor.

Since triplet–triplet energy transfer involves exchange between three donor and three acceptor spin levels, and the probability of the exchange depends upon the projection of the donor spin levels upon those of the acceptor,[58] the occurrence of the transfer will be reflected by the change in the relative triplet sublevel spin populations of the acceptor compared to those in the absence of donor. Thus, the transfer will affect the net spin polarization of the acceptor, with negligible effect on the radiative rate constants. The change in spin polarization is reflected by the strength and signs (increase or decrease) of the acceptor ODMR spectra.[59] In this way, it is possible to identify triplet–triplet energy transfer when the transition to the donor excited singlet state is of higher energy than the transition to the acceptor excited singlet state.

For the tryptophanyl-tyrosinate dipeptide, where the singlet excitation energy of the donor is lower than that of the acceptor, there is no ambiguity that the change in spin polarization of the acceptor is the result of triplet–triplet energy transfer alone since there is no direct excitation of the acceptor singlet manifold. By saturating a zero-field transition of the tyrosinate triplet state with microwave radiation, Rousslang and Kwiram[57] directly affected the phosphorescence intensity of tryptophan, monitoring in an emission wavelength region where the phosphorescence of tyrosinate is negligible (i.e., 520 nm; see Figure 1.3). Thus, this ODMR experiment is a direct demonstration of triplet–triplet energy transfer.

1.4.3. Disulfide Bonds and Sulfhydryl Groups

Quenching of tyrosine fluorescence by the single sulfur of methionine is inefficient,[60] with essentially no effect on the tyrosine quantum yield in model systems.[3] By contrast, the disulfide bridge of cystine has long been implicated in the fluorescence quenching of aromatic residues in polypeptides and proteins.[1, 3] The sulfhydryl group of cysteine also quenches tyrosine in model systems, but to a lesser extent.[3] Considerable effort has been made by a number of different research groups toward understanding the mechanism of disulfide quenching. While no conclusive picture has emerged, several important observations have been made regarding the physical properties of the disulfide bridge.

Cystine, in addition to tryptophan, tyrosine, and phenylalanine, absorbs in the near-ultraviolet region of the electromagnetic spectrum. This disulfide absorption band is thought to be a forbidden $\sigma^* \leftarrow n^*$ transition,[61, 62] which is reflected by its low oscillator strength; at 300 nm, the molar extinction coefficient is less than $50\ M^{-1}\ \mathrm{cm}^{-1}$.[13] This weak, near-ultraviolet absorption band of the disulfide bond overlaps the fluorescence emission bands of tryptophan, tyrosine, and phenylalanine. The observation that both tyrosine-sensitized photolysis and direct photolysis of dithioglycolic acid yield similar products led Shafferman and Stein[63] to propose energy transfer to the disulfide as a mechanism for the quenching of tyrosine fluorescence. If one considers the possibility that disulfide quenching proceeds via energy transfer, it is by definition a dynamic process. However, it should be recognized that it can be difficult to distinguish a highly efficient energy transfer from a static quenching process.

Cowgill[64] has cogently argued that the quenching of aromatic residues in proteins by disulfide bonds is neither collisional nor mediated through bonds. He has discounted energy transfer, collisional quenching, and hydrogen bond formation since the spectral overlap integral is small, bimolecular quenching with model compounds is not observed, and the disulfide group is a poor electron donor, respectively. As an alternative, he suggested that the sulfur-containing group facilitates deactivation of the aromatic excited state by increasing the coupling to vibrational levels of the sulfur group.

Another possibility is deactivation through increased intersystem crossing,[1] which would occur via enhanced spin–orbit coupling.[28] If intersystem crossing is enhanced, then the phosphorescence quantum yield of a tyrosine residue could increase, unless the disulfide interaction also promotes efficient deactivation of the triplet state. Whether the phosphorescence quantum yield increases or decreases, the phosphorescence lifetime will always decrease as a result of the mixing of the singlet-state and triplet-state wave functions during spin–orbit coupling, whether or not the triplet state is also deactivated nonradiatively. Although the decrease in the phosphorescence lifetime due to spin–orbit interaction is not easy to predict, in the case of spin–orbit coupling due to interaction with heavy atoms, such as iodide or bromide, the decrease is generally more than an order of magnitude.[28] The phosphorescence quantum yield of the model compound L-cystinyl-bis-L-tyrosine (Cys-Tyr Cys-Tyr) is much smaller than that of *N*-acetyltyrosinamide, but the phosphorescence lifetime of the tetrapeptide is only decreased by a factor of about two. This led Longworth[1] to argue that the role of the disulfide bridge is to facilitate internal conversion rather than to promote intersystem crossing. From the above discussion concerning the effects of spin–orbit coupling, however, it is difficult to exclude the possibility that disulfide quenching of tyrosine is in part a result of enhanced intersystem crossing since spin–orbit coupling can also enhance nonradiative decay.

The importance of comparing time-dependent and steady-state fluorescence measurements is well illustrated by the difficulty of resolving purely static from purely dynamic quenching. In either case, the basic relationship between the steady-state fluorescence intensity and quencher concentration is the same. The Stern–Volmer relationship[65] for static quenching due to formation of an intermolecular complex is

$$F_0/F = 1 + K_a[\mathrm{Q}] \tag{1.7}$$

where F_0 is the fluorescence intensity in the absence of quencher, F is the fluorescence intensity at a particular concentration of quencher, [Q], and K_a is the equilibrium association constant for complexation of the fluorophore with the quencher to form the dark complex. The corresponding relationship for dynamic quenching is

$$F_0/F = 1 + k_q\tau_0[\mathrm{Q}] \tag{1.8}$$

where k_q is the bimolecular collisional quenching constant, and τ_0 is the fluorescence lifetime in the absence of added quencher. We note that τ_0, as used here, should not be confused with the natural radiative lifetime, but is the fluorescence lifetime determined in the absence of quencher and includes all other nonradiative processes.

It has been pointed out by Eftink and Ghiron[60] that a general expression for the combination of static and dynamic quenching is

$$F_0/F = (1 + k_q\tau_0[\mathrm{Q}])e^{V[\mathrm{Q}]} \tag{1.9}$$

where V is a constant, representing an active volume with a radius slightly larger than the van der Waals contact distance between the quencher and the fluorophore. A typical reaction radius for values of V in the range of 1–3 M^{-1} is a distance of about 10 Å. It has been noted that, at low quencher concentrations, $e^{V[Q]}$ is approximately equal to $1 + V[Q]$, which has the same mathematical form (and dimensions) as the expression for static quenching (Eq. 1.7) with V and K_a interchangeable.[60, 66]

Other corrections, besides those for static interactions, are important for certain quenchers. For example, acrylamide quenching is often used to help determine the relative solvent accessibility of aromatic residue side chains. In addition to a correction for static quenching,[60, 66] acrylamide quenching data for tyrosine residues require both primary and secondary inner filter corrections since acrylamide absorbs both 280- and 305-nm light.[67]

By comparing time-resolved and steady-state fluorescence parameters, Ross *et al.*[68] have shown that in oxytocin, a lactation and uterine contraction hormone in mammals, the internal disulfide bridge quenches the fluorescence of the single tyrosine by a static mechanism. The quenching complex was attributed to an interaction between one $C^{\alpha}-C^{\beta}$ tyrosine rotamer and the disulfide bond. Swadesh *et al.*[69] have studied the dithiothreitol quenching of the six tyrosine residues in ribonuclease A. They carefully examined the steady-state criteria that are useful for distinguishing pure static from pure dynamic quenching by consideration of the Smoluchowski equation[70] for the diffusion-controlled bimolecular rate constant k_0,

$$k_0 = (4\pi NR/1000)(D_f + D_q) \tag{1.10}$$

where N is Avogadro's number, R is the distance of closest approach between the fluorophore and the quencher (sum of the molecular radii), and D_f and D_q are the diffusion coefficients of the fluorophore and the quencher, respectively. The diffusion-controlled bimolecular rate constant k_0, when multiplied by a constant for the efficiency of quenching, yields the bimolecular collisional quenching constant k_q [see Eq. (1.8)]. Approximating the diffusion coefficient by the Stokes–Einstein equation,

$$D = kT/6\pi\eta r \tag{1.11}$$

where k is Boltzmann's constant, T is the absolute temperature, η is the viscosity of the solvent, and r is the radius of an equivalent sphere for the molecule, it is clear that the molecular diffusion coefficients are proportional to the ratio T/η. Consequently, the Stern–Volmer quenching constant, $K_{sv} = k_q\tau_0$, is linearly proportional to T/η. Swadesh *et al.*[69] convincingly argue that in the limit of $T/\eta \to 0$, K_{sv} will approach zero. This dependence on viscosity and temperature can be taken as a hallmark of pure dynamic quenching. On the other hand, a static mechanism involving formation of a dark complex should exhibit only a $1/T$ temperature dependence for the logarithm of the equilibrium association constant. Using these two tests, Swadesh *et al.*[69] concluded that the disulfide quenching interaction was primarily static in nature. From the dithiothreitol quenching of the six tyrosines in native bovine pancreatic ribonuclease, the reduced and S-methylated form of the enzyme, and *N*-acetyltyrosine-*N*′-methylamide, and using 1 *M* as the standard state at 298 *K*, the average thermodynamic parameters obtained for complex formation were $\Delta G^0 = -3.3$ kcal/mol, $\Delta H^0 = -1.1$ kcal/mol, and $\Delta S^0 = 7.4$ eu. From the magnitude and positive sign of ΔS^0, it was suggested that hydrophobicity could be important in stabilizing the tyrosyl–disulfide complex. In addition, the negative ΔH^0 was interpreted in terms of polar interactions, which could also make a favorable energy contribution toward formation of the complex. Thus, the complexation reaction is slightly exothermic. If formation of the complex were due to hydrophobic interactions alone, then ΔH^0 would be positive.

1.4.4. Interactions with Ionizable Side Chains and Proton Acceptors

Cowgill, in his 1976 review,[3] notes that the protonated, charged forms of arginine, lysine, and histidine do not have an appreciable effect on tyrosine fluorescence. In their uncharged, basic forms, however, these amino acids can act as proton acceptors, and neutral primary amino groups are known to quench tyrosine fluorescence. In addition, Cowgill[3] has obtained data that show that the imidazole side chain is an effective quencher in its uncharged, basic form. This is an important consideration in proteins and polypeptides since the pK_a of histidine is physiologically relevant. It should be noted that other ionizable groups with pK_a values usually outside of the physiological range can be perturbed by their local environment and thus exhibit dramatically shifted pK_as which become physiologically important.

Longworth[1] and Cowgill[3] have reviewed the mechanism of tyrosine quenching by carboxylate side chains; they describe this process as largely due to collisional interactions. On the surface of proteins, this process is thought to involve transient hydrogen bonds; obviously, water will compete for formation of hydrogen bonds. Under these circumstances, aspartate or glutamate side chains would have to be very close to the phenol ring to be effective. On the basis of studies of tyrosine copolymers, Longworth[1] concluded that carboxylate quenching does not proceed via excited-state proton transfer, even though it has been observed that ionization of the phenol hydroxyl group leads to a dramatic reduction in the fluorescence quantum yield. Nevertheless, there are numerous reports of tyrosinate emission from both proteins and polypeptides, and proton transfer to acceptor side chains has been considered as a possible quenching mechanism for tyrosine (see Section 1.5.2).

Since excited-state proton transfer and a collisional interaction are both dynamic quenching mechanisms, and since a titratable proton is required for this deactivation of the excited state, we consider the two mechanisms to represent the same physical event in the case of an aromatic alcohol–proton acceptor quenching interaction. The important issue is distinguishing among three possible situations: (1) motion (diffusion) of a proton acceptor relative to the excited singlet-state alcohol, as the proton donor, resulting in an interaction (collision) that leads to excited-state proton transfer; (2) excited-state proton transfer via diffusion of the proton without movement of the proton donor or acceptor; and (3) a sufficiently close encounter leading to a ground-state hydrogen bond between the aromatic hydroxyl and the acceptor. Situation 1 could occur by either the normal, random dynamic motions of the protein, an induced structural perturbation during the reorientation about the excited-state dipole of tyrosine, or, in the case of external proton acceptors, a classical Stern–Volmer/Smoluchowski collision/diffusion process. Situation 2 could arise if the excited tyrosine residue and its local environment, which includes a neighboring acceptor group, are conformationally highly restricted. This generates an ion pair since the charges do not physically separate.

The hydrogen-bonded complex, listed above as situation 3, has some interesting outcomes upon singlet excitation of tyrosine. The hydrogen-bonded complex could still exist in the excited state. An example of this situation occurs in the binding of equilenin and 17β-dihydroequilenin, equine estrogens which have an aromatic structure similar to that of 2-naphthol, to the sex steroid-binding protein of human and rabbit sera.[71, 72] The result of formation of a ground-state hydrogen bond that is maintained in the excited state is a red shift of the estrogen excitation and emission spectra. The same magnitude of red shifts can be modeled by the interaction of 2-naphthol with triethylamine, a strong uncharged proton acceptor, in low-dielectric solvents. The red-shifted emission maximum, however, in these model systems depends on the degree of charge transfer, which is a function of the strength of the acceptor and the polarity and polarizability of the solvent.[21, 73, 74] Similar spectral shifts are observed for aqueous tyrosine in the presence of high buffer base concentrations.[22] In this case, the emission maximum is a function of the pK_a of the acceptor.

1.5. Emission from Polypeptides and Proteins

Most of the research on tyrosine fluorescence and phosphorescence in polypeptides and proteins has involved steady-state measurements. This is understandable when one considers that only recent developments have allowed high-precision decay measurements. Technical progress in instrumentation includes new pulsed light sources like synchrotrons and lasers as well as improved detection electronics, especially the microchannel plate photomultiplier tube. Data analysis methods have also improved (see the *Methods in Enzymology* volume entitled "Numerical Computer Methods").[75] The development of the global approach by Brand and co-workers[45,46] and the recent introductions of the linked-function analysis[46] and the distribution function concept[76–78] will all help tyrosine become a probe of protein structure in both steady-state and excited-state investigations. These improvements might also permit an analysis of tyrosine in the presence of tryptophan. For example, the decay-associated excitation spectrum of tyrosine has been resolved from that of a mixture of tyrosine and tryptophan model compounds as well as from that of two subtilisins.[79]

We will not present a comprehensive discussion in this chapter of all the proteins and polypeptides known to contain tyrosine but not tryptophan. Our purpose in this section is to review the important aspects of tyrosine emissions from proteins studied since the reviews by Longworth[1] and Cowgill,[3] as these reviews provide an excellent, comprehensive coverage of the older literature.

1.5.1. Fluorescence of Tyrosine

Tyrosine fluorescence emission in proteins and polypeptides usually has a maximum between 303 and 305 nm, the same as that for tyrosine in solution. Compared to the Stokes shift for tryptophan fluorescence, that for tyrosine appears to be relatively insensitive to the local environment, although neighboring residues do have a strong effect on the emission intensity. While it is possible for a tyrosine residue in a protein to have a higher quantum yield than that of model compounds in water, for example, if the phenol side chain is shielded from solvent and the local environment contains no proton acceptors, many intra- and intermolecular interactions result in a reduction of the quantum yield. As discussed below, this is evident from metal- and ion-binding data, from pH titration data, and from comparisons of the spectral characteristics of tyrosine in native and denatured proteins.

1.5.1.1. Nucleic Acid-Binding Proteins

Besides quenching from intramolecular interactions with peptide bonds and adjacent side chains, there can be interactions with other charged groups such as phosphate. Phosphate is well-known quencher of tyrosine fluorescence[80–84] since mono- and dianion phosphate ($H_2PO_4^-$ and HPO_4^{2-}) are good proton acceptors. A phosphate quencher can also be provided through an intramolecular interaction by a phosphorylated group in the same protein or by intermolecular interactions like the binding of histones to DNA.

1.5.1.1a. Histones. Tyrosine fluorescence of histones, which interact with DNA to form chromatin, has been found to be highly sensitive to the physical state of the protein. Both steady-state emission intensity and anisotropy measurements have been used to study the effects of ionic strength, urea denaturation, and pH on chromatin core particles, which include octamers of histone proteins H2a, H2b, H3, and H4 enwrapped by 145 base pairs of DNA.[85–89] One of the conformational changes that is related to the changes in the structure of chromatin upon transcription and replication of the DNA is thought to be similar to the low-salt transition of core particles. This low-salt transition, which occurs in the millimolar concentration range, has been shown by Libertini and Small[85, 86] to have only a slight effect upon the fluorescence emission intensity. The fluorescence anisotropy, however, increases with increasing salt concentration. A second, high-salt transition occurs above 1.4 *M* salt and affects both the fluorescence emission intensity and anisotropy. The low-salt transition is pH-dependent; the anisotropy showed a pH dependence that appeared to correlate with a conformational change with a pK_a near 7. The low-salt transition also involves cation binding, with divalent cations being more than an order of magnitude more effective than monovalent cations. The low-salt transition was interpreted in terms of a two-step mechanism involving interactions between dimers of H2a and H2b with a tetramer of H3 and H4.

Time-dependent fluorescence measurements have been made on tyrosine in calf thymus nucleosome core particles by Ashikawa *et al.*[87] Based on the salt dependence of the decay data, the tyrosines were divided into two classes. At 20 to 400 m*M* salt, about half of the tyrosine residues appear to be partially quenched, possibly by resonance energy transfer to DNA bases. The other half are thought to be statically quenched, possibly by hydrogen bonds; this quenching is partially eliminated at about 2 *M* salt. In view of the number of tyrosines per nucleosome core particle (estimated at 30), it is impossible to make a more detailed analysis of the decay data.

The fluorescence of purified histones has been studied by several different groups,[90–95] with the most detailed studies being on calf thymus histone H1. Histone H1, which binds to the outside of core particles, contains one tyrosine and no tryptophan. This protein exhibits a substantial increase in fluorescence intensity in going from a denatured to a folded state.[90] Collisional quenching studies indicate that the tyrosine of the folded H1 is in a buried environment.[91] Libertini and Small[94] have identified three emissions from this residue when in the unfolded state with peaks near 300, 340, and 400 nm. The 340-nm peak was ascribed to tyrosinate (*vide infra*), and several possibilities were considered for the 400-nm component, including room temperature phosphorescence, emission of a charge transfer complex, or dityrosine. Dityrosine has the appropriate spectral characteristics,[96] but would require formation of a covalent H1 dimer. The intensities of the three components are excitation wavelength dependent; with 295-nm excitation at pH 4, the 400-nm emission is primarily observed. The contributions of the three components also depend upon salt concentration and pH as the protein folds; at pH 7.4, the 400-nm component is essentially gone and the 340-nm emission is considerably reduced. Three fluorescence decay times of about 1, 2, and 4 ns were observed at 300 nm. While the decay constants are independent of salt concentration, the contribution of each decay constant is markedly affected in accord with the spectral results: at high salt concentration (0.5 *M*) the 4-ns component contributes over 90% of the total intensity. The two shorter decay constants, therefore, were attributed to denatured H1, and the 4-ns component was associated with folded H1.

Histone H1 from the fruit fly *Ceratitis capitata* has two tyrosine residues. Jordano *et al.*(92) have observed two differences from calf thymus H1: (1) the apparent quantum yield does not increase on protein folding; and (2) there is a pH- and conformation-dependent shoulder at 340 nm in the emission spectrum. This group has attributed this 340-nm emission to tyrosinate.(97) Their studies demonstrate that the folding of histone H1 from *C. capitata* is pH and ionic strength dependent. The possibility of tyrosinate formation at neutral pH is discussed in greater detail in Section 1.5.2.

The H2a–H2b histone dimer also has strong salt-dependent conformational properties, with a transition near 0.5 *M* NaCl.(93) Above 0.5 *M* NaCl, the tyrosine fluorescence emission becomes less quenchable by Cs^+, and the dimer structure becomes more compact.

In an investigation of the physical basis of the interaction of histones with DNA, De Petrocellis *et al.*(95) have examined the effect of phosphate ions on histone H1. Binding results have shown that there are high-affinity sites for phosphate ions. In addition, phosphate ions were found to perturb the absorption spectra of H1 and quench tyrosine fluorescence. Binding of the phosphate group resulted in positive difference absorption bands near 275 and 293 nm, which are similar to those produced at acid and alkaline pH, respectively.

During several stages of rat spermatogenesis, histones are replaced by other highly basic proteins. One of these testis-specific transition proteins, the testis protein TP, is a 54-residue polypeptide which is 19% lysine and 21% arginine and contains two tyrosines as the only aromatic amino acids. Singh and Rao(98) have shown that the tyrosine fluorescence is quenched when TP binds to DNA. They propose that the quenching is a result of intercalation, in the manner described in the review by Helene and Lancelot(99) for the tripeptide Lys-Tyr-Lys (*vide infra*). Although the association of TP and DNA is weak, Singh and Rao(98) find that TP prefers single-stranded DNA and might be acting as a DNA-melting protein.

The DNA of the thermophilic archaebacterium *Thermoplasma acidophilum* is stabilized by binding a histonelike protein, HTa. This homotetrameric protein has no tryptophan and only one tyrosine per subunit. To characterize the intrinsic fluorescence of the tetramer in the absence of DNA, Searcy *et al.*(100) carried out quenching and pH titration studies. Using tyrosine as a reference, steady-state quantum yields and fluorescence lifetimes were compared. It was concluded that only three of the four tyrosines were emissive; one tyrosine appeared to be completely quenched. This quenching was eliminated by denaturnation in 6 *M* guanidinium chloride. These results suggest that the native tetramer is not truly symmetric. However, since the HTa tyrosine absorption band is different from that of free tyrosine, an alternative model, consistent with the other data, would be equivalent tyrosine residues experiencing static quenching interactions. Each subunit also contains five phenylalanine residues. Resonance energy transfer from Phe to Tyr was demonstrated by analysis of excitation spectra and fluorescence lifetimes.(101)

1.5.1.1b. Model Peptides: The Tyrosine–Nucleic Acid Interaction. The perturbation of tyrosine fluorescence by binding to nucleic acids has been demonstrated in model peptides.(102–105) For example, Brun *et al.*(102) examined the fluorescence properties of oligopeptides bound to polynucleotides. The general structure of the oligopeptides was lysyl-X-lysine, where X was either tryptophan, tyrosine, or *O*-methyltyrosine. The results suggest that two kinds of complexes were formed, both as a result of electrostatic interactions between the two lysine side chains of the peptide and nucleic acid phosphate groups. Based on fluorescence lifetimes, the fluorescence quantum yield of one complex was essentially identical with that of the free peptide, but the other complex apparently was completely quenched. In the first complex, the aromatic residue appeared not to interact with the polynucleotide, whereas in the second complex it appeared that the aromatic ring was involved in stacking interactions. Comparison of fluorescence data for single-stranded and double-stranded polynucleotide/oligopeptide complexes indicated that the stacking interaction was favored in single-stranded polynucleotides.

In testing the possibility of proton transfer as a quenching mechanism of tyrosine in oligopeptide/polynucleotide complexes, Brun *et al.*(102) compared the fluorescence emission spectra of the tyrosine and *O*-methyltyrosine tripeptides. They noted that, in the complex, the *O*-methyltyrosine tripeptide had a unique secondary emission near 410 nm. Whether this emission is related to that observed by Libertini and Small(94) is an important question. While one must consider the possibility that two tyrosine side chains could be converted to dityrosine,(96) which has a fluorescence at 400 nm, another intriguing possibility is ambient temperature tyrosine phosphorescence. This could happen if the tyrosine side chain is in a rigid, protective environment, very effectively shielded from collisions with quenchers, particularly oxygen.

The fluorescence decay of the tripeptide lysyl-tyrosyl-lysine, measured by

Montenay-Garestier *et al.*,[105] is double-exponential, both in the absence and presence of either native or denatured DNA. The two decay times were interpreted as resulting from two peptide conformers. The relative weights (amplitudes) of the two lifetimes are affected only slightly by DNA binding. The main effect of binding (at 11 °C) is a reduction in the average lifetime; this decrease is largely reflected by a decrease in the shorter decay constant from about 1.1 to 0.6 ns. The decay parameters for the tripeptide in the native and denatured DNA complexes were very similar. The reduction in the average fluorescence lifetime in the complexes was significantly less than the reduction in the steady-state fluorescence quantum yields, and the reduction in the steady-state quantum yield was greater for the denatured DNA complex. These results can be explained by the formation of a static quenching interaction in the DNA/tripeptide complexes, with greater static quenching in the denatured DNA complex.

Three quenching mechanisms were considered for the interpretation of these data[105]: (1) stacking of tyrosine with the nucleic acid bases; (2) hydrogen bonding interactions with the hydroxyl group acting as a proton donor; and (3) energy transfer to the nucleic acid bases. In the case of the denatured DNA/tripeptide complex, the greater static quenching is consistent with stacking interactions. In the case of the native DNA/tripeptide complex, however, stacking interactions were discounted on the basis of NMR results from the literature. It is of interest that experiments with the *O*-methyltyrosine tripeptide also revealed strong quenching in the complex with native DNA, even though NMR indicated only limited stacking interactions. Thus, hydrogen bonding is not required for the quenching.

Two mechanisms were suggested for the reduction in the shorter lifetime upon binding[105]: (1) a binding-induced conformational change in the peptide that brings a quenching group nearer to the tyrosine ring; and (2) resonance energy transfer from the tyrosine to nucleic acid bases. If energy transfer is the quenching mechanism, then two kinds of complexes with different energy transfer efficiencies could explain not only the decrease in the short fluorescence decay component in native and denatured DNA complexes, but also the differential static quenching between the native and denatured DNA complexes. It was argued that a complex without stacking interactions would probably be less likely to undergo resonance energy transfer in comparison with a complex that forms as a result of stacking between the bases and the phenol side chain. It should be recognized, however, that it is difficult to exclude energy transfer in a complex without stacking interactions since the probability of transfer depends upon both the dipole orientations and the through-space interactions.

1.5.1.1c. Ribosomal Proteins. Lux *et al.*[106] have studied the fluorescence yields and lifetimes of the S8 and S15 ribosomal proteins from *Escherichia coli*, which are among five that bind directly to the ribosomal 16S

RNA. Both proteins contain no tryptophan; S8 contains three tyrosines, and S15 contains two tyrosines. The tyrosine emission of these two proteins represents a case in which the quantum yield is higher in the native than the denatured protein. The average fluorescence lifetime, however, was little affected by denaturation; no change was observed for S8, and the average lifetime decreased about 12% for S15. The collisional quenchers I^- and Cs^+ had essentially equivalent access to the tyrosines of both proteins, either in the native or denatured state, and comparison of the bimolecular quenching constants with that of free tyrosine suggested that the tyrosines were all well exposed. The mechanism for the reduction in quantum yield upon denaturation, apparently a static interaction, has not been elucidated.

1.5.1.1d. Specific Nucleic Acid-Binding Proteins. Tyrosine fluorescence has been used to study the protein–DNA interaction of the *lac* repressor protein of *Escherichia coli*,(107, 108) which is involved in regulation of the lactose operon. The *lac* repressor is a tetramer with 360 residues per subunit. Each subunit has a domain which is referred to as the short headpiece. It consists of the amino terminal part of the polypeptide chain and complexes with DNA. The fluorescence quenching of the four tyrosines in this domain was used by Schnarr *et al.*(108) to determine binding isotherms for natural DNA and poly(dG – dC). By comparing results from fluorescence with those from circular dichroism, the length of the polynucleotide binding site was estimated to be three to four base pairs. The binding interaction was salt-dependent and required protonation of a group with a pK_a of 6.7 ± 0.3; the imidazole of His-29 was postulated to be this group.

The gene V protein of fd bacteriophage is required for replication of viral DNA in infected *Escherichia coli* cells. Pretorius *et al.*(109) have shown that the gene V protein exists principally as a dimer at neutral pH and physiological salt concentrations (0.15 *M*). Higher concentrations of NaCl or $NaClO_4$ disrupt the dimer. While neither the tyrosine fluorescence nor the circular dichroism spectrum is affected by the monomer–dimer equilibrium, both optical signals are perturbed upon binding of the protein to fd-DNA or to poly(dT). Pretorius *et al.*(109) interpreted the quenching of the gene V protein tyrosine fluorescence as being consistent with reports in the literature that tyrosine is totally quenched on stacking with DNA. They concluded that three of the five tyrosines per gene V monomer were involved in stacking interactions with bases. This interpretation is consistent with the observation that the DNA base–base exciton circular dichroism band was reduced upon binding of the gene V protein, and it was pointed out that the DNA electronic transitions are appropriate for forming a base/tyrosine exciton band.

A single tyrosine is in the C-terminal portion of the transcription factor 1 (TF1), a type II procaryotic DNA binding protein encoded by *Bacillus subtilis* phage SPO1. Time-resolved fluorescence decay measurements yielded a single lifetime for the tyrosines in the unliganded dimer protein, and these results were interpreted in terms of a symmetric protein structure.[110] Based on quenching studies, the tyrosine appears to be on the surface of the protein but in a negatively charged environment. Binding of TF1 to various DNA sequences results in a decrease in tyrosine fluorescence. This quenching was ascribed to resonance energy transfer to the DNA bases since the tyrosine side chain does not appear to be in direct contact with the nucleotides.

A protein induced after coliphage N4 infection has been studied. Although it has one or two tryptophans, its intrinsic fluorescence is dominated by the ten tyrosines.[111] Tryptophan fluorescence is seen after denaturing the protein. Upon binding to single-stranded DNA, the tyrosine fluorescence is quenched. This signal has been used to demonstrate that the binding affinity is very dependent on salt concentration and is also very sensitive to the nucleotide sequence.

1.5.1.2. Calcium-Binding Proteins

The intrinsic fluorescence of tyrosine has been used extensively to study the biochemistry and physicochemical characteristics of several calcium-binding proteins, including calmodulin, troponin C, oncomodulin, the parvalbumins, and S100 proteins. Lux *et al.*[112] have recently examined some of these proteins and carefully compared the spectra from different species to assess how intramolecular interactions can affect tyrosine fluorescence spectra. They pointed out that while the emission maxima of S100b protein, ram testis calmodulin, and octopus calmodulin are similar (304 ± 1 nm), the bandwidths of the spectra show a strong dependence upon cation binding, particularly Ca^{2+} binding, and reflect whether the protein is denatured. Moreover, the extinction coefficients of the absorption bands near 275 nm are unusually high compared with that of free tyrosine. This appears to be due to hydrogen bond formation, since it is associated with static quenching of the fluorescence and high ground-state pK_a values for the phenol hydroxyl group.

1.5.1.2a. Calmodulin. The biophysical characteristics of calmodulin have been recently reviewed by Forsen *et al.*[113] Calmodulin is a regulatory protein found in all eucaryotic cells, which makes proteins to which it binds calcium-sensitive. It is the most studied member of the troponin C super family. It is heat stable, lacks cysteine and tryptophan residues (mammalian protein), and has more acidic than basic residues. The molecular weight of calmodulin is near 17,000. The low-resolution X-ray diffraction model is a dumbbell-shaped molecule with a central helix connecting two globular structures, each containing two metal ion-binding domains. The calcium binding sites are found in these four domains and are referred to as domains I, II, III, and IV. The two tyrosines, Tyr-99 and Tyr-138, are in domains III and IV, respectively, but appear to be in close proximity to one another since formation of dityrosine has been reported.[114] Calcium binding alters the protein conformation; these changes can be observed by monitoring various physical parameters including changes in tyrosine fluorescence.

The way in which calcium binding changes the tyrosine fluorescence of calmodulin was initially a controversial issue. In 1980, Seamon[115] examined the binding of Ca^{2+} and Mg^{2+} by NMR and concluded that both Tyr-99 and Tyr-138 were perturbed by the first two calcium ions. Although Tyr-138 was also perturbed by the binding of the fourth calcium, both residues appeared to be associated with high-affinity domains (III and IV).

Kilhoffer *et al.*,[116, 117] using Tb^{3+} to characterize the order of binding, came to the opposite conclusion: the two high-affinity calcium sites of calmodulin are domains I and II, with subsequent filling of domain III and then domain IV. Their results seem ambiguous, however, since the calcium and terbium data differed. Whereas the first two calciums bound resulted in a substantial increase in tyrosine fluorescence and the second two calciums had only a small effect, the first two ions of Tb^{3+} bound resulted in a small enhancement of tyrosine fluorescence. Furthermore, the second two terbium ions *quenched* the tyrosine fluorescence, perhaps due to tyrosine serving as an energy transfer donor for terbium.

To determine the influence of calcium binding upon the tyrosine fluorescence of calmodulin, Kilhoffer *et al.*[118] compared Ca^{2+} and Mg^{2+} binding in calmodulins from ram testis, which contains Tyr-99 and Tyr-138, and from octopus, which contains only Tyr-138. They found that the influence of calcium on tyrosine fluorescence was complete when two moles were bound per mole of protein. In addition, at physiological ionic strength (150 mM K^+), magnesium binding did not seem to exert a major influence on calmodulin conformation. In the absence of calcium, both species of protein had a low fluorescence quantum yield as the result of static quenching. In the case of the ram testis protein, the average fluorescence lifetime was about 1.3-fold shorter in the absence of calcium, but the steady-state quantum yield decreased about three-fold. In the case of the octopus protein, calcium had no effect on the fluorescence lifetime, but the steady-state quantum yield increased three-fold in the presence of calcium, indicating a change in static quenching of Tyr-138. When Kilhoffer *et al.*[118] analyzed the relative contributions of Tyr-99 and Tyr-138 to the total protein fluorescence, they assumed that the characteristics of Tyr-138 were the same for both the ram testis and octopus proteins. Using this assumption, they calculated that Tyr-99 exhibits a 2.5-fold increase in its steady-state quantum yield when calcium is present. Their main conclusion was that Ca^{2+} increases the quantum yield of calmodulin fluorescence because binding at domains I and II is coupled to domains III and IV, where the tyrosines are located, producing indirect fluorescence enhancement.

Kohse and Heilmeyer[119] reported in 1981 that rabbit skeletal muscle calmodulin had six calcium-binding sites at low ionic strength (1 m*M*). Investigating the competition of Ca^{2+} with Mg^{2+} at high ionic strength (0.18 *M*), they distinguished two high-affinity calcium-specific sites and two lower affinity sites which bind either calcium or magnesium ions. The remaining two sites were not observed in this experiment and were therefore interpreted as being specific for magnesium. At high ionic strength, magnesium enhanced the calcium affinity; this subsequent binding of calcium resulted in an enhancement of the tyrosine fluorescence intensity, which persisted even when the bound calcium was removed by chelating agents. Chelation of the magnesium was required to obtain the original intensity.

In 1982, Wang *et al.*[120] confirmed the observations by Kilhoffer *et al.*[116,117] of two high-affinity Tb^{3+} sites at domains I and II and two low-affinity Tb^{3+} sites at domains III and IV. Using terbium, Wallace *et al.*[121] also examined the order of filling the lanthanide binding sites and concluded that domains III and IV are indeed occupied subsequent to domains I and II. Domains I and II, however, appeared to have quite different affinities for terbium: one of these two sites was occupied first, and the other second. Wang *et al.*[122] then reexamined the relationship between calcium and terbium binding by stopped-flow kinetic studies and reported that, contrary to what was thought earlier, calcium and lanthanides in fact exhibit opposite preferences for the four metal-binding sites of calmodulin. Whereas sites III and IV are the high-affinity sites for calcium, sites I and II are the high-affinity sites for lanthanides. The resolution of the true high-affinity calcium sites in calmodulin dramatically and clearly demonstrates the point that great care must be taken when using lanthanides as site-specific probes to characterize calcium-binding sites in proteins.

The fluorescence of the two tyrosine residues in bovine testes calmodulin was investigated by Pundak and Roche.[123] Upon excitation at 278 nm, a second emission, in addition to tyrosine fluorescence, was observed at 330–355 nm, which they characterized as being due to tyrosinate fluorescence. The tyrosinate fluorescence appeared to be from Tyr-99, which has an anomalously low pK_a of about 7 for the phenol side chain. Pundak and Roche[123] reasoned that since tyrosinate emission is apparently not being seen in other species of calmodulin, it is possible that the bovine protein contains a carboxylate side chain in domain III which is amidated in other species. They further argued that the tyrosinate emission from bovine testes calmodulin arises from direct excitation of an ionized tyrosine residue. This tyrosinate fluorescence is discussed in more detail in Section 1.5.2.

The physical dimensions and dynamics of calmodulin have also been investigated by tyrosine fluorescence. To learn about the internal mobility of calmodulin, Lambooy *et al.*[124] and Steiner *et al.*[125] measured the steady-state fluorescence anisotropy of the tyrosine. Since the average correlation times derived from Perrin plots were about a factor of four shorter than the correlation time calculated for residues constrained within a rigid protein, they interpreted their anisotropy data as evidence for a fair degree of internal mobility for calmodulin.

To characterize the change in the conformation of calmodulin, which occurs in going from the calcium-free to the calcium-bound state, Steiner and Montevalli-Alibadi[126] measured the energy transfer distances between Tyr-99 and Tyr-138 in the presence and absence of Ca^{2+}. The measurement was made on mononitrotyrosine derivatives of calmodulin, using the nitrotyrosine residue as the acceptor of the energy from the singlet excited state of the unmodified residue. Since the two possible nitrotyrosine derivatives were made by selective chemical modification, it was possible to measure the energy transfer using either tyrosine residue as the donor or the acceptor. Steiner and Montevalli-Alibadi's[126] calculation of the energy transfer distance relied on an assumed value of $\frac{2}{3}$ for κ^2 (see Section 1.4.2), which corresponds to a random orientation between the donor and acceptor. They argued that the value of $\frac{2}{3}$ should be reasonable based on the low steady-state anisotropy observed by Lambooy *et al.*[124]

When Steiner and Montevalli-Alibadi[126] measured the efficiency of the resonance energy transfer, they used the fluorescence quenching of the unmodified tyrosine residue caused by the presence of the nitrotyrosine residue. The energy transfer distance measured in this way was about the same in the presence of calcium when either Tyr-99 or Tyr-138 was the donor (about 15.5 Å). The averaged data for both directions indicated little significant difference between the conformations of the calcium-free and the calcium-bound states. While it appears that nitration did not significantly alter the conformation of calmodulin based on circular dichroism spectra, other mechanisms can account for part of the fluorescence quenching of a donor in the presence of an acceptor. When an acceptor group is introduced into a protein or polypeptide, it is always possible that new, subtle interactions can occur which have the potential to provide new pathways for deactivation of the donor's excited state. This additional quenching would obviously bias the interpretation of energy transfer, and to establish whether or not these new pathways exist is not always straightforward; a fluorescence change does not necessarily imply a large structural perturbation.

Two independent groups have examined the tyrosine fluorescence intensity and anisotropy decay properties of calmodulin.[127,128] Both groups find that the binding of calcium ions causes an increase in the average fluorescence lifetime. Concomitantly, the number of exponentials needed to fit the anisotropy decay decreases. Both groups conclude that calcium binding results in protein conformational changes that restrict the motions of the phenol side chains and also affects their quantum yield. The agreement of the results, however, is only qualitative. The groups find different numbers of exponentials for both the intensity and anisotropy decays with or without calcium. Even when one takes into account the differences in temperature and other experimental conditions, the average fluorescence lifetimes and the observed rotational correlation times are quite different. Clearly, further experiments are needed.

These two groups each have made an important observation about calmodulin from its tyrosine fluorescence. Based on limiting anisotropy values determined from steady-state measurements on calmodulin and fragments of calmodulin, Gryczynski *et al.*[127] argued that resonance energy transfer occurs between the two tyrosine residues. This provides additional support for the idea that Tyr-99 and Tyr-138 are in close proximity to one another. Using circular dichroism to estimate α-helical content, Bayley *et al.*[128] have found that the amount of helix at neutral pH in the presence of saturating calcium is less than that calculated from the X-ray crystal model and that the conditions used to crystallize calmodulin cause an increase in the amount of α-helix. Furthermore, based on an isotropic polarization decay of calmodulin complexed with calcium, they find that the protein appears to adopt an essentially globular structure in solution. They postulated that the increased α-helical content is associated with residues 66–92, which connect the two globular regions that form the dumbbell observed in the X-ray crystal model, and that the two conformational states of the protein may have functional significance.

1.5.1.2b. Parvalbumin. The parvalbumins are calcium-binding proteins from the sarcoplasm. In lower vertebrates, the molecular weights are $11{,}000 \pm 2000$, and there are eight to ten phenylalanine residues in the polypeptide chain. Although tryptophan and tyrosine are generally absent, if either one is present, the other is usually not.[129–132] The three-dimensional X-ray diffraction model of one carp parvalbumin[133] consists of six helical regions, with two pairs of helices and their connecting loops forming two high-affinity calcium sites. Calcium binding strongly affects the intrinsic fluorescence. For example, Ca^{2+} binding to pike parvalbumin results in an increase in the quantum yield of its single tyrosine residue.[131] In addition, the quantum yield decreases as the pH is increased from 7 to 8. This effect has been interpreted as reflecting a pH-sensitive conformational transition. The removal of calcium increases the exposure of this tyrosine to collisional quenchers and decreases the stability of the protein against thermal denaturation.

Permyakov *et al.*[132] have compared the effects of calcium binding upon the steady-state and time-resolved fluorescence of two different species of fish parvalbumin, one with a single tryptophan (whiting) and one with a single tyrosine (pike). The fluorescence decays of both proteins were best fit by double exponentials either in the presence or absence of calcium. We focus here on the tyrosine results from pike parvalbumin. Calcium binding causes a 50% increase in the tyrosine steady-state fluorescence quantum yield and about a 10% increase in the mean (intensity-weighted) lifetime, which is defined by the relationship

$$\langle\tau\rangle = (\alpha_1\tau_1^2 + \alpha_2\tau_2^2)/(\alpha_1\tau_1 + \alpha\tau_2) \tag{1.12}$$

where the α_is are the amplitudes and the τ_is are the individual decay constants. The mean lifetimes were 3.65 ns and 3.35 ns in the presence and absence of calcium, respectively.

The mean lifetime can be compared with the commonly used average (amplitude-weighted) lifetime, defined by

$$\bar{\tau} = (\alpha_1\tau_1 + \alpha_2\tau_2)/(\alpha_1 + \alpha_2) \tag{1.13}$$

Calculating the average lifetimes from the data of Permyakov *et al.*,[132] one finds about a 20% increase in the average lifetime after binding of calcium. Comparison of these lifetime ratios with the steady-state quantum yield ratio could denote a static quenching interaction which is diminished upon the binding of calcium. The time-dependent parameters also imply a complex fluorescence. The longer lifetime was essentially unaffected by Ca^{2+} binding, increasing from 3.74 to 3.85 ns, while the shorter decay constant decreased from 1.40 to 0.85 ns. Our calculation of the relative amplitudes of the two decay components (which were used above to calculate the average lifetime) shows, in addition, an increase from 0.56 to 0.84 in the amplitude of the longer decay component upon binding of calcium. Thus, the contribution of the short decay component decreases as indicated by both the shorter lifetime and the smaller amplitude. Further investigations are required to clarify the relationship between Ca^{2+} binding and tyrosine fluorescence.

1.5.1.2c. Oncomodulin. Oncomodulin, first described by MacManus *et al.*,[134] is a 108-residue, parvalbumin-like tumor protein found in rats and humans. It is part of the troponin C super family, contains two tyrosine residues (Tyr-57, which is homologous to Tyr-99 in calmodulin, and Tyr-65), and has different conformational properties in the presence of Ca^{2+} compared to Mg^{2+}. The spectroscopy of rat oncomodulin, purified from Morris hepatoma 5123tc cells, was studied by MacManus *et al.*[135] Whereas calcium binding produces marked changes in the absorption, circular dichroism, and fluorescence excitation spectra, these spectra obtained in the presence of magnesium are remarkably similar to those of the metal-free protein. This is in contrast to rat parvalbumin, where Ca^{2+} and Mg^{2+} appear to induce the same conformational perturbations. The oncomodulin fluorescence emission was compared with that of free tyrosine, and the difference emission spectrum, with a maximum at 345 nm, appeared to be similar to that of tyrosinate (see Section 1.5.2). While the tyrosine emission increased in the presence of calcium, that associated with tyrosinate decreased. Based on the spectral perturbations, MacManus *et al.*[135] concluded that the stoichiometry of metal binding is two

Ca^{2+} and either one or two Mg^{2+} per molecule of protein. The uncertainty in the case of magnesium arises from the observation that binding of the second ion, if it occurs, has no effect on the protein.

1.5.1.2d. Troponin C. Johnson and Potter[136] have shown that the tyrosine fluorescence of troponin C increases upon binding of calcium. The maximal change occurs upon saturation of a class of high-affinity binding site(s). They also examined the circular dichroism spectrum as a function of calcium concentration, and from these results they interpreted the fluorescence data in terms of biphasic changes in the protein structure. A distinct conformational change occurs when Ca^{2+} binds to the high-affinity site(s), and a subsequent change occurs during saturation of the lower affinity site(s). Wang *et al.*[137] confirmed the existence of two classes of metal binding sites with lanthanide binding experiments. By comparing the binding of H^+, Ca^{2+}, and Mg^{2+}, Leavis and Lehrer[138] concluded that titration or complexation of the phenolic hydroxyl group inhibited proton transfer to nearby carboxyl residues. The increase in tyrosine fluorescence upon binding at the high-affinity calcium binding sites (III and IV) in the C-terminal domain has been used to demonstrate that the calcium regulatory activity in muscle is associated with the low-affinity calcium binding sites (I and II) in the N-terminal domain.[139]

Site II, a low-affinity calcium-specific site on troponin C, has been further investigated by synthesis of model peptides.[140, 141] To obtain a stronger fluorescence probe in these peptides, Phe-72 was replaced by tyrosine. Kanellis *et al.*[140] have measured the fluorescence decay kinetics of the Tyr-72 analogue; it is interesting to compare their decay data with those of Permyakov *et al.*[132] for parvalbumin. Whereas it appears that calcium binding relieves a static quenching of the tyrosine fluorescence decay of parvalbumin (Section 1.5.1.2b), calcium binding reduces a dynamic quenching interaction in the troponin C peptide. The fluorescence decay of the troponin C peptide is a double exponential of 0.67 and 2.14 ns in the absence of calcium and 0.82 and 2.73 ns in the presence of calcium. The values of 0.65 and 0.35 for the amplitudes of the shorter and longer decay components, respectively, are not significantly affected by Ca^{2+} binding.

1.5.1.2e. Intestinal Calcium-Binding Protein (ICaBP) and Brain S100b Protein. Bovine and porcine ICaBP are small proteins with molecular weights near 9000; both have a single tyrosine, and a structural model has been developed from X-ray diffraction data for the bovine protein.[142] O'Neil *et al.*[143] have shown that a 276-nm circular dichroism transition and the tyrosine fluorescence of porcine ICaBP are similarly affected by calcium binding. In addition, the tyrosine fluorescence is enhanced at low pH, and this transition has a pK_a of about 4.2. Based on the X-ray crystal[144] model and analysis of protein difference fluorescence emission spectra, O'Neil and Hofmann[145] suggested that the phenol moiety could hydrogen bond to Glu-38 in the ground state and that emission sometimes occurs from the completely deprotonated tyrosinate form due to the transient formation of the hydrogen bond. Chiba *et al.*[146, 147] found two sites that bind either calcium or terbium, but whereas the site affinities are essentially the same for calcium, they differ by more than an order of magnitude for terbium.[148]

This calcium-binding protein is also referred to as calbindin. The intensity and anisotropy decays of the tyrosine residue in wild-type and a mutant calbindin have been examined.[149] This particular mutant has had Ala^{14} and Asn^{21} deleted and Pro^{20} changed to glycine. For both proteins, the intensity decay parameters are similar, and the binding of Ca^{2+} does not perturb them. The intensity decays were fit to a sum of three exponentials. The two kinetic terms with relative amplitudes of 0.86 and 0.09 (wild type) were tentatively assigned to the hydrogen-bonded and uncomplexed forms of the phenol side chain, respectively. The third component, with an amplitude of 0.05, was ignored, even though this kinetic component contributed 33% of the total fluorescence intensity, which could raise some questions about this interpretation. In contrast to the intensity decay parameters, the anisotropy decay parameters are quite different for the two proteins. The rotational correlation times for the mutant are significantly shorter, suggesting an altered conformation in the environment of the tyrosine residue. This altered structure was also found to have a decreased affinity for calcium ion.

The brain S100b protein has homology with the ICaBP protein, including common structural features and a single tyrosine residue. An early report on the intrinsic fluorescence properties of S100b found an abnormal tyrosine emission spectrum[150]; this was later suggested to be tryptophan contamination.[151] Other similarities with ICaBP include a high pK_a for the phenol side chain and a small amount of a red-shifted component in the emission spectrum.[112] The binding of Ca^{2+} perturbs the absorption and fluorescence properties of the tyrosine residue. These spectral perturbations have been used to investigate various aspects of metal ion binding to S100b.[152–156]

1.5.1.2f. Other Calcium-Binding Proteins. Hauschka and Carr[157] used absorption, circular dichroism, and fluorescence spectroscopies to examine the conformation of osteocalcin, a 49-residue bone protein which binds calcium via three γ-carboxyglutamic acid residues. According to their data, metal-free osteocalcin exists mostly in a random coil conformation with a small amount (8%) of α-helical structure. The protein acquires more α-helical structure upon binding of calcium, and this conformational change results in a quenching of the tyrosine fluorescence.

A protein highly homologous to the S100 proteins has been isolated from Ehrlich ascites tumor cells; it has subsequently been shown to be nearly identical with human calcyclin. The fluorescence intensity from the three tyrosine residues is enhanced on the binding of Ca^{2+}.[158]

The p11 subunit of the calpactin I heterotetramer contains no tryptophan and two tyrosines. Although it is somewhat homologous to the S100 proteins, the tyrosine fluorescence is insensitive to Ca^{2+} binding.[159] The other subunit in calpactin I, p36, contains both tryptophan and tyrosine, and their fluorescence is affected by calcium ion. In addition, negatively charged phospholipids enhance the affinity of his heterotetramer for Ca^{2+}.

A protein of similar molecular weight to that of rat oncomodulin, rat and rabbit parvalbumins, S100, and the vitamin D-dependent calcium-binding proteins has been isolated from chicken gizzard smooth muscle. In this case, however, the fluorescence emission from the four tyrosine residues is quenched by Ca^{2+} binding.[160] The decrease in fluorescence intensity was used to suggest that there are two different classes of Ca^{2+} binding sites.

1.5.1.3. Mitochondrial Malate Dehydrogenase

The tyrosine fluorescence of porcine mitochondrial malate dehydrogenase (MDH) was initially described by Thorne and Kaplan[161] in 1963. MDH is a 70,000-dalton dimer of identical subunits, each with five tyrosine residues and no tryptophan. From titration, iodination, and nitration data, it appears that one or two of the tyrosines per dimer plays a role in the catalytic activity.[162] Wood *et al.*[163] have used the intrinsic tyrosine fluorescence to study the intersubunit interaction in the pH range between 4 and 7 and found that the apparent pK_a of the fluorescence increase was very similar to that for the change in the sedimentation coefficient due to the dissociation into monomers. Since the enzyme has a concentration-dependent dissociation, the apparent pK_a is slightly concentration dependent. Wood *et al.*[163] also measured the temperature dependence of the kinetics of the subunit reassociation in a pH jump experiment by following the change in both the specific activity and the intrinsic fluorescence. From these data, they obtained an activation energy of about 20 kcal/mol for the reassociation reaction. They then proposed a model for the enzyme which predicts two conformational states involving a *cis–trans* isomerization about one or more proline imino bonds: in the active form, the proline bond is *trans* and the tyrosine fluorescence is strong; in the inactive form, the proline bond is *cis* and the fluorescence is weak. Further, the isomerization pathway from the inactive to the active form occurs largely in the dimer after the deprotonation step and the subsequent reassociation, while the reverse pathway to the inactive form occurs in the monomer at acid pH.

Muller *et al.*[164] have examined the spectroscopy of the acid transition to understand better the role of tyrosine in the structure and biological function of MDH. Resolution of the protein absorption spectrum, using *N*-acetylphenylalanine ethyl ester in dioxane and *N*-acetyltyrosine ethyl ester in dioxane or 0.1 *M* phosphate buffer to model the effect of the local environments of the chromophoric groups, indicated that both the pig and the chicken enzymes have one strongly perturbed tyrosine (per subunit) in a hydrophobic environment. This residue has its absorption shifted about 4 nm to the red. Muller *et al.*[164] interpreted the red shift as resulting from hydrogen bond formation. They also compared the pK_a of the acid dissociation reaction for the chicken and pig enzymes using three different criteria: activity, fluorescence intensity, and ultraviolet absorption difference. The three techniques led to three different pK_a values, ranging from 5.8 to 5.25 for pig MDH and 5.3 to 4.45 for chicken MDH. To allow the three different pK_as to describe the same perturbation of the same tyrosine, they concluded that the differences in the pK_as were due to the inherent time scales of the physical processes being measured by the three techniques. The assumption that the three pK_as are involved with the same group is unfounded, especially in the case of enzyme activity. Numerous groups are titrated during a pH titration, and a change in biological activity will not necessarily correlate with a change in a spectroscopic parameter.

1.5.1.4. Protease Inhibitors

There has been considerable interest in certain protease inhibitors which contain tyrosine but no tryptophan. For example, there is a well-characterized X-ray diffraction crystal structure for bovine pancreatic trypsin inhibitor (BPTI),[165,166] which has provided a major impetus for theoretical[167] and experimental studies, including NMR[168] and fluorescence studies.[169–171] BPTI is a compact 58-residue protein with four tyrosines and three disulfide bridges. Kasprzak and Weber[169] examined the dynamics of the tyrosine residues by measuring steady-state fluorescence polarization as a function of temperature and viscosity, as well as in the presence of the quenchers citrate, acetate, and iodide. Their analysis of the polarization over the range −40 to +50 °C in 75% glycerol and as a function of glycerol concentration at 20 °C indicated several modes of motion. They pointed out, however, that while their results are consistent with the theoretical predictions for tyrosine motions in proteins made by Karplus and his collaborators,[167] the observed fluorescence depolarization could be the result of resonance energy transfer among the tyrosine residues.

With the development of multifrequency phase–modulation technology, Lakowicz and co-workers[171] were able to examine the time dependence of the anisotropy decay of BPTI. They noted that the intensity decay of the fluorescence is best fit by a biexponential decay law and that the anisotropy decay is also complex. At 25 °C and pH 6.5, correlation times of 39 ps and 2.25 ns were recovered from analysis of data obtained over the range 20 MHz to 2 GHz. The longer correlation time is close to that predicted for the overall rotational motion of a molecule of the size of BPTI. They indicated, however, that additional experiments need to be done to resolve whether the 39-ps component is the result of tyrosine side-chain torsional motions, as predicted from calculations, or the result of inter-residue energy transfer.

Nordlund *et al.*[172] have studied the fluorescence polarization decay of the single tyrosine in lima bean trypsin inhibitor (LBTI), a protein similar in size but with very little sequence homology to BPTI, using a signal-averaging streak camera. Although the fluorescence intensity decay appeared to be dominated by a 0.6-ns component, a slower decaying component is evident in the data (see Figure 2 of Ref. 151). The anisotropy decay is clearly resolved into two components: there is a fast correlation time of about 40 ps and a slower correlation time estimated to be 3 ns or longer. Thus, the BPTI and LBTI fluorescence anisotropy decay parameters are remarkably similar.

The single tyrosine in LBTI has an abnormally high pK_a of >11.5, most likely due to interactions with the protein.[173] Citrate fluorescence quenching studies of LBTI showed complex behavior. Based on simplifying assumptions, the quenching data also suggest that the tyrosine residue is shielded from solvent.[173]

1.5.1.5. Other Proteins

1.5.1.5a. Neurophysin. The neurophysins are highly homologous proteins that bind and transport the neurohypophyseal hormones oxytocin and vasopressin via the neurohypophyseal tract to the posterior lobe of the pituitary. Bovine neurophysins I and II are most commonly studied; they each have three phenylalanines, one tyrosine, and no tryptophan. Sur *et al.*[174] have shown that the fluorescence intensity of Tyr-49 in both neurophysin I and II is enhanced either by lowering the pH or by binding a hormone or hormone analogue. The pH effect demonstrates a pK_a of 4.6, suggesting that either of the nearby glutamate residues Glu-46 or Glu-47 are involved in a complex with Tyr-49. It was also shown that the single tyrosine residue in these hormones (see Section 1.5.1.6b) is effectively quenched upon binding to the neurophysins; the hormone tyrosine then appears to become an energy sink for the Tyr-49 in neurophysin through resonance energy transfer.

The neurophysins dimerize at higher concentrations. By investigating the rotational motions of Tyr-49 in highly viscous solvents, it has been shown that the local mobility of the phenol ring is not affected by the dimerization and that both tyrosine residues of the dimer appear to be equivalent.[175, 176] Upon ligand binding, however, the two tyrosine environments in the dimer become different. While the aromatic ring of the hormone is in a rigid environment upon binding, one Tyr-49 of the neurophysin dimer becomes very flexible and the other becomes restricted. From the energetics of these interactions, it was concluded that the first mole of ligand stabilizes the dimer while binding of the second ligand to the other subunit invokes the conformational changes.

Scarlata and Royer[177] have also examined and interpreted these data on the mobility of the Tyr-49 side chains. While they agreed that binding of ligand to equivalent sites in equivalent subunits causes the two tyrosines in the subunits to experience different environments, they argued that this difference is not due to the extent of coupling with neighboring residues but is more a result of the free space around the phenol ring. They also showed that ligand binding is stabilized by ring stacking, probably between the Tyr-2 of the hormone and a phenylalanine of the neurophysin, and that resonance energy transfer can occur between the ligand tyrosine and the neurophysin tyrosine. Furthermore, by correctly applying the order of free energy couplings between ligand binding and oligomerization, Scarlata and Royer[177] pointed out that the binding of the second ligand stabilizes the dimer and that the role of the first ligand is to change the affinity of neurophysin for the second.

1.5.1.5b. Ribonuclease A. The tyrosine residues of bovine pancreatic ribonuclease A (RNase A) have been characterized extensively in a number of investigations by Cowgill.[3] He classifies the six residues into two main groups (see Table 1.3). In the first group, the three residues at 25, 92, and 97 are type V, which means that they reside in a hydrophobic environment, are hydrogen bonded to peptide carbonyl groups, and are 100% quenched. In the second group, the residues are partially quenched by various mechanisms, depending on their intramolecular interaction; these have not been assigned to specific residues although two tyrosine residues are adjacent to disulfides. These mechanisms include type II, which denotes exposed residues that are quenched three- to fourfold by hydrated peptide carbonyl groups; type III, which denotes residues quenched by disulfide groups; and type VII, which denotes quenching by resonance energy transfer to other quenched tyrosines. The effects of the disulfide bridge upon the fluorescence decay of RNase A have been investigated by Barboy and Feitelson,[178] who showed that urea denaturation plus reduction of the disulfide bridges is required to effect a major change in the fluorescence properties. The mechanism of the disulfide quenching was investigated by Swadesh *et al.*[69] and was discussed earlier in Section 1.4.3.

RNase A has been used extensively as a model system for studies on protein folding. It has been established that in its unfolded state there are multiple forms of RNase A.[179, 180] The species of unfolded forms have been divided into two groups on the basis of refolding kinetics. These were identified by Garel and Baldwin[179] as fast-folding (U_f) and slow-folding (U_s) species. The U_f and U_s species have equilibrium populations of about 20% and 80%, respectively, and the equilibrium is thought to involve a proline *cis–trans* isomerization. Moreover, the tyrosine fluorescence appears to be sensitive to this isomerization.[181] Rehage and Schmid[182] have found that although both the U_f and U_s species are completely unfolded by several different criteria and they both have essentially identical absorption properties, the fluorescence yield of U_s is 20% higher than that of U_f. In interpreting their data, they argued that in the unfolded state U_f and U_s lack specific long-range interactions. Consequently, the observed difference in fluorescence must depend upon local interactions. If the *cis–trans* isomerization of proline produces the two different species, the fluorescence difference is probably due to Tyr-92, which is next to Pro-93. Recent studies strongly support the hypothesis that the fluorescence difference of U_s and U_f is associated with Tyr-92 and the conformation of Pro-93 in the fully unfolded, disulfide-intact protein.[183, 184] The refolding characteristics of pancreatic ribonucleases from sheep, red deer, and roe deer are similar to those of the bovine enzyme, even though they differ by 4 to 17 residues (out of 124) in the amino acid sequence, including some prolines.[185] It is interesting to note that the tyrosine fluorescence of the bovine and ovine proteins seems to originate from Tyr-76. This tyrosine has been replaced in the deer enzymes, and compared to the bovine proteins, these proteins have very low fluorescence.

Haas *et al.*[186] have examined the fluorescence decay of tyrosine due to different Tyr-Pro conformations in small peptides to elucidate further the nature of the fluorescence change associated with Tyr-92. These peptides have acetyl groups at the amino terminus and *N*-methylamide groups at the carboxyl terminus. They found that whereas the dipeptide fluorescence decay requires a double-exponential fit, that of the tripeptide Tyr-Pro-Asn can be fit by a single exponential. By comparison of the average fluorescence decay time and steady-state quantum yield of the peptide to that of *N*-acetyltyrosine-*N′*-methylamide, they found a relatively greater reduction in the steady-state quantum yield of the peptides. This is attributed to static quenching, which increased from 5% in the dipeptide to 25% in the tripeptide. The conformations of these peptides were also examined by NMR, but the results could be interpreted in terms of either *cis–trans* isomerization or other conformational isomerizations.

Haas *et al.*[186] also examined the Pro-Tyr sequence in relation to the more general question of initiation sites for protein folding. One common theme in protein folding concerns the role of aromatic residues. For example, Tulinsky *et al.*[187] postulated that aggregation of aromatic residues within the molecular structure of native chymotrypsin could play an important role in providing stability to the molecule. Coan *et al.*[188] have extended this concept by noting that tyrosine and tryptophan could act as initiators in the folding process since they have large, permanent dipole moments that provide for long-range orientation interactions with other dipoles, and they have relatively large, rigid contact surfaces that provide for stable short-range van der Waals interactions.

To investigate stabilizing interactions involving a tyrosine residue, Haas *et al.*[186] prepared the 105–124 tryptic peptide from performic acid-oxidized RNase A, which included the residues Pro-114 and Tyr-115. They measured the fluorescence decay kinetics of this and some shorter peptide fragments, which included the Pro-Tyr sequence, in water and in guanidine hydrochloride. These peptides had an acetyl group at the amino terminus and an *N*-methylamide group at the carboxyl terminus. At room temperature, Pro-Tyr and Asn-Pro-Tyr were found to have single exponential decays of 2.2 ns. Asn-Pro-Tyr-Val-Pro was found to decay as a sum of two exponentials with decay times of 2.2 and 0.9 ns and a value of 0.9 for the ratio of the amplitude of the longer decay component to that of the shorter one. The full 20-residue fragment had a double-exponential decay of 3.6 and 1.5 ns, with an amplitude ratio of 0.18 (longer:shorter). Haas *et al.*[186] also presented NMR data which they compared to their fluorescence results, and they concluded that in water at 25 °C there are locally ordered conformations in the 20-residue fragment that persist on a time scale intermediate between that of fluorescence (nanosecond) and NMR (millisecond).

1.5.1.5c. Apolipoproteins. Apolipoprotein A-II (apo A-II) is the second most abundant of the human high-density lipoproteins. It is a homodimer of 77 amino acid subunits, with each peptide chain containing no tryptophan and four tyrosines. Apo A-II and its association with lipid has been examined by spectroscopic methods, including circular dichroism, UV difference absorption on solvent perturbation, fluorescence quenching experiments, and fluorescence intensities as a function of both temperature and D_2O concentration.[189] The data suggest that two tyrosines are buried, two are exposed to solvent, and one of the exposed tyrosines becomes less solvated on binding lipid. A report on human apo A-IV also infers a burying of tyrosine on association with lipid.[190] However, since the experimental details for the fluorescence quenching experiments are lacking, and since apo A-IV also contains tryptophan, it is difficult to assess these results.

1.5.1.6. Peptide Hormones

There are several biologically important peptides which contain tyrosine but not tryptophan. These include small molecules with molecular weights of about 1000 or less. Molecules such as oxytocin, vasopressin, and tyrocidine A are cyclic, while others such as angiotensin II and enkephalin are linear. Schiller[191] has reviewed the literature up through 1984 on fluorescence of these and several other peptides. One major finding that has been reported recently is that the anisotropy and fluorescence intensity decays of many peptides are complex. This is especially evident in some of the tyrosine-containing peptides, and we expect that there will be considerable effort made over the next few years toward understanding the physical basis for these complex kinetics.

1.5.1.6a. Enkephalin. The conformation of the opioid peptide enkephalin [Tyr-Gly-Gly-Phe-Met (or -Leu); Met^5 and Leu^5 enkephalin, respectively] has been investigated by resonance energy transfer in the native peptide[56] and in Tyr^1-Trp^4 analogues, some of which were methylated to prevent formation of intramolecular hydrogen bonds.[191] Schiller[191] has concluded from these measurements that folded conformations need not be stabilized by $4 \rightarrow 1$ or $5 \rightarrow 2$ hydrogen bonds between amino and carbonyl groups of the peptide backbone or a hydrogen bond between the phenolic hydroxyl and a backbone carbonyl group.

The mobility of tyrosine in Leu^5 enkephalin was examined by Lakowicz and Maliwal,[170] who used oxygen quenching to measure lifetime-resolved steady-state anisotropies of a series of tyrosine-containing peptides. They measured a phase lifetime of 1.4 ns (30-MHz modulation frequency) without quenching, and they obtained apparent rotational correlation times of 0.18 ns and 0.33 ns, for Tyr^1 and the peptide. Their data analysis assumed a simple model in which the decays of the anisotropy due to the overall motion of the peptide and the independent motion of the aromatic residue are single exponentials and these motions are independent of each other.

Lakowicz *et al.*,[171] using multifrequency phase–modulation from 2 MHz to 2 GHz, have recently reexamined the intensity decay of Leu^5 enkephalin and were best able to fit their data with a triple exponential having time constants of 0.07, 0.32, and 1.36 ns with respective amplitudes of 0.12, 0.40, and 0.48. This gives a mean lifetime (Eq. 1.12) of 1.18 ns, somewhat shorter but more accurate than the previously measured 30-MHz phase lifetime of 1.4 ns. The anisotropy decay of the tyrosine fluorescence was best fit with a double exponential having rotational correlation times of 68 ± 20 and 247 ps. These correlation times are somewhat shorter than those obtained indirectly by the combination of oxygen quenching and steady-state anisotropy measurements, described above. The direct measurement, however, is expected to provide more accurate values for subnanosecond time-scale events.

1.5.1.6b. Oxytocin. The nine-residue cyclic hormone oxytocin has the sequence Cys-Tyr-Ile-Gln-Asn-Cys-Pro-Leu-Gly-NH_2, with the two cysteine residues joined by a disulfide bridge. The X-ray crystal structure model of the desamino analogue of oxytocin indicates that two of the $C^\alpha - C^\beta$ rotamers of Tyr-2 (rotamers I and II of Figure 1.5) could contact the disulfide bridge.[192]

It has long been known that the steady-state fluorescence of oxytocin is highly quenched.[1, 3] By comparing the time-resolved fluorescence (pH 3 and 5 °C) of oxytocin and its analogue desaminodicarbaoxytocin, in which the disulfide is replaced by an ethylene bridge, Ross *et al.*[68] found that the disulfide bridge is directly involved in the quenching and that the quenching involves a static mechanism (see Section 1.4.3). Whereas the fluorescence decay of oxytocin was a double exponential with time constants of about 0.7 and 1.9 ns and respective amplitudes of 0.75 and 0.25, that of the dicarba analogue could be fit by a triple exponential with either the identical time constants of oxytocin and an additional time constant of about 3 ns and respective amplitudes of 0.22, 0.08, and 0.70 or time constants of 0.78, 2.45, and 3.05 ns with respective amplitudes of 0.24, 0.08, and 0.68. The important feature of this result was that the amplitude ratio of the first two decay components of the dicarba analogue was 3:1, the same as for the decay components of oxytocin. The same 3:1 ratio was found for the NMR-determined populations of the $C^{\alpha}-C^{\beta}$ rotamers II and III in both the hormone and the analogue. On the basis of this correlation, Ross *et al.*[68] argued that rotamer I is interacting with the disulfide bridge in oxytocin, forming a nonradiative complex.

Lakowicz *et al.*[171, 193] examined the intensity and anisotropy decays of the tyrosine fluorescence of oxytocin at pH 7 and 25 °C. They found that the fluorescence decay was best fit by a triple exponential having time constants of 80, 359, and 927 ps with respective amplitudes of 0.29, 0.27, and 0.43. It is difficult to compare these results with those of Ross *et al.*[68] because of the differences in pH (3 vs. 7) and temperature (5° vs. 25 °C). For example, whereas at pH 3 the amino terminus of oxytocin is fully protonated, at pH 7 it is partially ionized, and since the tyrosine is adjacent to the amino terminal residue, the state of ionization could affect the tyrosine emission. The anisotropy decay at 25 °C was well fit by a double exponential with rotational correlation times of 454 and 29 ps. Following the assumptions described previously for the anisotropy decay of enkephalin, the longer correlation time was ascribed to the overall rotational motion of oxytocin, and the shorter correlation time was ascribed to torsional motion of the tyrosine side chain.

1.5.2. Fluorescence of Tyrosinate

A number of proteins that contain tyrosine but not tryptophan exhibit an emission band with a maximum in the wavelength region 315 to 350 nm at neutral pH, rather than at the expected wavelength near 305 nm. These red-shifted emissions cover the wavelength region that includes tyrosinate fluorescence, which in water has a maximum near 340 nm.[25] On the basis of this similarity, the red-shifted emission in these proteins lacking tryptophan has been identified variously as due to hydrogen bonding of the phenol hydroxyl group or due to tyrosinate formation in the excited state. In addition, the emission of the phenol excited-state dimer (excimer) also occurs near 340 nm.[194] Recently, a comparison of tyrosine absorption and fluorescence spectra of several proteins that lack tryptophan has been published in Russian by Khrapunov and Dragan.[195] According to the English translation of the abstract, they have classified tyrosine residues into three groups on the basis of their emission bands: hydrated phenols, hydrogen-bonded phenols, and phenols undergoing excited-state proton transfer forming tyrosinate. Our review of the following papers discusses issues about tyrosine and tyrosinate spectroscopy that can lead to misinterpretations of data.

For emission to occur from an ionized aromatic alcohol at neutral pH, it is clear that a specific mechanism must enhance proton transfer. The interaction promoting ionization could occur in either the ground state or the excited state. For example, the binding of equilenin to a steroid isomerase results in excitation and emission spectra that resemble those of the steroid analogue at high pH.[196] This is different from the hydrogen-bonding behavior seen on binding equilenin or 17β-dihydroequilenin to sex steroid binding protein.[71, 72] Rayner *et al.*[25] have shown that emission is observed near 350 nm for tyrosine at 293 K only if the pH is high enough for tyrosinate to exist in the ground state or if a high concentration of a strong proton-accepting buffer such as acetate is present under conditions where tyrosine is excited. As discussed earlier, excited-state proton transfer to bulk water will not be kinetically competitive with the other deactivation pathways. Consequently, excited-state proton transfer could become important for tyrosine only as the result of another deprotonation pathway due to the presence of a proton acceptor stronger than water.

It has been observed that ionization of the phenol hydroxyl group leads to a dramatic reduction in the fluorescence quantum yield.[1] Based on the ground-state and excited-state ionization constants for the tyrosine phenol ring, Rayner *et al.*[25] pointed out that it is extremely difficult to directly measure the quantum yields of the individual forms of tyrosine and tyrosinate because of the pK_as of the α-amino and α-carboxyl groups, although the value of 0.14 determined by Chen[197] for tyrosine near neutral pH is probably reasonable. Rayner *et al.*[25] argued that the pH dependence of the fluorescence quantum yields of tyrosine and tyrosinate does not fit a simple excited-state acid–base equilibrium model. Making a number of assumptions, they calculated a value of 0.16 for the fluorescence quantum yield of tyrosinate at neutral pH. Willis *et al.*[198] have measured a lifetime of 30 ps for tyrosinate. This implies that the quantum yield of tyrosinate is in fact much smaller than 0.16 and is in close agreement with the early assessment by Cornog and Adams[199] of 0.006 (adjusted for the revised quantum yield of tyrosine in water by Chen[197]). Recently, Willis and Szabo have examined the lifetime of tyrosinate in more detail.[22] By exciting only tyrosinate at 300 nm, between pH 9 and 11, they found that the ionization state of the α-amino group affected the lifetime. When the α-amino and α-carboxyl groups are both ionized, the fluorescence lifetime was found to be 69 ps. After deprotonation of the α-amino group with a pK_a of 9.7, the lifetime decreased to 26 ps, which agrees with their earlier work. Consequently, if tyrosinate emission is present in a protein sample, the expected lifetime is on the order of tens of picoseconds and the quantum yield should be very low.

In their study, Willis and Szabo also examined tyrosine at neutral pH in the presence of high buffer base concentrations.[22] Under these conditions, they did not observe tyrosinate, although a second emission band was observed that was of lower energy than that of tyrosine. This second emission was shown by decay-associated spectra (DAS) to be associated with a hydrogen-bonded species. The emission maxima of the DAS were dependent on the nature of the buffer ion, and the intensities were dependent on buffer concentration. The lifetime of this second emission was significantly longer than that expected for tyrosinate. In addition, this lifetime did not have a negative amplitude parameter as necessary for an excited-state reaction. Moreover, the excitation decay-associated spectrum (EDAS) of this component was that of hydrogen-bonded tyrosine, which is clearly distinct from that of tyrosinate.

Hasselbacher *et al.*[21] examined the possibility of tyrosinate being formed in a nonaqueous environment such as might be found in the interior of a protein. In this study, aromatic alcohols in cyclohexane and toluene, two low-dielectric solvents of differing polarity and polarizability, were titrated with the strong proton acceptor triethylamine (TEA). Ground-state hydrogen bond formation was demonstrated by absorption spectroscopy. A new fluorescence emission appeared on formation of the hydrogen-bonded complex that was to the red of the emission of the uncomplexed alcohol. While the fluorescence spectrum of the complex in cyclohexane was shifted only a few nanometers, the spectrum of the complex in toluene was shifted tens of nanometers and resembled that of the ionized alcohol in water. Linear combination of spectra (LINCS) of both the absorption and fluorescence spectra showed that only two species are present. Intensity decay studies revealed two lifetimes; based on several criteria, one lifetime could be assigned to the uncomplexed alcohol and the other lifetime to the hydrogen-bonded alcohol. The differences observed in several steady-state and time-resolved fluorescence parameters could be explained by the polarity and polarizability of the two solvent systems.

The above results for aqueous and nonaqueous conditions strongly suggest that the only way tyrosinate emission can be observed is if tyrosinate exists in the ground state and is directly excited. Ground-state tyrosinate can be identified by analysis of difference absorption spectra. If tyrosinate exists in the ground state, the protein absorption spectrum will have a significant contribution at 300 nm, and the extinction ratios at 280 and 295 nm will differ significantly from those predicted for the number of tryptophan and tyrosine residues in the protein.[200] The complete ionization of tyrosine in water produces an extinction difference spectrum with positive peaks near 240 and 300 nm; in a protein environment, these difference peaks may be shifted by a few nanometers.[201] Another way to demonstrate ground-state tyrosinate, especially since excited-state proton transfer is not likely to occur in the triplet state,[27] is to excite at 295–300 nm and look for tyrosinate phosphorescence (in the absence of oxygen). This can also be done in the presence of tryptophan because the phosphorescence spectra of tyrosinate and tryptophan are well resolved.[202] By contrast, phosphorescence or fluorescence resulting from excited-state ionization will have an excitation spectrum characteristic of tyrosine. Moreover, the decay kinetics of the product formed in the excited state will have a characteristic signature, specifically a kinetic component with a negative amplitude,[26] and short lifetime in the case of tyrosinate. In the presence of tryptophan, these processes will be difficult to recognize.

If the phenolic hydroxyl group is involved in a hydrogen bond in the ground state, then a tyrosine absorption spectrum shifted to the red should be observed.[17, 18] Lux *et al.*[112] have shown that the calcium-binding S100b protein from bovine brain and the calmodulins from ram testes and octopus have shifted tyrosine absorption spectra. The tyrosine fluorescence emission spectra are also slightly shifted to the red and have a broader bandwidth as compared to those of tyrosine, insulin, or ribonuclease. A "normal" tyrosine fluorescence is obtained upon denaturation, lowering the pH to 4, or raising the temperature to 50–60 °C. Furthermore, Lux *et al.*[112] noted that the binding of metal ions also perturbed the emission properties, although differently for the three proteins. They presented a strong argument that these spectral changes are all consistent with a hydrogen bond between the phenolic hydroxyl and an ionized carboxyl group. From the preceding discussions regarding tyrosine complexes with base buffers in water or aromatic alcohols complexed with TEA in organic solvents, the position of the spectrum of the hydrogen-bonded complex could reflect the polarizability of the environment as well as the acceptor strength of the interacting group(s).

The S100b protein has also been examined by time-resolved fluorescence.[156] The intensity decay was resolved into three exponentials. All three amplitudes and lifetimes were emission wavelength, pH, and metal ion dependent. Examination of these data suggests a problem with light scatter and stray light. The longest lifetime was assigned to tyrosinate since its relative amplitude appears to increase with increasing emission wavelength. This lifetime ranges from 3.6 to 14.5 ns, which is not characteristic of tyrosinate (*vide supra*). Thus, the intensity decay characteristics of the S100b protein should be reinvestigated.

In 1971, adrenodoxin, an iron–sulfur protein with a single tyrosine residue and no tryptophan, was shown to fluoresce at 331 nm upon 280-nm excitation at neutral pH.[203] On cooling from room temperature to 77 K, the emission maximum shifts to 315 nm. The redox state of the iron does not have any effect on the tyrosine emission. From these results, an exciplex between the excited singlet state of tyrosine and an unidentified group was suggested as the cause of the anomalous emission energy.[203] Later studies have shown that the excitation spectrum is a red-shifted tyrosine spectrum, that removal of the iron to form the apoprotein has no effect on the emission, and that heat, low pH, guanidine hydrochloride, urea, and LiCl all cause the emission to revert to 305 nm.[204-206] Based on a laser Raman study, it was found that the phenol hydroxyl group of the single tyrosine residue was hydrogen bonded to the carbonyl of a nearby carboxylic group.[207] Therefore, based on our considerations discussed above, it appears that there is a ground-state hydrogen-bonded complex in adrenodoxin that does *not* ionize in the excited state but remains as a temperature-sensitive complex, and this complex is obviously dependent on the conformation of the protein.

Plastocyanin from parsley, a copper protein of the chloroplast involved in electron transport during photosynthesis, has been reported to have a fluorescence emission maximum at 315 nm on excitation at 275 nm at pH 7.6.[208] Since the protein does not contain tryptophan, but does have three tyrosines, and since the maximum wavelength shifts back to 304 nm on lowering the pH to below 2, the fluorescence was attributed to the emission of the phenolate anion in a low-polarity environment. From this, one would have to assume that all three tyrosines are ionized. A closer examination of the reported emission spectrum, however, indicates that two emission bands seem to be present. If a difference emission spectrum is estimated (spectrum at neutral pH minus that at pH 2 in Figure 5 of Ref. 207), a "tyrosinate-like" emission should be obtained.

The tyrosine fluorescence from the H1 histone from the fruit fly *Ceratitis capitata* has been investigated.[92] This protein, which has two tyrosines, exhibits an emission at 303 nm and a shoulder at 340 nm. The intensity of this shoulder is dependent on the conformation of the protein and on the ionic state of a group with an apparent pK_a of 3.7. Thus, the emission of the second tyrosine residue, which is located on the surface of the insect H1 but is not present in the calf thymus H1, was attributed to excited-state proton transfer. It was noted that both the insect and calf thymus H1 proteins exhibit a red-shifted absorption on folding from the random coil state, which was interpreted as a result of transferring tyrosine from an aqueous to a nonaqueous environment. The difference absorption spectra[97] do not indicate ground-state tyrosinate but are what is expected for a hydrogen-bonded complex. In our view, based on the study by Willis and Szabo,[22] it is unlikely that the group with a pK_a of 3.7 forms the hydrogen bond since it would be too weak a base to shift the emission to 340 nm. However, the protein could undergo a conformational change controlled by a group with a pK_a of 3.7 that alters the ability of the hydroxyl group to interact with a stronger base in a polar, nonaqueous environment.

The absorption of the tyrosines in pig intestinal Ca^{2+}-binding protein is reported to be shifted to longer wavelengths; the intrinsic fluorescence, however, is in the normal energy region for tyrosine emission with a possibility of some emission from tyrosinate.[143, 145] These results can be equally well explained by a ground-state, hydrogen-bonded complex.

The H1 histone from calf thymus shows three different emission bands

(Section 1.5.1.1a), with the 340-nm peak being assigned to tyrosinate emission.[94] The 340-nm band is highly sensitive to salt concentrations: as the protein folds, this emission is lost. This band, however, was not sensitive to pH in the 4 to 7 region in the absence of salt.

Prendergast *et al.*[209] have examined a series of highly homologous, basic cytotoxic proteins, the purothionins, which contain a single tyrosine residue and no tryptophan. At a pH greater than 4, β-purothionin only has a single emission band with a maximum at 345 nm. At neutral pH, α_1- and α_2-purothionin exhibit two emissions, of equal intensity, at 308 and 345 nm. The α_1-purothionin displays an apparent pK_a in the 2 to 4 region for the loss/gain of intensity at 345/308 nm. Furthermore, denaturation of the native structure of the α_1- and α_2-purothionins resulted in emission at 303 nm. Consequently, Prendergast *et al.*[209] concluded that tyrosinate was being formed by intramolecular proton transfer. Since the absorption spectra are only slightly shifted to the red, which they interpreted as due to ground-state hydrogen bonding, and do not indicate any ground-state tyrosinate, they also concluded that the proton transfer is occurring in the excited state. At $-65\,°C$, the β-purothionin emission shifts from 345 to 323 nm, which was attributed to tyrosinate emission being solvent dependent. An alternate interpretation, however, is that the emission shift reflects the enthalpy, and hence the temperature dependence of the acceptor strength, of the hydrogen-bonded complex.

The tyrosinate fluorescence observed with bovine testes calmodulin is argued to be due to tyrosinate in the ground state.[123] Of the two tyrosine residues in this calmodulin, Tyr-99 apparently has a low pK_a near 7 for the formation of tyrosinate, which is most likely due to nearby side chains that are involved in calcium binding. These groups could then also account for the complex pH dependence of the 345-nm emission intensity. Besides the tyrosine and tyrosinate emissions at 305 and 345 nm, respectively, Pundak and Roche[123] also reported the existence of a third emission band between 312 and 320 nm. This band was similar in its pH and calcium dependence to the other residue, Tyr-138, and was speculated to be a result of a combination of contributions from the tyrosine and tyrosinate emissions. Since this band has its excitation profile shifted to the red, however, it could be that a hydrogen-bonded tyrosine exists in this calmodulin. Alternatively, it has also been found that the presence of the 345-nm emission depends upon the method of preparation (G. Sanyal, personal communication).

Rat oncomodulin, a parvalbumin-like tumor protein that has two tyrosine residues but no tryptophan, exhibits fluorescence emission at 301 and 345 nm.[135] Upon binding two moles of Ca^{2+} per mole of oncomodulin, the 301-nm intensity increases while the 345-nm band decreases. These results were explained in terms of acidic side chains involved in either binding Ca^{2+} or accepting a proton on excited-state generation of tyrosinate. The cloned gene product, however, has no emission at 345 nm, suggesting that the rat preparation was contaminated.[210]

An additional emission band near 350 nm has been observed for lima bean trypsin inhibitor (LBTI).[173] The authors discussed both the possibility of contamination by tryptophan and excited-state tyrosinate formation. Since this 350-nm emission has a tyrosine-like excitation spectrum that is slightly shifted compared to that of the major 302-nm emission, it is also possible that the tyrosine residue in a fraction of the LBTI molecules could be hydrogen bonded. This model is supported by the observations that the phenol side chain is shielded from solvent and has an anomalously high pK_a.

It has been reported that oxytocin, but not vasopressin, forms a stable intramolecular hydrogen bond with the Asn^5 carboxamide side chain in propylene glycol and that this leads to emission from tyrosinate.[211] Time-resolved studies assigned the longest lifetime (18.5 ns) to the emission from tyrosinate. Unfortunately, lack of experimental details and data, an unrealistic lifetime attributed to tyrosinate, and uncertainties concerning the purity of the solvents raise questions about this report.

Because tyrosinate fluorescence occurs at energies where tryptophan residues typically emit, its presence is likely to be masked in tryptophan-containing proteins. Nevertheless, fluorescence emission of tyrosinate has been identified by Longworth[212] in human serum albumin, a protein with a single tryptophan and 18 tyrosines. This emission was found by making several assumptions about relative quantum yields and absorption extinctions, normalizing emission spectra of the protein and model compounds taken at different excitation wavelengths, and calculating difference emission spectra. Another example of suspected tyrosinate emission in the presence of tryptophan has been reported by Pearce and Hawrot for binding-site fragments of the nicotinic acetylcholine receptor and their interaction with α-bungarotoxin.[213]

It is interesting to note that the first "demonstration" of tyrosinate fluorescence in a protein was made by Szabo *et al.*[214] with two cytotoxins from the Indian cobra *Naja naja*. While exhibiting different relative amounts of the two emission bands, both toxins had fluorescence at 304 and 345 nm, with the 304-nm band being greatly reduced on excitation at 290 nm. Since these proteins have three tyrosine residues and no tryptophan, it was concluded that the 345-nm emission band was due to tyrosinate. Furthermore, tyrosinate appeared to be formed in the excited state from a hydrogen-bonded ground-state complex based on the absorption spectra. Szabo subsequently reexamined these peptide samples and found that they were contaminated with tryptophan (A. G. Szabo, personal communication). While Szabo's approach to the demonstration of tyrosinate fluorescence was correct based on his initial data, his subsequent finding exemplifies an important caution: if tyrosinate emission is suspected, every effort must be made to demonstrate the *absence* of tryptophan. The presence of a single band on gel electrophoresis and the report that the protein does not contain tryptophan does not mean that the sample is necessarily pure in terms of fluorescence. Attempts should be made to determine the presence of indole by chemical methods. The sample should also be denatured and any disulfide bonds reduced. It should then be examined at low pH by difference and derivative absorption spectroscopies as well as by steady-state excitation and emission fluorescence spectra: under these conditions, neither tryptophan nor tyrosinate absorption or fluorescence bands should be detected. In addition, fluorescence lifetime measurements of the native and denatured protein should be compared. Particular attention should be made to the decay times since a lifetime on the picosecond time scale is expected for tyrosinate emission.[22, 198] Our personal experience in working with tyrosine-containing proteins has shown that special precautions are necessary when tyrosinate emission is suspected. Contaminating emissions in the tyrosinate spectral region can come from various sources, including dialysis tubing, apparently clean glassware, and buffer salts. In view of these considerations and the recent studies questioning the existence of tyrosinate emission at neutral pH or in a nonaqueous environment, it is our opinion that all of the proteins reported to have tyrosinate or hydrogen-bonded tyrosine emission should be reinvestigated.

1.5.3. Phosphorescence and ODMR of Proteins and Polypeptides

Phosphorescence and ODMR are additional spectroscopies that can be used to investigate intramolecular interactions that affect tyrosine residues in proteins and polypeptides.[215, 216] An example is tyrosine and tyrosinate in horse liver alcohol dehydrogenase.[202] The same approach has been used to study the role of tyrosine in the mechanism of action of carboxypeptidase B.[217, 218] In both these proteins, as in other proteins which contain both tyrosine and tryptophan, the tyrosine fluorescence is difficult to resolve from the tryptophan fluorescence. The tyrosine phosphorescence, however, is better resolved from the tryptophan phosphorescence since the high-energy edges of their emission bands are separated by about 50 nm; the high-energy edge of tyrosine phosphorescence begins near 350 nm whereas that of tryptophan typically begins near 400 nm.

In the case of horse liver alcohol dehydrogenase, a homodimeric enzyme, Subramanian *et al.*[202] used the relative phosphorescence of tyrosine and tryptophan to examine the effects of various ternary complexes known to selectively quench the fluorescence of the tryptophans of each subunit. One proposed quenching mechanism is the formation of a ground-state tyrosinate in a ternary complex at neutral pH.[201] This tyrosinate, by being a resonance energy transfer acceptor, could provide a means of specifically quenching the buried tryptophan which is at the subunit interface. Since tyrosinate can be selectively excited at wavelengths where tyrosine does not absorb (see Section 1.2), Subramanian *et al.*[202] argued that it should be possible to detect the tyrosinate directly through its phosphorescence. However, they were unable to detect any tyrosinate phosphorescence in ternary complexes at neutral pH. It is not possible to exclude completely ground-state formation of tyrosinate by monitoring phosphorescence emission, however, since either the triplet state or the singlet state could be deactivated by an efficient nonradiative process.

In the case of carboxypeptidase B, Shaklai *et al.*[217] compared the relative contributions to the protein phosphorescence from tyrosine and tryptophan for the apoenzyme, the zinc-containing metalloenzyme in the absence of substrate, the metalloenzyme in the presence of the substrate *N*-acetyl-L-arginine, and the metalloenzyme in the presence of the specific inhibitor L-arginine. The tyrosine:tryptophan emission ratio of the metalloenzyme was about a factor of four smaller than that of the apoenzyme. Binding of either the substrate or the inhibitor led to an increase in the emission ratio to a value similar to that of the apoenzyme. The change in the tyrosine: tryptophan phosphorescence ratio was attributed to an interaction between a tyrosine and the catalytically essential zinc. The emission ratio was also studied as a function of pH. The titration data are difficult to interpret, however, because a Tris buffer was used and the ionization of Tris is strongly temperature dependent. In general, the use of Tris buffers for phosphorescence studies should be avoided.

There are few reports of ODMR of tyrosine in proteins or polypeptides. In 1977, Ugurbil *et al.*[219] reported the phosphorescence and ODMR signals of tyrosine in two species of azurin, *Pseudomonas aeruginosa*, which contains both tryptophan and tyrosine, and *Pseudomonas fluorescens*, which contains only tyrosine. From their observations, they concluded that the tryptophan in *P. aeruginosa* did not completely quench the excited singlet state of tyrosine. The tyrosine phosphorescence emission of *P. aeruginosa* is shifted to the red by 4 nm compared with that of *P. fluorescens*, and their zfs differ slightly. Although the zfs of both of these copper-containing proteins appeared to be independent of the copper oxidation state, the zero-field parameter D increased slightly upon removal of the metal. The major effect of the oxidation state was on the spin polarization of the triplet state. That is, the intensities of the ODMR transitions were affected by the oxidation state.

In 1979, Ross *et al.*[220] measured the ODMR of tyrosine in glucagon and the derivative [12-homoarginine]glucagon to examine the effect of chemical modification of a lysine residue adjacent to Tyr-10 and Tyr-13. The guanidinated analogue had lower potency than glucagon in a fat cell hormone receptor assay. Since the tyrosine ODMR and other spectral properties of the polypeptide, including circular dichroism, were essentially identical, it was concluded that guanidination did not affect the conformation and that the loss of potency probably was due to a specific interaction between Lys-12 and the fat cell hormone receptor.

1.6. Tyrosine as an Excited-State Probe for Conformation and Dynamics

The major reasons for using intrinsic fluorescence and phosphorescence to study conformation are that these spectroscopies are extremely sensitive, they provide many specific parameters to correlate with physical structure, and they cover a wide time range, from picoseconds to seconds, which allows the study of a variety of different processes. The time scale of tyrosine fluorescence extends from picoseconds to a few nanoseconds, which is a good time window to obtain information about rotational diffusion, intermolecular association reactions, and conformational relaxation in the presence and absence of cofactors and substrates. Moreover, the time dependence of the fluorescence intensity and anisotropy decay can be used to test predictions from molecular dynamics.[167] In using tyrosine to study the dynamics of protein structure, it is particularly important that we begin to understand the basis for the anisotropy decay of tyrosine in terms of the potential motions of the phenol ring.[221] For example, the frequency of flips about the $C^{\alpha}-C^{\beta}$ bond of tyrosine appears to cover a time range from milliseconds to nanoseconds.[222]

Essentially nothing is known about tyrosine phosphorescence at ambient temperatures. In frozen solution, tyrosine residues have a phosphorescence decay of seconds. We would expect, however, a decay of milliseconds or shorter at ambient temperature. Observation of tyrosine phosphorescence from proteins in liquid solution will undoubtedly require efficient removal of oxygen. Nevertheless, it could be fruitful to explore ambient temperature measurements, since the phosphorescence decay could extend the range of observation of excited-state dynamics into the microsecond, or even millisecond, time range.

The important fluorescence and phosphorescence parameters that can be used to obtain information about physical structure are quantum yield, spectral linewidth and energy, and intensity and anisotropy decay times. These parameters can reveal quenching processes, excited-state reactions, and other dynamic changes that occur during the excited-state lifetime, thus providing an important part of the structural characterization of a protein or polypeptide in solution. Examples we have discussed here include detection of ionizable side chains and hydrogen bond and/or tyrosinate formation with the phenol hydroxyl group. In favorable cases, such as the calcium-binding proteins, the excited-state parameters can also provide information about biological function. Based on the current and improving state of the art of fluorescence and phosphorescence, we expect that tyrosine will prove increasingly valuable as an intrinsic spectroscopic probe.

Acknowledgments

We acknowledge the support of National Institutes of Health Grants HD/GM 17542 (J.B.A.R.), GM 39750 (J.B.A.R.), DK 39548 (W.R.L.), and DK 10080 (H.R.W.), the Jack Malamud Private Foundation (H.R.W.), and the Northwest Area Foundation Grant of Research Corporation (K.W.R.). Some of the work presented was supported by National Science Foundation Biological Instrumentation Award DMB-8516318 (J.B.A.R. and W.R.L.). We also thank Drs. Ludwig Brand, Gautam Sanyal, Arthur G. Szabo, and Kevin J. Willis for helpful discussions on tyrosine and tyrosinate fluorescence, Drs. Josef Eisinger and Carol A. Hasselbacher for critical readings of this chapter, and Dr. Steven Prestrelski for helping make some of the figures.

References

1. J. W. Longworth, Luminescence of polypeptides and proteins, in: *Excited States of Proteins and Nucleic Acids* (R. F. Steiner and I. Weinryb, eds.), pp. 319–484, Plenum Press, New York (1971).
2. S. V. Konev, *Fluorescence and Phosphorescence of Proteins and Nucleic Acids*, Plenum Press, New York (1967).
3. R. W. Cowgill, Tyrosyl fluorescence in proteins and model peptides, in: *Biochemical Fluorescence: Concepts 2* (R. F. Chen and H. Edelhoch, eds.), pp. 441–486, Marcel Dekker, New York (1976).
4. D. Creed, The photophysics and photochemistry of the near-UV absorbing amino acids—II. Tyrosine and its simple derivatives, *Photochem. Photobiol. 39*, 563–575 (1984).
5. P. Debye and J. O. Edwards, A note on the phosphorescence of proteins, *Science 116*, 143–144 (1952).
6. R. H. Steele and A. Szent-Gyorgyi, On excitation of biological substances, *Proc. Natl. Acad. Sci. U.S.A. 43*, 477–491 (1957).
7. G. Weber, Rotational Brownian motion and polarization of the fluorescence of solutions, *Adv. Protein Chem. 8*, 415–457 (1953).
8. D. Duggan and S. Udenfriend, The spectrofluorometric determination of tryptophan in plasma and of tryptophan and tyrosine in protein hydrolysates, *J. Biol. Chem. 223*, 313–319 (1956).
9. V. G. Shore and A. B. Pardee, Fluorescence of some proteins, nucleic acids and related compounds, *Arch. Biochem. Biophys. 60*, 100–107 (1956).
10. S. V. Konev, Fluorescence spectra and spectra of action of fluorescence in proteins, *Dokl. Akad. Nauk. SSSR 116*, 594–597 (1957).
11. Y. A. Vladimirov, Fluorescence of aromatic amino acids, *Dokl. Akad. Nauk. SSSR 116*, 780–783 (1957).
12. F. W. J. Teale and G. Weber, Ultraviolet fluorescence of the aromatic amino acids, *Biochem. J. 65*, 476–482 (1957).

13. G. H. Beaven and E. R. Holiday, Ultraviolet absorption spectra of proteins and amino acids, *Adv. Protein Chem. 7,* 319–386 (1952).
14. D. B. Wetlaufer, Ultraviolet spectra of proteins and amino acids, *Adv. Protein Chem. 17,* 303–390 (1962).
15. T. M. Hooker and J. A. Schellman, Optical activity of aromatic chromophores. I. *o, m,* and *p*-Tyrosine, *Biopolymers 9,* 1319–1348 (1970).
16. J. R. Platt, Classification of spectra of cata-condensed hydrocarbons, *J. Phys. Chem. 17,* 484–495 (1949).
17. G. C. Pimentel, Hydrogen bonding and electronic transitions: The role of the Franck–Condon principle, *J. Am. Chem. Soc. 79,* 3323–3326 (1957).
18. G. J. Brealey and M. Kasha, The role of hydrogen bonding in the $n \rightarrow \pi^*$ blue-shift phenomenon, *J. Am. Chem. Soc. 77,* 4462–4468 (1955).
19. D. A. Chignell and W. B. Gratzer, Solvent effects on aromatic chromophores and their relation to ultraviolet difference spectra of proteins, *J. Phys. Chem. 72,* 2934–2941 (1968).
20. S. Nagakura and M. Gouterman, The effect of H bonding on the near ultraviolet absorption of naphthol, *J. Phys. Chem. 26,* 881–886 (1957).
21. C. A. Hasselbacher, E. Waxman, L. T. Galati, P. B. Contino, J. B. A. Ross, and W. R. Laws, Investigation of hydrogen bonding and proton transfer of aromatic alcohols in non-aqueous solvents by steady-state and time-resolved fluorescence, *J. Phys. Chem. 95,* 2995–3005 (1991).
22. K. J. Willis and A. G. Szabo, The fluorescence decay kinetics of tyrosinate and tyrosine hydrogen bonded complexes, *J. Phys. Chem. 95,* 1585–1589 (1991).
23. I. Weinryb and R. F. Steiner, The luminescence of the aromatic amino acids, in: *Excited States of Proteins and Nucleic Acids* (R. F. Steiner and I. Weinryb, eds.), pp. 277–318, Plenum Press, New York (1971).
24. K. W. Rousslang, Optical detection of magnetic resonance in aromatic amino acids and proteins, Dissertation, University of Washington, Seattle, Washington (1976).
25. D. M. Rayner, D. T. Krajcarski, and A. G. Szabo, Excited-state acid–base equilibrium of tyrosine, *Can. J. Chem. 56,* 1238–1245 (1978).
26. W. R. Laws and L. Brand, Analysis of two-state excited-state reactions. The fluorescence decay of 2-naphthol, *J. Phys. Chem. 83,* 795–802 (1979).
27. C. A. Parker, *Photoluminescence of Solutions,* Elsevier, New York (1968).
28. S. P. McGlynn, T. Azumi, and M. Kinoshita, *Molecular Spectroscopy of the Triplet State,* Prentice-Hall, Englewood Cliffs, New Jersey (1969).
29. B. Smaller, E. C. Avery, and J. R. Remko, Triplet-state zero-field-splitting correlations in substituted molecules, *J. Chem. Phys. 46,* 3976–3983 (1967).
30. M. Ptak and P. Douzou, Examination of optically excited amino-acids by electron spin resonance at very low temperature, *Nature 199,* 1092 (1963).
31. T. Shiga and L. H. Piette, Triplet state studies of flavins by electron paramagnetic resonance—II, *Photochem. Photobiol. 3,* 223–230 (1964).
32. J. E. Maling, K. Rosenheck, and M. Weissbluth, Triplet ESR in L-tyrosine, *Photochem. Photobiol. 4,* 241–249 (1965).
33. J. Zuclich, Triplet-state electron paramagnetic resonance of the aromatic amino acids and proteins, *J. Chem. Phys. 52,* 3586–3591 (1970).
34. J. Zuclich, D. Schweitzer, and A. H. Maki, Optically detected magnetic resonance of the tryptophan phosphorescent state in proteins, *Photochem. Photobiol. 18,* 161–168 (1973).
35. A. L. Kwiram, Optical detection of magnetic resonance in molecular triplet states, in: *MTP International Review of Science, Ser. 1, Physical Chemistry 4* (C. A. McDowell, ed.), pp. 271–315, University Park Press, Baltimore (1972).
36. T.-T. Co, J. Hoover, and A. H. Maki, Dynamics of the tyrosine triplet state from magnetic resonance saturated phosphorescence decay measurements, *Chem. Phys. Lett. 27,* 5–9 (1974).

37. K. W. Rousslang and A. L. Kwiram, Triplet state decay and spin–lattice relaxation rate constants in tyrosinate and tryptophan, *Chem. Phys. Lett.* *39*, 226–230 (1976).
38. W. R. Laws, J. B. A. Ross, H. R. Wyssbrod, J. M. Beechem, L. Brand, and J. C. Sutherland, Time-resolved fluorescence and ^{1}H NMR studies of tyrosine and tyrosine analogues: Correlation of NMR-determined rotamer populations and fluorescence kinetics, *Biochemistry* *25*, 599–607 (1986).
39. J. P. Greenstein and M. Winitz, *Chemistry of the Amino Acids*, p. 498, Wiley, New York (1961).
40. J. Feitelson, On the mechanism of fluorescence quenching. Tyrosine and similar compounds, *J. Phys. Chem.* *68*, 391-397 (1964).
41. P. Gauduchon and P. Wahl, Pulse fluorimetry of tyrosyl peptides, *Biophys. Chem.* *8*, 87–104 (1978).
42. R. W. Cowgill, Fluorescence and protein structure X. Reappraisal of solvent and structural effects, *Biochim. Biophys. Acta* *133*, 6–18 (1967).
43. J. E. Tournon, E. Kuntz, and M. A. El Bayoumi, Fluorescence quenching in phenylalanine and model compounds, *Photochem. Photobiol.* *16*, 425–433 (1972).
44. I. H. Munro and N. Schwentner, Time resolved spectroscopy using synchrotron radiation, *Nucl. Instrum. Methods* *208*, 819–834 (1983).
45. J. M. Beechem, J. R. Knutson, J. B. A. Ross, B. W. Turner, and L. Brand, Global resolution of heterogeneous decay by phase/modulation fluorometry: Mixtures and proteins, *Biochemistry* *22*, 6054–6058 (1983).
46. J. R. Knutson, J. M. Beechem, and L. Brand, Simultaneous analysis of multiple fluorescence decay curves: A global approach, *Chem. Phys. Lett.* *102*, 501–507 (1983).
47. J. B. A. Ross, W. R. Laws, J. C. Sutherland, A. Buku, P. G. Katsoyannis, I. L. Schwartz, and H. R. Wyssbrod, Linked-function analysis of fluorescence decay kinetics: Resolution of side-chain rotamer populations of a single aromatic amino acid in small polypeptides, *Photochem. Photobiol.* *44*, 365–370 (1986).
48. P. B. Contino and W. R. Laws, Rotamer-specific fluorescence quenching in tyrosinamide: Dynamic and static interactions, *J. Fluorescence* *1*, 5–13 (1991).
49. R. S. Becker, *Theory and Interpretation of Fluorescence and Phosphorescence*, Wiley-Interscience, New York (1969).
50. K. W. Rousslang, unpublished results.
51. T. Förster, *Fluoreszenz Organischer Verbindungen*, Vandenhoeck and Ruprecht, Göttingen (1951).
52. D. L. Dexter, A theory of sensitized luminescence in solids, *J. Chem. Phys.* *21*, 836–850 (1953).
53. R. E. Dale, J. Eisinger, and W. E. Blumberg, The orientational freedom of molecular probes. The orientation factor in intramolecular energy transfer, *Biophys. J.* *26*, 161–194 (1979).
54. J. Eisinger, B. Feuer, and A. A. Lamola, Intramolecular singlet excitation transfer. Applications to polypeptides, *Biochemistry* *8*, 3908–3915 (1969).
55. J. Eisinger, Intramolecular energy transfer in adrenocorticotropin, *Biochemistry* *8*, 3902–3908 (1969).
56. M. Kupryszewska, I. Gryczynski, and A. Kawski, Intramolecular donor–acceptor separations in methionine- and leucine-enkephalin estimated by long-range radiationless transfer of singlet excitation energy, *Photochem. Photobiol.* *36*, 499–502 (1982).
57. K. W. Rousslang and A. L. Kwiram, Triplet–triplet energy transfer in the tryptophyl-tyrosinate dipeptide, *Chem. Phys. Lett.* *39*, 231–235 (1976).
58. M. A. El-Sayed, Optical pumping of the lowest triplet state and multiple resonance optical techniques in zero field, *J. Chem. Phys.* *54*, 680–691 (1971).
59. J. B. A. Ross, K. W. Rousslang, A. G. Motton, and A. L. Kwiram, Base interactions in the triplet states of NAD^{+} and NADH, *Biochemistry* *18*, 1808–1813 (1979).

60. M. R. Eftink and C. A. Ghiron, Fluorescence quenching studies with proteins, *Anal. Biochem. 114*, 199–227 (1981).
61. N. S. Kosower and E. M. Kosower, The glutathione–glutathione disulfide system, *Free Radicals Biol. 2*, 55–84 (1976).
62. D. Creed, The photophysics and photochemistry of the near-UV absorbing amino acids—III. Cystine and its simple derivatives, *Photochem. Photobiol. 39*, 577–583 (1984).
63. A. Shafferman and G. Stein, The effect of aromatic amino acids on the photochemistry of a disulfide: Energy transfer and reaction with hydrated electrons, *Photochem. Photobiol. 20*, 399–406 (1974).
64. R. W. Cowgill, Fluorescence and protein structure XI. Fluorescence quenching by disulfide and sulfhydryl groups, *Biochim. Biophys. Acta 140*, 37–44 (1967).
65. V. O. Stern and M. Volmer, On the quenching-time of fluorescence, *Physik. Zeitschr. 20*, 183–188 (1919).
66. W. R. Laws and P. B. Contino, Fluorescence quenching studies: Analysis of non-linear Stern–Volmer data, *Methods Enzymol. 210* (in press).
67. A. Follenius and D. Gerard, Acrylamide fluorescence quenching applied to tyrosyl residues in proteins, *Photochem. Photobiol. 38*, 373–376 (1983).
68. J. B. A. Ross, W. R. Laws, A. Buku, J. C. Sutherland, and H. R. Wyssbrod, Time-resolved fluorescence and ^{1}H NMR studies of tyrosyl residues in oxytocin and small peptides: Correlation of NMR-determined conformations of tyrosyl residues and fluorescence decay kinetics, *Biochemistry 25*, 607–612 (1986).
69. J. K. Swadesh, P. W. Mui, and H. A. Scheraga, Thermodynamics of the quenching of tyrosyl fluorescence by dithiothreitol, *Biochemistry 26*, 5761–5769 (1987).
70. M. V. Smoluchowski, Mathematical theory of the kinetics of the coagulation of colloidal solutions, *Z. Phys. Chem. 92*, 129–168 (1917).
71. A. Örstan, M. F. Lulka, B. Eide, P. H. Petra, and J. B. A. Ross, The steroid-binding site of human and rabbit sex steroid-binding protein of plasma: Fluorescence characterization with equilenin, *Biochemistry 25*, 2686–2692 (1986).
72. E. Casali, P. H. Petra, and J. B. A. Ross, Fluorescence investigation of the sex steroid binding protein of rabbit serum: Steroid and subunit dissociation, *Biochemistry 29*, 9334–9343 (1990).
73. N. Mataga and Y. Kaifu, Intermolecular proton transfer in the excited hydrogen-bonded complex in nonpolar solvent and fluorescence quenching due to hydrogen bonding, *J. Phys. Chem. 36*, 2804–2805 (1962).
74. A. Matsuzaki, S. Nagakura, and K. Yoshihara, Interactions of β-naphthol and β-naphthylamine in their excited singlet states with triethylamine, *Bull. Chem. Soc. Jpn. 47*, 1152–1157 (1974).
75. *Methods Enzymol. 210* (in press).
76. D. James and W. R. Ware, A fallacy in the interpretation of fluorescence decay parameters, *Chem. Phys. Lett. 120*, 455–459 (1985).
77. J. R. Alcala, E. Gratton, and F. G. Prendergast, Interpretation of fluorescence decays in proteins using continuous lifetime distributions, *Biophys. J. 51*, 925–936 (1987).
78. H. Szmacinski, R. Jayaweera, H. Cherek, and J. R. Lakowicz, Demonstration of an associated anisotropy decay by frequency-domain fluorometry, *Biophys. Chem. 27*, 233–241 (1987).
79. K. J. Willis, A. G. Szabo, J. Drew, M. Zuker, and J. M. Ridgeway, Resolution of heterogeneous fluorescence into component decay-associated excitation spectra, *Biophys. J. 57*, 183–189 (1990).
80. C. Helene, T. Montenay-Garestier, and J. L. Dimicoli, Interactions of tyrosine and tyramine with nucleic acids and their components. Fluorescence, nuclear magnetic resonance, and circular dichroism studies, *Biochim. Biophys. Acta 254*, 349–365 (1971).

81. O. Shimizu and K. Imakubo, New emission band of tyrosine induced by interaction with phosphate ion, *Photochem. Photobiol. 26*, 541–543 (1977).
82. O. Shimizu, J. Watanabe, and K. Imakubo, Effect of phosphate ion on fluorescence characteristics of tyrosine and its conjugate base, *Photochem. Photobiol. 29*, 915–919 (1979).
83. T. Alev-Behmoaras, J.-J. Toulme, and C. Helene, Quenching of tyrosine fluorescence by phosphate ions: A model study for protein–nucleic acid complexes, *Photochem. Photobiol. 30*, 533–539 (1979).
84. N. C. Verma, Fluorescence from L-tyrosine and its quenching by phosphate ions and deoxyribonucleic acid, *Indian J. Biochem. Biophys. 22*, 218–222 (1985).
85. L. J. Libertini and E. W. Small, Salt induced transitions of chromatin core particles studied by tyrosine fluorescence anisotropy, *Nucleic Acids Res. 8*, 3517-3534 (1980).
86. L. J. Libertini and E. W. Small, Effects of pH on low-salt transition of chromatin core particles, *Biochemistry 21*, 3327–3334 (1982).
87. I. Ashikawa, Y. Nishimura, M. Tsuboi, K. Watanabe, and K. Iso, Lifetime of tyrosine fluorescence in nucleosome core particles, *J. Biochem. (Tokyo) 91*, 2047–2055 (1982).
88. A. Mozo-Villarias, Fluorescence study of histone tyrosyl residues of DNA, *Biochem. Biophys. Res. Commun. 122*, 656–661 (1984).
89. L. J. Libertini and E. W. Small, Effects of pH on the stability of chromatin core particles, *Nucleic Acids Res. 12*, 4351–4359 (1984).
90. V. Giancotti, M. Fonda, and C. Crane-Robinson, Tyrosine fluorescence of two tryptophan-free proteins: Histones H1 and H5, *Biophys. Chem. 6*, 379–383 (1977).
91. V. Giancotti, F. Quadrifoglio, R. W. Cowgill, and C. Crane-Robinson, Fluorescence of buried tyrosine residues in proteins, *Biochim. Biophys. Acta 624*, 60–65 (1980).
92. J. Jordano, J. L. Barbero, F. Montero, and L. Franco, Fluorescence of histones H1. A tyrosinate-like fluorescence emission in *Ceratitis capitata* H1 at neutral pH values, *J. Biol. Chem. 258*, 315–320 (1983).
93. S. N. Khrapunov, A. I. Dragan, A. F. Protas, and G. D. Berdyshev, The structure of the histone dimer H2A–H2B studied by spectroscopy, *Biochim. Biophys. Acta 787*, 97–104 (1984).
94. L. J. Libertini and E. W. Small, The intrinsic fluorescence of histone H1. Steady-state and fluorescence decay studies reveal heterogeneous emission, *Biophys. J. 47*, 765–772 (1985).
95. L. De Petrocelis, G. Quagliarotti, L. Tomei, and G. Geraci, Structuring of H1 histone. Evidence of high-affinity binding sites for phosphate ions, *Eur. J. Biochem. 156*, 143–148 (1986).
96. R. Amado, R. Aeschbach, and H. Neukom, Dityrosine: *In vitro* production and characterization, *Methods Enzymol. 107*, 377–388 (1984).
97. R. Carallero, B. Fernandez, and F. Montero, Influence of carboxyl groups on conformation of histone H1 from *Ceratitis capitata*, *Int. J. Pept. Protein Res. 30*, 415–422 (1987).
98. J. Singh and M. R. S. Rao, Interaction of rat testis protein, TP, with nucleic acids in vitro, *J. Biol. Chem. 262*, 734–740 (1987).
99. C. Helene and G. Lancelot, Interactions between functional groups in protein–nucleic acid associations, *Prog. Biophys. Mol. Biol. 39*, 1–68 (1982).
100. D. G. Searcy, T. Montenay-Garestier, D. J. Laston, and C. Helene, Tyrosine environment and phosphate binding in the archaebacterial histone-like protein HTa, *Biochim. Biophys. Acta 953*, 321–333 (1988).
101. D. G. Searcy, T. Montenay-Garestier, and C. Helene, Phenylalanine-to-tyrosine energy transfer in the archaebacterial histone-like protein HTa, *Biochemistry 28*, 9058–9065 (1989).
102. F. Brun, J. J. Toulme, and C. Helene, Interactions of aromatic residues of proteins with nucleic acids. Fluorescence studies of the binding of oligopeptides containing tryptophan and tyrosine residues to polynucleotides, *Biochemistry 14*, 558–563 (1975).
103. R. Mayer, F. Toulme, T. Montenay-Garestier, and C. Helene, The role of tyrosine in the

association of proteins and nucleic acids. Specific recognition of single-stranded nucleic acids by tyrosine-containing peptides, *J. Biol. Chem. 254*, 75–82 (1979).
104. D. Porschke and J. Ronnenberg, The reaction of aromatic peptides with a double helical DNA. Quantitative characterization of a two step reaction scheme, *Biophys. Chem. 13*, 283–290 (1981).
105. T. Montenay-Garestier, M. Takasugi, and T. Le Doan, Fluorescence decay studies of peptide–nucleic acid complexes, in: *Nucleic Acids: the Vectors of Life* (B. Pullman and J. Jortner, eds.), pp. 305–315, Reidel, Dordrecht (1983).
106. B. Lux, D. Gerard, and G. Laustriat, Tyrosine fluorescence of S8 and S15 *Escherichia coli* ribosomal proteins, *FEBS Lett. 80*, 66–70 (1977).
107. F. Culard, M. Schnarr, and J. C. Maurizot, Interaction between the *lac* operator and the *lac* repressor headpiece: Fluorescence and circular dichroism studies, *EMBO J. 1*, 1405–1409 (1982).
108. M. Schnarr, M. Durand, and J. C. Maurizot, Nonspecific interaction of the *lac* repressor headpiece with deoxyribonucleic acid: Fluorescence and circular dichroism studies, *Biochemistry 22*, 3563–3570 (1983).
109. H. T. Pretorius, M. Klein, and L. A. Day, Gene V protein of fd bacteriophage. Dimer formation and the role of tyrosyl groups in DNA binding, *J. Biol. Chem. 250*, 9262–9269 (1975).
110. T. Härd, V. Hsu, M. H. Sayre, E. P. Geiduschek, K. Appelt, and D. K. Kearns, Fluorescence studies of a single tyrosine in a type II DNA binding protein, *Biochemistry 28*, 396–407 (1989).
111. G. Lindberg, S. C. Kowalczykowski, J. K. Rist, A. Sugino, and L. B. Rothman-Denes, Purification and characterization of the coliphage N4-coded single-stranded DNA binding protein, *J. Biol. Chem. 264*, 12700–12708 (1989).
112. B. Lux, J. Baudier, and D. Gerard, Tyrosyl fluorescence spectra of proteins lacking tryptophan: Effects of intramolecular interactions, *Photochem. Photobiol. 42*, 245–251 (1985).
113. S. Forsen, H. J. Vogel, and T. Drakenberg, Biophysical studies of calmodulin, in: *Calcium and Cell Function*, Vol. VI (W. Y. Cheung, ed.), pp. 113–157, Academic Press, New York (1986).
114. D. Malencik and S. R. Anderson, Dityrosine formation in calmodulin, *Biochemistry 26*, 695–704 (1987).
115. K. B. Seamon, Calcium- and magnesium-dependent conformational states of calmodulin as determined by nuclear magnetic resonance, *Biochemistry 19*, 207–215 (1980).
116. M.-C. Kilhoffer, J. G. Demaille, and D. Gerard, Terbium as luminescent probe of calmodulin calcium-binding sites. Domains I and II contain the high-affinity sites, *FEBS Lett. 116*, 269–272 (1980).
117. M.-C. Kilhoffer, D. Gerard, and J. G. Demaille, Terbium binding to octopus calmodulin provides the complete sequence of binding, *FEBS Lett. 120*, 99–103 (1980).
118. M.-C. Kilhoffer, J. G. Demaille, and D. Gerard, Tyrosine fluorescence of ram testis and octopus calmodulins. Effects of calcium, magnesium, and ionic strength, *Biochemistry 20*, 4407–4414 (1981).
119. K. P. Kohse and L. M. Heilmeyer, The effects of Mg^{2+} on the Ca^{2+}-binding properties and Ca^{2+}-induced tyrosine-fluorescence changes of calmodulin isolated from rabbit skeletal muscle, *Eur. J. Biochem. 117*, 507–513 (1981).
120. C. L. Wang, R. R. Aquaron, P. C. Leavis, and J. Gergely, Metal-binding properties of calmodulin, *Eur. J. Biochem. 124*, 7–12 (1982).
121. R. W. Wallace, E. A. Tallant, M. E. Dockter, and W. Y. Cheung, Calcium binding domains of calmodulin. Sequence of fill as determined by terbium luminescence, *J. Biol. Chem. 257*, 1845–1854 (1982).
122. C. L. Wang, P. C. Leavis, and J. Gergely, Kinetic studies show that Ca^{2+} and Tb^{3+} have different binding preferences toward the four Ca^{2+}-binding sites of calmodulin, *Biochemistry 23*, 6410–6415 (1984).

123. S. Pundak and R. S. Roche, Tyrosine and tyrosinate fluorescence of bovine testes calmodulin: Calcium and pH dependence, *Biochemistry 23*, 1549–1555 (1984).
124. P. K. Lambooy, R. F. Steiner, and H. Sternberg, Molecular dynamics of calmodulin as monitored by fluorescence anisotropy, *Arch. Biochem. Biophys. 217*, 517–528 (1982).
125. R. F. Steiner, P. K. Lambooy, and H. Sternberg, The dependence of the molecular dynamics of calmodulin upon pH and ionic strength, *Arch. Biochem. Biophys. 222*, 158–169 (1983).
126. R. F. Steiner and M. Montevalli-Alibadi, The determination of the separation of tyrosine-99 and tyrosine-138 in calmodulin: Radiationless energy transfer, *Arch. Biochem. Biophys. 234*, 522–530 (1984).
127. I. Gryczynski, J. R. Lakowicz, and R. F. Steiner, Frequency-domain measurements of the rotational dynamics of the tyrosine groups of calmodulin, *Biophys. Chem. 30*, 49–59 (1988).
128. P. Bayley, S. Martin, and G. Jones, The conformation of calmodulin: A substantial environmentally sensitive helical transition in Ca_4-calmodulin with potential mechanistic function, *FEBS Lett. 238*, 61–66 (1988).
129. E. A. Burstein, E. A. Permyakov, V. I. Emelyanenko, T. L. Bushueva, and J.-F. Pechere, Investigation of some physico-chemical properties of muscular parvalbumins by means of the luminescence of their phenylalanyl residues, *Biochim. Biophys. Acta 400*, 1–16 (1975).
130. E. A. Permyakov, V. V. Yarmolenko, V. I. Ememlanenko, E. A. Burstein, J. Closset, and C. Gerday, Fluorescence studies of the calcium binding to whiting (*Gadus merlangus*) parvalbumin, *Eur. J. Biochem. 109*, 307–315 (1980).
131. E. A. Permyakov, V. N. Medvedkin, L. P. Kalinichenko, and E. A. Burstein, Comparative study of physicochemical properties of two pike parvalbumins by means of their intrinsic tyrosyl and phenylalanyl fluorescence, *Arch. Biochem. Biophys. 227*, 9–20 (1983).
132. E. A. Permyakov, A. V. Ostrovsky, E. A. Burstein, P. G. Pleshanov, and C. Gerday, Parvalbumin conformers revealed by steady-state and time-resolved fluorescence spectroscopy, *Arch. Biochem. Biophys. 240*, 781–791 (1985).
133. R. H. Kretsinger and C. F. Nockolds, Carp muscle calcium-binding protein, *J. Biol. Chem. 248*, 3313–3326 (1973).
134. J. P. MacManus, D. C. Watson, and M. Yaguchi, The complete amino acid sequence of oncomodulin—a parvalbumin-like calcium-binding protein from Morris hepatoma 5123tc, *Eur. J. Biochem. 136*, 9–17 (1983).
135. J. P. MacManus, A. G. Szabo, and R. E. Williams, Conformational changes induced by binding of bivalent cations to oncomodulin, a parvalbumin-like tumour protein, *Biochem. J. 220*, 261–268 (1984).
136. J. D. Johnson and J. D. Potter, Detection of two classes of Ca^{2+} binding sites in troponin C with circular dichroism and tyrosine fluorescence, *J. Biol. Chem. 253*, 3775–3777 (1978).
137. C. L. Wang, P. C. Leavis, W. D. Horrocks, and J. Gergely, Binding of lanthanides to troponin C, *Biochemistry 20*, 2439–2444 (1981).
138. P. C. Leavis and S. S. Lehrer, Intrinsic fluorescence studies on troponin C, *Arch. Biochem. Biophys. 187*, 243–251 (1978).
139. Z. Grabarek, R.-Y. Tan, J. Wang, T. Tao, and J. Gergely, Inhibition of mutant troponin C activity by an intra-domain disulphide bond, *Nature 345*, 132–135 (1990).
140. P. Kanellis, J. Yang, H. C. Cheung, and R. E. Lenkinski, Synthetic peptide analogs of skeletal troponin C: Fluorescence studies of analogs of the low-affinity calcium-binding site II, *Arch. Biochem. Biophys. 220*, 530–540 (1983).
141. N. A. Malik, G. M. Anatharamaiah, A. Gawish, and H. C. Cheung, Structural and biological studies on synthetic peptide analogues of a low-affinity calcium-binding site of skeletal troponin C, *Biochim. Biophys. Acta 911*, 221–230 (1987).
142. D. M. E. Szebenyi, S. K. Obendorf, and K. Moffat, Structure of vitamin D-dependent calcium-binding protein from bovine intestine, *Nature 294*, 327–332 (1981).
143. J. D. O'Neil, K. J. Dorrington, D. I. Kells, and T. Hoffmann, Fluorescence and circular-

dichroism properties of pig intestinal calcium-binding protein ($M_r = 9000$), a protein with a single tyrosine residue, *Biochem. J. 207*, 389–396 (1982).
144. D. M. E. Szebenyi and K. Moffat, The refined structure of vitamin D-dependent calcium-binding protein from bovine intestine, *J. Biol. Chem. 261*, 8761–8777 (1986).
145. J. D. J. O'Neil and T. Hofmann, Tyrosine and tyrosinate fluorescence of pig intestinal Ca^{2+}-binding protein, *Biochem. J. 243*, 611–615 (1987).
146. K. Chiba, T. Ohyashiki, and T. Mohri, Quantitative analysis of calcium binding to porcine intestinal calcium-binding protein, *J. Biochem. (Tokyo) 93*, 487–493 (1983).
147. K. Chiba, T. Ohyashiki, and T. Mohri, Stoichiometry and location of terbium and calcium binding to porcine intestinal calcium-binding protein, *J. Biochem. (Tokyo) 95*, 1767–1774 (1984).
148. J. D. O'Neil, K. J. Dorrington, and T. Hofmann, Luminescence and circular-dichroism analysis of terbium binding by pig intestinal calcium-binding protein (relative mass = 9000), *Can. J. Biochem. Cell Biol. 62*, 434–442 (1984).
149. R. Rigler, J. Roslund, and S. Forsen, Side chain mobility in bovine calbindin D_{9k}, *Eur. J. Biochem. 118*, 541–545 (1990).
150. R. S. Mani, B. E. Boyes, and C. M. Kay, Physicochemical and optical studies on calcium- and potassium-induced conformational changes in bovine brain S-100_b protein, *Biochemistry 21*, 2607–2612 (1982).
151. J. Baudier and D. Gerard, The S-100_b protein: Tyrosine residues do not exhibit an abnormal fluorescence spectrum, *J. Neurochem. 40*, 1765–1767 (1983).
152. J. Baudier and D. Gérard, Ions binding to S100 proteins: Structural changes induced by calcium and zinc on S100a and S100b proteins, *Biochemistry 22*, 3360–3369 (1983).
153. J. Baudier, N. Glasser, and D. Gérard, Ions binding to S100 proteins, *J. Biol. Chem. 261*, 8192–8203 (1986).
154. J. Baudier and R. D. Cole, The Ca^{2+}-binding sequence in bovine brain S100b protein β-subunit, *Biochem. J. 264*, 79–85 (1989).
155. Y. Mely and D. Gérard, Structural and ion-binding properties of an S100b protein mixed disulfide: Comparison with the reappraised native S100b protein properties, *Arch. Biochem. Biophys. 279*, 174–182 (1990).
156. R. J. Turner, R. S. Roche, R. S. Mani, and C. M. Kay, Tyrosine and tyrosinate fluorescence of S100b. A time-resolved nanosecond fluorescence study. The effect of pH, Ca(II), and Zn(II), *Biochem. Cell Biol. 67*, 179–186 (1989).
157. P. V. Hauschka and S. A. Carr, Calcium-dependent alpha-helical structure in osteocalcin, *Biochemistry 21*, 2538–2547 (1982).
158. A. Filipek, C. W. Heizmann, and J. Kuźnicki, Calcyclin is a calcium and zinc binding protein, *FEBS Lett. 264*, 263–266 (1990).
159. C. Pigault, A. Follénius-Wund, B. Lux, and D. Gérard, A fluorescence spectroscopy study of the calpactin I complex and its subunits p11 and p36: Calcium-dependent conformational changes, *Biochim. Biophys. Acta 1037*, 106–114 (1990).
160. R. S. Mani and C. M. Kay, Isolation and characterization of a novel molecular weight 11000 Ca^{2+}-binding protein from smooth muscle, *Biochemistry 29*, 1398–1404 (1990).
161. C. J. R. Thorne and N. O. Kaplan, Physicochemical properties of pig and horse heart mitochondrial malate dehydrogenase, *J. Biol. Chem. 238*, 1861–1868 (1963).
162. H. B. Otwell, A. Y.-H. Tan, and M. E. Friedman, Implication of a tyrosyl residue at the active site of mitochondrial L-Malate:NAD^+ oxidoreductase, *Biochim. Biophys. Acta 527*, 309–318 (1978).
163. D. C. Wood, S. R. Jurgensen, J. C. Geesin, and J. H. Harrison, Subunit interactions in mitochondrial malate dehydrogenase, *J. Biol. Chem. 256*, 2377–2382 (1981).
164. J. Muller, M.-F. Manent, and G. Pfleiderer, Importance of tyrosine for structure and function of mitochondrial malate dehydrogenases, *Biochim. Biophys. Acta 742*, 189–196 (1983).

165. J. Walter and R. Huber, Pancreatic trypsin inhibitor. A new crystal form and its analysis, *J. Mol. Biol.* *167*, 911–917 (1983).
166. A. Wlodawer, J. Walter, R. Huber, and L. Sjolin, Structure of bovine pancreatic trypsin inhibitor, *J. Mol. Biol.* *180*, 301–329 (1984).
167. M. Karplus and J. A. McCammon, The internal dynamics of globular proteins, *CRC Crit. Rev. Biochem.* *9*, 293–349 (1981).
168. K. Wüthrich, *NMR of Proteins and Nucleic Acids*, Wiley-Interscience, New York (1986).
169. A. Kasprzak and G. Weber, Fluorescence depolarization and rotational modes of tyrosine in bovine pancreatic trypsin inhibitor, *Biochemistry* *21*, 5924–5927 (1982).
170. J. R. Lakowicz and B. Maliwal, Oxygen quenching and fluorescence depolarization of tyrosine residues in proteins, *J. Biol. Chem.* *258*, 4794–4801 (1983).
171. J. R. Lakowicz, G. Laczko, and I. Gryczynski, Picosecond resolution of tyrosine fluorescence and anisotropy decays by 2-GHz frequency-domain fluorometry, *Biochemistry* *26*, 82–90 (1987).
172. T. M. Nordlund, X.-Y. Liu, and J. H. Sommer, Fluorescence polarization decay of tyrosine in lima bean trypsin inhibitor, *Proc. Natl. Acad. Sci. U.S.A.* *83*, 8977–8981 (1986).
173. X.-Y. Liu, K. O. Cottrell, and T. M. Nordlund, Spectroscopy and fluorescence quenching of tyrosine in lima bean trypsin/chymotrypsin inhibitor and model peptides, *Photochem. Photobiol.* *50*, 721–731 (1989).
174. S. S. Sur, L. D. Rabbani, L. Libman, and E. Breslow, Fluorescence studies of native and modified neurophysins. Effects of peptides and pH, *Biochemistry* *18*, 1026–1036 (1979).
175. M. Rholam and P. Nicolas, Conformational flexibility of neurophysin as investigated by local motions of fluorophores. Relationships with neurohypophyseal hormone binding, *Biochemistry* *24*, 1928–1933 (1985).
176. M. Rholam, S. F. Scarlata, and P. Nicolas, Conformational flexibility of neurophysin as investigated by local motions of fluorophores. Relationships with neurohypophyseal hormone binding, *Biochemistry* *24*, 7853 (1985).
177. S. F. Scarlata and C. A. Royer, Ligand-induced asymmetry as observed through fluorophore rotations and free energy couplings: Application to neurophysin, *Biochemistry* *25*, 4925–4929 (1986).
178. N. Barboy and J. Feitelson, Fluorescence lifetime study of the denaturation of ribonuclease A, *Photochem. Photobiol.* *26*, 561–565 (1977).
179. J.-R. Garel and R. L. Baldwin, Both the fast and slow refolding reactions of ribonuclease A yield native enzyme, *Proc. Natl. Acad. Sci. U.S.A.* *70*, 3347–3351 (1973).
180. P. J. Hagerman, B. T. Nall, and R. L. Baldwin, A quantitative treatment of the kinetics of the folding of ribonuclease A, *Biochemistry* *15*, 1462–1473 (1976).
181. F. X. Schmid, A native-like intermediate on the ribonuclease A folding pathway. 1. Detection by tyrosine fluorescence changes, *Eur. J. Biochem.* *114*, 105–109 (1981).
182. A. Rehage and F. X. Schmid, Fast- and slow-refolding forms of unfolded ribonuclease A differ in tyrosine fluorescence, *Biochemistry* *21*, 1499–1505 (1982).
183. F. X. Schmid, R. Grafl, A. Wrba, and J. J. Beintema, Role of proline peptide bond isomerization in unfolding and refolding of ribonuclease, *Proc. Natl. Acad. Sci. U.S.A.* *83*, 872–876 (1986).
184. P. W. Mui, Y. Konishi, and H. A. Scheraga, Kinetics and mechanism of the refolding of ribonuclease A, *Biochemistry* *24*, 4481–4489 (1985).
185. H. Krebs, F. X. Schmid, and R. Jaenicke, Native-like folding intermediates of homologous ribonucleases, *Biochemistry* *24*, 3846–3852 (1985).
186. E. Haas, G. T. Montelione, C. A. McWherter, and H. A. Scheraga, Local structure in a tryptic fragment of performic acid oxidized ribonuclease A corresponding to a proposed polypeptide chain-folding initiation site detected by tyrosine fluorescence lifetime and proton magnetic resonance measurements, *Biochemistry* *26*, 1672–1683 (1987).

187. A. Tulinsky, R. L. Vandlen, C. N. Morimoto, N. V. Mani, and L. H. Wright, Variability in the tertiary structure of α-chymotrypsin at 2.8-Å resolution, *Biochemistry 12*, 4185–4192 (1973).
188. C. R. Coan, L. M. Hinman, and D. A. Deranleau, Charge-transfer studies of the availability of aromatic side chains of proteins in guanidine hydrochloride, *Biochemistry 14*, 4421–4427 (1974).
189. J. B. Massey and H. J. Pownall, Spectroscopic studies of the tyrosine residues of human plasma apolipoprotein A-II, *Biochim. Biophys. Acta 999*, 111–120 (1989).
190. R. B. Weinberg and M. K. Jordan, Effects of phospholipid on the structure of human apolipoprotein A-IV, *J. Biol. Chem. 265*, 8081–8086 (1990).
191. P. W. Schiller, Application of fluorescence techniques in studies of peptide conformations and interactions, in: *The Peptides*, Vol. 7 (S. Udenfriend, ed.), pp. 115–164, Academic Press, New York (1985).
192. S. P. Wood, I. J. Tickle, A. M. Treharne, J. E. Pitts, Y. Mascarenhas, J. Y. Li, J. Husain, S. Cooper, T. L. Blundell, V. J. Hruby, A. Buku, A. J. Fischman, and H. R. Wyssbrod, Crystal structure of deamino-oxytocin: Conformational flexibility and receptor binding, *Science 232*, 633–636 (1986).
193. J. R. Lakowicz, G. Laczko, and I. Gryczynski, Picosecond resolution of oxytocin tyrosyl fluorescence by 2 GHz frequency-domain fluorometry, *Biophys. Chem. 24*, 97–100 (1986).
194. S. S. Lehrer and G. D. Fasman, Excimer fluorescence in lipid phenol, *p*-ethylphenol, and anisole, *J. Am. Chem. Soc. 87*, 4687–4691 (1965).
195. S. N. Khrapunov and A. I. Dragan, Spectroscopy of molecular interactions of tyrosine chromophore. III. Classification of the state of tyrosine residues in protein composition according to their electronic spectra, *Biofizika 34*, 357–363 (1989).
196. T. C. M. Eames, R. M. Pollack, and R. F. Steiner, Orientation, accessibility, and mobility of equilenin bound to the active site of steroid isomerase, *Biochemistry 28*, 6269–6275 (1989).
197. R. F. Chen, Fluorescence quantum yields of tryptophan and tyrosine, *Anal. Lett. 1*, 35–42 (1967).
198. K. J. Willis, A. G. Szabo, and D. T. Krajcarski, The use of Stokes Raman scattering in time correlated single photon counting: Application to the fluorescence lifetime of tyrosinate, *Photochem. Photobiol. 51*, 375–377 (1990).
199. J. L. Cornog and W. R. Adams, The fluorescence of tyrosine in alkaline solution, *Biochim. Biophys. Acta 66*, 356–365 (1963).
200. H. Edelhoch, Spectroscopic determination of tryptophan and tyrosine in proteins, *Biochemistry 6*, 1948–1954 (1967).
201. W. R. Laws and J. D. Shore, Spectral evidence for tyrosine ionization linked to a conformational change in liver alcohol dehydrogenase ternary complexes, *J. Biol. Chem. 254*, 2582–2584 (1979).
202. S. Subramanian, J. B. A. Ross, L. Brand, and P. D. Ross, Investigation of the nature of enzyme–coenzyme interactions in binary and ternary complexes of liver alcohol dehydrogenase with coenzymes, coenzyme analogs, and substrate analogs by ultraviolet absorption and phosphorescence spectroscopy, *Biochemistry 20*, 4086–4093 (1981).
203. T. Kimura and J. J. Ting, Anomalous tyrosine emission at 331 nm in adrenal two iron and two labile-sulfur protein (adrenodoxin): A possible tyrosine exciplex, *Biochem. Biophys. Res. Commun. 45*, 1227–1231 (1971).
204. T. Kimura, J. J. Ting, and J. J. Huang, Studies on adrenal steroid hydroxylases. Anomalous fluorescence of a tyrosyl residue in adrenal iron–sulfur protein (adrenodoxin), *J. Biol. Chem. 247*, 4476–4479 (1972).
205. B. T. Lim and T. Kimura, Conformation-associated anomalous tyrosine fluorescence of adrenodoxin, *J. Biol. Chem. 255*, 2440–2444 (1980).

206. B. T. Lim and T. Kimura, Conformational prediction and spectral studies on adrenodoxin. The accessibility of the tyrosine at position 82 in the polypeptide, *J. Biol. Chem. 256*, 4400–4406 (1981).
207. E. Bicknell-Brown, B. T. Lim, and T. Kimura, Laser Raman spectroscopy of adrenal iron–sulfur apoprotein: The anomalous tyrosine residue at position 82, *Biochem. Biophys. Res. Commun. 101*, 298–305 (1981).
208. M. T. Graziani, A. F. Agro, G. Rotilio, D. Barra, and B. Mondovi, Parsley plastocyanin. The possible presence of sulfhydryl and tyrosine in the copper environment, *Biochemistry 13*, 804–809 (1974).
209. F. G. Prendergast, P. D. Hampton, and B. Jones, Characteristics of tyrosinate fluorescence emission in α- and β-purothionins, *Biochemistry 23*, 6690–6697 (1984).
210. C. M. L. Hutnik, J. P. MacManus, D. Banville, and A. G. Szabo, Comparison of metal ion-induced conformational changes in parvalbumin and oncomodulin as probed by the intrinsic fluorescence of tryptophan 102, *J. Biol. Chem. 265*, 11456–11464 (1990).
211. R. J. Turner, J. M. Matsoukas, and G. J. Moore, Tyrosinate fluorescence lifetimes for oxytocin and vasopressin in receptor-simulating environments: Relationship to biological activity and ^{1}H-NMR data, *Biochem. Biophys. Res. Commun. 171*, 996–1001 (1990).
212. J. Longworth, A new component in protein fluorescence, *Ann. N.Y. Acad. Sci. 366*, 237–245 (1981).
213. S. F. Pearce and E. Hawrot, Intrinsic fluorescence of binding-site fragments of the nicotinic acetylcholine receptor: Perturbations produced upon binding α-bungarotoxin, *Biochemistry 29*, 10649–10659 (1990).
214. A. G. Szabo, K. R. Lynn, D. T. Krajcarski, and D. M. Rayner, Tyrosinate fluorescence maxima at 345 nm in proteins lacking tryptophan at pH 7, *FEBS Lett. 94*, 249–252 (1978).
215. A. H. Maki and J. Zuclich, Protein triplet states, *Top. Curr. Chem. 54*, 115–163 (1975).
216. A. L. Kwiram and J. B. A. Ross, Optical detection of magnetic resonance in biologically important molecules, *Annu. Rev. Biophys. Bioeng. 11*, 223–249 (1982).
217. N. Shaklai, N. Zisapel, and M. Sokolovsky, The role of a tyrosyl residue in the mechanism of action of carboxypeptidase B: Luminescence studies, *Proc. Natl. Acad. Sci. U.S.A. 70*, 2025–2028 (1973).
218. N. Zisapel, N. Shaklai, and M. Sokolovsky, Metal-tyrosyl interaction in carboxypeptidase: Phosphorescence studies, *FEBS Lett. 51*, 262–265 (1975).
219. K. Ugurbil, A. H. Maki, and R. Bersohn, Study of the triplet state properties of tyrosines and tryptophan in azurins using optically detected magnetic resonance, *Biochemistry 16*, 901–907 (1977).
220. J. B. A. Ross, K. W. Rousslang, C. DeHaen, V. R. Lavis, and D. A. Deranleau, [12-Homoarginine]glucagon: Synthesis and observations on conformation, biological activity, and copper-mediated peptide cleavage, *Biochim. Biophys. Acta 576*, 372–384 (1979).
221. R. M. Levy and A. Szabo, Initial fluorescence depolarization of tyrosines in proteins, *J. Am. Chem. Soc. 104*, 2073–2075 (1982).
222. R. M. Levy and R. P. Sheridan, Combined effect of restricted rotational diffusion plus jumps on nuclear magnetic resonance and fluorescence probes of aromatic ring motions in proteins, *Biophys. J. 41*, 217–221 (1983).

2

Fluorescence and Dynamics in Proteins

A. P. Demchenko

2.1. Introduction

Recently, fluorescence spectroscopy has become one of the fundamental methods for the study of the structure and dynamics of microheterogeneous systems—colloid particles, micelles, liquid crystals, artificial and natural membranes, polymers, and biological macromolecules. It is being used increasingly in the field of protein research. This is because of the importance of studies at the molecular level for understanding biological function and the great diversity of protein molecules as well as the discrete nature of the structural forms of an individual protein. However, the most important reason is that fluorescence spectroscopy can be used to study those structural and dynamic properties of proteins which are directly related to such biological functions as specific binding (recognition), biocatalysis, membrane transport, and muscular motility.

The use of fluorescent probes has found wide application in studies of the structure and dynamics of proteins as well as in studies of other microheterogeneous systems. A major problem in gaining information concerning the structure (interactions in the ground state) and dynamics (processes in the electronic excited state) of aromatic fluorophores (probes) is that these properties may also be affected by the molecules and groups of atoms which surround the aromatic fluorophore and the dynamics of these surroundings. Apart from tryptophan[1,2] and tyrosine,†[1] amino acid residues occurring in practically all proteins, fluorescent coenzyme groups and their analogues (flavins[3] and nicotinamide[4] derivatives) as well as aromatic molecules possess suitable spectral properties that are introduced into the protein molecule as fluorescence probes are widely used in these investigations.[5,6]

† For a discussion of tyrosine fluorescence, see Chapter 1 in this volume.

A. P. Demchenko • A. V. Palladin Institute of Biochemistry of the Academy of Sciences, Kiev 252030, Ukraine.

Topics in Fluorescence Spectroscopy, Volume 3: Biochemical Applications, edited by Joseph R. Lakowicz. Plenum Press, New York, 1992.

The indole chromophore of tryptophan is the most important tool in studies of intrinsic protein fluorescence. The position of the maximum in the tryptophan fluorescence spectra recorded for proteins varies widely, from 308 nm for azurin to 350–353 nm for peptides lacking an ordered structure and for denatured proteins.[1] This is because of an important property of the fluorescence spectra of tryptophan residues, namely, their high sensitivity to interactions with the environment. Among extrinsic fluorescence probes, aminonaphthalene sulfonates are the most similar to tryptophan in this respect, which accounts for their wide application in protein research.[5]

The dynamics studied in protein molecules are associated with the relaxation and diffusion processes in the system consisting of the aromatic fluorophore and neighboring groups of atoms within the protein molecule. These processes affect the spectral, temporal, and polarization parameters of emission. The correlation of these parameters with the dynamics is ambiguous, and therefore difficulties arise in interpretation of the experimental results. For instance, a long-wavelength shift of the fluorescence spectrum could be induced by a variety of factors: (1) conformational changes in the protein molecule leading to an increase in the polarity of the fluorophore's environment; (2) an increase in the mobility of the fluorophore's environment and acceleration of the process of dipolar relaxation; and (3) reactions in excited states and, particularly, the formation of complexes (exciplexes). A decrease in fluorescence anisotropy occurs both in the case of protein and fluorophore rotation and in the case of energy transfer between the electronically excited fluorophores. The shortening of the fluorescence lifetime and the appearance of a nonexponential decay may be associated with either microheterogeneity of the fluorophore's environment in the ground state or different quenching processes in the excited state.

Such ambiguity and also the low structural resolution of the method require that the spectroscopic properties of protein fluorophores and their reactions in electronic excited states be thoroughly studied and characterized in simple model systems. Furthermore, the reliability of the results should increase with the inclusion of this additional information into the analysis and with the comparison of the complementary data. Recently, there has been a tendency not only to study certain fluorescence parameters and to establish their correlation with protein dynamics but also to analyze them jointly, to treat the spectroscopic data multiparametrically, and to construct self-consistent models of the dynamic process which take into account these data as a whole. Fluorescence spectroscopy gives a researcher ample opportunities to combine different parameters determined experimentally and to study their interrelationships (Figure 2.1). This opportunity should be exploited to the fullest.

In fluorescence studies on proteins, apart from spectral, temporal, and polarization measurements, it appears of importance to investigate the

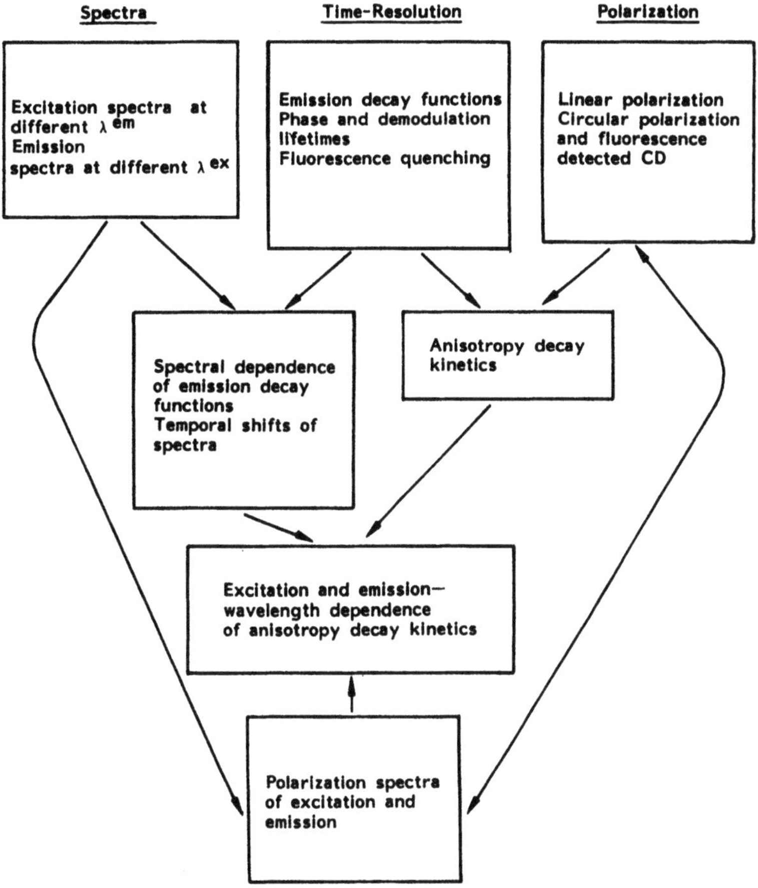

Figure 2.1. Combination of spectral, time-resolved, and polarization data for multidimensional analysis of spectroscopic information.

dependence of these data on experimental conditions, including temperature, solvent viscosity and dielectric constant, and concentration of collisional fluorescence quenchers.[1,7,8] These dependences should be analyzed in a consistent way in accord with the proposed model of the dynamic process. The construction of such models is a highly complicated problem, since both the photophysical processes determining the kinetics of emission and the dynamics themselves, that is, the kinetics of relaxational and diffusional motions in the protein molecule, are rather complex. There exist other difficulties associated with the heterogeneity of the emission parameters of fluorophores in structurally different environments within one protein molecule. At present, these difficulties are being overcome, not only by the study of simpler systems, but also by improvements in the methods for acquiring and analyzing spectroscopic information.

2.2. Dynamics in Proteins

2.2.1. Structural Hierarchy and Degrees of Mobility

The dynamics of protein molecules have been studied intensively using various experimental[9-11] and theoretical[12] approaches. Luminescence methods are widely applied in these investigations.[1,7,8] Modern concepts about the structure of proteins and their dynamics which have evolved from these investigations are presented briefly in this section.

The space-ordered structure of native proteins is formed by the arrangement of polypeptide chains and by its stabilization by noncovalent interactions and disulfide links.[13] A high degree of flexibility is typical of the polypeptide chain itself, since certain covalent bonds of the main chain and side groups allow rotation. Figure 2.2 presents the structures of a segment of a polypeptide chain and of one of the amino acid residues (tryptophan). The dihedral angles φ and ψ may vary, thus inducing changes in the spatial arrangement of the main chain groups. The C–N bond has partial double bond character. Therefore, each peptide linkage (CO–NH) and the C^{α} atom bonded to the carbonyl carbon are in the same plane. All the side chains have freedom of rotation around the bonds connecting the C^{α} and C^{β} atoms (dihedral angle χ^1) and the C^{β} and C^{γ} atoms (χ^2). Due to the partial double bond character of their bonds, aromatic rings are relatively planar and rigid, but their rotation with respect to the main chain is possible. Aliphatic side groups have additional mobility due to rotations around single bonds.

The polypeptide chain in the native protein is folded into a compact structure, which strongly limits the freedom of molecular movement. The arrangement in space of each atom in the protein molecule is fixed and does not change with time in the absence of thermal collisions with other atoms in a protein and solvent molecules. From the thermodynamic point of view, the

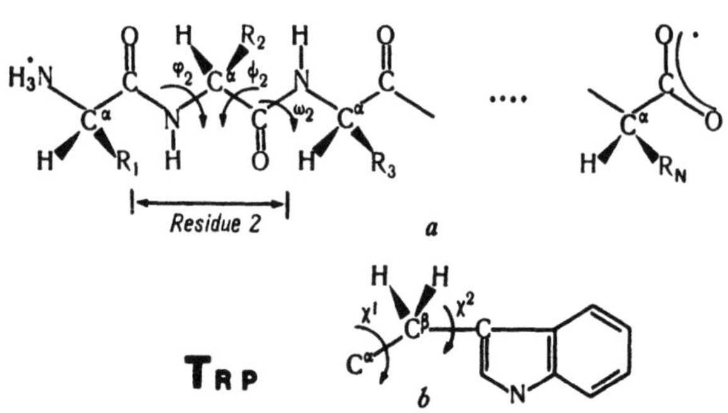

Figure 2.2. The structure of a segment of a polypeptide chain (a) and of the amino acid residue tryptophan (b). Rotations are possible around the bonds with dihedral angles φ, ψ, χ^1, and χ^2.

protein molecules are microphases, which are small systems, and structural fluctuations within them should be considerable. However, since these small systems are still highly organized, then it becomes possible for the extent of fluctuations to decrease along one coordinate and to increase along the others. This is determined by the structural features of protein molecules and primarily by their most rigid elements—periodic α-helices and β-sheets, which are stabilized by interactions between groups of the main chain. The side-chain groups also interact with each other and with the peptide groups of the main chain. This results in reduced mobility of some structures (for example, α-helical rods) and in relatively free mobility of others (for example, hinge-bending motions of large structural blocks or domains).

The spatial macrostructure of the native protein (the equilibrium location of the polypeptide main chain backbone and bulky side groups) is strictly determined. Individual protein molecules having the same sequence of amino acid residues do not differ in their three-dimensional structure, which is the equilibrium one and averaged in time. The activation energy of conformational transitions may be as high as several hundreds of kiloJoules per mole. Therefore, the extended fluctuations which are associated with the unfolding of the native macrostructure and transitions between conformations occur rather rarely.

However, the identicalness of protein molecules possessing the same macroconformation is not absolute. Within each structurally determined conformational macrostructure, there exists a microdisordering which is similar to that observed in amorphous solids and glasses.[11,14] It is associated with the presence of multiple relative minima of the free energy depending on small shifts and variations in orientation of certain groups within the limits of available space.

It should be noted that the condition of minimum free energy is realized at the level of sufficiently extended regions of the protein structure or even of the entire globule.[13] At the level of local interactions of atoms and groups, saturation by noncovalent (in particular, hydrogen) bonds is far from being complete.[15] The energy of hydrophobic interactions is slightly sensitive to the orientation of interacting groups. This leads to the formation of microstates that differ to a small extent in the orientations of atomic groups. These microstates may not differ, or may differ only slightly, in free energy. The activation energy of the small-scale fluctuations resulting in transitions between microstates is of the order of 10 kJ/mol, and these fluctuations may occur on the nanosecond or shorter time scales (Table 2.1).

Oscillations of atoms and their groups, typically occurring on a time scale of 10^{-12}–10^{-14} s, are much faster motions. However, there is evidently no sharp distinction between these motions and other slower motions. When damping and anharmonicity arise, the oscillations become diffusive and have the properties of transitions between microstates. It is natural to suppose (as

Table 2.1. Intramolecular Motions in Proteins and the Values of the Parameters that Characterize Them: Mass of Structural Element (m), Amplitude (A), Characteristic Time (τ), Activation Energy (E_a), and Cross-Correlation ($\langle \Delta X_1 \cdot \Delta X_2 \rangle$)

Motions	m (dalton)	A (nm)	τ (s)	E_a (kJ/mole)	$\langle \Delta X_1 \cdot \Delta X_2 \rangle$
Vibrations (with anharmonicity and damping)	10–10^2	$<5 \times 10^{-2}$	10^{-12}–10^{-14}	<10	Low
Small-scale diffusional motions	10–10^3	0.5–5×10^{-1}	10^{-7}–10^{-12}	10–10^2	Significant
Large-scale conformational transitions	10^3–10^5	0.5–5	$>10^{-6}$	10^2–10^3	Very high

is also confirmed by computer simulations of molecular dynamics) that the longer is the time scale of motions, the greater is their correlation with motions of other elements of the structure.

2.2.2. Distribution of Microstates

In all studies involving methods based on absorption or scattering of light, X rays, or neutrons, the characteristic time scales on which radiation interacts with the substance are many orders of magnitude shorter than those of atomic motions. Therefore, it is not the motions themselves but the disordering which arises due to molecular dynamics that should be investigated.

The distribution of microstates may be defined as the distribution of spatial dislocations, orientations, and interactions of groups of the main chain and side groups with respect to their most probable values.

X-ray diffraction analysis permits the root-mean-square shifts of atoms in protein molecules to be determined.[9, 16] The inhomogeneous character of fluctuations within globules as well as the possibilily of "freezing" of certain microstates when proteins are cooled down to definite glass-transition temperatures are observed in these studies. Mössbauer spectroscopy has shown that an increase in temperature changes the character of dynamic processes: new modes of motions become active.[11, 16] The distribution of microstates associated with the motions of atoms and groups in proteins affects the results of NMR studies[17, 18] and is manifested as very small uniform shifts in the absorption spectra of proteins with variations in temperature.[1] Nonexponential kinetics of ligand association with myoglobin and hemoglobin after their pulse photodissociation result from distributions of microstates.[19] The existence of several levels of such microstates and several levels of potential barriers which characterize transitions between them has been postulated from a detailed analysis of these kinetics.

The nature of the dynamic processes determining the distribution of microstates is under active discussion. The low adiabatic compressibility of proteins in the range of frequencies 0.5–10 MHz[20] shows that transitions between microstates are not related to high-amplitude motions or to changes in the arrangement of chain segments and protein domains. Instead, small-scale motions must be involved—oscillations having a strong anharmonicity and rapid decay, as well as activation processes associated with transitions through the potential barriers by a diffusion mechanism.[21] In this case diffusion is limited by steric effects and occurs in a medium with high and inhomogeneous viscosity.

If microstates lead to the existence of a distribution of energies of interaction between aromatic groups and neighboring groups of atoms, then the individual spectra of these groups in different microstates shift differently, which results in an inhomogeneous contour of the absorption band. The application of selective photoexcitation permits specific effects of the distribution of microstates on spectral, temporal, and polarization fluorescence properties to be observed.[22] Such effects have been observed in studies of proteins,[1, 8] and, as we show below, they may be used to obtain important information on dynamics.

2.2.3. Analysis of Motions Using Time-Resolved Methods

The characteristic times of motions of small molecules in solution are investigated by different methods. The values usually obtained are on the picosecond time scale, and only motions which are associated with high activation barriers are characterized by nanosecond times. In contrast, the motions within protein molecules are considerably slower. They include the motion of segments of various sizes surrounded by other protein groups with different packing density. This requires overcoming energetic and entropic activation barriers of different heights.

None of the methods currently used to study molecular dynamics can span the whole time range of motions of interest, from picoseconds to seconds and minutes. However, the structural resolution of a method is of equal importance. A method has to not only provide information about the existence of motions with definite velocities but also to identify what structural element is moving and what is the mechanism of motion. Computer simulation of molecular dynamics has proved to be a very important tool for the development of theories concerning times and mechanisms of motions in proteins.[12] In this approach, the initial coordinates and forces on each atom are input into the calculations, and classical equations of motions are solved by numerical means. The lengthy duration of the calculation procedure, even with powerful modern computers, does not permit the time interval investigated to be extended beyond hundreds of picoseconds. In addition, there are strong limitation on the size of the system (protein molecule). Some of the results from the computer simulation of dynamics in connection with the data obtained by fluorescence methods will be discussed below.

Apart from fluorescence, several other methods may be used to obtain time-resolved information. In the case of proteins containing an iron atom, Mössbauer spectroscopy allows the determination, in the iron binding site, of not only root-mean-square shifts of atoms but also the times over which such shifts occur. Detailed investigations of myoglobin have yielded relaxation times on the order of 10^{-8} s.[11, 16] Proton NMR spectroscopy can be used to study the rotational states of aromatic groups of phenylalanine and tyrosine (flips by 180°).[10, 17] Such motions are observed on the millisecond time scale. Paramagnetic and phosphorescent labels and probes are also used in the study of motions on this time scale.[23]

It should be noted that time resolution is not attained directly with conventional EPR and NMR techniques, and there are difficulties in directly recording the kinetics of processes. Light emission spectroscopic methods (fluorescence and phosphorescence) are suitable for such kinetic studies. Here the absorption of a quantum of light occurs on a time scale considerably shorter than that of any molecular motions, and the delay before emission occurs is on a time scale many orders of magnitude longer and coincides in time with the molecular motions studied. The time range under study is limited by the decay rate of the emission. Time-resolved studies may be conducted easily if the characteristic time of the studied process, τ, is of the same order as the fluorescence lifetime τ_F. If $\tau \ll \tau_F$, then such investigations are more complicated and are limited mainly by the possibility of resolving short decay components in emission. When $\tau \gg \tau_F$, direct kinetic measurements are not possible, and only certain limiting estimates may be obtained. Steady-state fluorescence parameters are functions integrated over the decay time which reflect indirectly the course of the dynamic process. Therefore, they may be useful in studies of motions in a more narrow time interval, that is, when $\tau \approx \tau_F$. The τ_F values of tryptophan residues in proteins vary between 3 and 5 ns, but they may be considerably lowered when the tryptophan residues interact with groups that are fluorescence quenchers.[1, 2] Extrinsic fluorescence probes, which are widely used in studies of proteins, exhibit fluorescence lifetimes on the order of nanoseconds (sometimes tens and hundreds of nanoseconds).[6]

At present, three main approaches to the analysis of intramolecular dynamics in proteins based on fluorescence studies are most commonly used (Figure 2.3).[24]

1. Quenching of fluorescence of tryptophan residues, coenzyme fluorophores, or extrinsic probes buried in the interior of proteins by collisional quencher molecules diffusing through the protein matrix.[7, 25–27]

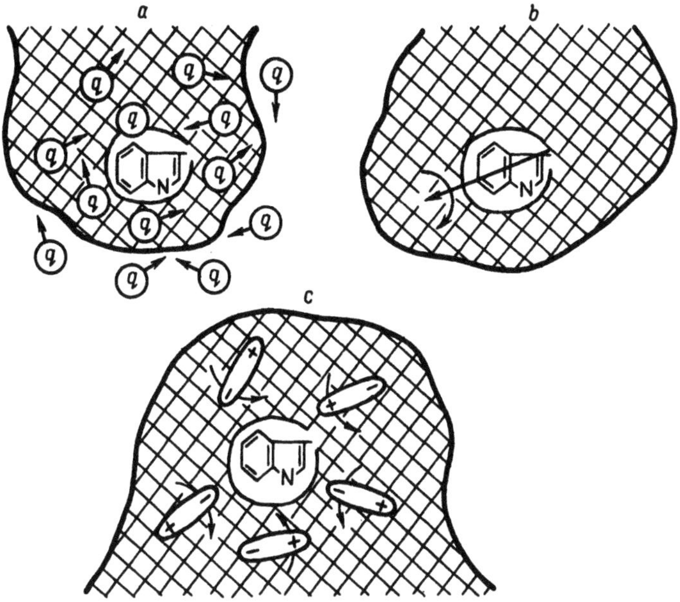

Figure 2.3. Scheme illustrating the main approaches to the study of protein dynamics by fluorescence techniques. (a) Fluorescence quenching. The quenching of the fluorescence a fluorophore buried inside the protein globule requires either diffusion of the quencher within the protein matrix or a local opening. (b) Fluorescence depolarization. Polarization of the fluorescence decreases as a result of fluorophore rotation. (c) Dipole-relaxational shifts and red-edge effects. The rotational motions of dipolar groups are studied. They are induced by the change in the chromophore dipole moment upon excitation.

2. Analysis of rotational mobility of fluorophores by observation of fluorescence depolarization with nanosecond time resolution[28] or by variation of the lifetime (by the action of quenchers).[9,29,30]
3. Observation of reorientational dynamics of dipolar groups surrounding the fluorophore in response to changes in the dipole moment of the fluorophore occurring upon electronic excitation. Such dynamics result in the appearance of spectral shifts with time,[1,31] in changes of fluorescence lifetime across the fluorescence spectrum,[7,32] and in a decrease in the observable effects of selective red-edge excitation.[1,24,33,34] The studies of these processes yield a very important parameter which characterizes dynamics in proteins—the reorientational dipolar relaxation time, τ_R.

In addition, the quenching of the fluorescence of fluorophore groups in protein molecules by neighboring groups[35] and its temperature dependence,[36] energy transfer of electronic excitation and its dependence on excitation wavelength,[1] the type of emission decay kinetics,[1,2] and changes in the half-width of the fluorescence spectrum with time[37] all depend on dynamic processes. The search for new effects related to dynamics in proteins is in progress. The different spectral manifestations of dynamics and the interrelationships between them will be considered in more detail below.

2.3. Decay and Quenching of Fluorescence

2.3.1. Emission Decay Kinetics

In this section, the information on structure and dynamics of proteins which may be obtained from direct observations of fluorescence decay will be considered. This type of information is afforded by methods which permit fluorescence decay kinetics to be followed with picosecond and nanosecond resolution.

Fluorescence lifetimes of tryptophan residues in proteins vary widely—from several picoseconds to 8–10 ns. This wide range of lifetimes may be related to differences arising from the interaction of tryptophan residues with their surroundings and to the fact that the surrounding groups of atoms may participate in reactions with the chromophore in the excited state. Fluorescence lifetimes of coenzyme groups and fluorescent probes associated with proteins vary over several orders of magnitude. These differences in fluorescence lifetimes are determined both by structure and dynamics, that is, by the interaction of chromophore groups with the environment and by variations in this interaction on the time scale of the emission.

The single-exponential decay kinetics, described by the equation

$$I(t) = I_0 \exp(-t/\tau_F) \tag{2.1}$$

is not observed for most chromophores in proteins with aromatic amino acid residues and their derivatives in solutions. In model experiments, it is seen only in special cases in which there is little interaction of the fluorophore with the environment, this interaction does not change upon electronic excitation, and no reactions occur in the excited state. In a more common case often observed in studies of proteins, the decay curve is nonexponential and may be described by a sum of several exponential terms:

$$I(t) = \sum_{i=1}^{n} f_i \tau_i^{-1} \exp(-t/\tau_i) \tag{2.2}$$

where n is the total number of independent components of the decay, and f_i is the fraction of light quanta contributing to the total emission of the ith component.

Nonexponential fluorescence decay would be expected to be observed in the following cases:

1. There are several conformational arrangements with different interactions of the fluorophores with surrounding groups of atoms. Such interactions may affect differently nonradiative deexcitation processes in the excited state, and the decay times for these conformational states will differ. If each of these states is characterized by single-exponential decay kinetics, then the number of constants f_i will correspond to the number of aromatic groups in the protein that are in conformationally different states.
2. Reactions take place in the excited state during the course of emission. This can lead to nonexponential decay kinetics even in the case of a single fluorophore in a single conformational state. Such excited-state reactions include dipolar relaxation of the environment, formation of exciplexes, and phototransfer of protons and electrons. In addition, certain fluorophores may undergo conformational rearrangements; examples include torsional vibrations of the indole ring of tryptophan[1,2] and closing and opening of nicotinic and adenine rings of NADH.[3] Deviations from the stationary diffusion pattern (transient effects) in fluorescence quenching[38,39] and excitation energy transfer between identical fluorophores[40] also lead to nonexponential decay.
3. There exists a distribution of microstates associated with the internal dynamics at the level of atomic groups. This may also result in nonexponential fluorescence decay if the transitions between microstates occur more slowly than the decay.

Proteins having one chromophore per molecule are the simplest and most convenient in studies of fluorescence decay kinetics as well as in other spectroscopic studies of proteins. These were historically the first proteins for which the tryptophan fluorescence decay was analyzed. It was natural to expect that, for these proteins at least, the decay curves would be single-exponential. However, a more complex time dependence of the emission was observed. To describe the experimental data for almost all of the proteins studied, it was necessary to use a set of two or more exponents.[2] The decay is single-exponential only in the case of apoazurin.[41] Several authors[41,42] explained the biexponentiality of the decay by the existence of two protein conformers in equilibrium. Such an explanation is difficult to accept without additional analysis, since there are many other mechanisms leading to nonexponential decay and in view of the fact that deconvolution into exponential components is no more than a formal procedure for treatment of nonexponential curves.

Recently, Alcala *et al.*[43] have applied a new version of the method of phase fluorimetry with variation of the modulation frequency and high time resolution. They have analyzed the fluorescence decay curves of proteins in terms of a continuous distribution of elementary fluorescence lifetimes τ_i. They estimated the width of the distribution function and could elucidate whether the distribution in a given case is mono- or polymodal. In several cases (ribonuclease T_1 and pancreatic phospholipase A_2), the distribution was found to be a bimodal function. With a decrease in temperature, the lifetime distribution became broader and shifted to longer lifetime values.

One can expect that the analysis of continuous distributions of electronic excited-state lifetimes will not only provide a higher level of description of fluorescence decay kinetics in proteins but also will allow the physical mechanisms determining the interactions of fluorophores with their environment in protein molecules to be elucidated. Two physical causes for such distributions of lifetimes may be considered:

1. *Static*: In this case, the distribution of lifetimes is due to the existence of a continuum of conformational microstates, each characterized by its own lifetime. For time-resolved fluorometric detection of heterogeneity in this case, it is necessary for the rate of transition between such microstates to be slower than that of emission.
2. *Dynamic*: In this case, the distribution of lifetimes is the result of the electronically excited tryptophan chromophore being perturbed by and colliding with the surrounding groups of atoms in the protein molecule and with solvent molecules.

One would expect that lowering the temperature or increasing the viscosity of the solvent would increase the width of the lifetime distribution, since both factors may affect the rate of transitions between microstates. If this rate is high as compared with the mean value of the fluorescence lifetime, the distribution should be very narrow, as for tryptophan in solution. When the rate of transitions between microstates is low, a wide distribution would be expected.

In protein molecules with two or more tryptophan residues, it is necessary to obtain first the fluorescence decay curves for the individual residues. For this purpose, additional spectroscopic information is necessary. One can use the dependence of the decay curves on emission wavelength, apply selective fluorescence quenchers, or selectively modify one of the tryptophan residues. The results of Brochon *et al.* for the *lac* repressor[(44)] and those of Beechem *et al.* for alcohol dehydrogenase[(45)] provide evidence in favor of such approaches.

The quantum yield of flavin fluorescence in proteins is very low in many cases, and the lifetimes are on the order of picoseconds. This is a result of the high electrophilicity of oxidized flavins, and their ability to quench fluorescence following electron transfer from the electron-rich groups of protein molecules. Lipoamide dehydrogenase, in which the isoalloxazine ring of flavin adenine dinucleotide (FAD) is evidently completely screened from the solvent, is among a small number of proteins that exhibit the intense emission of FAD. The fluorescence decay curves of this protein are nonexponential and are independent of the emission wavelength.[46] The same behavior is observed for the NADH decay kinetics in alcohol dehydrogenase[47] and glutamate dehydrogenase.[48]

In the majority of cases, fluorescent labels and probes, when studied in different liquid solvents, display single-exponential fluorescence decay kinetics. However, when they are bound to proteins, their emission exhibits more complicated, nonexponential character. Thus, two decay components were observed for the complex of 8-anilinonaphthalene-1-sulfonate (1,8-ANS) with phosphorylase[49] as well as for 5-diethylamino-1-naphthalenesulfonic acid (DNS)-labeled dehydrogenases.[50] Three decay components were determined for complexes of 1,8-ANS with low-density lipoproteins.[51] On the basis of only the data on the kinetics of the fluorescence decay, the origin of these multiple decay components (whether they are associated with structural heterogeneity in the ground state or arise due to dynamic processes in the excited state) is difficult to ascertain.

2.3.2. Fluorescence Quenching by Intrinsic Quenchers

Considerable variations of the quantum yield and lifetime and, probably, deviations from exponential decay as well result from the existence in protein molecules of atoms capable of quenching the fluorescence of chromophore groups in close proximity to them. A detailed analysis of the results of model studies on the interaction of indole and phenol chromophores with the functional groups present in proteins ($-COOH$, $-COO^-$, $-NH^+$, $-NH_2$, $-SH$, imidazole, and others) has shown that the rate constants of the dynamic quenching depend uniformly on the temperature, with an effective activation energy of 12 kJ/mol.[52] This fact, along with the high absolute values of the rate constants, has led to the conclusion that the rate of quenching in these cases is limited by diffusion, and the characteristic activation energy of the quenching itself is practically zero. This may indicate that fluorescence quenching by neighboring groups in proteins is limited by the frequency of active collisions of excited chromophores with nearby quencher groups. In this case the temperature dependence of the quenching of protein fluorescence should reflect the temperature-dependent rate of motions of the protein structure surrounding the aromatic group.

In the case in which the protein molecule contains only a single fluorophore, the equation for the fluorescence quantum yield Q will be

$$1/Q = 1 + k_2 + k_i(T) = a + b \cdot f(T) \tag{2.3}$$

where k_2 is the total rate constant for all nonradiative processes, whose rates are independent of the temperature, and the k_i are the temperature-dependent quenching constants.

It was shown[52] that the function $f(T)$ is not described properly by the exponent but corresponds perfectly to the temperature dependence of the T/η ratio:

$$1/Q = a + b \cdot \frac{T}{\eta} \tag{2.4}$$

where η is the viscosity of the solvent. This relationship is observed for emission of tyrosine and tryptophan residues,[36] as well as of coenzyme groups.[46]

Since the rate constants of bimolecular diffusion-limited reactions in isotropic solution are proportional to T/η, these data testify to the fact that the k_i values are linearly dependent on the diffusion coefficient D in water, irrespective of whether the fluorophores are present on the surface of the macromolecule (human serum albumin, cobra neurotoxins, proteins A and B of the neurotoxic complex of venom) or are localized within the protein matrix (ribonuclease C_2, azurin, L-asparaginase).[36] The linear dependence of the functions $1/Q$ and $1/\tau_F$ on τ/η indicates that the mobility of protein structures is correlated with the motions of solvent molecules, and this correlation results in similar mechanisms of quenching for both surface and interior sites of the macromolecule.

The motions of chromophore groups and of their environment that lead to temperature-dependent fluorescence quenching are those on the nanosecond time scale. Slower motions cannot manifest themselves in effects on the excited-state lifetime (this corresponds to the limit of high viscosity). On the other hand, if the motions are considerably faster (on the picosecond time scale), then they should give rise to static quenching.

2.3.3. Fluorescence Quenching by Extrinsic Quenchers

Small molecules that act as collisional quenchers may penetrate into the internal structure of proteins, diffuse, and cause quenching upon collision with the aromatic groups. Lakowicz and Weber[53] have shown that the interaction of oxygen molecules with buried tryptophan residues in proteins leads to quenching with unexpectedly high rate constants—from 2×10^9 to $7 \times 10^9\ M^{-1}\,s^{-1}$. Acrylamide is also capable of quenching the fluorescence of buried tryptophan residues, as was shown for aldolase and ribonuclease T_1.[54] A more hydrophobic quencher, trichloroethanol, is a considerably more efficient quencher of internal chromophore groups in proteins.[55]

Furthermore, the quenching of internal residues in proteins by ionic quenchers, although not strong, is quite detectable.[56] A double-quenching method was developed to separate fluorescence quenching parameters characteristic of solvent-exposed and buried fluorophores.[57] The method uses two types of quenchers simultaneously, one type penetrating and the other not penetrating into the protein matrix.

The possible mechanisms of quenching of internal residues may be discussed on the basis of two models of protein dynamics which were previously developed for interpretation of data on hydrogen–deuterium exchange.[58] One of the models suggests diffusion of the quencher inside the viscous protein matrix on the nanosecond time scale. In this case the quenching effect should have a small activation energy and depend slightly on the solvent viscosity, but strongly on the size and charge of the quencher. The other model suggests the existence of extensive fluctuations and local unfolding of the protein. In this case the elementary act of quenching occurs in the aqueous environment. According to such a kinetic model of quenching, the apparent quenching constant k_q^{app} may be written in the form

$$k_q^{\mathrm{app}} = \gamma k_d k_{\mathrm{or}} \tag{2.5}$$

where k_{or} is the equilibrium constant between the states of the protein with an open and a closed cavity, γ is the efficiency of the quenching reaction in the aqueous solution, and k_d is the diffusion-limited rate constant for collisions between the quencher and an exposed chromophore group of the protein. In this case not only nanosecond but also slower dynamics may affect the rate of the quenching process. Of importance is the fact that the lifetime of the open cavity in the protein in which diffusion of the quencher occurs should not be shorter than the lifetime of the excited state of the chromophore. In this model the activation energy of quenching will be determined by the energy of formation of defects within the protein structure and would be higher than for the diffusion model; the dependence on the size and the charge of the quencher would be lower. In addition, the rate of the quenching process may depend strongly on the viscosity of the solvent.

Low activation energy (12–16 kJ/mol) is normally observed in experiments on protein fluorescence quenching by oxygen.[53] However, higher values have also been found—for example, 40 kJ/mol for alcohol dehydrogenase[59] and 25–28 kJ/mol for cod parvalbumin.[60] In these cases some particular structural rigidity of the chromophore environment is indicated, and the diffusion of oxygen molecules requires deformation or breaking of several noncovalent bonds. In the case of alcohol dehydrogenase and alkaline phosphatase, a slight dependence of the efficiency of oxygen quenching on the viscosity of the medium is observed.[61] Taking into account the lack of charge and small size of the oxygen molecule and its ready solubility in nonpolar media, one may suggest that oxygen penetrates directly into the protein molecule, similarly to its diffusion in liquids, rather than by opening of the protein structure.

The question of what mechanisms is involved in the case of other quenchers is still unclear. For the quenching of aldolase and ribonuclease T_1 by acrylamide, the activation energy is rather high, 40–45 kJ/mol,[54] but the value in the case of cod parvalbumin[60] is lower (27 kJ/mol), being similar to that for oxygen quenching. According to Bushueva *et al.*,[56] the efficiency of fluorescence quenching by an ionic quencher (potassium iodide) and acrylamide in a number of proteins depends on the temperature and viscosity of the medium as a function of T/η. Eftink and co-workers found no dependence on the viscosity of the medium in experiments on the quenching of the fluorescence of ribonuclease T_1[54] and cod parvalbumin[60] tryptophan residues by acrylamide. The tryptophan residues in azurin, alcohol dehydrogenase,[62] and alkaline phosphatase[63] located inside the protein globule are practically not quenched by acrylamide. Thus, the mechanisms involved in the penetration of small molecules, except oxygen, into protein molecules and their diffusion on the nanosecond time scale may be complicated.

When analyzing fluorescence quenching of intraglobular chromophores from the standpoint of the diffusion mechanism, one must consider the possibility that at the moment of excitation the quencher molecules may be distributed differently inside the globule and in the surrounding solvent. Then when describing the kinetics, we must not only take into account the differences in the migration rate of the quencher but also estimate the rate constants of the penetration of the quencher into the protein molecule and its efflux. Gratton *et al.*[64] have developed a general model for fluorophore quenching inside protein globules by a diffusing quencher. At low concentrations of the quencher, the quenching process is dependent on the rate constants describing the entry of the quencher into the globule, its migration inside it, and its efflux from the globule. If the decay is single-exponential in the absence of quencher, then in its presence the decay becomes biexponential. At high concentrations of the quencher, the decay function becomes still more complicated. This model described well the results on fluorescence quenching of iron-free porphyrin in hemoglobin and myoglobin by oxygen.

Already in early work on the application of molecular oxygen as a fluorescence quencher, Vaughan and Weber[65] showed that the quenching rate constant of a pyrene derivative is decreased by two orders of magnitude on its binding to bovine serum albumin. A considerable drop in the efficiency of acrylamide quenching of the fluorescence label *N*-iodoacetyl-*N'*-(5-sulfo-1-naphthyl)ethylenediamine (IAEDANS) is observed on its binding to ribonuclease[66] and troponin.[67] The results of the above experiments have shown that the binding of fluorescence labels and probes to proteins may result in both a decrease in the access to the probe of quencher from the aqueous solvent and, evidently, immobilization of the probe binding site, which hampers the diffusion of the quencher through the protein structure.

Application of nanosecond time-resolved measurements should promote further development of the fluorescence quenching method. This approach allows quenching data to be analyzed in coordinates of fluorescence lifetimes; the static quenching component does not affect the result. As to nonstationary (transient) processes in quenching, it is these processes that may explain the nonexponential decay widely observed in protein fluorescence studies. The data obtained from fluorescence quenching studies reveal quite different and sometimes rather high levels of intramolecular dynamic processes on the nanosecond time scale. However, the exact relationship between the quenching parameters and the parameters describing protein dynamics cannot be extracted from these data. This is because not only are the local values of these parameters within the protein interior poorly defined, but also the dynamic mechanism leading to quenching is not always clear.

2.4. Rotation of Aromatic Groups

2.4.1. Fluorescence Polarization Studies with and without Time Resolution

If the aromatic group is bound tightly within the protein molecule, then one may obtain information on the rotational diffusion of the whole molecule from fluorescence polarization studies. Such investigations, which were started by Weber,[68] were widely popular in the 1960s and 1970s. Correlation times ϕ_D of macromolecule rotations were determined according to the Perrin equation:

$$1/r = 1/r_0 + \tau_F/r_0\phi_D \tag{2.6}$$

where r is the fluorescence anisotropy, and r_0 is the limiting value of the anisotropy observed in the absence of rotation. Since $\phi_D = V\eta/kT$, where V is the volume of the macromolecule and η is the viscosity of the solvent, Perrin plots showing the dependence of $1/r$ on T/η can be obtained by varying the temperature or the viscosity. Very often, however, the anisotropy of tryptophan fluorescence in proteins is observed to be lower than that which would correspond to the rotation of the whole molecule; deviations from the Perrin equation are observed at high viscosities and low temperatures.[69] This may occur for two reasons: (1) There exist intrinsic rotations of aromatic groups with respect to the protein globule; and (2) electronic excitation energy transfer occurs from one tryptophan residue to another (Figure 2.4).

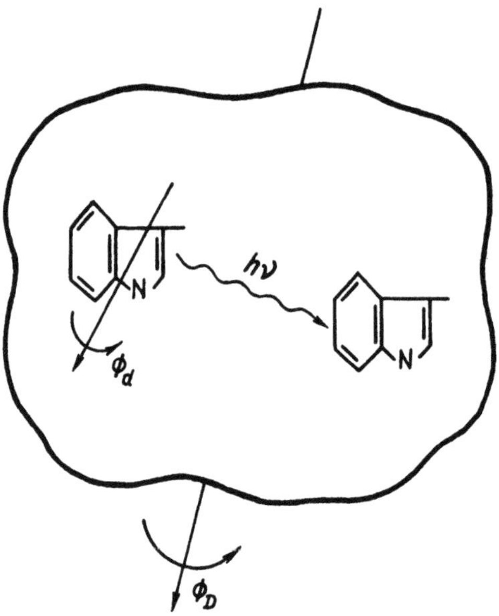

Figure 2.4. Schematic representation of processes which lead to fluorescence depolarization in proteins: rotation of the protein molecule as a whole with correlation time ϕ_D, rotation of the fluorophore with correlation time ϕ_d, and excitation energy transfer, represented by the wavy arrow.

The elucidation of the intramolecular dynamics of tryptophan residues became possible due to anisotropy studies with nanosecond time resolution. Two approaches have been taken: direct observation of the anisotropy kinetics on the nanosecond time scale using time-resolved[(28)] or frequency-domain[(70)] fluorometry, and studies of steady-state anisotropy for τ_F varying within wide ranges (lifetime-resolved anisotropy). The latter approach involves the application of collisional quenchers, oxygen[(29, 71)] or acrylamide.[(30)] The shortening of τ_F by the quencher decreases the mean time available for rotations of aromatic groups prior to emission.

In order to avoid complications caused by excitation energy transfer between tryptophan residues, most investigations have been performed with proteins containing one tryptophan residue per molecule. When studying protein solutions, there are difficulties in separating the effects of rotation of entire protein molecules and of the chromophores themselves relative to their environment in the protein matrix. It is usually assumed that intramolecular motions are more rapid and manifest themselves as short-lived components of anisotropy decay curves or in depolarization at short emission lifetimes.

Recent results show large variations in intramolecular rotations of tryptophan residues in proteins on the nanosecond time scale, ranging from complete absence of mobility to motions of considerable angular amplitudes. Among native proteins with internal tryptophan residues, wide angular amplitude rotations were observed only in studies of azurin,[(28, 29)] where the correlation time of the rapid component was $\tau_{\phi_1} = 0.51$ ns.[(28)] The existence of intramolecular tryptophan motions is confirmed by studies with azurin embedded into a polymeric matrix, which excludes the possibility of rotation of the entire molecule.[72] At the same time, the tryptophan residues of ribonuclease T_1, aldolase, and alcohol dehydrogenase[29] as well as of staphylococcal nuclease B[28] display no mobility relative to the protein matrix.

High intrinsic mobility of a single tryptophan residue is observed in basic myelin protein[28, 29] and monomeric melittin.[29, 70] One may postulate that azurin, on the one hand, and basic myelin protein, on the other hand, are two cases in which intramolecular rotation of tryptophan residues may occur for quite different reasons. In the case of azurin, the completely hydrophobic surroundings are not very dense.[73] The tryptophan residue in azurin does not form hydrogen bonds or exciplexes, which are known to be important in fixing the spatial orientation of tryptophan residues and lowering their mobility. Interaction of the indole group with hydrophobic protein groups is weaker and has no strict directionality in space. These conditions may give rise to the possibility of Brownian rotation of the indole ring. However, it should be emphasized that the case of azurin is unique, as are its spectral properties, it exhibits an extremely short-wavelength fluorescence maximum (308 nm) relative to that of any of the other proteins studied.

On the other hand, basic myelin protein and monomeric melittin are proteins which, by many criteria, are devoid of ordered structure in aqueous solutions. This results in freedom of rotation of tryptophan residues which are exposed to the solvent. Such a situation may exist for peptides without regular structure and for denatured proteins.

The most typical case is probably that in which tryptophan residues undergo rotations during the excited-state lifetime with a small angular amplitude (up to 30°).[30, 69] Similar motions are observed in protein–dye complexes.[74] We will consider below the qualitative models which have been put forward to describe such motions.

2.4.2. Models of Rotations

The most popular model describing small-angle rotational movements of aromatic rings is the model of torsional vibrations around the $C^\alpha-C^\beta$ and $C^\beta-C^\gamma$ bonds.[30, 70] This model has been used in simulations of motions by the methods of molecular dynamics.[75, 76] However, the results are not always satisfactory. In some cases, for example, for lysozyme,[77] the experimental data do not agree well with the results of simulations: the observed motions are slower and less extended than predicted.

Rotations of aromatic groups should be associated with a considerable shift of neighboring groups, especially those adjacent to the ring periphery.[35] On the basis of the discussion in Section 2.2, it can be concluded that such motions should be diffusional in character, limited in amplitude, and associated with viscous damping of vibrations. In this case, the rotation of the aromatic group itself becomes diffusive and proceeds by passing through the activation barriers separating the protein microstates.

It should be noted that the higher the angular amplitude of motion, the more extensive are the rearrangements required in the surroundings of the aromatic ring. Rotations of very small angular amplitude may occur with negligible activation energy[69] (they are probably responsible for the frequently observed[29] initial drop of the fluorescence anisotropy to 0.25–0.27 relative to the ultimate value of 0.3 at 300-nm excitation). Small-amplitude nanosecond motions should have activation energies of more than 10 kJ/mol. For wide-scale motions and flips by 180°, the activation energy is probably one order of magnitude higher. Evidently, this is the reason why they occur so slowly—on the millisecond and second time scales, as indicated by NMR data for phenylalanine and tyrosine rings.[10]

Both the model of torsional vibrations[69] and the diffusion model[78] follow from the concept that the electronically excited chromophore is at equilibrium with its environment. However, electronic excitation is known to lead to nonequilibrium energy states in the system comprised of the chromophores and surrounding groups, resulting in considerable strain and creation of a large excess of potential energy.[22] This excess potential energy would be expected to determine the course of dynamic processes, at least for short times before the chromophore groups and their environment reach equilibrium. However, if the surroundings of chromophore groups are of low mobility even on the nanosecond time scale, then the induced fluorophore motions arising under the effect of the reactive field and directed toward establishing dielectric equilibrium in the excited state should primarily determine the intramolecular mobility.

Brownian movements of the chromophore, similarly to dipolar relaxation, may proceed if the mobility of the groups of atoms surrounding the chromophore is sufficient. In model viscous media (e.g., in glycerol), the dipolar equilibrium is established by movement of the solvent molecules, the larger fluorophore molecule remains immobile on the nanosecond time scale, and the emission anisotropy is close to its limiting value. In protein molecules there are elements more massive and rigid than the fluorophore (for instance, macrodipoles of α-helical segments[79]), and equilibrium may be reached due to rotation of the fluorophore itself.[1] Therefore, it can be supposed that equilibrium Brownian rotation is activated with relaxation. As shown by Nemkovich *et al.*,[22] the dependence of the anisotropy decay kinetics on excitation wavelength may be used to reveal the effects of induced chromophore rotations associated with relaxation.

2.5. Fluorescence Spectroscopy of Molecular Relaxation

2.5.1. Dynamic Reorientation of Dipoles in the Fluorophore Environment

Fluorescence spectroscopy may be widely used for direct investigation of molecular movements occurring at the level of dipolar molecules and groups of atoms surrounding aromatic molecules or groups. The absorption of light results in disruption of the energy equilibrium existing between the fluorophore and its environment in the ground state. The time-dependent process of establishing a new equilibrium in the excited state (the relaxation process) can be studied. Such an equilibrium may or may not be reached, depending on the ratio between the fluorescence lifetime τ_R and the dipole-orientational relaxation time τ_F.

The dipole–dipole interactions of the fluorophore in the electronic excited state with the surrounding groups of atoms in the protein molecule or with solvent molecules give rise to considerable shifts of the fluorescence spectra during the relaxation process. These spectral shifts may be observed directly by time-resolved spectroscopic methods. They may be also studied by steady-state spectroscopic methods, but in this case additional data must be obtained by varying factors that affect the ratio between τ_F and τ_R.

The observed spectral shift depends both on the properties of the fluorophore itself (the vectorial difference between the dipole moments in the ground and the excited state, $\mu_e - \mu_g$) and also on properties of the environment interacting with it. The establishment of dielectric equilibrium with the environment occurs due to the following effects:

1. *Electronic polarization of the environment.* This effect is related to the square of the refractive index, n^2 (dielectric constant at the frequency of light). Here the spectral shift occurs instantly (10^{-15} s), and its evolution with time is not observed by the kinetic spectroscopic methods. The protein molecule is a medium with a relatively high electronic polarization ($n = 1.5 \div 1.6$).

2. *Reorientation of dipoles.* This effect, to a first approximation, is described by the static dielectric constant of the environment ε. However, a protein molecule is an environment with special dielectric properties. It contains in high concentrations fairly large electrical dipoles (the dipole moment of the peptide group is 3.6 D, which is twice as high as that of the water molecule), but their ability to reorient under the influence of an electric field is limited by steric effects. In addition, polar (and nonpolar) side groups may be arranged into clusters. Therefore, the results of calculations of the static dielectric constant vary within two orders of magnitude[80, 81] and are recognized to be unreliable.

Another aspect of dielectric relaxation in proteins should be considered. If in model solutions of aromatic molecules, dielectric relaxation occurs solely by virtue of orientational movements of dipoles (attainment of equilibrium solvation through translational movements is one to three orders of magnitude slower[82, 83]), then, in the protein molecule, relaxation owing to orientational and translational movements of surrounding fluorophore groups may be coupled and occur simultaneously. Probably, more general terms such as "structural relaxation" or "nuclear relaxation" could better describe this process.

3. *Charge transfer*. Dielectric equilibrium may be also established due to the shift of charges in the fluorophore's surroundings (transfer of protons and ions). Proton transfer is facilitated in systems with hydrogen bonding. If there is a chain of hydrogen bonds in the protein molecule, proton transfer may proceed with a low activation energy and lead to considerable redistribution of dipole moments.[84]

4. *Formation of specific complexes in the excited states (exciplexes)*.[1, 35, 52, 85] Exciplexes are complexes not present in the ground state that form due to the extensive redistribution of electron density that occurs upon excitation. Among exciplexes, there may be some whose formation does not require substantial nuclear rearrangements and thus occurs rather rapidly even at 77 K. The formation of exciplexes is accompanied by a spectral shift to longer wavelengths. It is postulated that the fluorescence from tryptophan in proteins in a variety of cases is fluorescence from tryptophan exciplexes.[35, 85] In studies of the effects of environmental dynamics on the spectra, the exciplexes may be considered as individual fluorophores.

Usually, the most general nonspecific effects of dipole-orientational and electronic polarization of the medium are discussed, and the results of the theory of relaxational shifts developed under the approximation of a continuous dielectric medium may be used.[86–88] The shift of the frequency of the emitted light with time is a function of the dielectric constant ε_0, the refractive index n, and the relaxation time τ_R:

$$\Delta\nu \sim (\mu_e - \mu_g)^2 \times \left(\frac{\varepsilon_0 - 1}{\varepsilon_0 + 2} - \frac{n^2 - 1}{n^2 + 2}\right) - [1 - \exp(-t/\tau_R)] \qquad (2.7)$$

Here the term involving ε_0 determines the spectral shift due to dipole–dipole interactions. This effect will be smaller, the greater the electronic polarization of the medium, which is expressed by the term involving n^2.

The description of the real process of dipole-orientational relaxation by one parameter τ_R is a first-order approximation which is far removed from reality even in studies with model solvents.[89] A set of relaxation times would exist in real systems. However, such an approximation is necessary since it allows rather simple models of relaxation to be developed and to be compared with the results of experiments. τ_R may be considered as a simple effective parameter characterizing the dynamic processes.

2.5.2. The Two-State Model of Relaxation

One may consider the relaxation process to proceed in a similar manner to other reactions in electronic excited states (proton transfer, formation of exciplexes), and it may be described as a reaction between two discrete species: initial and relaxed.[7, 90] In this case two processes proceeding simultaneously should be considered: fluorescence emission with the rate constant $k_F = 1/\tau_F$, and transition into the relaxed state with the rate constant $k_R = 1/\tau_R$ (Figure 2.5). The spectrum of the unrelaxed form can be recorded from solid solutions using steady-state methods, but it may be also observed in the presence of the relaxed form if time-resolved spectra are recorded at very short times. The spectrum of the relaxed form can be recorded using steady-state methods in liquid media (where the relaxation is complete) or using time-resolved methods at very long observation times, even as the relaxation proceeds.

According to the two-state model, the spectrum of the relaxed state has a mean frequency $\bar{\nu}_R$ and is shifted relative to the spectrum of the initial state, which has a mean frequency $\bar{\nu}_F$. If relaxation does not occur during the process of emission ($\tau_R \gg \tau_F$), the mean frequency of the fluorescence

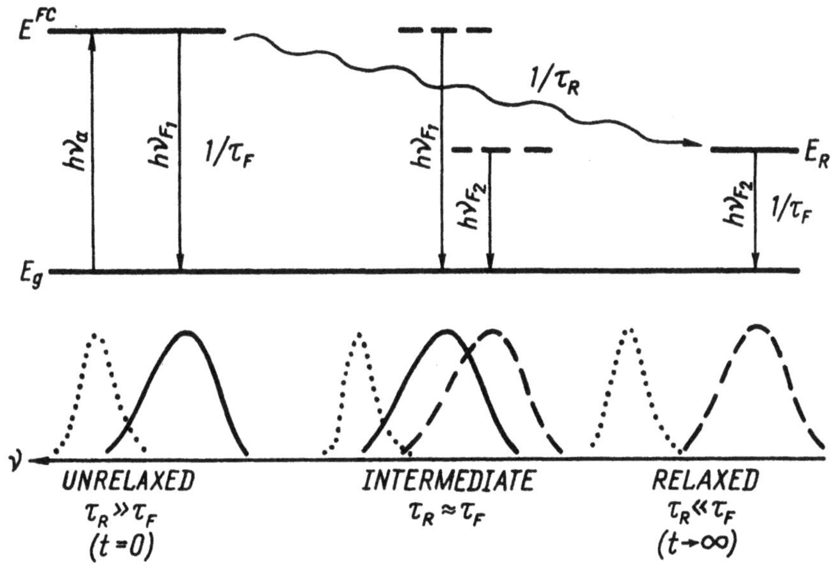

Figure 2.5. Energy level diagram (*top*) and spectra (*bottom*) illustrating the two-state model of relaxation. The energy of the absorbed quantum is $h\nu_a$, and the energies of the emitted quanta are $h\nu_{F_1}$ (unrelaxed) and $h\nu_{F_2}$ (relaxed). The fluorescence spectrum of the unrelaxed state (solid curve) is shifted relative to the absorption spectrum (dotted curve) due to the Stokes shift. The emission intensity from the unrelaxed state decreases and that from the relaxed state (dashed curve) increases as a result of relaxation.

spectrum is $\bar{\nu}_F$; when complete equilibrium is reached ($\tau_R \ll \tau_F$), it is $\bar{\nu}_R$. If $\tau_R = \tau_F$, then emission from the relaxed and unrelaxed states should contribute in equal proportions to the observed spectrum. Thereby, on the basis of this model, the fluorescence spectrum at the intermediate stages of relaxation should be considerably wider than at the initial and final stages.

This model permits τ_R to be determined using information on the fluorescence decay in a very simple way. If unrelaxed fluorophores are excited, the decay is exponential beyond the relaxation range and, in this range, consists of two components τ_1 and τ_2. These components will be simple functions of τ_R and τ_F. If we assume that emission on the short-wavelength side occurs only from the unrelaxed state and that the simultaneous loss of emitting quanta occurs due to relaxation, then the longer component, τ_1, equals τ_F, and the shorter one, τ_2, equals $1(1/\tau_R + 1/\tau_F)$. Unfortunately, this approach is difficult to apply when the decay is nonexponential, which is almost always the case with proteins (see Section 2.3.1.).

On the basis of the two-state model, the following explanation of the site-selective effects (the shifts of fluorescence spectra on excitation at the long-wavelength edge) may be put forward.[91] Red-edge excitation selects the relaxed species, that is, those in which the orientation of surrounding dipoles corresponds to the relaxed state. This model, however, does not explain why fluorescence spectra may be shifted to such a significant extent that they are observed at even lower energies than the spectra of the completely relaxed states. Time-resolved studies show that in this case the shift of the spectra with time toward the relaxed state is a shift to higher energies and shorter wavelengths. This phenomenon, which is called "up-relaxation,"[22] deserves special attention, and new relaxation models are needed to account for it.

2.5.3. Continuous Model of Relaxation

The continuous model of relaxation, suggested by Bakhshiev and Mazurenko,[86, 87] considers the relaxation as proceeding continuously and simultaneously with the emission. According to this model, the change in interaction energy and, correspondingly, the shift of the emission frequency in the course of relaxation is exponential. Emission is expressed by an exponential law (Figure 2.6).

According to the Bakhshiev–Mazurenko model, the emission intensity at frequency ν and time t after excitation, $I(\nu, t)$, may be expressed by the equation

$$I(\nu, t) = (1/\tau_F)\, I(\nu - \bar{\nu}) \exp(-t/\tau_F) \tag{2.8}$$

where the function $I(\nu - \bar{\nu})$ describes the spectral distribution of intensity in

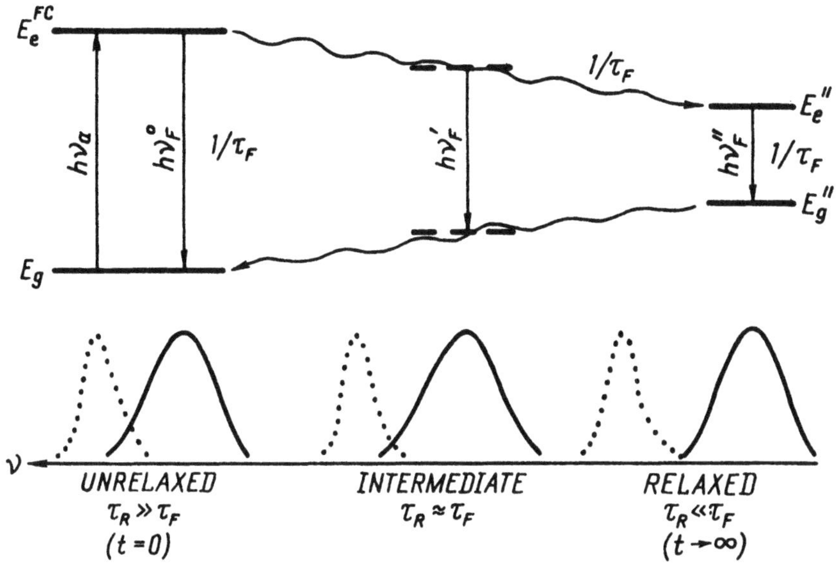

Figure 2.6. Energy level diagram (*top*) and spectra (*bottom*) illustrating the continuous model of relaxation. The energy of the emitted quanta decreases ($h\nu_F \to h\nu_F' \to h\nu_F''$) and the position of the fluorescence spectrum (solid curves) moves smoothly as a result of relaxation.

the emission process ($\bar{\nu}$ is the frequency of the "center of gravity" of the fluorescence spectrum, which may be replaced by the position of the maximum if the shape of the spectrum does not change much with relaxation). At a fixed time, $I(\nu, t)$ may be considered as the "instantaneous" fluorescence spectrum; at a fixed frequency ν, it may be considered as a function describing the decay kinetics.

If the limiting values of the spectral shift, $\bar{\nu}_0$ (at $t=0$) and $\bar{\nu}_\infty$ (at $t \to \infty$), are introduced, then the dependence of the frequency shift on the ratio between τ_R and τ_F may be expressed by the equation

$$\frac{\bar{\nu} - \bar{\nu}_\infty}{\bar{\nu}_0 - \bar{\nu}_\infty} = \frac{\tau_R}{\tau_R + \tau_F} \tag{2.9}$$

This equation is a good approximation to the description of the relaxational spectral shifts occurring with variations of τ_R and τ_F, which are brought about by temperature changes and effects of collisional fluorescence quenchers. Using this equation, τ_R can be easily determined if τ_F, $\bar{\nu}_0$, and $\bar{\nu}_\infty$ are known for the system (the chromophore and its environment) under study. The last two values may be obtained not only from time-resolved spectra but also from steady-state spectra at the lowest ($\bar{\nu}_0$) and highest ($\bar{\nu}_\infty$) temperatures. The latter measurement is difficult to achieve with such labile species as protein molecules in solution. In determining $\bar{\nu}_0$, one may shorten τ_F by making use of collisional quenchers.[92]

The Bakhshiev–Mazurenko model predicts the following dependence of the fluorescence decay curves on the emission wavelength (Figure 2.7). The decay is exponential at the maximum of the fluorescence band. In the short-wavelength region, an additional short-lived positive exponent is observed, caused by the loss of excited-state species in the process of relaxation (as the excited-state species relax, they emit light at longer wavelengths), and in the long-wavelength region, there is an additional negative exponent, which describes the increase in the number of excited fluorophores emitting in this region of the spectrum due to relaxation. Such a dependence was observed experimentally in studies of model systems.[93] The mean lifetime τ_F should increase with wavelength across the fluorescence spectrum. Relaxation affects the results of phase fluorometry as well: the phase angle and modulation depth become functions of the emission wavelength.[7, 87]

It should be noted that a number of experimental observations do not agree with the Bakhshiev–Mazurenko model: (1) the time-dependent range of relaxational shifts of spectra is considerably wider than that described by Eq. (2.9), which may be associated with the existence of a distribution of relaxation times[89, 94]; (2) the bandwidth of the fluorescence spectrum varies significantly during relaxation[93]; (3) substantial deviations from exponential

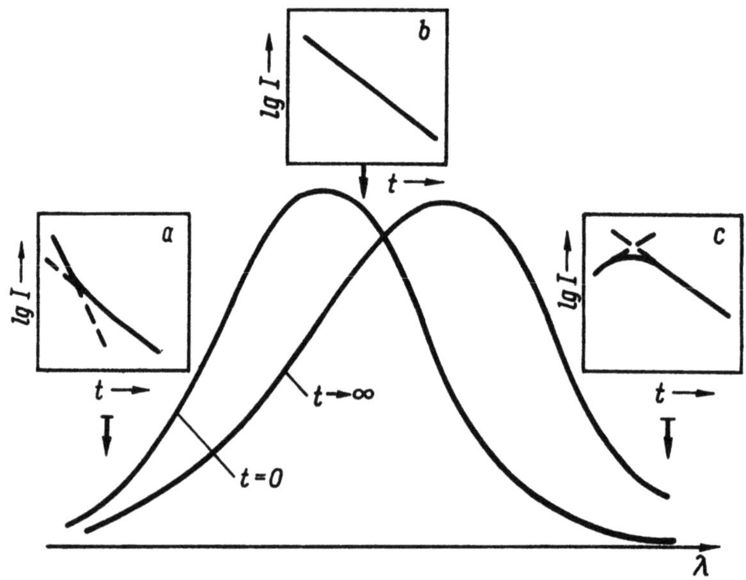

Figure 2.7. The fluorescence spectra from unrelaxed ($t = 0$) and relaxed ($t \to \infty$) states and the emission decay curves at the short-wavelength edge (a), the maximum (b), and the long-wavelength edge (c) of the spectrum.

decay are observed[95]; and (4) the model does not suggest the existence of any dependence of the fluorescence spectra on the excitation wavelength or of "red-edge" effects.

The difficulties encountered in the application of the above simple models of dipole-orientational relaxation to the interpretation of the experimental data necessitate the development of more complex models. In a realistic description of the relaxation process, two approaches may be taken: one that makes allowance for the distribution of fluorophores in fluorescence lifetimes, and one that makes allowance for their distribution in initial energies of interaction with the environment. The latter approach is more promising, since it allows new experimental data on the excitation wavelength dependence of fluorescence spectra, as well as on the influence of relaxation on this dependence, to be obtained and analyzed.

2.5.4. Site-Photoselection Model

As has been reported,[1,8,22] specific effects at the "red edge" in fluorescence spectra and at the "blue edge" in excitation spectra are observed in systems with delayed relaxation ($\tau_R > \tau_F$). They are also manifested in the time and polarization properties of fluorescence and in excitation energy transfer. A shift of fluorescence spectra toward longer wavelengths with an increase in the excitation wavelength at the long-wavelength (red) edge, is the basic and most easily observable effect.

The analysis of these phenomena requires the use of more complicated models which take into account the fact that at the moment of excitation individual aromatic molecules in the ensemble under study may interact differently with their environment.[96,97] The existence of a distribution of fluorophores differing in such interactions leads to inhomogeneous broadening of the spectra. Upon excitation by light whose energy is insufficient to excite all the fluorophores in the ensemble, there occurs a selection of those species whose spectral properties differ from the average ones. These properties and their changes with time may characterize the relaxation process.[1,24,33,98]

Consider the energy level diagram in the case of inhomogeneously broadened spectra (Figure 2.8). The energy E of any ground or excited level may be represented in the form $E = E_0 + \Omega(W_{dd})$, where E_0 is the energy level in the absence of interaction, and the energy of solvation is represented by the distribution $\Omega(W_{dd})$ of the energy of dipolar interactions. This distribution is different in the ground and the excited state, since the electronic structure changes significantly upon excitation, and the orientation of the surrounding dipoles which is energetically favored in the ground state may be unfavorable in the excited state, and vice versa.

If excitation is by light of very low energy, then there will be photo-

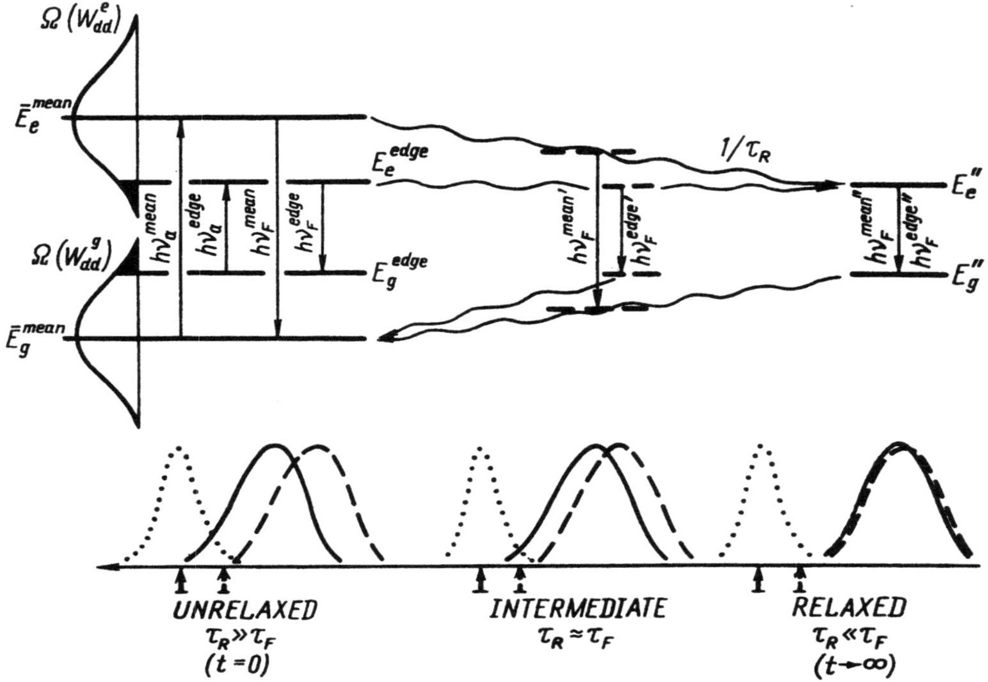

Figure 2.8. Energy level diagram (*top*) and spectra (*bottom*) illustrating the site-photoselection model of relaxation. The mid-band excitation ($h\nu_a^{\text{mean}}$) results in fluorescence with energy $h\nu_F^{\text{mean}}$. Excitation by low-energy quanta $h\nu_a^{\text{edge}}$ leads to a fluorescence spectrum shifted to longer wavelengths, $h\nu_F^{\text{edge}}$. Upon relaxation, there is both a decrease in the mean excited-state energy level and the establishment of a new energy distribution. The excitation wavelength dependence of the spectra disappears. Excitation wavelengths are indicated by arrows on the absorption spectra (dotted curves). For simplicity, the distribution in interaction energy of the relaxed state is not shown.

selection of only those fluorophores whose interaction with the environment (the solvation energy) is the lowest in the ground and the highest in the excited state. Photoselection of mainly those chromophores which can absorb the low-energy light occurs, whereas other chromophores with a higher energy gap between the ground and the excited state are not excited. Therefore, chromophores having a ground-state energy not lower than a certain E_g^{edge} and an excited-state energy not higher than a certain E_e^{edge} (see shaded areas in Figure 2.8) are excited at the red edge. If dipole-orientational relaxation does not occur during the lifetime τ_F (i.e., $\tau_R \gg \tau_F$), the energy of the emitted light will also be lower than $h\nu_F^{\text{mean}}$. Thereby, the red-edge excitation effect appears: the fluorescence spectrum is situated at longer wavelengths than in the usual case of excitation at the band maximum. This effect should be observed only when dipolar relaxation occurs slowly and does not distort the distribution of fluorophores in interaction energy.

Now consider the case in which rapid structural relaxation takes place in the medium ($\tau_R \ll \tau_F$). This should lead to rapid redistribution of interaction energies with the environment, which corresponds to a new distribution of electron density in the excited state. As a result, a fluorophore excited at the red edge "forgets" about it. The averaging of the properties of all fluorophores in the system occurs more rapidly than the emission process. An energy distribution of the emitted light also exists in this case, but it is dynamic and does not depend on the excitation wavelength. Thus, it is possible to gain information about structurally (dipole-orientationally) nonequilibrium electronic states by a rather simple method, that is, by observation of the dependence of the stationary fluorescence spectra on the excitation wavelength.

Studies of the red-edge effects permit very important information on the rate of relaxation to be obtained. In the case of relaxation processes for which $\tau_R \approx \tau_F$, the effects should vanish with time, and this can be observed by recording the excitation wavelength dependence of time-resolved spectra or decay functions at different emission wavelengths. In steady-state spectra, the red-edge effects must depend on the factors affecting the ratio between τ_R and τ_F. If we assume that the kinetics of the relaxation of the spectra for individual fluorophores are adequately described by the Bakhshiev–Mazurenko equation (Eq. 2.9) for both mean and edge excitation, then a relation between the parameters characterizing the red-edge effect and the relaxation properties of the chromophore's environment can be obtained:

$$\frac{\bar{\nu}^{\text{mean}} - \bar{\nu}^{\text{edge}}}{\bar{\nu}_0^{\text{mean}} - \bar{\nu}_0^{\text{edge}}} = \frac{\tau_R}{\tau_R + \tau_F} \tag{2.10}$$

Here $\bar{\nu}^{\text{mean}}$ and $\bar{\nu}^{\text{edge}}$ are the frequencies of the fluorescence spectra excited at fixed excitation frequencies at the maximum and at the red edge, respectively, and $\bar{\nu}_0^{\text{mean}} - \bar{\nu}_0^{\text{edge}}$ is the limiting value of the red-edge effect, given by the difference $\bar{\nu}^{mean} - \bar{\nu}^{edge}$ observed for very short times or at low temperatures (no effect of relaxation). It is assumed that τ_F does not depend on the excitation frequency. (This may not necessarily be the case, since fluorescence quenching may also be site-selective.) However, the main advantage of this approach consists in the fact that ν_∞ does not appear in Eq. (2.10), since the fluorescence spectrum of the completely relaxed state does not depend on the excitation conditions. Thus, to obtain τ_R, it is sufficient to record the difference between the fluorescence spectra at two fixed excitation frequencies, $\bar{\nu}^{mean}$ and $\bar{\nu}^{edge}$, and the limiting value of this difference, $\bar{\nu}_0^{mean} - \bar{\nu}_0^{edge}$. The latter may be measured by applying collisional fluorescence quenchers, lowering the temperature, or recording time-resolved spectra at "early" times.

The results of the determination of τ_R in a model viscous solvent (glycerol) from the temperature-dependent shifts of spectra obtained in our laboratory by N. V. Shcherbatska are illustrated in Figure 2.9. The temperature-dependent positions of the fluorescence maxima of the probe 2-(*p*-toluidinyl)naphthalene-6-sulfonate (2,6-TNS) in glycerol at different excitation wavelengths are similar, but the amplitude of the spectral changes decreases with the transition to long-wavelength edge excitation. Both the curves themselves and their differences are sigmoid functions which may be described by Eq. (2.10). At their bend point, which is observed at +5 °C, τ_R and τ_F must be equal. The value of τ_F for 2,6-TNS in this system is about 6 ns. At this temperature the dielectric relaxation time of glycerol is on the order of nanoseconds.[99] Therefore, the spectroscopic estimates agree with the data from dielectric measurements. We observed agreement between the dependence of τ_R on

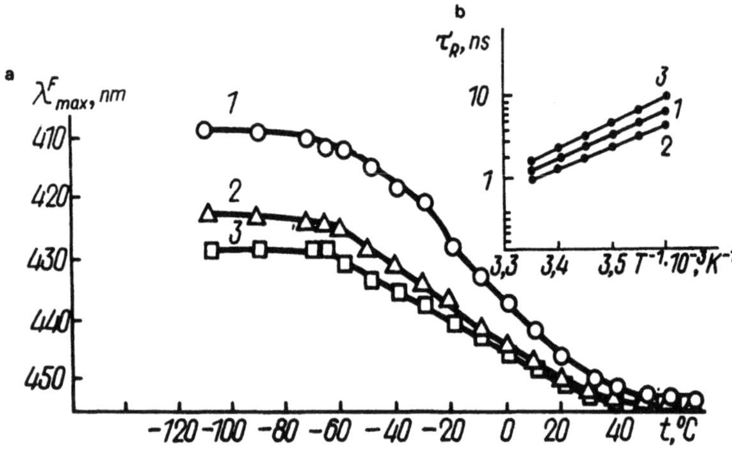

Figure 2.9. (a) The dependence of the position of the fluorescence spectrum maximum on temperature for 2,6-TNS in glycerol for excitation wavelengths of 360 (1), 390 (2), and 410 nm (3). (b) The dependence of log τ_R on $1/T$, calculated from Eq. (2.9) with the data for 360 nm (1) and from Eq. (2.10) with the data for 360 and 390 nm (2) and 360 and 410 nm (3).

temperature obtained from the relaxational shift of the spectra (Eq. 2.9) and that obtained from the red-edge effect (Eq. 2.10).

Analogous results for the temperature dependence of τ_R have been obtained in fluorescence studies of indole and tryptophan in glycerol.[24, 33] Therefore, the above approach may be considered to be adequate for the description of the dynamics of the model viscous media.

2.6. Molecular Relaxation and Dynamics of Dipoles in the Protein Globule

As shown above, the intrinsic fluorescence spectra of proteins as well as coenzyme groups and probes shift within very wide ranges depending on their environment. Since the main contribution to spectral shifts is from relaxational properties of the environment, the analysis of relaxation is the necessary first step in establishing correlations of protein structure with fluorescence spectra. Furthermore, the study of relaxation dynamics is a very important approach to the analysis of the fluctuation rates of the electrostatic field in proteins, which is of importance for the understanding of biocatalytic processes and charge transport.[8] Here we will discuss briefly the most illustrative results obtained by the methods of molecular relaxation spectroscopy.

2.6.1. Relaxational Shift of Steady-State Spectra

The limiting short-wavelength (unrelaxed) and long-wavelength (relaxed) positions of spectra may be obtained by variation of τ_R and τ_F using Eq. (2.9). Applying these data, it is possible to determine τ_R using the information on the position of the fluorescence spectra and the lifetime τ_F under the experimental conditions of interest. In steady-state spectroscopy, τ_R and τ_F may be varied in two ways: either by changing the temperature or by introducing dynamic quenchers of fluorescence. The necessary condition is that the structure of the fluorophore's environment (and for proteins this means their conformation) should not be changed by these variations.

There are substantial difficulties in the interpretation of temperature-dependent shifts of protein spectra because of the thermal lability of proteins and the possibility of temperature-dependent conformational transitions. Low-temperature studies in aqueous solutions revealed that for many of the proteins investigated the observed shifts of the fluorescence spectra within narrow temperature ranges were probably the result of cooperative conformational transitions, and not of relaxational shifts.[100] Spectral shifts have also been observed for proteins in glass-forming solvents,[101] but here there arise difficulties associated with the possible effects of viscous solvents on the protein dynamics.

Short-wavelength spectral shifts may also be observed for proteins under conditions of dynamic quenching by oxygen[29] and acrylamide.[102] However, the existing data do not yield reliable estimates of τ_R. If, under the initial conditions, $\tau_R \approx \tau_F$, then the quenching experiments do not allow $\bar{\nu}_\infty$ to be determined and Eq. (2.9) cannot be applied.

2.6.2. Time-Resolved Spectra

The discussion of the mechanisms and models of the relaxation process given in Section 2.5 shows that the application of time-resolved methods produces substantial advantages in accessing dynamical information, but it does not allow the complete pattern of the dynamic process to be obtained. The analysis of the experimental results requires that a particular dynamic model be assumed. Information on the dynamics is obtained from studies of the dependence of emission intensity on two parameters: the frequency (or the wavelength) of emission and on time. The function $I(\nu^{em}, t)$ may be investigated by two types of potentially equivalent experiments:

1. Measurement of the decay kinetics $I(t)$ in different regions of the fluorescence spectrum. If relaxation (or any reaction in the excited state) is absent, $I(t)$ does not depend on ν^{em}, whereas in its presence, the spectral dependence illustrated in Figure 2.7 is observed.

2. Analysis of the instantaneous fluorescence spectra corresponding to different times after excitation. Such spectra may be obtained by several means: directly from a pulse excitation experiment by scanning the spectrum after introducing time discrimination, or by constructing the spectra on the basis of data on the emission wavelength dependences of the decay curves[7] or of the results of phase–modulation measurements.[103] If emission and relaxation occur simultaneously, then within the range of initial times it is possible to observe the spectrum of the unrelaxed state; at longer times, spectra shifted more and more toward longer wavelengths are recorded.[89]

The results obtained show that the dipole-relaxational motions in protein molecules are really very retarded as compared to such motions in the environment of aromatic molecules dissolved in liquid solvents (where they occur on a time scale of tens of picoseconds).[82] Dipole-relaxational motions on the nanosecond time scale have been observed for a variety of proteins. For example, such motions were recorded for apohemoglobin and bovine serum albumin[104, 105] labeled with the fluorescent probe 2,6-TNS.

In studies of intrinsic protein fluorescence from tryptophan residues, a dependence of the decay kinetics on the emission wavelength typical of dipolar relaxation was observed for chicken pepsinogen,[106] with a component with negative amplitude at the long-wavelength edge of the fluorescence spectrum being detected. However, in further investigations of the emission decay of

17 other proteins, eight of which contained a single tryptophan residue, such a negative component was not detected.[107] These results do not rule out the possibility of dipolar relaxation since it may be masked by other processes. Since the fluorescence decay kinetics are not exponential and contain a short-lived positive component, the negative component at the long-wavelength edge may be masked. An increase of the mean fluorescence lifetime on passing to the long-wavelength emission edge may be considered as an indication of the existence of relaxation on the nanosecond time scale.

2.6.3. Red-Edge Excitation Spectroscopy

Sometimes the time scale of dipolar relaxation may be beyond the nanosecond time range, and in this case no relaxation effects will be seen in the time-resolved spectra. Thus, the problem arises of ascertaining whether the dynamics is faster or slower than the emission rate. The red-edge excitation method suggested recently may be used in this case.[1, 8, 24] This method involves a study of the dependence of the fluorescence spectra on the excitation wavelength. According to the theory of this method (Section 2.5.3), excitation at the long-wavelength absorption band edge results in shifts of the fluorescence spectra toward longer wavelengths. This occurs if the relaxation rate is comparable to or slower than the rate of fluorescence decay.

The theory predicts that if $\tau_R \gg \tau_F$, then the shift of the fluorescence spectra for a fixed difference in the excitation wavelengths should be maximal and independent of the factors influencing the ratio between τ_R and τ_F (temperature, collisional quenchers). At $\tau_R \ll \tau_F$, this shift should not occur. If the emission and relaxation take place simultaneously ($\tau_R \approx \tau_F$), then using Eq. (2.10) we may determine τ_R (Figure 2.9).

A variety of results obtained in studies of dipolar relaxation in the environment of the fluorescence probe 2,6-TNS are illustrated in Figure 2.10. In the model viscous medium (glycerol at 1 °C), the fluorescence spectra exhibit a marked dependence on the excitation wavelength. When λ^{ex} varies from 360 to 400 nm, the shift of the fluorescence spectrum maximum is 10 nm with a certain decrease of the half-width. In media with low viscosity, for instance, in ethanol (Figure 2.10a), this effect is never observed.

The results of studies of proteins complexed with 2,6-TNS (Figure 2.10b) show the existence of considerable effects of red-edge excitation. Thus, with a shift in the excitation wavelength from 360 nm to 400 nm, the shift of the fluorescence spectra for the complex of 2,6-TNS with β-lactoglobulin is 14 nm, with β-casein, 9.7 nm, with bovine serum albumin, 8 nm, and with human serum albumin, 13 nm.[98] Considerable shifts of the spectra upon excitation at the long-wavelength edge are observed for the complex of 2,6-TNS with melittin[108] and apomyoglobin[91] and also for the complex of the

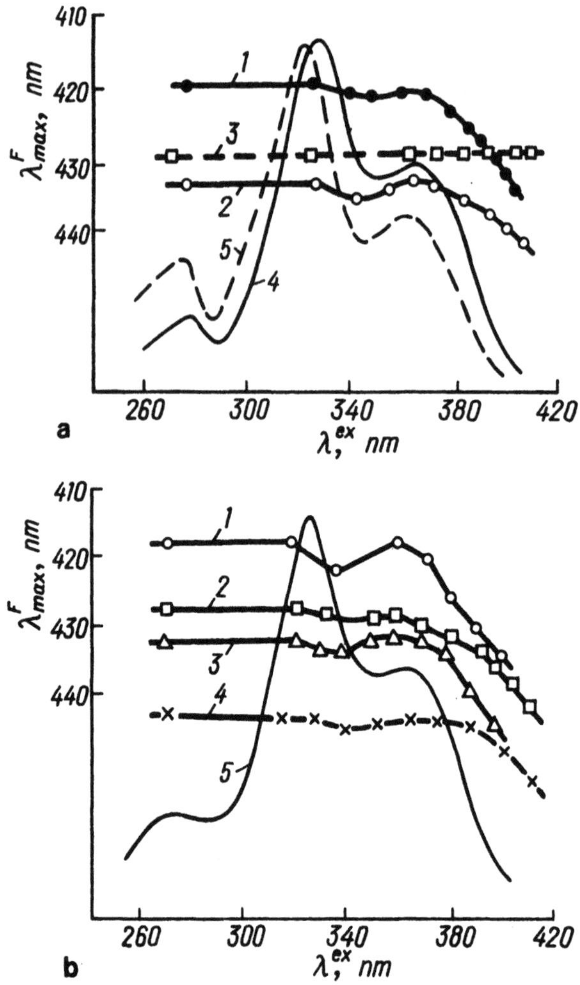

Figure 2.10. The dependence of the position of the fluorescence spectrum maximum on excitation wavelength for 2,6-TNS in model media (a) and in complexes with proteins (b). (a) 2,6-TNS (3×10^{-5}) M in glucose glass at 20 °C (1), glycerol at +1 °C (2), and 80% aqueous ethanol at 20 °C (3). Excitation spectra are for glycerol (4) and 80% ethanol (5). (b) 2,6-TNS in complexes with β-lactoglobulin (1), tetrameric melittin (2), human serum albumin (3), and lysozyme (4) at 20 °C. Excitation spectrum (5) is for human serum albumin.

probe 6-propionyl-2-(*N*,*N*-dimethylamino)naphthalene (PRODAN) with apomyoglobin.[109]

A comparison with the results of model studies indicates that the behavior of these probes bound to proteins differs fundamentally from their behavior in liquid media in which the position of their fluorescence spectra with ordinary excitation is similar to that for the protein complexes. In the latter case, the red-edge effect is always absent.

Heterogeneity of probe binding in the ground state cannot be responsible for this effect, since the absorption spectra of the probes are practically unchanged on binding with protein, and thus photoselection of different binding sites cannot occur. Furthermore, on going from ordinary to red-edge excitation the half-width of the fluorescence spectra decreases to some extent, whereas it would be expected to increase in the case of heterogeneity of the spectral properties of the dye itself or its binding sites. The form of the dependences of the position of the maximum and the half-width of the spectra on the excitation wavelength is the same for the probe associated with proteins and in the model solid and viscous media.

The fluorescent probe 2,6-TNS and other similar aminonaphthalene derivatives (1,8-ANS, DNS) were considered to be indicators of the polarity of protein molecules, and they were assumed to be bound only to hydrophobic sites on the protein surface. The detection of considerable spectral shifts with red-edge excitation has shown that the reason for the observed short-wavelength location of the spectra of these probes when complexed to proteins is not the hydrophobicity of their environment (or, at least, not only this) but the absence of dipole-relaxational equilibrium on the nanosecond time scale. Therefore, liquid solvents with different polarities cannot be considered to simulate the environment of fluorescent probes in proteins.

What is the extent of relaxational mobility during the fluorescence lifetime? The investigations that have been conducted reveal a very small temperature dependence of the red-edge effect for the complexes of 2,6-TNS with β-lactoglobulin and human serum albumin in the range 1–45 °C[93] and with melittin in the range 5–45 °C.[108] The results of studies on apomyoglobin showed almost no change between 4 and 20 °C.[91] Data obtained in our laboratory for lysozyme indicate that the magnitude of the red-edge effect does not vary either with temperature or in the presence of a quencher (0.1 *M* CsCl). Therefore, binding of the probe to the protein leads to immobilization of its environment, and $\tau_R \geqslant \tau_F$. However, this should not be considered to be a general rule. The fluorescence maximum for the complex of 1,8-ANS with aldolase has been observed to be located at 480 nm with practically no dependence on either temperature or the excitation wavelength (Yu. V. Chumachenko, unpublished results). Evidently, in this case the site of probe binding has high mobility on the nanosecond time scale.

The fact that observation of the red-edge effects requires a high concentration of immobile dipoles creating a sufficiently wide energy distribution by their interactions with the fluorophore is also an argument against the concept of "hydrophobic binding." These dipoles cannot be attributed to hydration water, which would give rise to relaxation times of 10^{-10}–10^{-11} s,[110] but rather to the polypeptide chain segments and side groups of protein. Thus, in the binding sites of fluorescent probes in protein molecules, there is a sufficiently high static microdisordering of the structure to give rise to such distributions in the energy of dipolar interactions, which do not differ substantially from the distributions occurring in the disordered solid and viscous polar media.

Considerable red-edge effects exhibiting a dependence on the viscosity of the medium are observed for model solutions of indole and tryptophan[33] (Figure 2.11a), which permits this approach to be applied to studies of the dynamics of the environment of tryptophan residues in proteins. In discussion

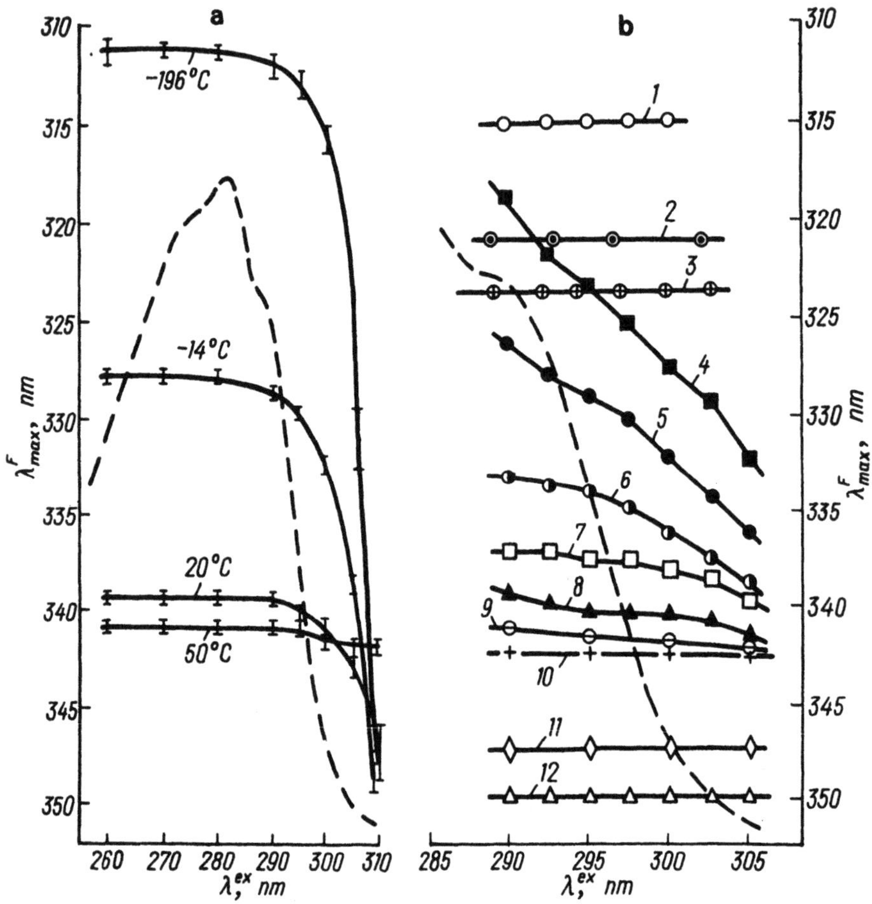

Figure 2.11. The dependence of the position of the fluorescence spectrum maximum on excitation wavelength for tryptophan in a model medium (glycerol) at different temperatures (a) and single-tryptophan proteins (b). 1, Whiting parvalbumin, pH 6.5 in the presence of Ca^{2+} ions; 2, ribonuclease T_1, pH 6.5; 3, ribonuclease C_2, pH 6.5; 4, human serum albumin, pH 7.0, $+10^{-4}$ *M* sodium dodecyl sulfate; 5, human serum albumin, pH 3.2; 6, melittin, pH 7.5, +0.15 *M* NaCl; 7, protease inhibitor IT-AJ from *Actinomyces janthinus*, pH 2.9; 8, human serum albumin, pH 7.0; 9, β-casein, pH 7.5; 10, protease inhibitor IT-AJ, pH 7.0; 11, basic myelin protein, pH 7.0; 12, melittin in water. The dashed line is the absorption spectrum of tryptophan.

the results of studies on the red-edge effects in UV fluorescence, we begin with the simplest case of protein molecules containing a single tryptophan residue. Here macroheterogeneity of tryptophan residues arising from different surroundings may be decreased or almost excluded with preservation of microheterogeneity associated with the intramolecular dynamics. Studies conducted by varying the excitation wavelength λ^{ex} from 290 to 310 nm have revealed that the position of the UV fluorescence spectrum of several such proteins may depend or not depend on the excitation wavelength. It was found that for proteins with a fluorescence maximum that is considerably shifted toward shorter wavelengths, such as azurin ($\lambda^F_{max} = 308$ nm), whiting parvalbumin with excess calcium ions ($\lambda^F_{max} = 315$ nm), and ribonuclease T_1 ($\lambda^F_{max} = 321$ nm) and C_2 ($\lambda^F_{max} = 323$ nm), the fluorescence spectra do not depend on the excitation wavelength, and the red-edge effects are not observed. This may be due to the hydrophobic environment of tryptophan residues in these proteins and to insufficient dipole-orientational broadening of the spectra. Proteins with several tryptophan residues per molecule that exhibit this type of spectroscopic behavior include bacteriorhodopsin ($\lambda^F_{max} = 315$ nm) and aldolase quenched by NADH ($\lambda^F_{max} = 313$ nm). In these cases it is impossible to obtain information on the intramolecular dynamics using the above approach.

There are also no red-edge effects for proteins emitting in the most extreme long-wavelength range of the spectrum, such as melittin at low ionic strength ($\lambda^F_{max} = 350$ nm), protease inhibitor IT-AJ ($\lambda^F_{max} = 342$ nm), myelin basic protein ($\lambda^F_{max} = 347$ nm), and β-casein ($\lambda^F_{max} = 341$ nm). This, evidently, is due to the exposure of the tryptophan residues to the rapidly relaxing water solvent. This should also result in the absence of red-edge effects for all denatured proteins.

Red-edge effects have been recorded for a number of proteins whose fluorescence spectra are in the intermediate range of 325–340 nm (Figure 2.11b). Here there is a characteristic shift of the spectra, which, with increase of the excitation wavelength to 305 nm, does not reach any limiting value. A comparison of the results obtained for excitation at 305 and at 295 nm shows that significant shifts of the spectra are observed in the case of protease inhibitor at pH 2.9 (2 nm) and human serum albumin at pH 7.0 (2 nm). Larger shifts take place in the case of melittin at high ionic strength (6 nm) and also of albumin in the F-form at pH 3.2 (7 nm) and the albumin–sodium dodecyl sulfate complex (11 nm). The shifts are not accompanied by noticeable changes in the shape of the spectrum.

These data may be explained in terms of the above mechanism of the long-wavelength shift of fluorescence spectra for red-edge excitation. The properties of the environment of the tryptophan residues in the proteins studied are such that during the lifetime of the excited state, structural relaxation of the surrounding dipoles fails to proceed. Studies of the dependence of the magnitude of the red-edge effect on temperature in the range 1–45 °C have shown that rapid (nanosecond time scale) motions in the environment of the tryptophan residues in the proteins investigated are activated only to an insignificant extent with the increase in temperature. In this range of temperatures in the model viscous medium (glycerol) the rate of dipole-orientational relaxation changes by an order of magnitude and the magnitude of the red-edge effect for indole and tryptophan is reduced by half. In this case the condition $\tau_R \approx \tau_F$ is fulfilled. One may think that in the case of the proteins studied, $\tau_R \gg \tau_F$, and the magnitude of the red-edge effect is close to its limiting value. It should be noted that the environment of the tryptophan residues in proteins may be heterogeneous, and the characteristic time scales of the motions of certain groups may vary by many orders of magnitude. Therefore, even with the absence of a temperature dependence of the effect in the range 1–45 °C, the magnitude of the effect is only some fraction of what it would be in a completely immobile environment.

Let us consider in greater detail the temperature dependence of the position of the maximum in the fluorescence spectrum of melittin (Figure 2.12). Three characteristic temperature regions may be distinguished. At $T < 30$ °C the spectrum does not depend on the temperature with excitation at both 280 nm and 305 nm; in this case the red-edge effect is maximal. Evidently, the condition $\tau_R \gg \tau_F$ holds in this region. In the range 30 to 50 °C there is a

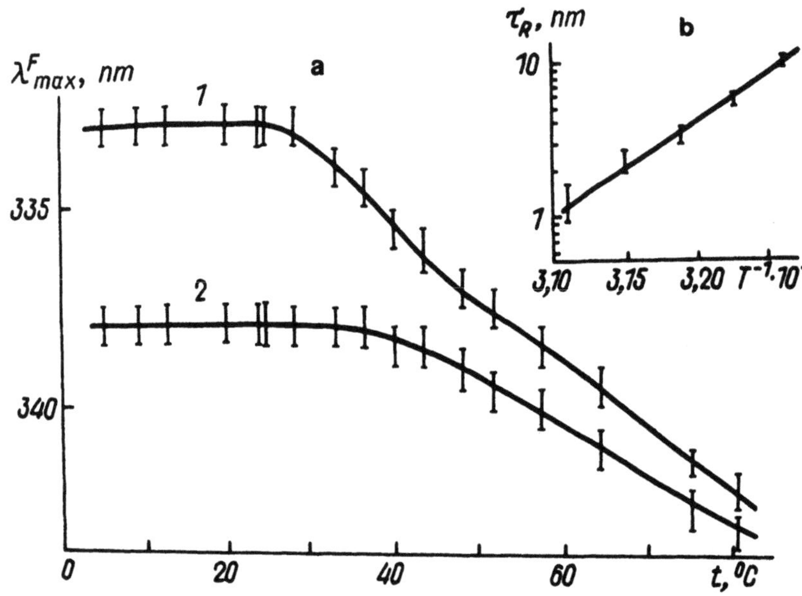

Figure 2.12. (a) The dependence of the position of the fluorescence spectrum maximum of tetrameric melittin on temperature at excitation wavelengths 280 (1) and 305 nm (2). (b) The dependence of log τ_R on $1/T$, calculated from these data using Eq. (2.10).

sharp decrease in the magnitude of the red-edge effect for excitation at 280 nm due to a temperature-dependent spectral shift that may be associated with dipole-orientational structural relaxation on the nanosecond time scale. Using Eq. (2.10), with $\lambda_F^{\text{mean}} = 333$ nm, $\lambda_F^{\text{edge}} = 338$ nm, and $\tau_F = 3.1$ ns at 27 °C, and assuming that τ_F decreases with increasing temperature in proportion to the change in the quantum yield, we have determined the temperature dependence of τ_R. A linear dependence of $\log \tau_R$ on $1/T$ is observed, and the activation energy determined from the slope is 57 kJ/mol. Extrapolation of this function to temperatures below this intermediate range leads to the value $\tau_R = 40$ ns at 25 °C. In the high-temperature range ($T > 50$ °C), in which the fluorescence spectra at 280- and 305-nm excitation are shifted simultaneously, a temperature-dependent protein conformational transition probably occurs.

In using the method of the red-edge shift in UV fluorescence spectroscopy, we should take into account the possibility of emission not only of tryptophan but also of tyrosine residues. In many tryptophan-containing proteins, tyrosine fluorescence is not observed. However, it is considerable in serum albumin, and the decrease in its intensity is responsible for the long-wavelength shift of the spectra recorded at $\lambda^{\text{ex}} < 290$ nm. At $\lambda^{\text{ex}} > 292$ nm, the tyrosine component should be completely absent.

A more complicated problem is associated with the structural heterogeneity of tryptophan residues in the ground state and the possibility at the red edge of selective photoexcitation of a structural form (structurally determined environment of tryptophan) whose absorption spectrum is shifted to a considerable extent toward longer wavelengths. When analyzing such a possibility, we must take into account that the shifts of absorption and fluorescence spectra are determined by quite different factors. Variations of the electronic polarization of the medium[1, 85] are manifested mainly in shifts of the absorption spectra, and there is a long-wavelength shift on going to a more polarizable medium. Shifts of the fluorescence spectra are determined to a considerably higher extent by dipolar interactions. Thus, if two structural forms exist in a protein molecule, then photoselection of a form with a longer wavelength absorption at the red edge should give rise to a shorter wavelength emission. We have observed quite a different effect. Photoselection in this case involves the selective excitation of chromophore microstates whose energy of dipolar interaction with the environment corresponds more closely to the highest interaction energy in the electronic excited state.

Among multitryptophan proteins emitting light around 330 nm, we have observed the largest red-edge effect (estimated from the difference between the maxima of the fluorescence spectra obtained at 290- and 305-nm excitation) for papain in the active and inactive forms (13 and 10 nm, respectively). Large shifts were also observed for rabbit muscle asparagyl- and valyl-RNA synthetases (8 nm). For rabbit aldolase A, the observed shift was 6 nm, for skeletal muscle myosin, 4.5 nm, for chymotrypsin, 2.5 nm, and for carbonic anhydrase, 2 nm. No shift was observed for trypsin and β-lactoglobulin.[1] This evidences the importance of the red-edge shift parameter for detection of differences in the dynamic properties of the environment of the tryptophan residues in these proteins.

2.7. Conclusion and Future Prospects

Let us turn our attention back again to the scheme illustrating various versions of the joint application of fluorescence parameters (Figure 2.1) and consider the possibilities for constructing more general and more definite models of protein dynamics. These models can be suggested and confirmed or rejected by comparing predicted behavior with the results of spectroscopic experiments of different kinds.

Both red-edge excitation spectroscopy and time-resolved spectroscopy are applicable to studies of dipole-orientational relaxation, and ambiguity in the interpretation of the results obtained by one of these methods may be resolved by a comparison with the results of the other. Using nanosecond time-resolved spectroscopy, it is difficult to study relaxation which occurs on a time scale that is either much shorter or much longer than the excited-state lifetime, which is usually on the order of nanoseconds. However, if nanosecond relaxation of the spectra is recorded, then it may be associated not only with dipole-orientational relaxation but also with other time-dependent processes leading to a decrease in the energy of the electronic excited state (for instance, changes in the solvation shell due to translational diffusion, isomerization, production of exciplexes, or directed excitation energy transfer between identical fluorophores.[1] Meanwhile, cases in which $\tau_R \geqslant \tau_F$ may be easily and definitely characterized by red-edge excitation spectroscopy, but here there may be difficulties associated with the structural heterogeneity in the ground state. Since the dipole-orientational relaxation is, evidently, the most rapid process that involves a shift of the atomic nuclei, then it would primarily lead to a redistribution of microstates. Therefore, if there is good agreement between the value of τ_R determined from the red-edge effect and that obtained from the nanosecond time-resolved spectra, the involvement of dipolar relaxation is indicated; if these values do not agree, then another mechanism is required to explain the spectral kinetics. Such an approach was applied in studies of the dynamics in phospholipid bilayer membranes using the probe 2,6-TNS.[111] Red-edge excitation spectroscopy may be combined directly in the same experiment with nanosecond time-resolved spectroscopy.

The results of fluorescence polarization studies of proteins were discussed above. Time-resolved anisotropy measurements often permit, without any additional variation of experimental conditions, intramolecular rotations to be distinguished from rotation of the whole protein molecule and characterized, and they sometimes provide information about the accessible rotation angle of the aromatic group. Studies of the wavelength dependence of the fluorescence anisotropy allow data to be obtained which may be analyzed and compared with a proposed dynamic model. Thus, if the time course of the anisotropy decay parallels the spectral kinetics, then the steady-state anisotropy would increase sharply on the short-wavelength side of the fluorescence spectrum. This is connected with a decrease in the effective fluorescence lifetime.[86]

There should exist a correlation between the two time-resolved functions: the decay of the fluorescence intensity and the decay of the emission anisotropy. If the fluorophore undergoes intramolecular rotation with some potential energy and the quenching of its emission has an angular dependence, then the intensity decay function is predicted to be strongly dependent on the rotational diffusion coefficient of the fluorophore.[112] It is expected to be single-exponential only in the case when the internal rotation is fast as compared with an averaged decay rate. As the internal rotation becomes slower, the intensity decay function should exhibit nonexponential behavior.

Of considerable interest is the fact that not only the steady-state anisotropy but also its kinetics depend on the excitation wavelength. In this case another red-edge effect connected with site photoselection may be observed. Dipole-orientational relaxation may occur not only by rotation of the dipoles surrounding the fluorophore but also by rotation of the aromatic group itself. If red-edge excitation results in the photoselection of fluorophores whose energy of interaction with the environment already corresponds to that in the excited state, then the relaxation-associated rotation should not be observed and the rotation that occurs should be completely Brownian in character.[22]

A comparative analysis of different functions of fluorescence parameters is necessary to elucidate whether or not excitation energy transfer between identical fluorophores occurs. In a protein molecule containing several tryptophan residues, these residues may be situated sufficiently closely for effective energy transfer to occur. In cases of incomplete dipolar relaxation in the excited state, such energy transfer affects not only the polarization, but also the spectral and temporal properties of emission. The reason is that in transfer the emission properties are not averaged; instead, there is a directed energy flow from the donor with a shorter wavelength emission spectrum to the acceptor with a longer wavelength absorption spectrum. Such energy transfer may be observed from both the nanosecond spectral kinetics and the kinetics of anisotropy. If we observe such effects and have grounds to consider that no mobility is responsible for them (for instance, by lowering the temperature), then they should be attributed to tryptophan–tryptophan energy transfer. The observation of the Weber red-edge effect[113] (a fall of polarization for long-wavelength edge excitation) may serve as additional evidence of directed transfer.[1]

Thus, at present, fluorescence spectroscopy is capable of providing direct information on molecular dynamics on the nanosecond time scale and can estimate the results of dynamics occurring beyond this range. The present-day multiparametric fluorescence experiment gives new opportunities for interpretation of these data and construction of improved dynamic models. A further development of the theory which would provide an improved description of the dynamics in quantitative terms with allowance for the structural inhomogeneity of protein molecules and the hierarchy of their internal motions is required.

It should be noted that the dynamics studied by fluorescence methods is the dynamics of relaxation and fluctuations of the electric field. Dipole-orientational processes may be directly related to biological functions of proteins, in particular, charge transfer in biocatalysis and ionic transport. One may postulate that, irrespective of the origin of the charge balance disturbance, the protein molecule responds to these changes in the same way, in accordance with its dynamic properties. If the dynamics of dipolar and charged groups in proteins does play an important role in protein functions, then fluorescence spectroscopy will afford ample opportunities for its direct study.

References

1. A. P. Demchenko, *Ultraviolet Spectroscopy of Proteins*, Springer-Verlag, Berlin (1986).
2. J. M. Beechem and L. Brand, Time-resolved fluorescence of proteins, *Annu. Rev. Biochem. 54*, 43–71 (1985).
3. A. J. W. G. Visser, in: *Pyridine Nucleotide Coenzymes: Chemical, Biochemical and Medical Aspects* (D. Dolphin, R. Poulson, and O. Avramovic, eds.), Vol. 2A, pp. 163–183, John Wiley and Sons, New York (1987).
4. A. J. W. G. Visser, in: *Excited State Probes in Biochemistry and Biology* (A. G. Szabo and L. Masotti, eds.), Plenum Press, New York (1987).
5. L. Brand and J. R. Gohlike, Fluorescence probes for structure, *Annu. Rev. Biochem. 41*, 843–868 (1972).
6. Iu. A. Vladimorov and G. E. Dobretsov, *Fluorescence Probes in the Studies of Biological Membranes*, Nauka, Moscow (1980).
7. J. R. Lakowicz, *Principles of Fluorescence Spectroscopy*, Plenum Press, New York (1983).
8. A. P. Demchenko, *Luminescence Spectroscopy and Dynamics of Protein Structure*, Naukova Dumka, Kiev (1988).
9. G. A. Petsko and D. Ringe, Fluctuations in protein structure from X-ray diffraction, *Annu. Rev. Biophys. Bioeng. 13*, 331–371 (1984).
10. G. Wagner, Internal mobility in globular proteins, *Comments Mol. Cell. Biophys. 1*, 261–280 (1982).
11. V. I. Gol'danskii, Yu. F. Krupyanskii, and V. N. Fleurov, Rayleigh scattering of Mössbauer radiation (RSMR) data, hydration effects and glass-like dynamical model of biopolymers, *Phys. Scr. 33*, 527–540 (1986).
12. J. A. McCammon, Protein dynamics, *Rep. Prog. Phys. 47*, 1–46 (1984).
13. G. E. Schulz and R. H. Schirmer, *Principles of Protein Structure*, Springer-Verlag, New York (1979).

14. D. L. Stein, A model of protein conformational substates, *Proc. Natl. Acad. Sci. U.S.A. 82*, 3670–3672 (1985).
15. E. N. Baker and R. E. Hubbard, Hydrogen bonding in globular proteins, *Prog. Biophys. Mol. Biol. 44*, 97–179 (1984).
16. H. Hartmann, F. Parak, W. Seigelmann, G. A. Petsko, D. R. Ponzi, and H. Frauenfelder, Conformational substates in a protein: Structure and dynamics of metmyoglobin at 80 K, *Proc. Natl. Acad. Sci. U.S.A. 79*, 4967–4971 (1982).
17. R. J. P. Williams, Energy states of proteins, enzymes and membranes, *Proc. Roy. Soc., B 200*, 353–389 (1978).
18. J. Thomas and G. N. LaMar, Heme orientational heterogeneity in deuterohemin-reconstituted horse and human hemoglobin characterized by proton NMR spectroscopy, *Biochem. Biophys. Res. Commun. 119*, 640–654 (1984).
19. H. Frauenfelder and R. D. Young, Protein dynamics and ligand binding, *Comments Mol. Cell. Biophys. 3*, 347–372 (1986).
20. B. Gavish, E. Gratton, and C. J. Hardy, Adiabatic compressibility of globular proteins, *Proc. Natl. Acad. Sci. U.S.A. 80*, 750–754 (1983).
21. F. Parak and E. W. Knapp, A consistent picture of protein dynamics, *Proc. Natl. Acad. Sci. U.S.A. 81*, 7088–7092 (1984).
22. N. A. Nemkovich, A. N. Rubinov, and V. I. Tomin, Inhomogeneous broadening of electronic spectra of dye molecules in solution, in: *Topics in Fluorescence Spectroscopy* (J. R. Lakowicz, ed.), Vol. 2, pp. 367–428, Plenum Press, New York (1992).
23. G. I. Likhtenstein, A. V. Kulikov, A. I. Kotelnikov, and L. A. Levchenko, Methods of physical labels—a combined approach to the study of microstructure and dynamics in biological systems, *J. Biochem. Biophys. Meth. 12*, 1–28 (1986).
24. A. P. Demchenko, Fluorescence analysis of protein dynamics, *Essays Biochem. 22*, 120–157 (1986).
25. J. R. Lakowicz and G. Weber, Quenching of protein fluorescence by oxygen. Detection of structural fluctuations in proteins on the nanosecond time scale, *Biochemistry 12*, 4171–4179 (1973).
26. M. R. Eftink and C. A. Ghiron, Fluorescence quenching studies with proteins, *Anal. Biochem. 114*, 199–227 (1981).
27. M. R. Eftink, Fluorescence quenching: Theory and applications, in: *Topics in Fluorescence Spectroscopy* (J. R. Lakowicz, ed.), Vol. 2, pp. 53–120, Plenum Press, New York (1992).
28. J. Munro, J. Pecht, and L. Styer, Subnanosecond motions of tryptophan residues in proteins, *Proc. Natl. Acad. Sci. U.S.A. 76*, 56–60 (1979).
29. J. R. Lakowicz, B. P. Maliwal, H. Cherek, and A. Batter, Rotational freedom of tryptophan residues in proteins and peptides, *Biochemistry 22*, 1741–1752 (1983).
30. M. Eftink, Quenching-resolved emission anisotropy studies with single and multitryptophan-containing proteins, *Biophys. J. 43*, 323–334 (1983).
31. A. Grinvald and J. Z. Steinberg, Fast relaxation processes in a protein revealed by the decay kinetics of tryptophan fluorescence, *Biochemistry 13*, 5170–5177 (1974).
32. J. R. Lakowicz and H. Cherek, Dipolar relaxation in proteins on the nanosecond time scale observed by wavelength-resolved phase fluorometry of tryptophan residue, *J. Biol. Chem. 255*, 831–834 (1980).
33. A. P. Demchenko and A. S. Ladokhin, Red-edge-excitation fluorescence spectroscopy of indole and tryptophan, *Eur. Biophys. J. 15*, 369–379 (1988).
34. A. P. Demchenko, Red-edge-excitation spectroscopy of single-tryptophan proteins, *Eur. Biophys. J. 16*, 121–129 (1988).
35. E. A. Burstein, Intrinsic protein luminescence as a tool for studying fast structural dynamics, *Mol. Biol. (Moscow) 17*, 455–467 (1983).
36. T. L. Bushueva, E. P. Busel, and E. A. Burstein, Relationship of thermal quenching of

protein fluorescence to intramolecular structural mobility, *Biochim. Biophys. Acta 534*, 141–152 (1978).
37. V. F. Kamalov, A. C. Ladokhin, and B. N. Toleutaev, Nanosecond intramolecular dynamics of melittin, *Dokl. Akad. Sci. USSR 296*, 742–745 (1987).
38. W. R. Ware and J. C. Andre, The influence of diffusion on fluorescence quenching, in: *Time-Resolved Fluorescence Spectroscopy in Biochemistry and Biology* (J. R. Lakowicz, ed.), pp. 383–392, Plenum Press, New York (1983).
39. R. W. Wijnaendts van Resandt, Picosecond transient effect in the fluorescence quenching of tryptophan, *Chem. Phys. Lett. 95*, 205–208 (1983).
40. V. M. Agranovich and M. D. Galanin, *The Electronic Excitation Energy Transfer in Condensed Media*, Nauka, Moscow (1978).
41. A. G. Szabo, T. M. Stepanik, D. M. Wayner, and N. M. Young, Conformational heterogeneity of the copper binding site in azurin. A time-resolved fluorescence study, *Biophys. J. 41*, 233–244 (1983).
42. E. A. Permyakov, A. V. Ostrovsky, E. A. Burstein, P. G. Pleshanov, and C. Gerday, Parvalbumin conformers revealed by steady-state and time-resolved fluorescence spectroscopy, *Arch. Biochem. Biophys. 240*, 781–791 (1985).
43. J. R. Alcala, E. Gratton, and F. G. Prendergast, Interpretation of fluorescence decays in proteins using continuous lifetime distributions, *Biophys. J. 51*, 925–936 (1987).
44. J.-C. Brochon, P. Wahl, M. Charlier, J. C. Maurizot, and C. Helene, Time-resolved spectroscopy of the tryptophanyl fluorescence, *Biochem. Biophys. Res. Commun. 79*, 1261–1271 (1977).
45. J. M. Beechem, J. R. Knutson, J. B. A. Rose, B. W. Turner, and L. Brand, Global resolution of heterogeneous decay by phase modulation fluorometry: Mixtures and proteins, *Biochemistry 22*, 6054–6058 (1983).
46. A. J. W. G. Visser, H. J. Grande, and C. Veeger, Rapid relaxation processes in pig heart lipoamide dehydrogenase revealed by subnanosecond resolved fluorometry, *Biophys. Chem. 12*, 35–49 (1980).
47. A. Gafni and L. Brand, Fluorescence decay studies of reduced nicotinamide adenine dinucleotide in solution and bound to liver alcohol dehydrogenase, *Biochemistry 15*, 3165–3171 (1976).
48. J.-C. Brochon, P. Wahl, J. M. Jallon, and M. Iwatsubo, Pulse fluorimetry study of beef liver glutamate dehydrogenase–reduced nicotinamide adenine dinucleotide phosphate complexes, *Biochemistry 15*, 3259–3265 (1976).
49. M. S. Tung and R. F. Steiner, The use of nanosecond fluorimetry in detecting conformational transitions of an allosteric enzyme, *Biopolymers 14*, 1933–1949 (1975).
50. G. Hoenes, M. Hauser, and G. Pjeiderer, Dynamic total fluorescence and anisotropy decay study of the dansyl fluorophore in model compounds and enzymes, *Photochem. Photobiol. 43*, 133–137 (1986).
51. S. P. Spragg and R. W. Wijnaendts van Resandt, The temperature dependence of the fluorescence decay of low-density lipoproteins, *Biochim. Biophys. Acta 792*, 84–91 (1984).
52. E. A. Burstein, Luminescence of protein chromophores (model studies), in: *Biophysica*, Vol. 6, pp. 1–214, VINITI, Moscow (1976).
53. J. R. Lakowicz and G. Weber, Quenching of protein fluorescence by oxygen. Detection of structural fluctuations in proteins on the nanosecond time scale, *Biochemistry 12*, 4171–4179 (1973).
54. M. R. Eftink and C. A. Ghiron, Exposure of tryptophanyl residues and protein dynamics, *Biochemistry 16*, 5546–5551 (1977).
55. M. R. Eftink, J. L. Zajicek, and C. A. Ghiron, A hydrophobic quencher of protein fluorescence: 2,2,2-trichloroethanol, *Biochim. Biophys. Acta 491*, 473–481 (1977).
56. T. L. Bushueva, E. P. Busel, and E. A. Burstein, Some regularities of dynamic accessibility

of buried fluorescent residues to external quenchers in proteins, *Arch. Biochem. Biophys. 204*, 161–166 (1980).
57. B. Somogyi, S. Papp, A. Rosenberg, I. Seres, J. Matko, G. R. Welch, and P. Nagy, A double-quenching method for studying protein dynamics: Separation of the fluorescence quenching parameters characteristic of solvent-exposed and solvent-masked fluorophores, *Biochemistry 24*, 6674–6679 (1985).
58. C. K. Woodward and B. D. Hilton, Hydrogen exchange kinetics and internal motions in proteins, *Annu. Rev. Biophys. Bioeng. 8*, 99–128 (1979).
59. K. A. Hagaman and M. R. Eftink, Fluorescence quenching of Trp-314 of liver alcohol dehydrogenase by oxygen, *Biophys. Chem. 20*, 201–207 (1984).
60. M. R. Eftink and K. A. Hagaman, Fluorescence quenching of the buried tryptophan residue of cod parvalbumin, *Biophys. Chem. 22*, 173–180 (1985).
61. D. B. Calhoun, J. M. Vanderkooi, G. V. Woodrow, and S. W. Englander, Penetration of dioxygen into proteins studied by quenching of phosphorescence and fluorescence, *Biochemistry 22*, 1526–1533 (1983).
62. M. R. Eftink and L. A. Selvidge, Fluorescence quenching of liver alcohol dehydrogenase by acrylamide, *Biochemistry 21*, 117–125 (1982).
63. D. B. Calhoun, J. M. Vanderkooi, and S. W. Englander, Penetration of small molecules into proteins studied by quenching of phosphorescence and fluorescence, *Biochemistry 22*, 1533–1540 (1983).
64. E. Gratton, D. M. Jameson, G. Weber, and B. Alpert, A model of dynamic quenching of fluorescence in globular proteins, *Biophys. J. 45*, 789–794 (1984).
65. W. M. Vaughan and G. Weber, Oxygen quenching of pyrenebutyric acid fluorescence in water. A dynamic probe of the microenvironment, *Biochemistry 9*, 464–473 (1970).
66. M. Jullien, J.-R. Garel, F. Merola, and J.-C. Brochon, Quenching by acrylamide and temperature of a fluorescence probe attached to the active site of ribonuclease, *Eur. Biophys. J. 13*, 131–138 (1986).
67. P. C. Leavis, E. Gowell, and T. Tao, Fluorescence lifetime and acrylamide quenching studies of the interactions between troponin subunits, *Biochemistry 23*, 4156–4161 (1984).
68. G. Weber, Rotational Brownian motion and polarization of the fluorescence of solutions, *Adv. Protein Chem. 8*, 415–459 (1953).
69. K. K. Turoverov and I. M. Kuznetsova, Polarization of intrinsic fluorescence of proteins. 2. The studies of intramolecular dynamics of tryptophan residues, *Mol. Biol. (Moscow) 17*, 468–475 (1983).
70. J. R. Lakowicz, G. Laszko, I. Gryczynski, and H. Cherek, Measurement of subnanosecond anisotropy decays of protein fluorescence using frequency-domain fluorometry, *J. Biol. Chem. 261*, 2240–2245 (1986).
71. J. R. Lakowicz and G. Weber, Nanosecond segmental mobilities of tryptophan residues in proteins observed by lifetime-resolved fluorescence anisotropies, *Biophys. J. 32*, 591–601 (1980).
72. E. Gratton, R. Alcala, G. Marriott, and F. Prendergast, Fluorescence studies of protein dynamics, Preprint, University of Illinois, ILL-(EX)-85/53 (1985).
73. K. K. Turoverov, I. M. Kuznetsova, and V. N. Zaitsev, The environment of the tryptophan residue in *Pseudomonas aeruginosa* azurin and its fluorescence properties, *Biophys. Chem. 23*, 79–89 (1985).
74. A. J. Cross and G. R. Fleming, Influence of inhibitor binding on the internal motions of lysozyme, *Biophys. J. 50*, 507–512 (1986).
75. J. A. McCammon, P. G. Wolynes, and M. Karplus, Picosecond dynamics of tyrosine side chains in proteins, *Biochemistry 18*, 927–942 (1979).
76. T. Ichiye and M. Karplus, Fluorescence depolarization of tryptophan residues in proteins: A molecular dynamics study, *Biochemistry 22*, 2884–2894 (1983).

77. E. Gratton, J. R. Alcala, and G. Marriott, Rotations of tryptophan residues in proteins, *Biochem. Soc. Trans. 14*, 835–838 (1986).
78. R. M. Levy and R. P. Sheridan, Combined effect of restricted rotational diffusion plus jumps of nuclear magnetic resonance and fluorescence probes of aromatic ring motions in proteins, *Biophys. J. 41*, 217–221 (1983).
79. W. G. J. Hol, The role of the α-helix dipole in protein function and structure, *Prog. Biophys. Mol. Biol. 45*, 149–196 (1985).
80. A. Warshel and S. T. Russell, Calculations of electrostatic interactions in biological systems and in solutions, *Quart. Rev. Biophys. 17*, 283–422 (1984).
81. B. H. Honig, W. L. Hubbell, and R. F. Flewelling, Electrostatic interactions in membranes and proteins, *Annu. Rev. Biophys. Chem. 15*, 163–194 (1986).
82. T. Okamura, M. Sumitani, and K. Yoshihara, Picosecond dynamics Stokes shift of α-naphthylamine, *Chem. Phys. Lett. 94*, 339–343 (1983).
83. Y. T. Mazurenko and V. S. Udaltsov, Spectral relaxations of fluorescence. 2. Three-component solutions, *Opt. Spektrosk. 45*, 255–263 (1978).
84. R. Lindeman and G. Zundel, Proton transfer in and polarizability of hydrogen bonds coupled with conformational changes in proteins. II. IR investigation of polyhistidine with various carboxylic acids, *Biopolymers 17*, 1285–1301 (1978).
85. R. Lumry and M. Hershberger, Status of indole photochemistry with special reference to biological application, *Photochem. Photobiol. 27*, 819–840 (1978).
86. Y. T. Mazurenko and N. G. Bakhshiev, The influence of orientational dipolar relaxation on spectral, temporal and polarizational properties of luminescence in solutions, *Opt. Spektrosk. 28*, 905–913 (1970).
87. N. G. Bakhshiev, *Spectroscopy of Intermolecular Interactions*, Nauka, Leningrad (1972).
88. G. van der Zwan and J. T. Hynes, Time-dependent fluorescence solvent shifts, dielectric friction and nonequilibrium solvation in polar solvents, *J. Phys. Chem. 89*, 4181–4188 (1985).
89. Y. T. Mazurenko and V. S. Udaltsov, Spectral relaxations of fluorescence. 3. Kinetics of spectra of polar solutions with distributed dielectric relaxation time, *Opt. Spectrosc. (Engl. transl.) 45*, 765–767 (1978).
90. J. R. Lakowicz and A. Balter, Resolution of initially excited and relaxed states of tryptophan fluorescence by differential-wavelength deconvolution of time-resolved fluorescence decays, *Biophys. Chem. 15*, 353–360 (1982).
91. J. A. Lakowicz and S. Keating-Nakamoto, Red-edge excitation of fluorescence and dynamic properties of proteins and membranes, *Biochemistry 23*, 3013–3021 (1984).
92. J. R. Lakowicz and D. Hogen, Dynamic properties of the lipid–water interface of model membranes as revealed by lifetime-resolved fluorescence emission spectra, *Biochemistry 20*, 1366–1373 (1981).
93. R. P. DeToma, J. H. Easter, and L. Brand, Dynamic interactions of fluorescence probes with the solvent environment, *J. Am. Chem. Soc. 98*, 5001–5007 (1976).
94. A. I. Kotelnikov, G. I. Likhtenstein, V. P. Fogel, and G. B. Postnikova, On the interpretation of relaxational shift of luminescence spectra of chromophores in proteins and model systems, *J. Appl. Spectrosc. (USSR) 40*, 564–568 (1984).
95. E. Bismuto, D. M. Jameson, and E. Gratton, Dipolar relaxations in glycerol: A dynamic fluorescence study of 4,2′-(dimethylamino)-6′-naphthylcyclohexanecarboxylic acid (DANSA), *J. Am. Chem. Soc. 109*, 2354–2357 (1987).
96. R. B. Macgregor and G. Weber, Fluorophores in polar media. Spectral effects of the Langevin distribution of electrostatic interactions, *Ann. N. Y. Acad. Sci. 366*, 140–154 (1981).
97. Y. T. Mazurenko, Statistics of solvation and solvatochromy, *Opt. Spektrosk. 55*, 471–478 (1983).
98. A. P. Demchenko, On the nanosecond mobility in proteins. Edge excitation fluorescence red

shift of protein-bound 2-(*p*-toluidinylnaphthalene)-6-sulfonate, *Biophys. Chem. 15*, 101–109 (1982).
99. G. E. McDuffie and T. A. Litovitz, Dielectric relaxation in associated liquids, *J. Chem. Phys. 37*, 1699–1705 (1962).
100. E. A. Permyakov and E. A. Burstein, Relaxation processes in frozen aqueous solution of proteins; temperature dependence of fluorescence parameters, *Stud. Biophys. 51*, 91–103 (1975).
101. G. I. Likhtenstein and A. I. Kotelnikov, The studies of fluctuational intramolecular mobility in proteins by physical labels, *Mol. Biol. (Moscow) 17*, 505–519 (1983).
102. M. R. Eftink and C. A. Ghiron, Exposure of tryptophanyl residues in proteins. Quantitative determination by fluorescence quenching studies, *Biochemistry 15*, 672–680 (1976).
103. E. Gratton, D. M. Jameson, and R. D. Hall, Multifrequency phase and modulation fluorometry, *Annu. Rev. Biophys. Bioeng 13*, 105–124 (1984).
104. L. Brand and J. R. Gohlike, Nanosecond time-resolved fluorescence spectra of a protein–dye complex, *J. Biol. Chem. 246*, 2317–2324 (1971).
105. A. Gafni, R. P. DeToma, R. E. Manrow, and L. Brand, Nanosecond decay studies of a fluorescence probe bound to apomyoglobin, *Biophys. J. 17*, 155–168 (1977).
106. A. Grinvald and I. Z. Steinberg, Fast relaxation processes in a protein revealed by the decay kinetics of tryptophan fluorescence, *Biochemistry 13*, 5170–5177 (1974).
107. A. Grinvald and I. Z. Steinberg, The fluorescence decay of tryptophan residues in native and denatured proteins, *Biochim. Biophys. Acta 427*, 663–678 (1976).
108. A. P. Demchenko, Fluorescence molecular relaxation studies of protein dynamics. The probe binding site of melittin is rigid on the nanosecond time scale, *FEBS Lett. 182*, 99–102 (1985).
109. R. B. Macgregor and G. Weber, Estimation of the polarity of the protein interior by optical spectroscopy, *Nature 319*, 70–73 (1986).
110. I. D. Kuntz and W. Kauzmann, Hydration of proteins and polypeptides, *Adv. Protein Chem. 28*, 239–345 (1974).
111. A. P. Demchenko and N. V. Shcherbatska, Nanosecond dynamics of the charged fluorescent probes at the polar interface of the membrane phospholipid bilayer, *Biophys. Chem. 22*, 131–143 (1985).
112. F. Tanaka and N. Mataga, Fluorescence quenching dynamics of tryptophan in proteins. Effect of internal rotation under potential barrier, *Biophys. J. 51*, 487–495 (1987).
113. G. Weber and M. Shinitsky, Failure of energy transfer between identical aromatic molecules on excitation at the longwave edge of the absorption spectrum, *Proc. Natl. Acad. Sci. U.S.A. 65*, 823–830 (1970).

3

Tryptophan Phosphorescence from Proteins at Room Temperature

Jane M. Vanderkooi

3.1. Background

Long-lived luminescence from protein-containing materials was reported many years ago. Debye and Edwards reported that a bluish light was emitted from proteins at cryogenic temperatures after illumination.[1] Work in the 1950s established the relationship between fluorescence and the long-lived phosphorescence for the aromatic amino acids in proteins.[2-4]. Konev in his classic work *Fluorescence and Phosphorescence of Proteins and Nucleic Acids* summarized this early history.[5]

Although protein phosphorescence was in fact observed earlier than fluorescence, fluorescence of proteins is now widely used, whereas phosphorescence receives much less attention. The reason for this is that until recently it was thought that protein phosphorescence could only be observed in frozen samples, thereby limiting its use. The early literature provides clues that this need not be the case. Beccari reported in 1746 that phosphorescence was observed from a cold hand after it had been exposed to the sunlight.[6] A comprehensive coverage of the early sightings of phosphorescence is found in the book by Harvey.[7]

In spite of this long fascination with luminescence, it was only within the last 25 years that protein phosphorescence at room temperature was convincingly documented. Hastings and Gibson in 1967 reported that a long-lived emission centered at 430 nm could be observed for luciferase and other proteins in the absence of oxygen.[8] In 1974, Saviotti and Galley showed that room temperature phosphorescence could be observed from liver alcohol dehydrogenase and alkaline phosphatase by the emission of resolved spectra characteristic of tryptophan phosphorescence.[9] The spectra, as well as a long

Jane M. Vanderkooi • Department of Biochemistry and Biophysics, School of Medicine, University of Pennsylvania, Philadelphia, Pennsylvania 19104.

Topics in Fluorescence Spectroscopy, Volume 3: Biochemical Applications, edited by Joseph R. Lakowicz. Plenum Press, New York, 1992.

lifetime, provided convincing evidence that the observed long-lived emission occurred from the tryptophan triplet state.

It is now clear that in the absence of molecular oxygen most proteins phosphoresce in aqueous solutions at ambient temperature.[10] In this chapter we discuss the use of phosphorescence of tryptophan to study proteins, with emphasis on measurements at room temperature. Comparisons between phosphorescence and the more commonly used fluorescence spectroscopy are made. Comprehensive reviews of protein luminescence have been written by Longworth.[11,12] A discussion on the use of phosphorescence at room temperature for the study of biological materials was given by Horie and Vanderkooi.[13]

3.2. Triplet State Formation and Disappearance

3.2.1. Energy Diagram

A modified Jablonski energy diagram, in Figure 3.1, shows the relationship between the ground state and the excited singlet and triplet states, where S_0 represents the ground state and S_1 and T_1 refer to the excited singlet and triplet states, respectively. By definition, fluorescence is the light emitted from the singlet state, and phosphorescence is the light emitted from the triplet state.

Several routes are possible to populate the triplet state. The triplet excited state can, in principle, be directly excited from the ground state, but a low extinction coefficient associated with the S_0 to T_1 transition (reflected in the long lifetime) makes direct excitation an inefficient process for tryptophan. The triplet state can be thermally populated, but for tryptophan the large energy gap between the ground state and the triplet state makes this process unfavorable. Energy transfer from a higher state can also populate the

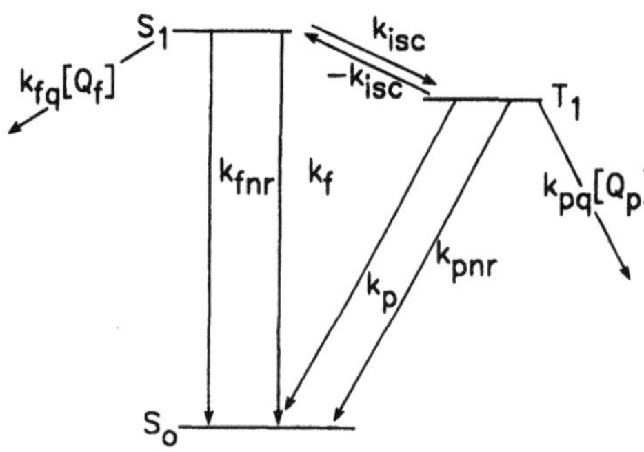

Figure 3.1. Energy diagram. S_0, S_1, and T_1 refer to the ground, excited singlet, and excited triplet state, respectively.

triplet level. In some cases the excited triplet state can be chemically generated in proteins.[14]

Ordinarily, however, the population of the triplet state is achieved through excitation into the singlet manifold which is followed by intersystem crossing to the excited triplet state. Therefore, in considering the yield of phosphorescence we must consider the processes involved in formation and disappearance of both the singlet and triplet states.

Referring to the diagram, four rates must be considered in the decay of S_1:

$$d[S_1]/dt = -(k_f + k_{\rm fnr} + k_{\rm fq}[{\rm Q}_f] + k_{\rm isc})[S_1] \tag{3.1}$$

In this we have assumed that the back reaction, $k_{-\rm isc}$, is negligible. Because the absorption of light is so rapid compared with the decay, it is further assumed that absorption is instantaneous. The observed singlet lifetime is

$$1/\tau_f = k_f + k_{\rm fq}[{\rm Q}_f] + k_{\rm isc} + k_{\rm fnr} \tag{3.2}$$

The triplet decay will also be governed by four rates:

$$d[T_1]/dt = -(k_p + k_{\rm pq}[{\rm Q}_p] + k_{\rm pnr})[T_1] + k_{\rm isc}[S_1] \tag{3.3}$$

The selection rules for quenching are different for fluorescence and phosphorescence. Hence, in Eqs. (3.1) and (3.3), quenchers of phosphorescence, $[{\rm Q}_p]$, and fluorescence, $[{\rm Q}_f]$, are distinguished because molecules that quench one may not quench the other with the same efficiency.

Making the assumption that the rate of intersystem crossing is fast relative to phosphorescence emission, the decay of phosphorescence will be exponential and the observed lifetime for phosphorescence, for most conditions, will be governed only by three rates as given by Eq. (3.4):

$$1/\tau_p = k_{\rm nr} + k_p + k_{\rm pq}[{\rm Q}_p] \tag{3.4}$$

When quantum yield of phosphorescence, Φ_p, is considered, $k_{\rm isc}$, the rate of intersystem crossing, must also be included:

$$\Phi_p = k_p \Phi_{\rm isc}/(k_p + k_{\rm pq}[{\rm Q}_p] + k_{\rm nr}) \tag{3.5}$$

where the quantum yield of intersystem crossing, $\Phi_{\rm isc}$, is

$$\Phi_{\rm isc} = k_{\rm isc}/(k_f + k_{\rm fq}[{\rm Q}_f] + k_{\rm isc} + k_{\rm fnr}) \tag{3.6}$$

3.2.2. General Considerations of Phosphorescence Yield

The phosphorescence lifetime of tryptophan at 77 K is 6.5 s in aqueous solution. The long lifetime is characteristic of $^1(n, \pi^*) \leftrightarrow {}^3(n, \pi^*)$ or $^1(\pi, \pi^*) \leftrightarrow {}^3(\pi, \pi^*)$ transitions.[15]

The long lifetime has important consequences on the decay rates. First, we consider what affects the nonradiative rates (k_{nr}) which change the yields of fluorescence and phosphorescence. The nonradiative decay rate is often enhanced in molecules which have flexible constituents (the so-called "loose-bolt" effect). Therefore, both fluorescence and phosphorescence yields are generally larger for rigid molecules than flexible molecules. For the same reason, a rigid environment will increase the emission yields; hence both fluorescence and phosphorescence yields often increase with increasing viscosity.

The long lifetime also has important consequences for the effect of specific quenching between the chromophore and surrounding quenching species. The probability of bimolecular collisions is related to the duration of the excited state. *The triplet excited state molecule is more susceptible than the singlet excited molecule to quenching simply because it has more time to interact with the surroundings.*

The intersystem crossing process has opposite effects on the yields of fluorescence and phosphorescence since it depletes the singlet state and populates the triplet state. It is commonly known that heavy ions, such as iodide and bromide, increase intersystem crossing by spin–orbit coupling.[16,17] For proteins, fluorescence can be quenched as phosphorescence yield is enhanced.[5,18,19] However, although the phosphorescence yield is increased, the lifetime is decreased. This effect arises because spin–orbit coupling, which increases the intersystem crossing rate from S_1 to T_1, also increases the conversion rate from T_1 to S_0.

Tryptophan at 77 K in rigid solution has a phosphorescence quantum yield of 0.17[20] and a lifetime of ~6 s. These values at 77 K are relatively invariant from protein to protein and do not vary significantly between buried and exposed tryptophans.[21,22] If one assumes that the intersystem crossing yield is a constant, a calculation of the quantum yield of indole phosphorescence can be roughly estimated from the lifetimes. The phosphorescence yield is related to lifetime by

$$\Phi_p = \tau_p(0.17/6) \qquad (3.7)$$

A tryptophan with a lifetime of 1 s has a quantum yield of $\sim 3 \times 10^{-2}$, and when the lifetime is 1 ms, then the quantum yield is $\sim 3 \times 10^{-5}$.

3.2.3. Measurement of Phosphorescence

The appearance and disappearance of the triplet state can be measured by light emission or by absorption change. The absorption change arises because the ground and triplet states have different absorption spectra. The absorption spectrum of tryptophan in the triplet state is red shifted in comparison with that of the ground state.[23,24] Use of emission has an advantage in that, by definition, it is measuring the triplet state. Photochemical products which have absorption similar to that of the triplet state molecule may complicate triplet absorption measurements. On the other hand, phosphorescence from tryptophans with quantum yields less than 10^{-6} is difficult to measure experimentally because of background luminescences. For such proteins, transient absorption measurement may be the method of choice.

Most commercially available phosphorimeters measure phosphorescence intensity and spectra by exciting the sample with a pulse of light and measuring the light intensity after a delay, thereby eliminating fluorescence. Similarly, phosphorescence lifetimes are usually measured by monitoring light intensity as a function of time after a flash of light. As with fluorescence lifetime measurement, a phase method can also be used. This method uses modulated exciting light and monitors the phase delay in modulation in the emission. This method has recently been applied for determining triplet lifetimes and anisotropy.[25]

3.3. Tryptophan Phosphorescence Emission from Proteins

3.3.1. Comparison of Fluorescence and Phosphorescence Emission Spectra

Figure 3.2 shows the fluorescence and phosphorescence emission spectrum from tobacco mosaic virus coat protein. These spectra are fairly typical of the tryptophan emission spectra observed from proteins at room temperature.

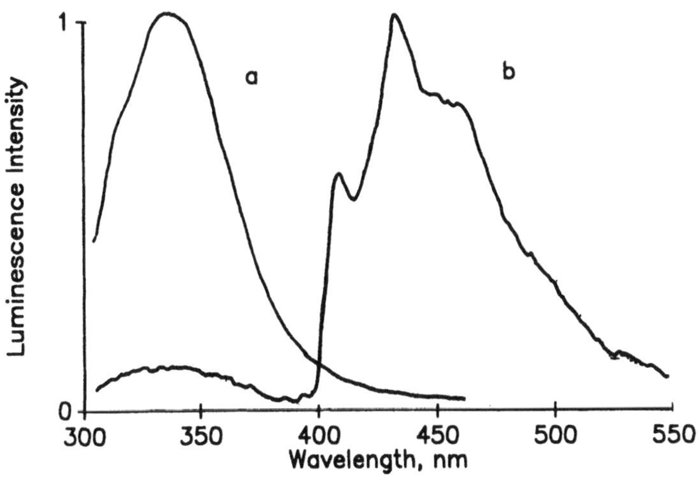

Figure 3.2. Luminescence from tobacco mosaic virus. (a) Fluorescence; (b) phosphorescence. Conditions described in Ref. 88. Spectra courtesy of J. W. Berger.

Differences between the spectra of fluorescence and phosphorescence are immediately obvious. For all tryptophans in proteins the phosphorescence spectrum, even at room temperature, is structured, while the fluorescence emission is not. (Even at low temperatures the fluorescence emission spectrum is usually not structured. The notable exceptions include α-amylase and aldolase,[5,26] protease, azurin[27,28] and ribonuclease T_1, staphylococcal endonuclease, elastase, tobacco mosaic virus coat protein, and *Drosophila* alcohol dehydrogenase[12].)

The broad spectrum of fluorescence, compared with phosphorescence, is attributed to a larger dipole moment in the excited singlet state as compared with the triplet state. This results in greater interaction with the environment and produces a larger spectral shift depending upon environment. For instance, the fluorescence emission maximum of tryptophan in aqueous solution is 350 nm, compared with 300 nm for tryptophan in butanol at 20°C.[5] This compares with a 0–0 transition for phosphorescence at 404 nm for indole in ethylene glycol/water and 408–410 nm in a hydrophobic environment. However, because the spectra of phosphorescence are resolved, it is often possible to distinguish the emission of individual tryptophans in the spectrum by distinguishing the 0–0 emission. In contrast, resolution of fluorescence from individual tryptophans is often obscured by the broad fluorescence emission band.

At 77 K the position of the 0–0 band is generally blue shifted for exposed tryptophans and red shifted for buried tryptophans. Along with a shift in wavelength to the red, the phosphorescence lifetime decreases.[28] The single tryptophan of human serum albumin shows red-shifted phosphorescence and $D - L$ triplet zero-field splitting, indicating that it is in a hydrophobic environment.[29]

The width of the 0–0 line in single-tryptophan proteins at 77 K has been interpreted to reflect inhomogeneous broadening arising because the protein exists as a distribution of conformations.[30–34] The width of the 0–0 band of liver alcohol dehydrogenase is $\sim 500\ \mathrm{cm}^{-1}$ at 22°C.[10,31,35] The widths of the 0–0 transition for other proteins are somewhat greater. In many cases for the spectra taken at room temperature, low-resolution optics were used (as in Figure 3.2), and hence the published spectra may overestimate the width of the emission band.

3.3.2. Delayed Fluorescence

Another feature of the spectrum shown in Figure 3.2 is a long-lived emission of the same wavelength as the fluorescence emission spectrum. This so-called "delayed fluorescence" is very weak relative to phosphorescence, and care must be taken to ensure that the long-lived fluorescence emission does not correspond to fluorescence induced by the extended tail of the lamp flash. Several mechanisms are possible to account for repopulation of the singlet state from the triplet state.[36] Tryptophan is reported to show delayed fluorescence due to thermal repopulation of the singlet state (called phosphorescence in the Polish and Russian literature).[5] In this case, delayed fluorescence arises from thermal repopulation of the singlet state from the triplet state and depends upon the energy gap between the T_1 and S_1 states. The following relationship describes this mechanism:

$$k_{-\mathrm{isc}} = A \exp(-E/RT) \tag{3.8}$$

where A is a frequency factor, and E is the activation energy, equal to the energy difference between the S_1 and T_1 states. Because the fluorescence emission spectrum is broad, it is hard to judge the precise position of the 0–0 emission, but we can estimate it to be around 310 nm, while the phosphorescence emission is around 410 nm. This corresponds to an energy gap between T_1 and S_1 of over 20 kcal, a very large value, making this path very unlikely. Another mechanism for population of singlet states from the triplet states is by triplet–triplet transfer ("annihilation"), whereby two excited-state triplets react to form an excited singlet state and a ground-state molecule, which has been observed for aromatic amino acids in crystals.[37] A third possible mechanism for delayed fluorescence is electron recombination, in which an electron is ejected from the molecule and then recombines to form an excited state. Emission from the electron recombination process can last up to an hour. The origin of the delayed fluorescence from the tobacco mosaic virus protein (Figure 3.2) has not yet been determined.

3.3.3. Lifetime of Tryptophan Phosphorescence in Proteins

A large number of proteins have now been reported to phosphoresce at room temperature. A recent survey of 40 proteins revealed that about 75% of them showed lifetimes longer than 1 ms.[10]

The phosphorescence lifetimes of various proteins at room temperature are given in Table 3.1. Some variability in the lifetimes reported from lab to lab is evident, possibly due to different enzyme preparation, removal of oxygen (see below), or other conditions. Nevertheless, when measured under the same conditions, it is apparent that the tryptophan lifetimes vary dramatically from protein to protein. Alkaline phosphatase exhibits the longest lifetime from a protein in solution with a lifetime of 1.5–1.7 s at 22°C, approaching the lifetime of 5.5 s at 77 K. The lifetime of free indole in solution is ~15–30 μs at ~22°C.[38,39] Therefore, in the absence of other quenching mechanisms, the lower limit for the phosphorescence lifetime of a fully exposed tryptophan moiety in a protein should be about 20 μs.

Table 3.1. Phosphorescence Lifetimes from Tryptophan-Containing Proteins at Ambient Temperatures under Anaerobiosis

Protein	Lifetime (ms)	Reference
Actin, rabbit muscle	15	88
Albumin, bovine serum	0.9	10
Albumin, human serum	0.9	10
	0.5	24
Albumin, chicken egg	15	10
Alcohol dehydrogenase	134	9
	300	35
	400	54
Aldolase	45	10
Alkaline phosphatase	800	9
Escherichia coli	1500	10
	1700	54
Asparaginase	50	10
ATPase, sarcoplasmic reticulum	30	75
Azurin	400	10
Cellulase	200	10
Coat protein, tobacco mosaic virus	25	85
Edestin, hemp seed	500	10
Glucose oxidase	120	10
Glutamate dehydrogenase		
Bovine liver	1.2	72
Escherichia coli	0.65	72
Keratin	1400	47
Lactic dehydrogenase	25	10
β-Lactoglobulin	15	10
Myosin, rabbit muscle	100	10
Nuclease, micrococcal	400	10
Parvalbumin (with calcium)	5	10
Protease		
Streptomyces griseus	100	10
Bacillus subtilis	10	10
Ribonuclease T_1	14	10
Streptokinase	50	10
Trypsin, bovine pancreas	1.4	73

The range of six orders of magnitude for lifetimes of tryptophan phosphorescence in proteins at room temperature is larger than for fluorescence. The lower limit for fluorescence lifetime is about 0.5 ns, while the upper limit is ~8 ns.[21] Typical values range from 3 to 5 ns.

From an experimental viewpoint, the wide range of phosphorescence lifetimes is advantageous for the study of proteins. It means that it should be possible to select one tryptophan from a population of emitting tryptophans by varying gate and delay times for signal acquisition.

3.3.4. What Affects the Phosphorescence Lifetime?

Conceptually, we can separate "environmental" effects (k_{nr}) and "specific" quenching mechanisms (k_{pq}), where k_{pq} is due to the presence of a quenching moiety within the protein. Specific quenching effects of externally added quenchers are discussed in Section 3.3.5.

3.3.4.1. Effect of Environment on Phosphorescence

Strambini and Gonnelli[40] have studied the effect of viscosity on the phosphorescence of tryptophan, 1-methyltryptophan, *N*-methyltryptophan-amide, and tryptophan-containing peptides. Over a viscosity range of 10^9 to 10^4 poise, the decay rate of the excited triplet state changed by a factor of 100. This change was insensitive to polarity of the solvent. It was also insensitive to proton exchange at the ring nitrogen since *N*-methyltryptophanamide showed the same viscosity dependence as the derivatives in which the nitrogen was acylated.

Out-of-plane vibrations which increase k_{nr} will decrease the observed lifetime. Therefore, based upon the viscosity dependence of phosphorescence, an attractive hypothesis is that the long lifetime of tryptophan in the protein may reflect the rigidity of the tryptophan site. The relationship between out-of-plane motion of the tryptophan and phosphorescence yield can be examined by comparing fluorescence anisotropy with phosphorescence lifetime. Fast (i.e., subnanosecond) segmental motion is reported for the tryptophan in monellin[41,42] and melittin,[42,43] both of which exhibit short (<0.5 ms) phosphorescence lifetimes. The tryptophan of ribonuclease T_1 is immobilized on the nanosecond time scale,[41,43] and one of the two tryptophans of liver alcohol dehydrogenase is immobilized[42,44]. These two proteins show long phosphorescence lifetimes. The single tryptophan of azurin from *Pseudomonas aeruginosa*, at position 48, is located in the core of a β-barrel structure, surrounded by hydrophobic side chains. It is immobilized on the picosecond and nanosecond time scales[45] and also exhibits a long phosphorescence lifetime, $\sim$400 ms.[10]

It has been observed that for some proteins the room temperature phosphorescence lifetimes are increased in D_2O. The phosphorescence lifetime of liver alcohol dehydrogenase is 300 ms in H_2O and 500 ms in D_2O.[10] Phosphorescence lifetimes are often dramatically increased by exchanging hydrogen with deuterium. The reason for this is that decay rates are affected by overtones of the C–H or N–H stretch. In the case of tryptophan in proteins, we exchange only the N−H, not the C−H, so that this mechanism may not have much effect. Indeed, the model study of Strambini and Gonnelli[40] would suggest that exchange of the enamine proton of tryptophan with deuterium does not affect the lifetime. An isotope effect on the quenching by an intrinsic ionizable group could also produce a D_2O effect.[46] An alternative reason for the effect of D_2O on the lifetime could be the difference in the hydration of proteins, which produces subtle differences in structure.

Room temperature phosphorescence can be observed from dried proteins. Sheep wool keratin[47] has a phosphorescence lifetime of 1.4 s. Six lyophilized proteins were shown to exhibit phosphorescence at room temperature.[48] The spectra were diffuse, and the lifetime was non-single-exponential, which the authors interpreted as due to inhomogeneous distribution of tryptophans. As the protein was hydrated, the phosphorescence lifetime decreased. This decrease occurred over the same range of hydration where the tryptophan fluorescence becomes depolarized. Hence, these results are consistent with the idea that rigidity of the site contributes to the lifetimes.

For single-tryptophan proteins there is some correlation between blue-shifted fluorescence emission maximum and phosphorescence lifetime (Table 3.2). Another correlation is that three of the proteins which exhibit phosphorescence, azurin, protease (subtilisin Carlsberg), and ribonuclease T_1, are reported to show resolved fluorescence emission at 77 K. Both blue-shifted emission spectra and resolved spectra are characteristic of indole in a hydrocarbon-like matrix.

In summary, it appears that phosphorescence at room temperature is a function of "burial" or "rigidity" of the site, but, as for all excited states, the competing nonradiative pathways are influenced by the polarizability, polarity, and mobility of the local environment.

Table 3.2. Correlation between Fluorescence Emission Maximum and Phosphorescence Lifetime in Single-Tryptophan Proteins[a]

Protein	Fluorescence maximum (nm)	Phosphorescence lifetime (ms)
Azurin	305	400
Protease (subtilisin Carlsberg)	310	10
Parvalbumin	320	5
Ribonuclease T_1	325	14
Melittin	340	<0.5
Monellin	345	<0.5
Parvalbumin (no calcium)	350	<0.5

[a] Ref. 10.

3.3.4.2. Specific Quenching Mechanisms

Some tryptophans do not exhibit phosphorescence because of quenching by specific sites from within the protein. The absence of phosphorescence could be due to quenching of either the singlet state or the triplet state. For example, in horse heart cytochrome *c* the tryptophan is adjacent to the heme, and its fluorescence is quenched by Förster transfer to the heme. Since the singlet state is populating the triplet state, the lack of observable phosphorescence is likely to be due to an unpopulated triplet state. Another example where the redox center of the protein interacts with the tryptophan excited states is found in azurin. The copper(II) quenches both the singlet and triplet states.(28)

Other groups within the protein may affect excited states. Disulfide bonds quench the excited states of tryptophan. For instance, at 77 K the phosphorescence lifetime of native lysozyme is low, 1.4 s; reduction of the disulfide bonds or denaturation gave the typical phosphorescence lifetime of 5.6 s.(49) Therefore, the absence of phosphorescence at room temperature from this protein is likely to be due to quenching of both the singlet and the triplet state.

Other groups may cause shortening of the lifetime. The phosphorescence of parvalbumin is quenched by free tryptophan with a quenching rate constant of about $10^5\ M^{-1}\ s^{-1}$ (D. Calhoun, unpublished results). A more extensive survey of proteins or model compounds with known distances between tryptophans is needed to study how adjacent tryptophans affect the lifetime. It should be noted that at low temperature the phosphorescence lifetime of poly-L-tryptophan is about the same as that of the monomer.(12) This does not necessarily mean that in a fluid solution tryptophan–tryptophan interaction could not take place. Thermal fluctuations in the polypeptide chain may transiently produce overlap in the π orbitals between neighboring tryptophans, thus resulting in quenching.

3.3.5. Phosphorescence Quenching by External Molecules

3.3.5.1. Quenching Equation

Fluorescence quenching has proven to be a powerful means to determine location of tryptophans. Small organic molecules, such as acetone, acrylamide, and amino acids, have been used to quench fluorescence of tryptophans which are exposed to the solvent.(50, 51) These molecules apparently quench by close interaction and so provide a tool to determine the surface accessibility of tryptophan in a protein.

The same rationale which is used for fluorescence quenching can be used for phosphorescence quenching. Because the lifetime of phosphorescence is so long, slow processes that are out of the time range for fluorescence can be detected.

The Stern–Volmer[52] equation relates fluorescence intensity and the quenching rate constant, k_q:

$$F_0/F = 1 + \tau_0 k_q[Q_f] \tag{3.9}$$

where F_0 is the fluorescence intensity in the absence of quencher, and F is the fluorescence intensity in the presence of quencher at a given concentration, $[Q_f]$.

Making the assumption that k_{isc} is fast, we can modify Eq. (3.9) to yield:

$$\tau/\tau_0 = 1 + \tau_0 k_q[Q_p] \tag{3.10}$$

which would apply to phosphorescence lifetimes τ and τ_0 in the presence and absence of quencher, respectively. Note that the ratio of phosphorescence intensities does not equal the ratio of lifetimes because quenchers can increase the intersystem crossing rate.

3.3.5.2. Oxygen, NO, and CO Quenching

Experimentally, oxygen quenching represents the most serious problem for phosphorescence measurement of biological samples. If the diffusion-limited quenching rate constant of oxygen is $10^9\ M^{-1}\ s^{-1}$, then for a molecule with a lifetime of 1 s, the concentration of oxygen that will reduce the lifetime by 10% is 0.11 n*M*. Such low concentrations are often difficult to achieve. If τ_0 is 1 ms, the concentration which will reduce the lifetime by 10% is 0.11 μM, a concentration that is more readily attainable. For lifetimes typical of fluorescence, say, 1 ns, the concentration for 10% quenching would be 0.11 *M*. The concentration of oxygen in aqueous buffer at 20°C equilibrated with air is only around 0.25 m*M*; hence, one does not need to deoxygenate the sample before measurement of fluorescence.

The original claim by Saviotti and Galley[9] that phosphorescence can be observed in oxygenated medium appears to have been due to the decomposition of oxygen by the UV lamp.[35,53,54] However, there still appears to be a discrepancy in the quenching constants for oxygen observed for proteins in different laboratories. Reexamination of the quenching of phosphorescence of alkaline phosphatase and liver alcohol dehydrogenase gave bimolecular rate constants for oxygen quenching of 1.2×10^6 and $3.6 \times 10^7\ M^{-1}\ s^{-1}$, respectively.[54,55] This is much less than the respective values of $\sim 10^9$ and $6 \times 10^8\ M^{-1}\ s^{-1}$ reported by Calhoun *et al.*[35] and the value of $1.4 \times 10^8\ M^{-1}\ s^{-1}$ for alcohol dehydrogenase reported by Barboy and Feitelson.[56]

The differences lie in the difficulty of making oxygen measurements at low concentrations. On one hand, the oxygen may be consumed by the lamp, and, on the other hand, addition of partially deoxygenated buffer (such as the procedure used by Calhoun *et al.*[35]) may inadvertently allow more oxygen to be added than intended. A solution to the experimental difficulties may be to include in the sample another soluble dye whose oxygen dependency of the triplet lifetime is known.[57] This approach was recently taken by Calhoun *et al.*[58] The k_q for alkaline phosphatase and alcohol dehydrogenase was 1.2×10^6 and $3.6 \times 10^7\ M^{-1}\ s^{-1}$, respectively, comparing favorably with results found by Strambini.[55]

The low rate constants for oxygen quenching obtained for alkaline phosphatase and alcohol dehydrogenase modify the conclusion that O_2 can diffuse through proteins in general, which was suggested by studies of oxygen quenching of tryptophan fluorescence of 14 proteins by the work of Lakowicz and Weber.[59] Of these proteins, only azurin is now known to have a buried and no exposed tryptophan. The oxygen quenching constant measured for azurin phosphorescence, $2 \times 10^7\ M^{-1}\ s^{-1}$,[58] compares with $1 \times 10^8\ M^{-1}\ s^{-1}$ determined for fluorescence quenching.[59] The difference can perhaps arise due to a statistical factor for phosphorescence quenching by oxygen ranging between $\frac{1}{9}$ and $\frac{5}{9}$.[60,61]

Two other diatomic molecules, CO and NO, quench tryptophan phosphorescence. Strambini reported that the k_q for NO quenching of alkaline phosphatase is $8.2 \times 10^5\ M^{-1}\ s^{-1}$, which is about the same as that reported for oxygen.[55] The k_q of CO for alkaline phosphatase is $2 \times 10^4\ M^{-1}\ s^{-1}$ and for liver alcohol dehydrogenase is $1 \times 10^5\ M^{-1}\ s^{-1}$.[58] The value for CO quenching does not indicate restricted motion since the quenching constant for *N*-acetyltrytophanamide (NATA) was $3 \times 10^6\ M^{-1}\ s^{-1}$, about three orders of magnitude less than if every collision is effective, and about three orders of magnitude reduced from the quenching constant for oxygen.

3.3.5.3. *Other Quenchers*

Paramagnetic molecules, electron-dense molecules, and electron donors or acceptors are expected to quench phosphorescence. The quenching constants of NATA and parvalbumin for five quenchers are given in Table 3.3. The viscosity of 85% glycerol is approximately 20 times that of water at 22°C. Assuming the applicability of Stokes' law, the quenching rate is expected to be about $10^9\ M^{-1}\ s^{-1}$ for the free molecule in water. The quenching rate in parvalbumin is less than for the free NATA, indicating restricted access to the tryptophan in the protein.

A survey of the quenching constants for a series of proteins was made using one quencher, nitrite (Table 3.4).[58] It is noted that the phosphorescence lifetime τ_0 is not correlated with the quenching constant. For example, the

Table 3.3. Survey of Quencher Molecules of NATA and Parvalbumin Phosphorescence[a]

Quencher	k_q at 22°C ($M^{-1}\,s^{-1}$) NATA[b]	Parvalbumin[c]
CO	3×10^6	2.5×10^5
CS_2	2.2×10^8	4×10^7
Ethanethiol	1.6×10^7	3.2×10^6
Nitrite	6.8×10^7	2.1×10^6
Methyl vinyl ketone	4.4×10^7	4.9×10^5
Cinnamamide	4×10^7	2.4×10^7

[a] Ref. 58.
[b] In 85% glycerol.
[c] In 0.1 M NaCl, 0.01 M Tris, pH 7.0.

phosphorescence lifetimes of liver alcohol dehydrogenase and azurin are about the same, whereas the quenching constants differ by a factor of ~100. This indicates that the lifetimes observed from proteins in solution in the absence of added quenchers are not solely dependent upon residual impurities in the solution. The quenching constant varied greatly from protein to protein—often by about five orders of magnitude.

3.3.5.4. How Does an Externally Added Quencher Quench a Buried Tryptophan?

Many lines of evidence suggest that proteins undergo structural fluctuations.[(62–65)] A question is how a molecule in solution can interact with a

Table 3.4. Survey of Protein Phosphorescence Quenching by NO_2[a,b]

	k_q ($M^{-1}\,s^{-1}$)	τ_0 (ms)
NATA	6.8×10^7	0.5
Parvalbumin	2.1×10^6	5.2
Lactoglobulin	2×10^6	13
Ribonuclease T_1	5×10^5	14
Liver alcohol dehydrogenase	4.7×10^4	350
Aldolase	5.8×10^3	40
Pronase	2.5×10^3	700
Edestin	1.4×10^3	400
Azurin	7.4×10^2	400
Alkaline phosphatase	5.0×10^1	1600

[a] Ref. 58.
[b] Conditions: NATA solution contained 85% glycerol. Proteins were dissolved in aqueous buffers.

buried moiety in a protein. The goal of the experiment described in Table 3.4 using externally added quenchers is to learn how interaction occurs. The following three models for quenching are usually considered in interpreting results.

1. There is a thermally activated structural rearrangement of the protein such that the tryptophan is transiently exposed. The following scheme describes this mechanism:

$$\text{protein (closed)} \underset{k_c}{\overset{k_{op}}{\rightleftharpoons}} \text{protein (open)} \xrightarrow{k_d} \text{quench}$$

This is the so-called "gating" model. The following relationship between k_d and k_q would hold:

$$k_q = (k_{op}/k_c)\, k_d \tag{3.11}$$

where k_d is the diffusion-limited rate constant. The more general expression has been derived by Somogyi *et al.*[66]

Since the opening reaction would depend upon the properties of the protein, this model predicts that the quenching could vary dramatically for different proteins and that the observed quenching would be directly proportional to k_d. Therefore, changes in viscosity and size of the molecule would affect k_q according to the Stokes relationship.

2. There is a thermally activated structural rearrangement of the protein such that channels appear and the quencher molecules are able to penetrate the protein—the "penetration" model.[67] This model distinguishes between external diffusion, $k_d(\text{ext})$, and diffusion within the protein as follows:

$$\text{Q} + \text{protein} \overset{k_d(\text{ext})}{\rightleftharpoons} [\text{Q–protein}] \overset{k_d(\text{int})}{\rightleftharpoons} \text{quench}$$

The observed quenching rate constant, k_q, according to this model will be a function of the internal diffusion of the quencher, $k_d(\text{int})$, and is given by

$$k_q = k_d(\text{int})\, k_d(\text{ext})/[k_d(\text{int}) + k_d(\text{ext})] \tag{3.12}$$

In this model, whether k_q is a function of the solvent viscosity depends upon the relative magnitudes of $k_d(\text{int})$ and $k_d(\text{ext})$. If $k_d(\text{int}) \ll k_d(\text{ext})$, then k_q will depend upon viscosity; if $k_d(\text{int}) > k_d(\text{ext})$, the structural fluctuations in the protein allowing penetration of the quencher determine the magnitude of k_q, and change in bulk viscosity may not affect this rate. Simulation of protein penetration behavior suggests that the penetration rate should be extremely sensitive to the size and charge of the quencher.[65]

3. Quenching occurs by long-range interactions. In this model, the quencher molcule does not need to touch the tryptophan, but molecules at the surface of the protein can quench tryptophans some distance away—the "long-range transfer" model. The following equation would describe the results:

$$k_q = k_t k_d/(k_t + k_d) \tag{3.13}$$

The long-range transfer model is formally the same as the second model, but now k_t does not require penetration but instead indicates transfer at a distance. When $k_t < k_d$, the reaction rate will not depend upon k_d. This model can therefore account for the viscosity dependence. Since penetration is not required, this model also predicts that quenching will not be critically dependent upon size and charge of the molecule.

There are many examples of long-range triplet–triplet interactions involving indole-type molecules.[68,69] Triplet state porphyrins in proteins have been shown to be quenched by long-range electron exchange reactions.[70,71] Because the triplet state is so long, "forbidden" processes may become significant and therefore long-range interactions must be considered. Triplet quenching by electron exchange has been demonstrated by Vanderkooi *et al.*[71a]

3.3.6. Phosphorescence Lifetimes to Measure Conformational Changes in Proteins

It is clear that the wide range of protein phosphorescence lifetimes is due to various specific quenching mechanisms (k_q) or due to flexibility of the tryptophan site, thereby affecting k_{nr}. It also follows that phosphorescence will be very sensitive to conformational fluctuations since subtle changes in distance or orientation relative to a specific quenching moiety within the protein will affect the lifetimes dramatically. The phosphorescence emission from protein tryptophan remains relatively unexplored in terms of investigation of dynamic structure–function relationships.

3.3.6.1. Temperature-Dependent Conformational Changes

The phosphorescence lifetimes have been examined for many protein systems as a function of temperature. In the early work oxygen was not removed from the sample.[72,73] In these works the lifetimes are dominated by quenching by oxygen, and so the temperature dependencies probably represent temperature-dependent oxygen diffusion.

Domanus *et al.*[74] proposed that the ratio of phosphorescence intensity to lifetime, (P/τ), of tryptophan phosphorescence as a function of temperature be used to distinguish heterogeneity in emission from multitryptophan proteins. Since different tryptophans within one protein show different temperature-dependent quenching and since lifetime measurements can be arranged to select for a long-lived component, this ratio is particularly sensitive to heterogeneity. The ratio of intensity to lifetime showed steplike transitions as a function of temperature for multitryptophan proteins, but this ratio remained constant for free tryptophan or for the single-tryptophan protein myelin basic protein. They interpreted these results as indicating that large variations exist in the rate of fluctuations in the structure surrounding individual tryptophans in the protein. A similar stepwise decrease in P/τ was observed for glutamate dehydrogenase as a function of temperature.[75]

Kai and Imakubo[76] found that the temperature at which emission from the "exposed" tryptophan is no longer observed appears to be characteristic of the protein, having values of 180 K for trypsin, 200 K for aldolase, and 230 K for alkaline phosphatase.

Bismuto *et al.*[77] compared the phosphorescence from both tuna and sperm whale apomyoglobin. The emission occurs from a tryptophan in the A helix. The temperature dependence of lifetime and the position of the 0–0 vibrational band differ as a function of temperature for the two proteins. The authors interpreted their results to indicate that the microenvironment of the tryptophan in sperm whale apomyoglobin possesses a higher degree of internal flexibility than that in the tuna protein.

3.3.6.2. Effect of Substrate on Phosphorescence of the Sarcoplasmic Reticulum ATPase

Vanderkooi *et al.*[78] examined the phosphorescence from tryptophan in sarcoplasmic reticulum vesicles and the purified Ca transport ATPase at room temperature in deoxygenated solutions. The phosphorescence decay is multi-exponential; the lifetime of the long-lived component of phosphorescence is ~22 ms. Addition of ATP or vanadate decreased the phosphorescence yield. The Ca^{2+}-ATPase of the sarcoplasmic reticulum alternates between two conformations, called E_1 and E_2, during Ca^{2+} transport. The observations were interpreted to indicate that either the binding of vanadate or phosphate to the phosphorylation site of the ATPase or the induced shift in the conformation from the E_1 to the E_2 state produced the phosphorescence quenching.

3.3.6.3. Denaturation of Proteins

For horse liver alcohol dehydrogenase, denaturation by guanidine hydrochloride resulted in a decrease in phosphorescence lifetime parallel with loss of activity.[79] With urea as a denaturant, the decrease in phosphorescence lifetime appeared cooperative, and it is suggested that the denaturant loosened intramolecular interactions (such as hydrogen bonds), resulting in greater fluidity of the tryptophan environment.[80]

3.4. Phosphorescence Anisotropy and Rotational Motion

3.4.1. Phosphorescence Anisotropy

Phosphorescence, like fluorescence, is based upon dipolar interactions and therefore is polarized.

Experiments involving anisotropy of phosphorescence or of the absorption of the triplet state rely upon the same principles as the measurement of fluorescence anisotropy. All are based upon the photoselection of molecules by polarized light and the randomization of polarization due to Brownian motion occurring on the time scale of the excited state. Anisotropy is defined as

$$A = (I_{\|} - I_{\perp})/(I_{\|} + 2I_{\perp}) \tag{3.14}$$

where $I_{\|}$ is the intensity in the parallel direction, and $I_{\perp}$ is the intensity in the perpendicular direction. Details of the measurement of rotational diffusion and anisotropy are described by Jovin *et al.*[81] and Cherry.[82]

The value of A in the absence of motion is referred to as the limiting anisotropy, A_0. This value is related to the relative direction of the absorption and emission dipoles. For tryptophan, A_0 is negative, which indicates that the absorption and emission dipoles are approximately perpendicular to each other.[83]

The limiting value of A_0 is never achieved in practice, and partial depolarization can result from molecular motion. For a chromophore which moves with the motion of a rigid spherical macromolecule to which it is attached, the observed anisotropy will decay exponentially as a function of ϕ, the rotational correlation time, according to

$$A(t) = A_0 e^{-t/\phi} \tag{3.15}$$

In the case that the macromolecule is nonspherical or sidechain or segmental motions occur, then the anisotropy will decay as a sum of exponential functions. The work of Kinosita *et al.*[83] deals with the case in which there are restricted motions. The anisotropy decay function becomes

$$A(t) = A_\infty + (A_0 - A_\infty)\, e^{-t/\phi} \tag{3.16}$$

Anisotropy of phosphorescence then becomes a powerful tool to study the overall rotation of large biological macromolecules and to study segmental motions which occur in these structures.

3.4.2. Anisotropy to Study Proteins

Strambini and Galley have used tryptophan anisotropy to measure the rotation of proteins in glassy solvents as a function of temperature. They found that the anisotropy of tryptophan phosphorescence reflected the size of globular proteins in glycerol buffer in the temperature range −90 to −70°C.[84,85] Tryptophan phosphorescence of erythrocyte ghosts depolarized discontinuously as a function of temperature. These authors interpreted the complex temperature dependence to indicate protein–protein interactions in the membrane.

The rotational mobility of human low-density (LDL) and very-low-density (VLDL) lipoproteins was studied as a function of viscosity and temperature in the range of −90 to −50°C.[86] The rotational behavior for LDL is represented by a single correlation time, consistent with the overall rotation of a spherical rigid particle as the source of the phosphorescence depolarization. For VLDL, internal peptide motions dominate the depolarization profile.

The phosphorescence anisotropy of liver alcohol dehydrogenase was studied in crystals and in solution.[87] The phosphorescence, arising from the tryptophan in the coenzyme-binding domain, showed no depolarization in the triplet lifetime. This result could be accounted for by segmental motion. Such a finding would indicate an immobilization of the tryptophan. The tryptophan in the catalytic site of glutamate dehydrogenase, an enzyme which shows a similar peptide conformation to that of alcohol dehydrogenase, is also immobilized.[87]

Berger and Vanderkooi[88] studied the depolarization of tryptophan from tobacco mosaic virus. The major subunit of the coat protein contains three tryptophans. The phosphorescence decay is non-single-exponential. At 22°C the lifetime of the long component decays with a time constant of 22 ms, and at 3°C the lifetime is 61 ms. The anisotropy decay is clearly not single-exponential and was consistent with the known geometry of the virus.

3.5. Tryptophan Phosphorescence from Cells

The long-lived phosphorescence of the tryptophan in alkaline phosphatase is unusual. Horie and Vanderkooi examined whether its phosphorescence could be detected in *E. coli* strains which are rich in alkaline phosphatase.[89] They observed phosphorescence at 20°C with a lifetime of 1.3 s, which is comparable to the lifetime of purified alkaline phosphatase (1.4 s). Long-lived luminescence was not observed from strains deficient in alkaline phosphatase. The temperature dependence of tryptophan phosphorescence in the living cells was slightly different from that for the purified enzyme, indicating an environmental effect.

Mazhul *et al.*[90] have reported that long-lived luminescence could be detected in intact human erythrocytes and white blood cells at ambient temperature. They have shown by emission spectra and pH dependency that this emission arises from tryptophan. The emission was not single-exponential, suggesting that more than one population of tryptophan emitted. Identification of the emitting species has not yet been conclusively made, but the white blood cell protein content is about 10% actin, a protein known to phosphorescence.[91]

3.6. Conclusions

Phosphorescence is readily detectable from most types of proteins at room temperature. Tryptophan phosphorescence lifetimes and yields are very sensitive to environment, and therefore phosphorescence is sensitive to conformational changes in proteins. Fundamental questions concerning exactly what parameters affect lifetime and spectra of tryptophan in proteins remain still to be answered.

The long lifetime of phosphorescence allows it to be used for processes which are slow—on the millisecond to microsecond time scale. Among these processes are the turnover time of enzymes and diffusion of large aggregates or smaller proteins in a restricted environment, such as, for example, proteins in membranes. Phosphorescence anisotropy is one method to study these processes, giving information on rotational diffusion. Quenching by external molecules is another potentially powerful method; in this case it can lead to information on tryptophan location and the structural dynamics of the protein.

Acknowledgment

This work was supported by NIH grants GM 34448 and GM 36393.

References

1. P. Debye and J. O. Edwards, A note on the phosphorescence of proteins, *Science 116*, 143–144 (1952).
2. W. F. J. Teale and G. Weber, Ultraviolet fluorescence of the aromatic amino acids, *Biochem. J. 65*, 476–482 (1957).
3. D. Duggan and S. Udenfriend, The spectrophotofluorometric determination of tryptophan in plasma and of tryptophan and tyrosine in protein hydrolysates, *J. Biol. Chem. 223*, 313–319 (1956).
4. V. G. Shore and A. B. Pardee, Fluorescence of some proteins, nucleic acids and related compounds, *Arch. Biochem. Biophys. 60*, 100–107 (1956).

5. S. V. Konev, *Fluorescence and Phosphorescence of Proteins and Nucleic Acids*, Plenum Press, New York (1967).
6. J. B. Beccari, De quam plurinis phosphoris nune prinun detectis, *Commen. Acad. Bonon. 2*, 136–179 (1746).
7. E. N. Harvey, *A History of Luminescence*, pp. 305–365, The American Philosophical Society, Philadelphia (1957).
8. J. W. Hastings and Q. H. Gibson, The role of oxygen in the photoexcited luminescence of bacterial luciferase, *J. Biol. Chem. 242*, 720–726 (1967).
9. M. L. Saviotti and W. C. Galley, Room temperature phosphorescence and the dynamic aspects of protein structure, *Proc. Natl. Acad. Sci. U.S.A. 71*, 4154–4158 (1974).
10. J. M. Vanderkooi, D. B. Calhoun, and S. W. Englander, On the prevalence of room temperature protein phosphorescence, *Science 236*, 568–569 (1987).
11. J. W. Longworth, in: *Time-Resolved Fluorescence Spectroscopy in Biochemistry and Biology* (R. B. Cundall and R. E. Dale, eds.), pp. 651–725, Plenum Press, New York (1983).
12. J. W. Longworth, in: *Excited States of Proteins and Nucleic Acids* (R. F. Steiner and G. Weinryb, eds.), pp. 319–484, Plenum Press, New York (1971).
13. T. Horie and J. M. Vanderkooi, Use of phosphorescence at room temperature for the study of biological molecules, *Life Sci. Rep. 2*, 141–178 (1983).
14. C. C. C. Vidigal, A. Faljoni-Alario, N. Duran, K. Zinner, Y. Shimizu, and G. Cilento, Electronically excited species in the peroxidase catalyzed oxidation of indoleacetic acid. Effect upon DNA and RNA, *Photochem. Photobiol. 30*, 195–198 (1979).
15. S. K. Lower and M. A. El-Sayed, The triplet state and molecular electronic processes in organic molecules, *Chem. Rev. 66*, 199–241 (1966).
16. S. P. McGlynn, T. Azumi, and M. Konoshita, *Molecular Spectroscopy of the Triplet State*, Prentice-Hall, Englewood Cliffs, New Jersey (1969).
17. J. B. Birks, *Photophysics of Aromatic Molecules*, Wily-Interscience, New York (1970).
18. S. S. Lehrer, Solute perturbation of protein fluorescence. The quenching of the tryptophyl fluorescence of model compounds and of lysozyme by iodide ion, *Biochemistry 10*, 3254–3263 (1971).
19. R. F. Steiner, Varying luminescence behavior of the different tryptophan residues of papain, *Biochemistry 10*, 771–778 (1971).
20. F. Bishai, E. Kuntz, and L. Augenstein, Intra- and intermolecular factors affecting the excited states of aromatic amino acids, *Biochim. Biophys. Acta 140*, 381–394 (1967).
21. J. W. Longworth, C. L. McLaughlin, and A. Solomon, Luminescence studies on Bence–Jones proteins and light chains of immunoglobulins and their subunits, *Biochemistry 14*, 2953–2959 (1976).
22. J. Domanus, G. B. Strambini, and W. Galley, Heterogeneity in the thermally-induced quenching of the phosphorescence of multi-tryptophan proteins, *Photochem. Photobiol. 34*, 15–21 (1980).
23. L. I. Grossweiner, Metastable states of photoexcited ovalbumin and constituents, *J. Chem. Phys. 24*, 1255–1256 (1956).
24. B. Hicks, M. White, C. A. Ghiron, R. R. Kuntz, and W. A. Volker, Flash photolysis of human serum albumin: Characterization of the indole triplet absorption spectrum and decay at ambient temperature, *Proc. Natl. Acad. Sci. U.S.A. 75*, 1172–1175 (1978).
25. P. B. Garland, Phase and modulation optical spectroscopic methods for determining triplet lifetimes and slow rotational diffusion coefficients, *Biochem. Soc. Trans. 15*, 838–839 (1986).
26. V. P. Bobrovich and S. V. Konov, Luminescence characteristics of amylase in the crystalline state, *Dokl. Akad. Nauk SSSR 155*, 197–200 (1964).
27. E. A. Burstein, V. A. Permyakov, S. A. Yashin, S. A. Burkhanov, and A. Finazzi Agro, The fine structure of luminescence spectra of azurin, *Biochim. Biophys. Acta 491*, 155–159 (1977).
28. K. Ugurbil, A. H. Maki, and R. Bersohn, study of the triplet state properties of tyrosines and

tryptophan in azurins using optically detected magnetic resonance, *Biochemistry 16*, 901–907 (1977).
29. C. F. Beyer, W. A. Gibbons, L. C. Craig, and J. W. Longworth, Heterogeneous tryptophan environments in the cyclic peptides tyrocidines B and C, *J. Biol. Chem. 249*, 3204–3211 (1974).
30. K. L. Bell and H. C. Brenner, Phosphorescence and optically detected magnetic resonance study of the tryptophan residue in human serum albumin, *Biochemistry 21*, 799–804 (1982).
31. W. C. Galley, Heterogeneity in protein emission spectra, in: *Concepts of Biochemical Fluorescence* Vol. 2 (R. F. Chen and H. Edelhoch, eds.), pp. 409–439, Marcel Dekker, New York (1976).32. S.-Y. Mao and A. H. Maki, Comparative phosphorescence and optically detected magnetic resonance studies of fatty acid binding to serum albumin, *Biochemistry 26*, 3576–3582 (1987).
33. M. V. Hershberger, A. H. Maki, and W. C. Galley, Phosphorescence and optically detected magnetic resonance studies of a class of anomalous tryptophan residues in globular proteins, *Biochemistry 19*, 2204–2209 (1980).
34. R. M. Purkey and W. C. Galley, Phosphorescence studies of environmental heterogeneity for tryptophyl residues in proteins, *Biochemistry 9*, 3569–3574 (1970).
35. D. B. Calhoun, J. M. Vanderkooi, G. V. Woodrow III, and S. W. Englander, Penetration of dioxygen into proteins studied by quenching of phosphorescence and fluorescence, *Biochemistry 22*, 1526–1532 (1983).
36. C. A. Parker, *Photoluminescence of Solutions*, pp. 97–127, Elsevier, Amsterdam (1968).
37. M. E. McCarville and S. P. McGlynn, Delayed luminescence of organic mixed crystals—XI. Amino acids and proteins, *Photochem. Photobiol. 10*, 171–181 (1969).
38. D. V. Bent and E. Hayon, Excited state chemistry of aromatic amino acids and related peptides. III. Tryptophan, *J. Am. Chem. Soc. 97*, 2612–2619 (1975).
39. C. Pepmiller, E. Bedwell, R. R. Kuntz, and C. A. Ghiron, A flash photolysis study of 1-methylindole, *Photochem. Photobiol. 38*, 273–280 (1983).
40. G. B. Strambini and M. Gonnelli, The indole nucleus triplet-state lifetime and its dependence on solvent microviscosity, *Chem. Phys. Lett. 115*, 196–200 (1985).
41. I. Munro, I. Pecht, and L. Stryer, Subnanosecond motions of tryptophan residues in proteins, *Proc. Natl. Acad. Sci. U.S.A. 76*, 56–60 (1979).
42. J. R. Lakowicz, B. P. Maliwal, H. Cherek, and A. Balter, Rotational freedom of tryptophan residues in proteins and peptides, *Biochemistry 22*, 1741–1752 (1983).
43. S. Georghiou, M. Thompson, and A. H. Mukhopadhyay, Melittin–phospholipid interaction. Evidence for melittin aggregation, *Biochim. Biophys. Acta 642*, 429–432 (1981).
44. J. A. B. Ross, C. J. Schmidt, and L. Brand, Time-resolved fluorescence of the two tryptophans in horse liver alcohol dehydrogenase, *Biochemistry 20*, 4369–4377 (1981).
45. J. W. Petrich, J. W. Longworth, and G. R. Fleming, Internal motion and electron transfer in proteins: A picosecond fluorescence study of three homologous azurins, *Biochemistry 26*, 2711–2722 (1987).
46. M. Nakanishi, M. Kobayashi, M. Tsuboi, C. Takasaki, and N. Tamiya, Electronic spectroscopy and deuteration kinetics of tyrosine and tryptophan residues: An application to the study of erabutoxin b. *Biochemistry 19*, 3204–3208 (1980).
47. I. H. Leaver, On the room temperature phosphorescence of wool keratin, *Photochem. Photobiol. 27*, 439–443 (1978).
48. G. B. Strambini and E. Gabellieri, Intrinsic phosphorescence from proteins in the solid states, *Photochem. Photobiol. 39*, 725–729 (1984).
49. J. E. Churchich, Luminescence properties of muramidase and reoxidized muramidase, *Biochim. Biophys. Acta 92*, 194–197 (1964).
50. M. K. Eftink and C. A. Ghiron, Review of fluorescence quenching studies with proteins, *Anal. Biochem. 114*, 199–227 (1981).
51. D. B. Calhoun, J. M. Vanderkooi, G. R. Holtom, and S. W. Englander, Protein fluorescence

quenching by small molecules: Protein penetration versus solvent exposure, *Proteins 1*, 109–115 (1986).

52. O. Stern and M. Volmer, Über die abklingunszeit der fluoreszenz, *Physik. Zeitschr. 20*, 183–188 (1919).
53. D. B. Calhoun, J. M. Vanderkooi, and S. W. Englander, Penetration of small molecules into proteins studied by quenching of phosphorescence and fluorescence, *Biochemistry 22*, 1533–1539 (1983).
54. G. B. Strambini, Singular oxygen effects on the room-temperature phosphorescence of alcohol dehydrogenase from horse liver, *Biophys. J. 43*, 127–130 (1983).
55. G. B. Strambini, Quenching of alkaline phosphatase phosphorescence by O_2 and NO, *Biophys. J. 52*, 23–28 (1987).
56. N. Barboy and J. Feitelson, Quenching of tryptophan phosphorescence in alcohol dehydrogenase from horse liver and its temperature dependence, *Photochem. Photobiol. 41*, 9–13 (1985).
57. J. M. Vanderkooi, G. Maniara, T. J. Green, and D. F. Wilson, An optical method for measurement of dioxygen concentration based upon quenching of phosphorescence, *J. Biol. Chem. 262*, 5476–5482 (1987).
58. D. B. Calhoun, W. W. Wright, S. W. Englander, and J. M. Vanderkooi, The quenching of room temperature protein phosphorescence by added small molecules, *Biochemistry 27*, 8466–8474 (1988).
59. J. R. Lakowicz and G. Weber, Quenching of protein fluorescence by oxygen. Detection of structural fluctuations on proteins on the nanosecond time scale *Biochemistry 12*, 4171–4179 (1973).
60. O. L. J. Gijzeman, F. Kaufman, and G. Porter, Oxygen quenching of aromatic triplet states in solution, *J. Chem. Soc., Faraday Trans. 2, 69*, 708–720 (1973).
61. J. Saltiel and B. W. Atwater, Spin-statistical factors in diffusion-controlled reactions, *Adv. Photochem. 14*, 1–90 (1988).
62. F. R. N. Gurd and M. Rothgeb, Motions in proteins, *Adv. Protein Chem. 33*, 73–165 (1979).
63. S. W. Englander and N. R. Kallenbach, Hydrogen exchange and structural dynamics of proteins and nucleic acids, *Q. Rev. Biophys. 16*, 521–655 (1984).
64. M. Karplus and J. A. McCammon, The internal dynamics of globular proteins, *CRC Crit. Rev. Biochem. 9*, 293–349 (1981).
65. F. M. Richards, Packing defects, cavities, volume fluctuations and access to the interior of proteins, *Carlsberg Res. Commun. 44*, 47–63 (1979).
66. B. Somogyi, J. A. Norman, and A. Rosenberg, Gated quenching of intrinsic fluorescence and phosphorescence of globular proteins, *Biophys. J. 50*, 55–61 (1986).
67. E. Gratton, D. M. Jameson, and G. Weber, Model of dynamic quenching of fluorescence in globular proteins, *Biophys. J. 45*, 789–794 (1984).
68. N. J. Turro, *Modern Molecular Photochemistry*, pp. 296–361, Benjamin/Cummings, Menlo Park, California (1978).
69. J. R. Miller, J. A. Peeples, M. J. Schmitt, and G. L. Closs, Long-distance fluorescence quenching by electron transfer in rigid solutions, *J. Am. Chem. Soc. 104*, 6488–6493 (1982).
70. H. E. Zemel and B. M. Hoffman, Long-range triplet–triplet energy transfer within metal-substituted hemoglobins, *J. Am. Chem. Soc. 103*, 1192–1201 (1981).
71. H. Koloczek, T. Horie, T. Yonetani, H. Anni, G. Maniara, and J. M. Vanderkooi, Interaction between cytochrome *c* and cytochrome *c* peroxidase: Excited-state reactions of zinc- and tin-substituted derivatives, *Biochemistry 26*, 3142–3148 (1987).

71a. J. M. Vanderkooi, S. W. Englander, S. Papp, W. W. Wright, and C. S. Owen, Long-range electron exchange measured in proteins by quenching of tryptophan phosphorescence, *Proc. Natl. Acad. Sci. USA*, 5099–5103 (1990).

72. L. Augenstein and J. Nag-Caudhur, Energy transfer in Proteins, *Nature 203*, 1145–1146 (1964).

73. E. Kuntz, F. Bishai, and L. Augenstein, Quantitative emission spectroscopy in media where appreciable light scattering occurs, *Nature 212*, 980–983 (1966).
74. J. Domanus, G. B. Strambini, and W. C. Galley, Heterogeneity in the thermally-induced quenching of the phosphorescence of multi-tryptophan proteins, *Photochem. Photobiol. 34*, 15–21 (1980).
75. G. B. Strambini, P. Cioni, and R. A. Felicioli, Characterization of tryptophan environments in glutamate dehydrogenases from temperature-dependent phosphorescence, *Biochemistry 26*, 4968–4975 (1987).
76. Y. Kai and K. Imakubo, Temperature dependence of the phosphorescence lifetimes of heterogeneous tryptophan residues in globular proteins between 293 and 77 K, *Photochem. Photobiol. 29*, 261–265 (1979).
77. E. Bismuto, G. B. Strambini, and G. Irace, Temperature dependence of phosphorescence parameters of phylogenetically distant apomyoglobins, *Photochem. Photobiol. 45*, 741–744 (1987).
78. J. M. Vanderkooi, S. Papp, T. Samoriski, S. Pikula, and A. Martonosi, Tryptophan phosphorescence of the Ca^{2+}-ATPase of sarcoplamic reticulum, *Biochim. Biophys. Acta 957*, 230–236 (1988).
79. G. B. Strambini and M. Gonnelli, Effects of urea and guanidine hydrochloride on the activity and dynamical structure of equine liver alcohol dehydrogenase, *Biochemistry 25*, 2471–2476 (1986).
80. M. Gonnelli and G. B. Strambini, The rate of equine liver alcohol dehydrogenase denaturation by urea: Dependence on temperature and denaturant concentration, *Biophys. Chem. 24*, 161–167 (1986).
81. T. M. Jovin, M. Bartholdi, W. L. C. Vaz, and R. H. Austin, Rotational diffusion of biological macromolecules by time-resolved delayed luminescence (phosphorescence, fluorescence) anisotropy, *Ann. N. Y. Acad. Sci. 366*, 176–196 (1981).
82. R. J. Cherry, Measurement of protein rotational diffusion in membranes by flash photolysis, *Methods Enzymol. LIX*, 47–61 (1978).
83. K. Kinosita, Jr., S. Kawato, and A. Ikegami, Dynamic structure of biological and model membranes: Analysis by optical anisotropy decay measurement, *Adv. Biophys. 17*, 147–203 (1984).
84. G. B. Strambini and W. C. Galley, Detection of slow rotational motions of proteins by steady-state phosphorescence anisotropy, *Nature 260*, 554–555 (1976).
85. G. B. Strambini and W. C. Galley, Time-dependent phosphorescence anisotropy measurements of the slow rotational motions of proteins in viscous solution, *Biopolymers 19*, 383–394 (1980).
86. H. Kim and W. C. Galley, Rotational mobility associated with the protein moiety of human serum lipoproteins from tryptophan phosphorescence anisotropy measurements, *Can. J. Biochem. Cell Biol. 61*, 46–53 (1983).
87. G. B. Strambini and E. Gabellieri, Phosphorescence anisotropy of liver alcohol dehydrogenase in the crystalline state. Apparent glass-like rigidity of the coenzyme-binding domain, *Biochemistry 26*, 6527–6530 (1987).
88. J. W. Berger and J. M. Vanderkooi, Intrinsic phosphorescence anisotropy measurements of the tobacco mosaic virus, work in progress.
89. T. Horie and J. M. Vanderkooi, Phosphorescence of alkaline phosphatase of *E. coli in vitro* and *in situ*, *Biochim. Biophys. Acta 670*, 290–297 (1981).
90. V. M. Mazhul, Y. S. Ermolaev, and C. V. Konev, Tryptophan phosphorescence at room temperature: New method for the study of the structural composition of biological membranes and proteins in cells, *Zh. Prikl. Spectrosk. 32*, 903–907 (1980).
91. T. Horie and J. M. Vanderkooi, Phosphorescence of tryptophan from parvalbumin and actin in liquid solutions, *FEBS Lett. 147*, 69–73 (1982).

4

Fluorescence Studies of Nucleic Acids: Dynamics, Rigidities, and Structures

J. Michael Schurr, Bryant S. Fujimoto, Pengguang Wu, and Lu Song

4.1. Introduction

The broad field of nucleic acid structure and dynamics has undergone remarkable development during the past decade. Especially in regard to dynamics, modern fluorescence methods have yielded some of the most important advances. This chapter concerns primarily the application of time-resolved fluorescence techniques to study the dynamics of nucleic acid/dye complexes, and the inferences regarding rotational mobilities, deformation potentials, and alternate structures of nucleic acids that follow from such experiments. Emphasis is mainly on the use of time-resolved fluorescence polarization anisotropy (FPA), although results obtained using other techniques are also noted. This chapter is devoted mainly to free DNAs and tRNAs, but DNAs in nucleosomes, chromatin, viruses, and sperm are also briefly discussed.

The reader is referred to other reviews for detailed discussions of the electronic states and luminescence of nucleic acids and their constituents,[1] fluorescence correlation spectroscopy,[2] spectroscopy of dye/DNA complexes,[3] and ethidium fluorescence assays.[4, 5] A brief review of early work on DNA dynamics[6] as well as a review of tRNA kinetics and dynamics[7] have also appeared. The diverse and voluminous literature on the use of fluorescence techniques to assay the binding of proteins and antitumor drugs to nucleic acids and on the use of fluorescent DNA/dye complexes in cytometry and cytochemistry lies entirely outside the scope of this chapter.

J. Michael Schurr, Bryant S. Fujimoto, Pengguang Wu, and Lu Song • Department of Chemistry, University of Washington, Seattle, Washington 98195.

Topics in Fluorescence Spectroscopy, Volume 3: Biochemical Applications, edited by Joseph R. Lakowicz. Plenum Press, New York, 1992.

4.2. Rotational Dynamics of DNA

4.2.1. Background

The rotational Brownian motions of a high-molecular-weight DNA span an enormous range of time scales from subnanosecond wobble of the bases, or intercalated dyes, to the slowest Rouse–Zimm coil-deformation mode. The Langevin relaxation time[8] of the latter varies with molecular weight (M_r) as $M_r^{1.6}$,[9] and it approaches 1 s for $M_r = 10^8$. Figure 4.1 exhibits the pertinent time scales of the presently assigned rotational relaxation processes and the techniques that have been employed for measurements in different time ranges.

The fluorescence quantum yield of native DNA ($\phi_f = 4 \times 10^{-5}$)[10, 11] is much too small and its fluorescence lifetimes ($\tau_1 \approx 10$ ps, $\tau_2 \approx 65$ ps)[12] are far too short to be useful for studying its rotational Brownian dynamics, so one must employ an extrinsic probe. Most commonly used is ethidium dye. Upon

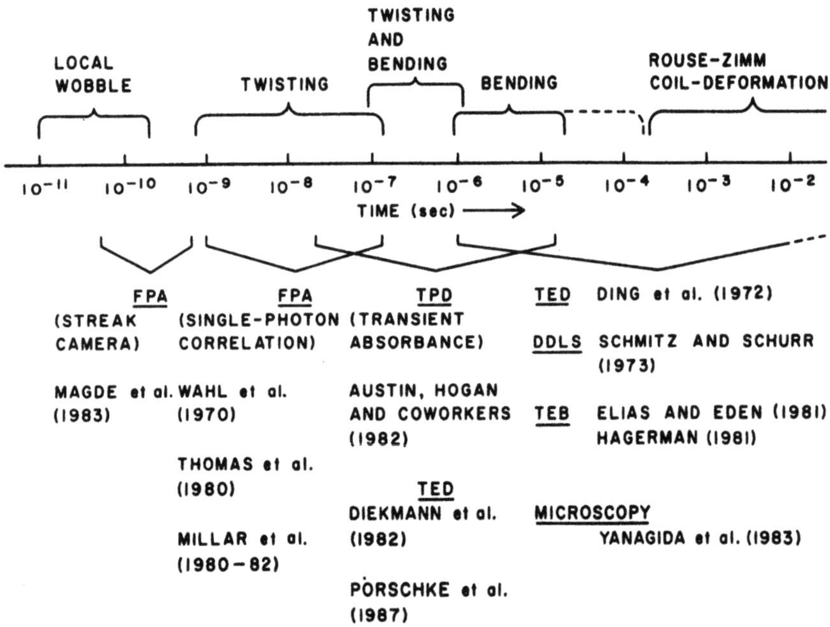

Figure 4.1. Time scales for rotational motions of long DNAs that contribute to the relaxation of the optical anisotropy $r(t)$. Experimental methods used to study these motions in different time ranges are also indicated along with the authors and dates of some early work in each case. FPA, Fluorescence polarization anisotropy (Refs. 15, 18–20, and 87); TPD, transient photodichroism (Refs. 28 and 62); TEB, transient electric birefringence (Refs. 26 and 27); DDLS, depolarized dynamic light scattering (Ref. 116); TED, transient electric dichroism (Refs. 25, 115, and 130); Microscopy, time-resolved fluorescent microscopy (Ref. 176).

intercalation into DNA, its fluorescence lifetime increases from 1.7–1.8 ns to 21.8 ± 1 ns, with a corresponding increase in quantum yield. The mean residence time of ethidium in an intercalation site exceeds 0.01 s, when the NaCl concentration is less than or equal to 1.0 *M*.[13,14] Hence, the dye remains bound to the DNA for vastly longer times than its fluorescence lifetime. In these and certain other regards, ethidium is an almost ideal extrinsic probe for studying the rotational dynamics of DNA at times shorter than 150 ns.[15]

The rotational relaxation of DNA from 1 to 150 ns is due mainly to Brownian torsional (twisting) deformations of the elastic filament. Partial relaxation of the FPA on a 30-ns time scale was observed and qualitatively attributed to torsional deformations already in 1970.[15] However, our *quantitative* understanding of DNA motions in the 0- to 150-ns time range has come from more accurate time-resolved measurements of the FPA in conjunction with new theory and has developed entirely since 1979. In that year, the first theoretical treatments of FPA relaxation by spontaneous torsional deformations appeared,[16,17] and the first commercial synch-pump dye laser systems were delivered. Experimental confirmation of the predicted FPA decay function and determination of the torsional rigidity of DNA were first reported in 1980.[18] Other labs[19–21] subsequently reported similar results, although their anisotropy formulas were not entirely correct, and they did not so rigorously test the predicted decay function or attempt to fit likely alternatives. The development of new instrumentation, new data analysis techniques, and new theory and their application to different DNAs in various circumstances have continued to advance this field up to the present time.

Potential uncertainties in regard to sample quality have been minimized in our laboratory by the preparation and study of clean, monodisperse (in composition and size) samples of native linear and supercoiled DNAs and restriction fragments, as well as short synthetic DNAs of specified sequence and fractionated narrow-size distributions of alternating synthetic polynucleotides. Sample characterization by other techniques, including fluorescamine tests for contaminating proteins and polyamines, gel electrophoresis, dynamic light scattering (DLS),[8,22] and sedimentation, has often been crucial for a proper interpretation of the results. Especially, DLS at large scattering vector has corroborated nearly all of the important and/or unexpected changes in FPA dynamics. As DLS is independent of the extrinsic probe and reflects motions of all parts of the molecule, not just at the site of the probe, it provides an extremely valuable independent measurement, which is also quite sensitive to the torsional rigidity.[18–24]

The (bending) persistence length of DNA, which is proportional to its bending rigidity, has been extensively studied for well over two decades, usually via its effect on the overall dimensions of the DNA coil. This work has been reviewed elsewhere.[22] Direct observation of dynamic bending over short distances has come from comparatively recent studies using transient electric dichroism[25] and birefringence,[26,27] which are insensitive to torsional motions. Although FPA does not provide a long enough time window to yield much useful information about bending dynamics, a transient photodichroism (TPD) technique provides the same kind of dynamical information out to times as long as 20 μs.[28] The theory presented herein for FPA is directly transferable to TPD,[29] as well as to phosphorescence anisotropy.

4.2.2. Pertinent Questions and Problems

Before delving into theoretical and experimental details, it is useful to consider some of the motivations for research in this area. The kinds of problems and questions addressed by such experiments can be classified into several categories from physical to biological, as follows.

4.2.2.1. Brownian Dynamics

At present, the Brownian motions of isolated *rigid* macromolecules are quite well understood. The challenge now is to understand the Brownian *deformations* of *nonrigid* macromolecules and to ascertain the time scales on which the *coupled motions* of their subunits relax various experimental signals.

A question is paramount importance is whether simple coupled Langevin equations and generalized diffusion equations for such motions are valid at nanosecond times, and in the presence of strong direct forces. Brownian motions of the DNA subunits, which are coupled by elastic twisting and bending forces, are a particular example of diffusion of strongly interacting species. Both translational and rotational motions of strongly interacting diffusers in many different situations are presently under intensive investigation in numerous laboratories.[22,30] The results already obtained have deepened our understanding considerably and inspire some confidence in the general validity of simple Langevin and diffusion theories for interacting species.[22,30]

A particular question of interest is whether the DNA torsional motions observed on the nanosecond time scale are overdamped, as predicted by simple Langevin theory, and as observed for Brownian motions on longer time scales, or instead are underdamped, so that damped oscillations appear in the observed correlation functions. A related question is whether the solvent water around the DNA exhibits a normal constant viscosity on the nanosecond time scale, or instead begins to exhibit viscoelastic behavior with a time-, or frequency-, dependent complex viscosity. In brief, are the predictions for a simple elastically deformable filament in a solvent with normal constant viscosity obeyed by DNA on the nanosecond time scale?

4.2.2.2. *Longitudinal Diffusion of Overdamped Solitons*

It is conceivable that diffusion of kinks, or overdamped solitons, along the DNA could act to relax the FPA with a time dependence similar to that predicted for torsional deformation.[31, 32] High levels of intercalated dyes would be expected to alter both the equilibrium population of kinks and their mobility along the DNA. Hence, this question is addressed by examining the effect of intercalating dyes on the torsional dynamics.

4.2.2.3. *Mechanism of Spontaneous Deformation*

The possible role of spontaneous transient opening, or disruption, of the local double-helical structure in determining the long-range torsional and flexural rigidities, and Brownian dynamics, of the DNA filament is an intriguing and important question.[23] At one extreme, the DNA can be imagined to bend and twist by progressive deformation of its native structure with virtually no contribution from spontaneous opening, and at the other to remain quite stiff, largely unbent and untwisted, except for occasional spontaneously denatured, or disrupted, regions, where the bulk of the twisting and bending takes place. Both smooth and segmental models for bending deformation are illustrated in Figure 4.2. The segmental model merits consideration, because it has been proposed for DNA bending.[33] Kinked structures of different types,[34, 35] which would likely be sites of major rigidity weaknesses,[35] have also been proposed. The existence of a definite, localized fluctuated state through which hydrogen exchange occurs has recently been demonstrated.[36, 37] Although the probability (K) of that fluctuated state in duplex DNAs with 10 and 12 base pairs (bp) ($K \gtrsim 10^{-5}$) is most likely too

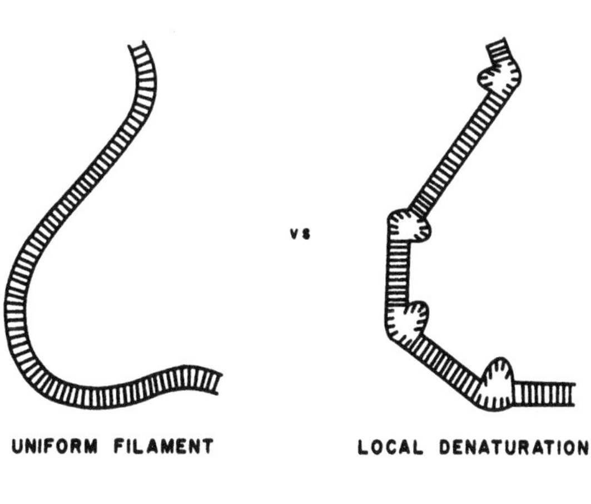

Figure 4.2. Illustration of two extreme models for the bending (and/or twisting) of DNA. Uniform deformation of the native structure on the left is contrasted with bending (and/or twisting) exclusively at sites of local denaturation in an otherwise stiff filament on the right. [From *Biopolymers 22*, 2273–2321 (1983), Ref. 23, with permission.]

low to affect the Brownian dynamics,[36, 37] the probability of that state was found to be substantially larger ($K \gtrsim 10^{-4}$, estimated for 25 °C) in a 14-bp duplex, and larger still ($K \gtrsim 10^{-3}$, estimated for 25 °C) in a 16-bp duplex linked at the ends by TTTT loops.[38] Recent evidence also indicates that this fluctuated state is most probably not the unstacked open state of optical melting theory.[38] Nevertheless, a localized fluctuated state with a probability $K \gtrsim 10^{-3}$ at 25 °C and a standard enthalpy change in the range 20–25 kcal/mol[38] is a potential site of a major rigidity weakness that could facilitate segmental deformation. Whether torsional deformation is smooth or segmental is, therefore, a nontrivial question that can in principle be answered from the time course of the FPA relaxation.[17, 18, 39] The temperature dependences of the bending[23] and twisting[40] rigidities also bear critically on this same issue.

4.2.2.4. *Magnitude of the Torsional Rigidity and Anisotropy of Deformation*

The layerlike structure of the base pairs in DNA suggests the possibility that the restoring torque for twisting might be much smaller than the corresponding restoring torque for bending or, equivalently, that the long-range torsional rigidity might be much smaller than the long-range bending rigidity.[23] For a macroscopic rod composed of typical bulk polymeric material (Poisson ratio $\sigma = 0.5$), the torsional rigidity would be two-thirds of the bending rigidity, but for a highly anisotropic material, the torsional rigidity could be very much smaller. For DNA, this question is addressed by comparing the magnitude of the torsional rigidity obtained from the FPA with the bending rigidity obtained from persistence length measurements.[23]

DNA is permanently twisted, as well as bent, in many, if not all, of its natural states, including supercoiled plasmids, chromatin, and condensed DNAs in sperm and phage heads. It also appears to be significantly deformed in *specific* complexes with various proteins, including RNA polymerase,[41] catabolite activator protein (CAP),[42–47] *lac* repressor,[48] and 434 protein.[49] The magnitude of the torsional rigidity is an essential ingredient for estimating the energetics of many or all of these deformed natural states and DNA/protein complexes. In fact, models have been proposed in which a sequence-dependent bending rigidity or torsional rigidity of the DNA is the critical factor in determining the relative affinity of the 434 protein for different DNA sequences.[50, 218]

4.2.2.5. *Water in the DNA Grooves*

Ordering of water in the DNA grooves is a topical subject.[51, 52] Whether water in the DNA grooves behaves like normal water or instead rotates, as if rigidly attached to DNA for times as long as 120 ns, can be directly ascertained by measuring the hydrodynamic radius for azimuthal rotation of the DNA around its symmetry axis.[53] This question can also be indirectly addressed by a phosphorescence spectroscopic technique.[54]

4.2.2.6. The Twisting Potential

The effective potential governing torsional deformations could conceivably be quite anharmonic, so that overwinding is much more strongly resisted than underwinding for finite deformations. This question is addressed by examining the dependence of the torsion constant on temperature[40] and on superhelix density.

4.2.2.7. Secondary Structure

The twisting and bending rigidities of a given DNA depend sensitively on its secondary structure. These rigidities therefore provide a novel probe for differences, or changes, in secondary structure. Given a change in the *long-range* torsional rigidity, there arises the question of whether the change in secondary structure is local, introducing segmental character into the motion, or global. Both types of change in secondary structure have been observed, as will be detailed subsequently. Of special interest are the effects of polyamines, pH, Mg^{2+} ions, NaCl concentration, temperature, bound proteins, and intercalating ligands on the magnitude and uniformity of the torsional rigidity of DNA. A related question concerns the effect of base composition on torsional rigidity. The popular belief that GC-rich sequences should be stiffer against torsion is explicitly tested by FPA.

The rather small amount of DNA (20 μg) required for measurement of the torsional rigidity by FPA places this technique, along with circular dichroism (CD), as one of the more sensitive indicators of changes in secondary structure. In contrast to CD, however, the FPA is largely independent of tertiary structure. In this sense, it provides perhaps the most unambiguous indication of changes in secondary structure of supercoiled DNAs. Partly for this reason, FPA measurements have detected changes in the secondary structure of supercoiled DNAs which have previously gone unnoticed.[55, 56] In view of the possible role of (allosteric) secondary structure transitions in the regulation of gene activity, their characterization is very likely a matter of fundamental biological importance.

The torsional mobility of DNA in viruses,[57] in sperm,[58] in chromosomes,[21, 59–61] and in core particles[21, 60, 62, 63] or condensed by polyamines[21, 57, 64] and the accessibility of DNA to intercalating dyes are significant problems that are also addressed using FPA and TPD techniques.

4.2.2.8. *Effect of Intercalating Dyes*

Whenever an extrinsic probe is used, one must be concerned with artifacts arising from perturbations of the macromolecular rigidity and dynamics by that probe. It has been proposed that ethidium binds intercalatively at the site of a so-called β-kink, which was further suggested to exhibit a tenfold smaller torsional rigidity than that of normal DNA.[35] If so, then ethidium would be reporting the dynamics at major torsional rigidity weaknesses, rather than the dynamics of normal DNA, and would yield anomalously low torsional rigidities. This raises the general question of how the torsion potential between an intercalated dye and a base pair differs from that between two base pairs. This question is addressed by FPA studies of the torsional dynamics as a function of bound intercalator up to very high binding ratios (dye/bp).[53] To avoid depolarization by excitation transfer, it is necessary to employ intercalators that do not engage in excitation transfer with the ethidium probe.

One would still like to examine the effect of ethidium on the torsional rigidity and dynamics at high binding ratios. One would also like to test the Förster theory for excitation transfer between bound ethidium molecules, since it has been questioned.[65] This is possible in principle by deconvoluting the effects of depolarization by excitation transfer on the FPA, as will be shown subsequently. DLS also provides crucial information on this same question.

4.2.2.9. *Free Energy of Supercoiling*

There exists a serious (twofold) discrepancy between the free energies of supercoiling estimated by the ligation[66–69] and dye-binding[70, 71] methods,[53] as will be described in greater detail. One possible explanation is that intercalated dyes themselves might substantially alter the twisting and bending rigidities, thereby violating one of the underlying assumptions of the dye-binding method, namely, that the rigidities *per se* are unaffected by intercalator binding. It is also conceivable that excess Mg^{2+} in the ligation medium could directly affect the structure and rigidity of the DNA. FPA measurements of the torsional rigidity of linear and supercoiled DNAs provide the means to investigate such possibilities, and perhaps ultimately pinpoint the origins of this discrepancy. The equilibrium binding constants for linear DNA/chloroquine complexes and the supercoiling free energy for circular DNA/chloroquine complexes can also be obtained by resolving the contributions of free and intercalated ethidium to the fluorescence decay, as will be described.

4.2.2.10. *Allosteric Transitions of Secondary Structure Induced by Superhelical Stress*

As an initially relaxed DNA is progressively strained by increasing superhelical stress, it is conceivable that the secondary structure is not just simply strained, but instead becomes *non-simply* strained in the sense that it undergoes an allosteric transition to one or more alternate secondary structures, which may exhibit different twisting and bending rigidities.[55, 72] The existence of two or more nearly iso(free)-energetic secondary structures would enable DNA to function as a kind of long-range switch, thereby facilitating communication between distant specifically bound proteins that are involved in genetic information processing and providing an important control element for regulation of gene activity. FPA provides a sensitive probe for any allosteric transitions that may be induced by changes in superhelical stress (or other means).[55, 56, 72] Evidence for such transitions from FPA and other techniques is described subsequently.

4.2.3. Theory

Throughout this chapter it is assumed that the intensity of the polarized exciting pulse is sufficiently low that only a small fraction of the fluorophores are ever excited.[29] High light intensities are treated elsewhere.[73, 74] The following subsection presents some very general and basic theory that is not specifically directed toward rotational relaxation of DNA. The reader may wish to skip directly to the final result in Eq. (4.15), or even skip this subsection entirely.

4.2.3.1. *Reorientation of Coordinate Frames and Vectors by Arbitrary Reorientation Mechanisms*

We present here some very general exact results, which hold for arbitrary reorientation mechanisms of any molecule in an equilibrium *isotropic* fluid (but not a liquid crystal). A coordinate frame (R) is rigidly attached to the molecule of interest. Its orientation in the laboratory frame (L) is defined by the Euler rotation $\Phi = (\alpha\beta\gamma)$ that carries a coordinate frame from coincidence with the laboratory frame L to coincidence with the molecular frame R.[29] The conditional probability per unit Euler "volume" $[G(\Phi t \mid \Phi_0 0)]$ that a molecule with orientation Φ_0 at time $t=0$ will move to Φ at time t must depend *only* on the Euler rotation $\Delta = \Phi\Phi_0^{-1}$ (i.e., rotate *first* by Φ_0^{-1}, then by Φ) that carries R from its position at $t=0$ to its position at t, but not on the absolute orientation of R in the laboratory frame. Thus, $G(\Phi t|\Phi_0 0)$ can be expanded in the complete set of Wigner rotation functions $D^l_{mn}(\Delta)$. For a given Euler rotation, Φ, these rotation functions are defined by[75, 76]

$$D^l_{mn}(\Phi) = e^{in\alpha} d^l_{mn}(\beta) e^{im\gamma} \tag{4.1}$$

where the $d^l_{mn}(\beta)$ are given by Wigner[75] and Edmonds.[76] The rotation functions also obey the orthogonality relation[75, 76]

$$\int d\Phi\, D^{L*}_{NM}(\Phi)\, D^l_{mn}(\Phi) = [(8\pi^2)/(2l+1)]\, \delta_{L,l}\, \delta_{M,m}\, \delta_{N,n} \tag{4.2}$$

and exhibit the matrix multiplicative property for successive Euler rotations $(\Phi_2 \Phi_1)$,[75, 76]

$$D^l_{mn}(\Phi_2 \Phi_1) = \sum_{p=-l}^{l} D^l_{mp}(\Phi_2)\, D^l_{pn}(\Phi_1) \tag{4.3}$$

Moreover, $D^l_{pn}(\Phi_0^{-1}) = D^{l*}_{np}(\Phi_0)$. From these considerations, one obtains

$$G(\Phi t|\Phi_0 0) = \sum_{l=0}^{\infty} \sum_{m} \sum_{n\,(-l)}^{(l)} \sum_{p} b^l_{mn}(t)\, D^l_{mp}(\Phi)\, D^{l*}_{np}(\Phi_0) \tag{4.4}$$

where the $b^l_{mn}(t)$ are coefficients that in general depend on the details of the motion. Using Eqs. (4.2) and (4.4) and the reality of $G(\Phi t|\Phi_0 0) = G(\Phi t|\Phi_0 0)^*$, it follows that[74]

$$\begin{aligned}\langle D^L_{MN}(\Phi)\, D^{l*}_{mn}(\Phi_0)\rangle &\equiv (1/8\pi^2) \int d\Phi \int d\Phi_0\, G(\Phi t|\Phi_0 0)^*\, D^L_{MN}(\Phi)\, D^{l*}_{mn}(\Phi_0) \\ &= \delta_{L,l}\, \delta_{N,n} \langle D^l_{M0}(\Phi)\, D^{l*}_{m0}(\Phi_0)\rangle \end{aligned} \tag{4.5}$$

which is independent of the value of the index $N=n$. This relation is important in the subsequent discussion.

Consider the vector that lies along the z-axis of the molecular frame. It has orientation $\Omega_L = (\alpha\beta)$ in the laboratory frame. The conditional probability per unit solid angle, $g(\Omega t|\Omega_0 0)$, that this vector with orientation Ω_0 at time $t=0$ will move to Ω at time t is obtained from $G(\Phi t|\Phi_0 0)$ by integrating over the angles γ and γ_0 of Φ and Φ_0, respectively. Thus,

$$\begin{aligned} g(\Omega t|\Omega_0 0) &= \int d\gamma \int d\gamma_0\, G(\Phi t|\Phi_0 0) \\ &= \sum_{l=0}^{\infty} c_l(t) \sum_{p=-l}^{l} Y_{lp}(\Omega)\, Y^*_{lp}(\Omega_0) \end{aligned} \tag{4.6}$$

wherein $Y_{lp}(\alpha\beta) = [(2l+1)/4\pi]^{1/2} D^l_{0p}(\alpha\beta\gamma)$ is a spherical harmonic,[75, 76] and $c_l(t) = b^l_{00}(t)\, 2\pi(2l+1)/2$ is independent of p. It follows from either Eq. (4.5) or (4.6) that

$$\langle Y_{LN}(\Omega)\, Y^*_{ln}(\Omega_0)\rangle = \delta_{L,l}\, \delta_{N,n} \langle Y_{l0}(\Omega)\, Y^*_{l0}(\Omega_0)\rangle \tag{4.7}$$

Ω is just the instantaneous orientation in the lab frame of *any* vector fixed in the molecule. Thus, under equilibrium averaging (for an isotropic fluid), the correlation functions for different spherical harmonic functions of the same vector are orthogonal, and independent of the index $n = N$![17, 29, 74, 77–79] This conclusion, which is valid for arbitrary reorientation mechanisms, seems not to be widely known.

Let $\omega_L(t)$ denote the orientation of the *absorption* transition dipole in the lab frame, and $\omega_R(t) = [\mu(t), \nu(t)]$ its orientation in the molecular frame. Likewise, let $\Omega_L(t)$ denote the orientation of the *emission* dipole in the lab frame, and $\Omega_R(t) = [\varepsilon(t), \zeta(t)]$ its orientation in the molecular frame. Using the transformation property of the Wigner rotation functions,

$$Y_{ln}[\omega_L(t)] = \sum_{p=-l}^{l} D^l_{pn}(\Phi)\, Y_{lp}[\omega_R(t)] \tag{4.8}$$

and the corresponding relation for $Y_{ln}[\Omega_L(t)]$ together with Eq. (4.5), it is found that[74]

$$\langle Y_{ln}[\omega_L(0)]\, Y^*_{LN}[\Omega_L(t)]\rangle = \delta_{l,L}\, \delta_{n,N} \langle Y_{l0}[\omega_L(0)]\, Y^*_{l0}[\Omega_L(t)]\rangle \tag{4.9}$$

Evidently, correlation functions for different spherical harmonic functions of two different vectors in the *same molecule* are also orthogonal under equilibrium averaging for an isotropic fluid. Thus, if the excitation process "photoselects" particular *lm* components of the (solid) angular distribution of absorption dipoles, then only those same *lm* components of the (solid) angular distribution of emission dipoles will contribute to observed signal, regardless of the other *lm* components that may in principle be detected, and vice versa. The result in this case is likewise independent of the index $n = N$. Equation (4.7) is just the special case of Eq. (4.9) when the two dipoles coincide.

4.2.3.2. Fluorescence Polarization Anisotropy

The sample is illuminated at $t = 0$ by an infinitely short pulse delivering I photons/cm^2 polarized along the lab z-axis. The subsequent rate of emission of (lab) z-polarized photons is[73]

$$I_{\parallel}(t) = N\langle P[\omega_L(0)]\, E_{\parallel}[\Omega_L(t)]\rangle \tag{4.10}$$

where

$$P[\omega_L(0)] = \left(\frac{aI}{3}\right)\left\{1 + 2\left(\frac{4\pi}{5}\right)^{1/2} Y_{20}^*[\omega_L(0)]\right\} \tag{4.11}$$

is the probability that a fluorophore with orientation $\omega_L(0)$ of its absorption dipole absorbs a photon, and

$$E_{\parallel}[\Omega_L(t)] = \frac{1}{3} E(t) \left\{1 + 2\left(\frac{4\pi}{5}\right)^{1/2} Y_{20}[\Omega_L(t)]\right\} \tag{4.12}$$

is the subsequent rate of emission by its emission dipole. N is the number of molecules in the sample, a is the cross section for absorption of a photon polarized along the absorption dipole, and $E(t)$ is the average rate of emission by any given fluorophore. Likewise, the rate of emission of (lab) x-polarized photons is

$$I_{\perp}(t) = \langle P[\omega_L(0)]\, E_{\perp}[\Omega_L(t)]\rangle \tag{4.13}$$

where

$$E_{\perp}[\Omega_L(t)] = \frac{1}{3} E(t) \left(1 - \left(\frac{4\pi}{5}\right)^{1/2} Y_{20}[\Omega_L(t)] + 3\left(\frac{2\pi}{15}\right)^{1/2} \{Y_{22}[\Omega_L(t)] + Y_{2,-2}[\Omega_L(t)]\}\right) \tag{4.14}$$

The averages in Eqs. (4.10) and (4.13) are simplified using Eq. (4.9). The general result for the optical anisotropy is[74, 79, 80]

$$r(t) = \frac{I_{\parallel}(t) - I_{\perp}(t)}{I_{\parallel}(t) + 2I_{\perp}(t)} = \frac{2}{5} 4\pi \langle Y_{20}^*[\omega_L(0)]\, Y_{20}[\Omega_L(t)]\rangle = \frac{2}{5} \langle P_2[\hat{\mu}_A(0) \cdot \hat{\mu}_E(t)]\rangle \tag{4.15}$$

where $\hat{\mu}_A(t)$ and $\hat{\mu}_E(t)$ are instantaneous unit vectors along the absorption and emission dipoles, respectively. For the case of parallel emission and absorption dipoles $[\omega_L(t) = \Omega_L(t)]$, this becomes

$$r(t) = \tfrac{2}{5} 4\pi \langle Y_{20}^*[\Omega_L(0)]\, Y_{20}[\Omega_L(t)]\rangle = \tfrac{2}{5} \langle P_2(\hat{\mu}(0) \cdot \hat{\mu}(t))\rangle \tag{4.16}$$

where $\hat{\mu}(t)$ is an instantaneous vector along the transition dipole. The second lines of Eqs. (4.15) and (4.16) are obtained using the Addition Theorem for spherical harmonics. Both forms of Eqs. (4.15) and (4.16) are useful in different contexts. Equations (4.15) and (4.16) reflect the photoselection of the $Y_{20}[\omega_L(0)]$ component of the angular distribution of absorption dipoles by the exciting pulse [cf. Eq. (4.11)].

The motion of the absorption and emission dipoles in the molecular frame R is now assumed to be statistically independent of the motion of the R frame in the laboratory frame L. In a deformable molecule, the R frame may be attached to some small part of the molecule, which can be regarded as locally rigid. In this case, motion of the R frame occurs as a consequence of molecular deformation, as well as overall (uniform) rotation of the molecule. In such a case, statistical independence of the motion of a dipole in the R frame and the motion of the R frame itself is not guaranteed. However, with this assumption, Eq. (4.15) becomes

$$r(t)=\frac{2}{5}4\pi\sum_{p}\sum_{q}\langle Y^{*}_{2p}[\omega_R(0)]\,Y_{2q}[\Omega_R(t)]\rangle\cdot\langle D^{2*}_{p0}(\Phi_0)\,D^{2}_{q0}(\Phi)\rangle \qquad (4.17)$$

Further simplification is possible when the motion of the R frame is, on the average, cylindrically symmetric about its z-axis, as shown in the next section.

Equations (4.15)–(4.17) and subsequent theoretical expressions for $r(t)$ are the *true* anisotropy, which is defined here as the fluorescence response to an instantaneous light pulse when measured by an instrument with infinitely rapid temporal response. In a real experiment this is convoluted with the instrument response function, as discussed in a later section.

Most of the results presented in this section, including Eqs. (4.15)–(4.17), are not valid when the equilibrium state of the fluid exhibits global orientational order, for example, a global director. However, they do apply to an isotropic suspension of locally anisotropic objects, such as vesicles or liposomes, which may exhibit a local director, provided that long-range orientational correlations do not extend over a significant fraction of the volume sampled in the experiment.

The photoinduced *absorbance* anisotropy in a TPD experiment relaxes according to the same correlation function as in Eq. (4.16).[29] Effects of spatial variations in the excitation and probe beams, and chromophore concentration, have been treated and shown not to alter the final result.[29] NMR dipolar relaxation rates are expressed in terms of Fourier transforms of the correlation functions, $4\pi\langle Y_{2m}[\Omega(0)]\,Y_{2m}[\Omega(t)]\rangle$, where $\Omega(t)$ denotes the orientation of a particular internuclear vector. In view of Eq. (4.7), these correlation functions are independent of the index m, hence formally the same as in Eq. (4.16). For the analysis of NMR relaxation data, it is necessary also to evaluate Fourier transforms of the correlation functions. Methods to accomplish this in the case of deformable DNAs have been developed and applied to analyze a variety of data.[81–83]

The convention employed here for the rotation functions matches that of Wigner[75] and Edmonds[76] and differs from that employed previously in this laboratory,[23, 29, 81, 82, 84, 85] which corresponds to defining all Euler rotations in a negative sense. This change in convention alters none of the observable correlation functions.

4.2.3.3. Macromolecules with Mean Local Cylindrical Symmetry[29]

The deformable macromolecule is regarded as a linear, but by no means always straight, array of $N+1$ rigid cylindrical rods or disks with appropriate lengths and radii, as illustrated in Figure 4.3. Each rod is coupled to its neighbors by a twisting and bending potential. A coordinate frame is rigidly attached to each rod, with the z-axis taken along the local symmetry axis. It is assumed that the fluorescent probe is part of, or attached to, a particular rod. Each elementary rod undergoes in time t mean squared angular displacements about its *body-fixed* x, y, and z axes. These mean squared displacements are rigorously defined by

$$\langle \Delta_j(t)^2 \rangle = \left\langle \left(\int_0^t \omega_j(t')\, dt' \right)^2 \right\rangle, \qquad j = x, y, z \tag{4.18}$$

where $\omega_j(t)$ is the instantaneous angular velocity about the body-fixed j-axis of the rod, and the angular brackets denote an average over all trajectories of the given rod. The $\langle \Delta_j(t)^2 \rangle$ may differ from one rod to the next. Results applicable to an arbitrary rod are discussed first. Throughout this discussion, it should be understood that the unstrained macromolecule is neither bent nor twisted, so that instantaneously bent or twisted configurations represent spontaneous thermal fluctuations away from the equilibrium geometry.

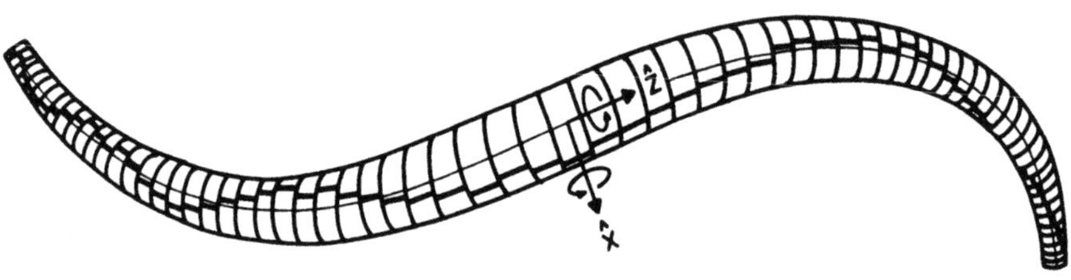

Figure 4.3. The rod (or disk) model for torsion and flexure of DNA. The DNA is modeled as a string of rods (or disks) connected by Hookean twisting and bending springs which oppose, respectively, torsional and flexural deformations. The instantaneous $\hat{z}$ and $\hat{x}$ axes of a subunit rod around which the mean squared angular displacements $\langle \Delta_j(t)^2 \rangle$, $j = x, z$, take place are indicated. The filament is assumed to exhibit mean local cylindrical symmetry in the sense that $\langle \Delta_x(t)^2 \rangle = \langle \Delta_y(t)^2 \rangle$ for any pair of transverse $\hat{x}$- and $\hat{y}$-axes. Twisting: $\langle \Delta_z(t)^2 \rangle$ = mean squared angular displacement about *body-fixed* $\hat{z}$-axis. Tumbling (bending): $\langle \Delta_x(t)^2 \rangle$ = mean squared angular displacement about *body-fixed* $\hat{x}$-axis $= \langle \Delta_y(t)^2 \rangle$ (assumed).

The phrase "mean local cylindrical symmetry" is understood to imply the following two assumptions[29]:

$$\langle \Delta_x(t)^2 \rangle = \langle \Delta_y(t)^2 \rangle \tag{4.19}$$

$$\left\langle \int_0^t \omega_i(t')\,dt' \int_0^t \omega_j(t'')\,dt'' \right\rangle = \delta_{ij} \langle \Delta_j(t)^2 \rangle \tag{4.20}$$

Equation (4.19) merely asserts that the same average dynamics takes place around any two transverse axes of a given rod. Equation (4.20) asserts that correlations between *different* angular velocity components of the same rod are negligible. Such correlations arise only from hydrodynamic couplings in this model. A perfectly cylindrical object, such as the rod itself, cannot generate hydrodynamic self-couplings between its different angular velocity components. Thus, any correlations between different angular velocity components of the same rod must arise from hydrodynamic interactions that are mediated by other instantaneously noncollinear rods. Such hydrodynamic interactions are second, or higher, order and rather long-range, provided that bending between adjacent rods is comparatively slight. These two circumstances, and also the relatively short range of the hydrodynamic interactions for local rotation, ensure that Eq. (4.20) is an extremely good approximation.

It is assumed that inertial and memory effects in the usual sense can be ignored. However, coupling of the angular degrees of freedom of the $N+1$ rods by elastic restoring forces leads to a *nonstationary* Markov process for rotational diffusion, so the time ($t=0$) of the initial photoselection event has special significance.[29] It is also assumed that the fluid in which the macromolecule resides exhibits an isotropic equilibrium state. Under these conditions, the conditional probability density for an infinitesimal change in the orientation of a rod between t and $t+dt$ is independent of its *orientation* at time t, although in general it depends on t.

The orientation of the coordinate frame R fixed in a particular rod with respect to the laboratory frame L is specified by the Euler rotation, $\Phi = (\alpha\beta\gamma)$, as before. Under the preceding assumptions, a diffusion equation for the probability density $[f(\Phi, t)]$ is derived,[29] namely,

$$\frac{\partial f(\Phi, t)}{\partial t} = \left\{ \frac{1}{2} \frac{d\langle \Delta_x(t)^2 \rangle}{dt} (L^2 - M_\gamma^2) + \frac{1}{2} \frac{d\langle \Delta_z(t)^2 \rangle}{dt} M_\gamma^2 \right\} f(\Phi, t) \tag{4.21}$$

where $L^2 = -\hat{L}^2/\hbar^2$ and $M_\gamma^2 = -\hat{M}_\gamma^2/\hbar^2$ are proportional to the quantum-mechanical operators for total squared angular momentum ($\hat{L}^2$) and squared angular momentum around the body-fixed z-axis ($\hat{M}_\gamma^2$), respectively. Explicit formulas are given elsewhere.[29] The diffusion operator in braces in Eq. (4.21) is the same as that for rigid cylinders, except that the diffusion coefficient for rotation around a transverse axis is replaced by $D_\perp(t) = \frac{1}{2} d\langle \Delta_x(t)^2 \rangle / dt$

and that for rotation around the symmetry axis is replaced by $D_{\|}(t) = \frac{1}{2}d\langle \Delta_z(t)^2\rangle/dt$. When the macromolecule is a completely rigid cylinder, $\langle \Delta_x(t)^2\rangle = 2D_{\perp}t$ and $\langle \Delta_z(t)^2\rangle = 2D_{\|}t$ so the effective diffusion coefficients $D_{\perp}(t) = D_{\perp}$ and $D_{\|}(t) = D_{\|}$ are constants, independent of the time. However, for a deformable macromolecule with mean local cylindrical symmetry, the effective diffusion coefficients are time-dependent at small times and become constant only after the internal deformational coordinates have diffused from their initial values to their equilibrium distributions.(29)

The conditional probability density that a rod with orientation Φ_0 at $t=0$ will move to Φ at the time t is just the solution of Eq. (4.21) subject to the initial condition $f(\Phi, 0) = \delta(\alpha - \alpha_0)\,\delta(\cos\beta - \cos\beta_0)\,\delta(\gamma - \gamma_0)$. The solution is almost trivial, because the diffusion operator is formally identical to the Hamiltonian for a symmetric top, the eigenfunctions of which are known to be the Wigner rotation functions.(75, 76) The eigenvalues are readily expressed in terms of $\langle \Delta_x(t)^2\rangle$ and $\langle \Delta_z(t)^2\rangle$. The conditional probability density is(29)

$$G(\Phi t \mid \Phi_0 0) = \sum_{l=0}^{\infty} \sum_{m=-l}^{l} \sum_{n=-l}^{l} \frac{(2l+1)}{8\pi^2} \exp\left\{-\frac{1}{2}[l(l+1) - m^2]\langle \Delta_x(t)^2\rangle\right\}$$
$$\times \exp\left[-\frac{m^2}{2}\langle \Delta_z(t)^2\rangle\right] D^{l*}_{mn}(\Phi_0)\, D^{l}_{mn}(\Phi) \tag{4.22}$$

Using $G(\Phi t \mid \Phi_0 0)$ from Eq. (4.22) and the orthogonality of the Wigner rotation functions (Eq. 4.2), it is found that(29)

$$\langle D^{L}_{MN}(\Phi_0)\, D^{l*}_{mn}(\Phi)\rangle = \delta_{l,L}\,\delta_{m,M}\,\delta_{n,N}(2l+1)^{-1}$$
$$\times \exp\{-\tfrac{1}{2}[l(l+1) - m^2]\langle \Delta_x(t)^2\rangle\}$$
$$\times \exp\{-\tfrac{1}{2}m^2\langle \Delta_z(t)^2\rangle\} \tag{4.23}$$

Equation (4.23) may be incorporated directly into Eq. (4.17) to obtain the central result(29)

$$r(t) = \frac{2}{5}\sum_{n=0}^{2} I_n(t)\, F_n(t)\, C_n(t) \tag{4.24}$$

wherein the twisting correlation functions are defined by

$$C_n(t) = \langle \exp[-n^2\langle \Delta_z(t)^2\rangle/2]\rangle_R \tag{4.25}$$

the tumbling correlation functions are defined by

$$F_n(t) = \langle \exp[-(6 - n^2)\langle \Delta_x(t)^2\rangle/2]\rangle_R \tag{4.26}$$

and the internal correlation functions are defined by

$$I_n(t) = \frac{4\pi}{5} [\langle Y^*_{2n}[\omega_R(0)] \, Y_{2n}[\Omega_R(t)] \rangle \\ \times (1 - \delta_{n,0}) \langle Y^*_{2,-n}[\omega_R(0)] \, Y_{2,-n}[\Omega_R(t)] \rangle] \qquad (4.27)$$

The outer angle brackets $\langle \cdots \rangle_R$ in $F_n(t)$ and $C_n(t)$ imply an average over the different rods to which the fluorophore is bound. It has been assumed that the motions in the different factors in Eq. (4.24) are statistically independent. Equation (4.24) is expected to be rather generally valid for deformable macromolecules with mean local cylindrical symmetry. Relaxation of the FPA by rotation of the rods around their symmetry axes is contained in $C_n(t)$. Likewise, relaxation of the FPA by rotation, or end-over-end tumbling, of the rods about their transverse axes is contained in $F_n(t)$. Motion of the transition dipole with respect to the frame of the rod in which it is attached is contained in $I_n(t)$. Further progress requires the evaluation, or estimation, of $I_n(t)$, $F_n(t)$, and $C_n(t)$ for particular models.

Equation (4.24) differs significantly from the corresponding anisotropy expression of Barkley and Zimm,[16] which was shown to be incorrect whenever $\langle \Delta_x(t)^2 \rangle > 0$, or $F_n(t) < 1.0$.[29] Unfortunately, that incorrect expression was used in some early data analyses.[19, 28] The data of Hogan *et al.*[28] have been reanalyzed using the correct anisotropy expression in Eq. (4.24).[39]

An interesting aspect of Eq. (4.24) is that, even though $F_n(t)$ and $C_n(t)$ must represent genuine Brownian motions, $I_n(t)$ may represent non-Brownian, even cyclic, motions of the antenna in the rod frame. Motion of the coordinates $\omega_R(t)$ and $\Omega_R(t)$ has been assumed to be statistically independent of the rod motions, but the nature of their trajectories has not yet been specified.

4.2.3.4. Internal Correlation Functions

Several special cases are considered here.

4.2.3.4a. Rigidly Bound Fluorophore. In the case of a rigidly bound fluorophore, $\omega_R(0) = (\mu_0, 0)$ and $\Omega_R(t) = (\varepsilon_0, \zeta_0)$ are both independent of t.

$$I_0(t) = (\tfrac{3}{2}\cos^2 \mu_0 - \tfrac{1}{2})(\tfrac{3}{2}\cos^2 \varepsilon_0 - \tfrac{1}{2}) \qquad (4.28a)$$

$$I_1(t) = 3 \sin \mu_0 \cos \mu_0 \cos \zeta_0 \sin \varepsilon_0 \cos \varepsilon_0 \qquad (4.28b)$$

$$I_2(t) = \tfrac{3}{4} \sin^2 \mu_0 \cos 2\zeta_0 \sin^2 \varepsilon_0 \qquad (4.28c)$$

4.2.3.4b. Overdamped Brownian Libration of the Fluorophore Transition Moment in a Harmonic Potential Well; Absorption and Emission Dipoles

Parallel. In this case, which is treated elsewhere,[29, 81, 83] one has $\mu_0 = \varepsilon_0$, $\nu_0 = \zeta_0 = 0$, and

$$I_0(t) = \tfrac{1}{4}\{\tfrac{1}{4} + \tfrac{3}{2}\cos(2\varepsilon_0)\exp[-2\langle\delta\varepsilon^2\rangle] + \tfrac{9}{8}\cos(4\varepsilon_0) \times \exp[-4\langle\delta\varepsilon^2\rangle(1+e^{-t/\tau_\varepsilon})] + \tfrac{9}{8}\exp[-4\langle\delta\varepsilon^2\rangle(1-e^{-t/\tau_\varepsilon})]\} \quad (4.29a)$$

$$I_1(t) = \tfrac{3}{8}\{\exp[-4\langle\delta\varepsilon^2\rangle(1-e^{-t/\tau_\varepsilon})] - \cos(4\varepsilon_0)\exp[-4\langle\delta\varepsilon^2\rangle(1+e^{-t/\tau_\varepsilon})]\} \times \exp[-\langle\delta\zeta^2\rangle(1-e^{-t/\tau_\zeta})] \quad (4.29b)$$

$$I_2(t) = \tfrac{3}{16}\{1 - 2\cos(2\varepsilon_0)\exp(-2\langle\delta\varepsilon^2\rangle) + \tfrac{1}{2}\exp[-4\langle\delta\varepsilon^2\rangle(1-e^{-t/\tau_\varepsilon})] + \tfrac{1}{2}\cos(4\varepsilon_0)\exp[-4\langle\delta\varepsilon^2\rangle(1+e^{-t/\tau_\varepsilon})]\} \times \exp[-4\langle\delta\zeta^2\rangle(1-e^{-t/\tau_\zeta})] \quad (4.29c)$$

Here $\langle\delta\varepsilon^2\rangle = k_B T/G$ and $\langle\delta\zeta^2\rangle = k_B T/g \sin^2 \varepsilon_0$ are the rms angular displacements of the transition dipole in, respectively, the polar and azimuthal harmonic potential wells with corresponding force constants G and g.[29, 81–84] Also, τ_ε and τ_ζ are the corresponding relaxation times for motion in those wells. Equations (4.29) are valid only when $\langle\delta\varepsilon^2\rangle \ll \varepsilon_0$, due to approximations invoked in the derivation. Thus, Eqs. (4.29) are not valid for $\varepsilon_0 = 0$, except when $\langle\delta\varepsilon^2\rangle = 0$ also. Practically, it is not possible to distinguish four constants characterizing the internal motion, so one typically assumes that the motion is either *isotropic*[29, 39, 81–83] (i.e., $\langle\delta\varepsilon^2\rangle = \langle\delta\zeta^2\rangle \sin^2 \varepsilon_0$) or purely polar (i.e., $\langle\delta\zeta^2\rangle = 0$) or purely azimuthal (i.e., $\langle\delta\varepsilon^2\rangle = 0$). Isotropic motion of the transition dipole in the rod frame means that the probability of an angular deflection away from the minimum-energy orientation, or vector, at $(\varepsilon_0, 0)$ is invariant (or symmetric) with respect to rotation around that vector. Using Eq. (4.29c), $I_2(t)$ is calculated for isotropic motion with $\langle\delta\varepsilon^2\rangle = \langle\delta\zeta^2\rangle \sin^2 \varepsilon_0 = 0.0149\ \text{rad}^2 = (7°)^2$ and $\tau_\varepsilon = \tau_\zeta = 100$ ps and plotted versus t in Figure 4.4. The internal correlation functions do not in general

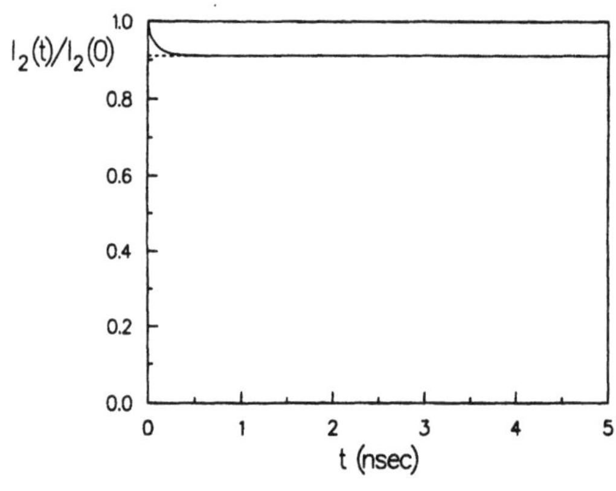

Figure 4.4. Relative internal correlation function $I_2(t)/I_2(0)$. The ratio is calculated using Eqs. (4.29) with the values $\varepsilon_0 = 70.5$, $\tau_\varepsilon = \tau_\zeta = 100$ ps, and $\langle\delta\varepsilon^2\rangle = \langle\delta\zeta^2\rangle \sin^2 \varepsilon_0 = 0.0149$, which corresponds to an rms angular libration of 7° in both the polar and azimuthal directions. The dashed line is the value of the ratio at $t = \infty$.

relax to zero, but exhibit a finite asymptote. Equations (4.29a–c) are valid for anisotropic as well as isotropic internal motion.

4.2.3.4c. Isotropic Internal Motion of the Transition Dipole. For the special case of *isotropic* internal motion, regardless of the particular form of the potential, it is rigorously found that

$$I_0(\infty) = \text{ARF}(\tfrac{3}{2}\cos^2 \varepsilon_0 - \tfrac{1}{2})^2 \tag{4.30a}$$

$$I_1(\infty) = \text{ARF}\ 3\cos^2 \varepsilon_0 \sin^2 \varepsilon_0 \tag{4.30b}$$

$$I_2(\infty) = \text{ARF}(\tfrac{3}{4}) \sin^4 \varepsilon_0 \tag{4.30c}$$

where the amplitude reduction factor is given by

$$\text{ARF} = \langle P_2(\cos \xi)\rangle^2 \tag{4.30d}$$

and ξ is the deflection of the transition dipole vector away from its equilibrium (minimum-energy) orientation.[83] Equations (4.30) apply *after* the internal motions have *completely* relaxed. The effect of rapid *isotropic* internal motions, then, is to reduce all three internal correlation functions *by the same factor*, as might have been expected. Numerical investigations have confirmed this conclusion also for Eqs. (4.29) under conditions where they are valid.

Most FPA studies to date on DNA have lacked sufficient time resolution to observe directly the relaxation of the internal correlation functions. Instead, the initial anisotropy r_0 is taken as an adjustable parameter. Equations (4.30) show that such a procedure is completely valid for anisotropic diffusors (i.e., $\langle \Delta_z(t)^2\rangle \neq \langle \Delta_x(t)^2\rangle$), provided the rapid internal motion of the transition dipole is isotropic. It has not yet been ascertained whether the internal motion actually is isotropic, so this must be assumed.[83] A recent claim[86] that large amplitudes of polar wobble are required to fit both the small amplitude of initial FPA relaxation[87] and the linear dichroism[88] has been refuted.[83]

4.2.3.4d. Kinetic Jump Models. Kinetic models in which the transition dipole hops between discrete orientations have been developed.[74, 89–93] An error in the treatment of King and Jardetzky[91] was corrected.[90] Such models are frequently employed to interpret NMR relaxation data on polynucleotides,[90, 94–103] possibly because the authors are unfamiliar with the results for libration of a vector in a harmonic potential well. In any case, the common failure to take account also of collective twisting and bending deformations in those analyses[90, 94–103] has invariably resulted in spuriously large estimates for the angular amplitudes of local hopping motions.[81–83]

4.2.3.4e. Restricted Diffusion Models. Models in which the internal motion proceeds by restricted diffusion, or wobbling in a cone, have also been developed.[90, 74]

4.2.3.4f. Non-Brownian Motions of the Absorption and Emission Dipoles in the Rod Frame. One can imagine *non-Brownian* processes in which the transition dipole undergoes regular, or cyclic, motions in the rod frame. For example, a transition dipole attached to an actively transcribing RNA polymerase would exhibit progressive net rotation in one direction around the helix axis. Although much too slow for FPA experiments, such motion could in principle be detected by TPD techniques. For simplicity, we assume that the polar angles of the absorption and emission dipoles remain fixed, but that a driven regular motion $\rho(t)$ and a Gaussian random process $\delta(t)$ are superimposed on the azimuthal motions of those dipoles. In that case, $\nu(t)=\rho(t)+\delta(t)$ and $\zeta(t)=\zeta_0+\rho(t)+\delta(t)$. These expressions for $\zeta(t)$ and $\nu(t)$ are inserted in Eq. (4.27), and the average over $\delta(t)$ performed to yield

$$I_0(t)=(\tfrac{3}{2}\cos^2\mu_0-\tfrac{1}{2})(\tfrac{3}{2}\cos^2\varepsilon_0-\tfrac{1}{2}) \tag{4.31a}$$

$$I_1(t)=3\sin\mu_0\cos\mu_0\sin\varepsilon_0\cos\varepsilon_0\langle\cos[\zeta_0+\rho(t)-\rho(0)]\rangle_{\text{in}} \times\exp[-\langle\delta(t)^2\rangle/2] \tag{4.31b}$$

$$I_2(t)=\tfrac{3}{4}\sin^2\mu_0\sin^2\varepsilon_0\langle\cos\{2[\zeta_0+\rho(t)-\rho(0)]\}\rangle_{\text{in}} \times\exp[-4\langle\delta(t)^2\rangle/2] \tag{4.31c}$$

where the subscript (in) denotes an average over the initial coordinate $[\rho(0)]$ of the regular motion. Use has been made of the relation $\langle\exp[in\,\delta(t)]\rangle=\exp[-n^2\langle\delta(t^2)\rangle/2]$ for a Gaussian random process.

When the regular motion is simply uniform rotation of the absorption and emission dipoles with angular velocity ω around the helix axis, one has $\rho(t)-\rho(0)=\omega t$. For the corresponding random motion, one might have $\langle\delta(t)^2\rangle=2Dt$, where D is the effective diffusion coefficient for Brownian rotation of the transition dipole around the helix axis. When these expressions are incorporated in Eqs. (4.31) and (4.24), the latter becomes a generalization of a relation recently derived using a more cumbersome approach.[104]

When the driven regular motion is harmonic azimuthal libration of the absorption and emission dipoles with angular frequency ω and amplitude A about their equilibrium positions 0 and ζ_0, respectively, one has $\rho(t)-\rho(0)=A\cos(\omega t+\phi)-A\cos\phi$, where ϕ denotes the initial phase of the libration. This can be inserted in Eqs. (4.31), and an average over the uniform distribution of ϕ performed. When $A\ll 1.0$, one obtains $\langle\cos\{\zeta_0+[A\cos(\omega t+\phi)]-[A(\cos\phi)]\}\rangle_{\text{in}}\cong\cos\zeta_0[1-(A^2/2)(1-\cos\omega t)]$. In this case, one might have also $\langle\delta(t)^2\rangle=\langle\delta(0)^2\rangle(1-e^{-t/\tau})$, where $\langle\delta(0)^2\rangle$ is the mean squared amplitude of the random process, and τ its relaxation time. Insertion of these expressions into Eqs. (4.31) and (4.24) yields the anisotropy. We emphasize that such regular motions are not exhibited by equilibrium systems, as they require continuous dissipation of free energy.

4.2.3.5. *Twisting Correlation Function*

Each rod m is assumed to be coupled to its neighbors on either side by Hookean torsion springs and to obey a simple Langevin equation of the form

$$J\ddot{\phi}_m + \gamma\dot{\phi}_m + \alpha(-\phi_{m-1} + 2\phi_m - \phi_{m+1}) = T_m(t) \quad (4.32)$$

with appropriate modifications for the first ($m = 1$) and the last ($m = N + 1$) rods.[17] In Eq. (4.32) ϕ_m is the angular displacement of the mth rod around its symmetry axis, α is the torsion spring constant between rods, γ is the friction factor per rod for rotation around the symmetry axis, and $T_m(t)$ is the random Brownian torque, which is assumed to fluctuate rapidly compared to any deformational relaxation of the filament. J is the moment of inertia of a rod, which is neglected in treating the deformational normal modes, and cancels out in the treatment of the uniform axial spinning mode.[17] Equation (4.20) predicts a set of overdamped torsion normal modes, each of which makes a separate contribution to $\langle \Delta_z(t)^2 \rangle = \langle [\phi_m(t) - \phi_m(0)]^2 \rangle$ for the mth rod.[17, 29] For a linear DNA with two free ends, the relaxation time of the lth normal mode ($2 \leqslant l \leqslant N + 1$) is

$$\tau_l = \gamma / \{4\alpha \sin^2[(l-1)\pi/2(N+1)]\} \quad (4.33)$$

its wavelength in rods is given by[17, 53]

$$\lambda_l = \pi/\arcsin[(\gamma/4\alpha\tau_l)^{1/2}] = 2(N+1)/(l-1) \quad (4.34)$$

and its equilibrium mean squared amplitude is finite, namely,

$$\langle d_l^2 \rangle = k_B T / \{4\alpha \sin^2[(l-1)\pi/2(N+1)]\} \quad (4.35)$$

The expression for the projection of the lth normal mode onto ϕ_m of the mth rod is presented elsewhere.[17, 29] Besides these deformational normal modes, there is also a uniform axial spinning mode in which all of the rods rotate in phase like a speedometer cable. The contribution of that mode to $\langle \Delta_z(t)^2 \rangle$ is just $2k_B Tt/(N+1)\gamma$, as expected for diffusion in one dimension with the friction factor $(N+1)\gamma$ of the entire filament.

This simple model of rigid rods connected by Hookean torsion springs has been criticized as unrealistic, because it does not reflect the atomic structure of a real DNA. However, this objection misses an essential point, namely, that the correlation functions obtained for this simple model are also valid for a much wider class of models over the observable time domain. The reason is as follows. The earliest time at which depolarization due to twisting can be distinguished from wobble is about 0.5 ns.[39, 87] The wavelength of the torsion normal mode with $\tau_l = 0.5$ ns is $\lambda_l = 41$ bp [estimated from Eq. (4.34) using $\alpha = 5.2 \times 10^{-12}$ dyn-cm and $\gamma = 6.2 \times 10^{-23}$ dyn-cm s). Thus, the torsion potential need not be Hookean between base pairs, but only when averaged over 5 to 10 bp! The central limit theorem of statistics ensures that the equilibrium distribution, $P(\phi_n - \phi_1)$, of the relative angular displacement, $(\phi_n - \phi_1) = (\phi_n - \phi_{n-1}) + (\phi_{n-1} - \phi_{n-2}) + \cdots + (\phi_2 - \phi_1)$, between rods n and 1 will be Gaussian for large n, regardless of the distribution for *adjacent* rods; namely, $p(\phi_n - \phi_{n-1}) \propto \exp[-V(\phi_n - \phi_{n-1})/k_B T]$, which depends on the potential V and need not be Gaussian. A Gaussian distribution $P(\phi_n - \phi_1)$ implies an *effective long-range* potential that is quadratic, or Hookean in $(\phi_n - \phi_1)$! For n in the range 5–10, $P(\phi_n - \phi_1)$ is expected to be nearly Gaussian for almost any reasonable choice of $V(\phi_2 - \phi_1)$, including even a square well. Thus, over most or all of the accessible domain of observation, almost any potential between neighboring rods is expected to give the same functional form of the decay, which is characterized simply by the long-range effective torsional rigidity. Due to their comparatively long wavelength, the torsional deformations observed in the FPA are "macroscopic" in the sense that the discrete molecular structure and detailed interatomic forces play no role except to determine the effective torsion constant for the long-range deformations and the hydrodynamic radius.

The neglect of hydrodynamic interactions between rods in Eq. (4.35) was originally a matter of some concern. However, Allison subsequently demonstrated that their neglect introduces no significant error into the predicted correlation functions at times longer than 0.2 ns.[105, 106]

It is conceivable that the twisting motion experiences internal friction, by which is meant the occurrence of bumps or barriers in the potential surface along which the DNA deforms. This would cause γ to exhibit a temperature (T) dependence differing from that due to the viscosity of water. Experimental results[40] give no indication of such anomalous T dependence, as shown subsequently.

From the set of Eqs. (4.32), the general twisting correlation functions for each rod, and the average over all rods, have been obtained using well-established techniques[17, 107] for linear[17, 29] and circular[82] DNAs and for linear DNAs with one or both ends clamped.[59] General anisotropy expressions have also been given for linear DNAs wrapped around the equator of a sphere with zero, one, or both ends clamped to the sphere.[29] Lengthy summations preclude the use of any of these general expressions in routine data analysis of high-molecular-weight DNAs. Fortunately, it has been possible in most cases to perform at least part of the summation analytically and to obtain a sequence of accurate analytical approximations, each valid in a particular time zone, that span the complete time course of the decay. The formulas for $C_n(t)$ for a linear DNA, averaged over all rods, and their domains of validity are as follows[29]:

(i) Initial Exponential Decay Zone $(0 \leqslant t \leqslant \gamma/\alpha)$

$$C_n(t) = \exp\left[-n^2 \frac{k_B T}{\alpha}\left(\frac{\alpha t/\gamma}{1+\alpha t/\gamma}\right)\right] \tag{4.36}$$

(ii) Intermediate Zone $(\gamma/4\alpha \leqslant t \leqslant \tau_{31} = \gamma/\{4\alpha \sin^2[30\pi/2(N+1)]\})$

$$C_n(t) = \exp\{-n^2 k_B T[t/(\pi\alpha\gamma)]^{1/2}\} \tag{4.37}$$

(iii) Post-Intermediate Zone $(\tau_{31} \leqslant t \leqslant \tau_4 = \gamma/\{4\alpha \sin^2[3\pi/2(N+1)]\})$

$$\begin{aligned} C_n(t) = (N+1)^{-1} &\sum_{m=1}^{(N+2)/2} (2-\delta_{m,(N+2)/2}) \\ &\times \exp[-n^2 k_B T[t/(\pi\alpha\gamma)]^{1/2} (1+e^{-b_m/t} \\ &- (\pi b_m/t)^{1/2} \operatorname{erfc}[(b_m/t)^{1/2}])] \end{aligned} \tag{4.38}$$

where $b_m \equiv (2m-1)^2 \gamma/4\alpha$ and erfc( ) is the complement of the error function. Equation (4.38) is applicable only when $N+1$ is odd.

(iv) Pre-Uniform Mode Zone $(\tau_4 \leqslant t \leqslant 6\tau_2 = 6\gamma/\{4\alpha \sin^2[\pi/2(N+1)]\})$.

$$\begin{aligned} C_n(t) = (N+1)^{-1} &\exp[-n^2 k_B T t/(N+1)\gamma] \\ &\times \sum_{m=1}^{N+1} \exp\left[\frac{-n^2 k_B T}{\alpha}\left(\frac{N+1}{3} - \frac{(2m-1)}{2} + \frac{(2m-1)^2}{4(N+1)}\right)\right. \\ &+ \frac{n^2 k_B T}{\alpha} \frac{1}{4(N+1)} \\ &\left.\times \sum_{l=2}^{7} \frac{\{\cos[(2m-1)(l-1)\pi/(N+1)] + 1\} e^{-t/\tau_l}}{\sin^2[(l-1)\pi/2(N+1)]}\right] \end{aligned} \tag{4.39}$$

Equation (4.39) is valid for all integral $N+1$.

(v) Uniform Mode Zone $(t \geqslant \tau_{\mathrm{un}} = 6\tau_2)$

$$C_n(t) = B_n(\infty) \exp[-n^2 k_B T t/(N+1)\gamma] \tag{4.40}$$

where

$$B_n(\infty) = (n^2 Z)^{-1/2} \exp(-n^2 Z/3)(\pi^{1/2}/2) \operatorname{erf}[(n^2 Z)^{1/2}] \tag{4.41}$$

and $Z \equiv k_B T(N+1)/4\alpha$. Equation (4.25) is valid for all integral $N+1$.

This sequence of analytical expressions has been computationally tested against the exact expression. There are no visually resolvable differences, nor are there any visually apparent discontinuities in $C_n(t)$ or its slope at the boundaries of these different zones. The theoretically predicted $C_1(t)$ for a DNA with 2001 bp is plotted over several time spans of about 100 ns in Figure 4.5. If the elementary rod length is 1 bp, the motion passes out of the Initial Exponential Decay Zone into the Intermediate Zone at $t = \gamma/\alpha = 1.6 \times 10^{-11}$ s, before any significant rotational relaxation has occurred. At all subsequent times the motion is indistinguishable from that of a continuum filament with the same long-range torsional rigidity. Although appreciable amplitude remains at 1 μs, that largely dies out by 5 μs, and $C_1(t)$ is entirely negligible by 40 μs, which is close to the start of the Uniform Mode Zone ($\tau_{un} = 39$ μs). Thus, the residual amplitude of the uniform mode is practically zero for such a long DNA. The relaxation time of the uniform mode in $C_1(t)$ is $\tau_1 = (k_B T/(N+1)\gamma)^{-1} = 3.0$ μs. Thus, the uniform spinning mode of such a long DNA contributes little to the total relaxation of $C_1(t)$ at 1 μs, and virtually nothing to the relaxation at 150 ns. The superposition of both $C_1(t)$ and $C_2(t)$ terms in $r(t)$ is illustrated for DNAs of different length in Figure 4.6. The anisotropy $r(t)$ is calculated from Eqs. (4.24) and (4.37)–(4.41) assuming $F_n(t) = 1.0$ (no tumbling) and Eqs. (4.28a–c) (rigidly bound fluorophore) with $\varepsilon_0 = 70.5°$. The curve for 2001 bp is still rather close to the Intermediate Zone formula at 150 ns, even though its own Intermediate Zone ends at $\tau_{31} = 7.3$ ns. In general, the Inter-

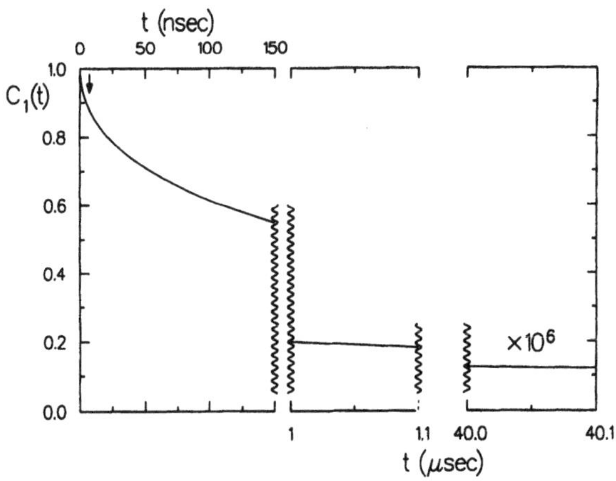

Figure 4.5. Twisting correlation function $C_1(t)$ versus t. $C_1(t)$ is calculated using Eqs. (7.37–7.41) for a 2001-base pair-long piece of DNA with $\alpha = 3.8 \times 10^{-12}$ dyn-cm, $\tau = 6.15 \times 10^{-23}$ dyn-cm s, and $T = 293.15$ K. The Initial Exponential Decay Zone (Eq. 4.36) ends at 16 ps and is not shown. The Intermediate Zone formula (Eq. 4.37) and the Post-Intermediate Zone formula (Eq. 4.38) are used in the left section of the figure. The arrow marks the dividing line between the two ranges at $\tau_{31} = 7.3$ ns. The middle section of the figure is the Pre-Uniform Mode Zone (Eq. 4.39) range which extends from $\tau_4 = 730$ ns to $6\tau_2 = 39$ μs. At the right is the Uniform Mode Zone [Eqs. (4.40 and (4.41)], which is valid for all times greater than $6\tau_2$. As can be seen from Figure 4.6, the Intermediate Zone formula is still a good approximation for times well beyond τ_{31}.

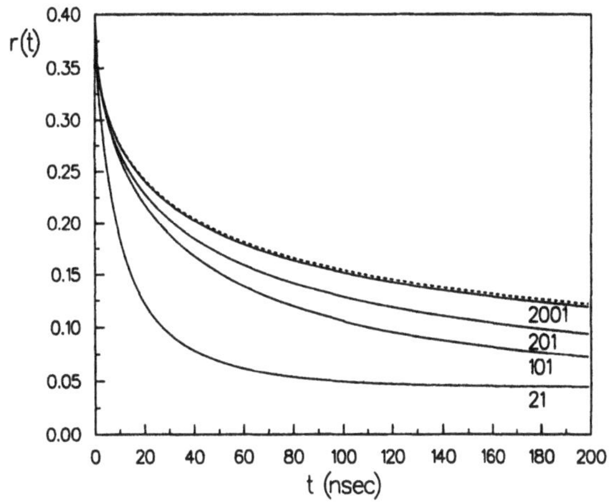

Figure 4.6. The anisotropy $r(t)$ due to *twisting only* versus t. The curves are calculated for different values of $N+1$ (as indicated in the figure) using Eqs. (4.24) and (4.37)–(4.41) with $\alpha = 3.8 \times 10^{-12}$ dyn-cm, $\gamma = 6.15 \times 10^{-23}$ dyn-cm s, $T = 293.15$ K, and Eqs. (4.28a–c) with $\varepsilon_0 = \mu_0 = 70.5°$. The dashed line is $r(t)$ calculated using only the Intermediate Zone formula.

mediate Zone formula is followed sufficiently closely out to τ_8 that its use over that time span introduces negligible error into the best-fit parameters. For this 2001-bp filament, $\tau_8 = 134$ ns. For DNAs with $N+1 \leqslant 1000$ bp, significant error appears in the best-fit parameters when data extending to 120 ns are fitted using the Intermediate Zone formula.

The unusual time dependence of the Intermediate Zone decay is a direct consequence of the dispersion relation, $1/\tau_l \sim \sin^2[(l-1)\pi/2(N+1)] \sim (l-1)^2/(N+1)^2$, and may be regarded as the characteristic signature for such a spectrum of collective modes.

The residual amplitude of the Uniform Mode $[B_n(\infty)]$ falls rapidly with increasing filament length. Using $\alpha = 5.0 \times 10^{-12}$ dyn-cm, we calculate $B_2(\infty) \cong 0.021$ for $N+1 = 1000$ bp. Thus, for this and all longer lengths, the relaxation of $C_2(t)$ is essentially complete in the earlier zones, so negligible amplitude remains in the Uniform Mode Zone. Conversely, $B_n(\infty)$ increases toward 1.0 as the filament length decreases. For example, using $\alpha = 5.0 \times 10^{-12}$ dyn-cm and $N+1 = 10$, we estimate $B_2(\infty) \cong 0.95$. In this case, relaxation of the twisting deformations causes only a very small reduction in the residual amplitude of the uniform mode. This very strong length dependence of the residual amplitude is a distinguishing feature of *collective* deformations. In contrast, the relaxation of *local* internal motions acts to reduce $I_n(t)$ in a manner that is independent of length. For this reason, the separate contributions of collective deformations and local internal motions must be distinguished if results obtained for long DNAs are to be reliably extrapolated to short DNAs. The vast majority of analyses of NMR relaxation data on DNAs with 43 to 600 or more bp[90, 94–103] have not made this distinction, so their conclusions are not transferable to much shorter DNAs.[81–83]

As the torsion constant α approaches infinity, the relaxation times τ_l all approach 0, and the amplitude of the uniform mode, $B_n(\infty)$, approaches 1.0, in which case $C_n(t)$ is precisely the result for a rigid (against twist) rod.

For DNAs with less than 1000 bp, significant deviations from Intermediate Zone behavior are expected (and observed) in the time range of the FPA experiment (0.5 to 150 ns). Unfortunately, many data, especially on synthetic DNA samples, where much or all of the preparation exists as fragments with lengths substantially less than 1000 bp, have been analyzed using the inappropriate Intermediate Zone formula.[19, 21, 61, 108] This work needs to be repeated on preparations with known length distributions using the proper twisting correlation functions. For DNAs with $N+1 \geqslant 2000$ bp and *uniform* torsional rigidities, the decay of $C_n(t)$ is virtually indistinguishable from Intermediate Zone behavior up to 150 ns, even though the relaxation may progress somewhat into the Post-Intermediate Zone. In this case, one can determine only the product $\alpha\gamma$ of the torsion constant and friction factor, but not either factor separately, as is evident from Eq. (4.37). However, for DNAs sufficiently short that appreciable amplitude $[B_n(\infty)]$ remains in the uniform mode, it is possible to determine γ alone, as is evident from Eq. (4.40). Historically, long DNAs were studied first, and α was calculated using an assumed value of γ. The value of γ measured for short restriction fragments[109] is very close to that assumed in this laboratory, corresponding to a hydrodynamic radius $R_H = 12$ Å, but significantly (17–27%) smaller than those assumed in other laboratories, corresponding to $R_H = 13$ to 13.5 Å.

Similar sequences of accurate analytical approximations to $C_n(t)$ have been presented for circular DNAs[82] and linear DNAs with both ends clamped,[29, 63] but are not reproduced here. At present, no complete sequence of accurate analytical approximations is available for linear DNAs with only one end clamped.

Though derived for DNA, and nucleosome core particles, these twisting correlation functions are potentially useful for the analysis of other filamentous biopolymers. Such analyses of optical anisotropy data for actin filaments have already been carried out.[110, 111]

4.2.3.6. *Tumbling Correlation Function*

Theoretical and experimental elucidation of the tumbling correlation function has proved more difficult than in the case of the twisting correlation function. The tumbling dynamics of a long inextensible filament with persistence length (P) much less than its contour length (L) is an inherently nonlinear problem that does not admit a straightforward normal-mode analysis. The difficulties are compounded by the long range ($\sim 1/r$) of the hydrodynamic interactions for rotations of a filament, or rod, around a transverse axis. The force per unit length on the rod due to deformation is[112]

$$\mathbf{F}(s) = -\kappa \frac{\partial^4 \mathbf{r}}{\partial s^4} + T(s, t) \frac{\partial^2 \mathbf{r}}{\partial s^2} \tag{4.42}$$

where $\mathbf{r}(s)$ is the spatial position of the point s along the contour, κ is the bending rigidity, and $T(s, t)$ is the tension at the contour point s. $T(s, t)$ can be identified with the Lagrange multiplier associated with the constraint $(\partial\mathbf{r}/\partial s) \cdot (\partial\mathbf{r}/\partial s) = 1.0$, which prevents contraction or extension of the filament. For a long filament with $L \gg P$, $T(s, t)$ will vary along s, and also fluctuate in time, taking both positive and negative values. The tension is expected to be associated mainly with large-amplitude bending fluctuations with wavelengths of two or more persistence lengths. Unfortunately, our quantitative understanding of the tension and its role in bending dynamics is negligible at present.

Barkley and Zimm[16] (BZ) formulated a normal-mode analysis of bending by considering only small deviations of the helix axis from its average orientation (along $\hat{\mathbf{z}}$) and by neglecting $T(s, t)$. Their theory should be valid for short-wavelength motions in weakly bending filaments ($L \lesssim P$), where all motion is transverse to a common axis and significant tension is not developed. It has been argued that this theory should apply also to the circumstance $L \gg P$ at very short times. However, for such filaments the deviation from linearity is extreme over large distances, so even at early times there might be significant tension and variations therein at different points along the contour. Whether this has any dynamically important consequences is not known.

For the *central* point (or subunit) of a filament of length L, the BZ result[16] can be written as

$$\langle \Delta_x(t)^2 \rangle = 2D_\perp t + 2\left(\frac{P}{L}\right) \sum_{\substack{n \geqslant 2 \\ \text{(even)}}}^{\infty} (1 - e^{-t/T_n})/\kappa_n^2 \tag{4.43}$$

where

$$1/T_n = \frac{k_B T P \kappa_n^4}{4\pi\eta} \left[K_0(\kappa_n b) + \left(\frac{\kappa_n b}{2}\right) K_1(\kappa_n b) \right] \tag{4.44}$$

is the relaxation time of the nth bending normal mode ($n = 1, 2, 3, 4, 5, \ldots, \infty$), $\kappa_n = (2n+1)\pi/2L$, $b \cong 12.5$ Å is the hydrodynamic radius for motion transverse to the helix axis, and $D_\perp$ is the average rotational diffusion coefficient for a "frozen" equilibrium ensemble of variously bent DNAs. The sum over even n in Eq. (4.43) can be evaluated approximately by integration to yield the more compact form[16]

$$\langle \Delta_x(t)^2 \rangle = D_\perp t + \tfrac{1}{2} B(t) t^{1/4} \tag{4.45}$$

where

$$B(t) \equiv \frac{4(k_B T)^{1/4}\,\Gamma(3/4)}{\pi^{5/4} P^{3/4} \eta^{1/4} [K_0(z_m) + (z_m/2)\,K_1(z_m)]} \tag{4.46}$$

and z_m is the solution of the transcendental equation

$$z_m = \frac{b(5.024\pi\eta/k_B TPt)}{K_0(z_m) + (z_m/2)\,K_1(z_m)} \tag{4.47}$$

In Eqs. (4.43)–(4.47), use is made of the well-known relation[23] $P = \kappa/k_B T$, where κ is the bending rigidity. Equation (4.45) differs from the corresponding BZ result by the inclusion of the $D_\perp t$ term. Normally, this is negligible on the fluorescence time scale, but for sufficiently short filaments it could make a significant contribution. It appears with the coefficient 1.0 instead of 2.0 because the correction term from the Euler–McLaurin summation formula actually cancels out one of the two $D_\perp t$ terms in Eq. (4.43). Equation (4.43) or (4.45) is then inserted in Eq. (4.26) to obtain the BZ tumbling correlation function for the filament.

The unusual time dependence in Eq. (4.45) is a direct consequence of the dispersion relation, $1/T_n \sim \kappa_n^4$ in Eq. (4.44), and may be regarded as the characteristic signature for such a spectrum of collective modes.

For the jth subunit in a weakly bending filament ($L \lesssim P$) at long times, after the bending modes have all relaxed to their equilibrium mean squared displacements, one has[109]

$$\langle \Delta_x(t)^2 \rangle = 2D_\perp t + 2\langle \eta_j^2 \rangle \tag{4.48}$$

where η_j is the angular displacement of the jth bond vector from the average end-to-end vector. Thus, $2\langle \eta_j^2 \rangle$ represents the contribution of *complete relaxation* of all bending deformations to the mean squared angular displacement of the jth rod around its x-axis. The general expression for $\langle \eta_j^2 \rangle$ is given elsewhere.[109] At such long times, the tumbling correlation function becomes

$$\langle F_n(t) \rangle_R = D_n(\infty) \exp[-(6-n^2)\,D_\perp t] \tag{4.49}$$

where the reduced amplitude, $D_n(\infty)$, of the uniform mode represents the effects of complete relaxation of all bending modes. The correct expression, *including the average over all rods*, is[109]

$$D_n(\infty) = Z_n^{-1/2} \exp(-Z_n/3)(\pi^{1/2}/2)\,\mathrm{erf}(Z_n^{1/2}) \tag{4.50}$$

where $Z_n = (6-n^2)L/4P$. For just the central subunit (c), or central point, the corresponding formula is

$$D_n^c(\infty) = \exp[-(6-n^2)\langle \eta_c^2 \rangle] \tag{4.51}$$

where $\langle \eta_c^2 \rangle = (0.08333)L/P$. Using the BZ theory (Eq. 4.43), a similar formula is obtained for $D_n^c(\infty)$, but in that case $\langle \eta_c^2 \rangle_{BZ} = (0.06065)L/P$.[109] This BZ result arises from the use of approximate eigenvalues and eigenvectors of the potential energy operator in the evaluation of $\langle \eta_c^2 \rangle$ and is consequently inexact.[109] The expressions for $D_n(\infty)$ and $D_n^c(\infty)$ also become substantially incorrect for discrete filaments, when the rms angle between bond vectors of adjacent subunits exceeds 18°.[109]

$F_1(t)$ is illustrated in Figure 4.7 for a 69-bp DNA. The uniform mode decays for both the centrally bound dye and the average over all binding sites are shown. Comparison between the BZ result and the exact result is also indicated.

These analytical results can be compared with the Brownian dynamics simulations of Allison and McCammon, who determined $-\ln\langle P_2[\hat{\mu}(0)\cdot\hat{\mu}(t)]\rangle = 3\langle \Delta_x(t)^2 \rangle$ for the central bond vectors of discrete wormlike coils with $L/P = 0.5$, $P = 600$ Å[113] and with $L/P = 2.36$, $P = 400$ Å.[114] The subunit radius was chosen to provide agreement with $D_\perp$ for DNA. For $L/P = 0.5$, the longest bending relaxation time of the central subunit (T_2) was found to be somewhat (~30%) longer than that predicted using Eq. (4.44),[39, 113] while the simulated $\langle \eta_c^2 \rangle$ was found to be in good agreement with the inexact

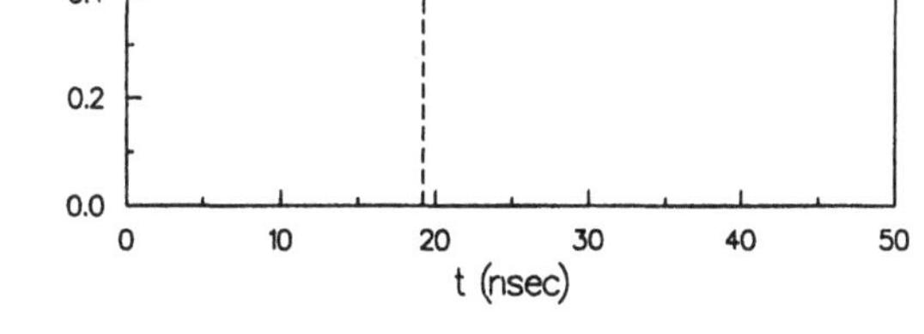

Figure 4.7. Tumbling correlation function $F_1(t)$ versus t. The $F_1(t)$ are calculated for a 69-bp fragment with a persistence length of 500 Å at 20 °C. The vertical dashed line is drawn at $6T_2 = 19.2$ ns, where T_n is given in Eq. (4.44). The curve to the left of the vertical line is the Barkley–Zimm formula in Eqs. (4.45)–(4.47). For times longer than $6T_2$, $F_1(t)$ is exponential with a decay time equal to the end-over-end tumbling time. The amplitude is reduced due to the internal flexural motions, and the three curves to the right of the vertical dashed line represent three different estimates of the residual amplitude in $F_1(t)$ after the internal flexural motions have relaxed. Equation (4.51), which applies to just the central subunit, yields the solid and dotted lines. The solid line uses the B2 result $\langle \eta_c^2 \rangle = (0.06065)L/P$. The dotted line uses the exact result $\langle \eta_c^2 \rangle = (0.08333)L/P$. The dashed line was calculated using Eq. (4.50), which averages the exact result over all rods in the DNA fragment.

$\langle \eta_c^2 \rangle_{BZ}$. A recent analysis[109] shows that the use of rather long bond vectors (31.8 Å), corresponding to about 9 bp, in the simulation necessitates a large rms angle (18.7°) between those bond vectors, so the expression for $\langle \eta_c^2 \rangle$ following Eq. (4.51) is not expected to apply. When shorter bond vectors (3.4 Å), corresponding to 1 bp, are employed in the analytical theory, but the mean squared angular displacement at $t = \infty$ is calculated for the *average* of the central nine bond vectors, that is, $\langle \bar{\eta}^2 \rangle = \langle [(\frac{1}{9}) \sum_{l=-4}^{4} \eta_{c+l}]^2 \rangle$, the resulting $\langle \bar{\eta}^2 \rangle$ agrees very well with $\langle \eta_c^2 \rangle$ from the simulation.[109] This strongly suggests that the agreement between the simulation and $\langle \eta_c^2 \rangle_{BZ}$ may be an artifact due to the use of long bond vectors in the former and that use of shorter bond vectors would probably yield results more in agreement with $\langle \eta_c^2 \rangle$ in Eq. (4.51) than $\langle \eta_c^2 \rangle_{BZ}$.

A recent normal-mode theory[216] for the flexure dynamics of weakly bending finite filaments ($L \lesssim P$) improves on the Barkley–Zimm theory (*as applied to finite filaments*) in three important respects: (1) The hydrodynamic interactions are appropriate for the finite length; (2) the Langevin equations of motion are solved exactly; and (3) the mean squared amplitudes of the normal coordinates are evaluated exactly. This theory requires considerably more elaborate computation than Eqs. (4.44) and (4.45), but requires orders of magnitude less computer time than a Brownian dynamics simulation. For $L/P = 0.5$, the $\langle \Delta_x(t)^2 \rangle$ calculated from this theory agrees with the Brownian dynamics simulations within the statistical errors in the latter.[113] The longest bending time T_2 from this theory exceeds that from Eq. (4.44) by the factor 1.33. In view of these results, we believe that the BZ theory somewhat underestimates both the relaxation time T_2 and the mean squared amplitudes of the *longer* bending modes for a filament with $L/P = 0.5$ and sufficiently short bond vectors. For shorter wavelength bending modes, both the mean squared amplitudes and relaxation times from the BZ theory should be rather more accurate.

For $L/P = 2.36$, the simulated $\langle \Delta_x(t)^2 \rangle$ agrees very well with the BZ result in Eq. (4.45) out to 200 ns, at which time the longest contributing bending mode ($T_2 = 595$ ns) is only about 30% relaxed.[114] In regard to this good agreement, it must be emphasized that the simulation applies for an intermediate-size filament containing only 2.36 persistence lengths ($P = 400$ Å). The $\langle \Delta_x(t)^2 \rangle$ for such a filament must constitute an *upper limit* to the $\langle \Delta_x(t)^2 \rangle$ that would be obtained for an infinitely long filament with the same P. From its agreement with the simulation, one may conclude that $\langle \Delta_x(t)^2 \rangle_{BZ}$ similarly provides an *upper bound* for the actual $\langle \Delta_x(t)^2 \rangle$ of an infinite filament, at least up to $t = 200$ ns!

The simulation procedure was generalized by Allison *et al.*[209] to treat both symmetric and asymmetric anisotropic bending, as well as permanent bends. Symmetric anisotropic bending is found to have little effect for times longer than a few nanoseconds, provided the long-range persistence length[85] is not altered. The bending normal modes that relax in a few nanoseconds extend over several full turns of the helix, or more, which evidently averages out the effects of the anisotropy.

For bending modes with wavelengths longer than two persistence lengths, approximations in the BZ theory, including the neglect of the tension term, become rapidly invalid. At present, there is no reliable theoretical guide for this regime. The empirical electric birefringence decay for 587-bp restriction fragments at times longer than 1 μs gives[26]

$$F_0(t) = 0.31 \exp[-t/(7.8 \times 10^{-6})] + 0.69 \exp[-t/(45 \times 10^{-6})] \quad (4.52)$$

after correction to 20 °C.[82] The ~8-μs component has been observed also in transient electric dichroism[115] and depolarized dynamic light scattering.[116] It corresponds to the rotational relaxation time for a rigid rod with $b = 12.5$ Å and $L = 800$ Å, which is between 1 and 2 persistence lengths. The relations

$$F_1(t) = [F_0(t)]^{5/6} \quad (4.53)$$

$$F_2(t) = [F_0(t)]^{2/6} \quad (4.54)$$

which are exactly valid for very long filaments, are proposed to estimate $F_1(t)$ and $F_2(t)$ from $F_0(t)$.[39, 82] A recent Brownian dynamics simulation shows that use of Eqs. (4.53) and (4.54) in Eq. (4.24) is a good approximation even for a 209-bp restriction fragment.[209] A very crude estimate of $F_0(t)$ for larger DNAs on still longer time scales of the Rouse–Zimm coil deformation modes has also been given.[82] These Rouse–Zimm modes, which begin to relax at about 100 μs, are far to slow to have any significant effect on the FPA, but are crucial for the determination of NMR R_2 relaxation rates of large DNAs.[82]

As the persistence length P approaches infinity, the bending relaxation times T_n all approach zero, and the amplitude of the uniform mode approaches 1.0, in which case $F_n(t)$ is precisely the result for a rigid (against flexure) rod.

4.2.3.7. *An Alternate Theoretical Approach*

Yamakawa and co-workers have formulated a discrete helical wormlike chain model that is mechanically equivalent to that described above for twisting and bending.[79, 111, 117] However, their approach to determining the dynamics is very different. They do not utilize the mean local cylindrical symmetry to factorize the terms in $r(t)$ into products of correlation functions for twisting, bending, and internal motions, as in Eq. (4.24). Instead, they solve a large complex set of nonlinear coupled equations that contain the effects of both twisting and bending motions by invoking a preaveraging approximation, which, they note, is rather severe. Results are obtained in the form of extended sums of exponential decays, in which the decay constants are eigenvalues of large matrices. The formal simplicity that arises from retaining the mean squared angular displacements about the individual body-fixed axes in the exponents in Eqs. (4.25) and (4.26) is entirely absent in their treatment. In fact, their formulation is so complex and computationally intensive that it is entirely impractical for routine fitting of large quantities of data. On the other hand, their formulation (though not their approximate results) in principle contains the effects of coupling of large-amplitude twisting and bending motions. Although such coupling is *implicitly* contained in Eqs. (4.24)–(4.27), it is *not* present in the specific models that we use to determine $\langle \Delta_z(t)^2 \rangle$ and $\langle \Delta_x(t)^2 \rangle$. However, it is doubtful whether any *linear* treatment, including that of Yamakawa and co-workers, is valid in a regime where the amplitudes of twisting and bending are so large that these motions are strongly coupled.

Yamakawa and co-workers[79] compare theoretical curves for various input parameters with the experimental data of Millar *et al.*[19, 20] over a limited time span up to 80 ns. For $t \geqslant 40$ ns, the theoretical curves deviate significantly and progressively farther from the experimental values. As they note, the fit is not nearly so good, even over this limited time range, as fits achieved using the Intermediate Zone formula for $C_n(t)$. By using the latter formula in Eq. (4.24), excellent fits with small reduced chi-squared, $\chi_r^2 \leqslant 1.1$, can routinely be obtained for times as long as 120 ns or more. Not only does the theory of Yamakawa and co-workers not match the data so well, but their optimum hydrodynamic radius, 15.8–18.1 Å, is much too large, substantially exceeding that recently measured for rotation around the symmetry axis (12.0 ± 0.6 Å)[109] and those less precise values determined from transient electrooptic[25–27] and steady-state transport measurements (12.5–13.0 Å). To estimate the torsion constant reliably, within a factor of two, requires rigorous least-squares fitting of the data over at least one, and preferably several, time spans, which Yamakawa and co-workers did not do, as well as an accurate independent value of the friction factor γ, which they did not have. It also appears as if their optimum torsion constant must vary significantly with the time span of the data fitted, although they did not investigate that possibility. In view of these problems, their reported torsional rigidities could easily differ from the correct values by a factor of two or more. For such reasons, we do not pursue this theory further, except to note that it yields torsional rigidities in a reasonable range.

For much more flexible chains, the approximations invoked by Yamakawa and co-workers seem to be less severe, and agreement with the experimental FPA data is rather good.[118, 119]

4.2.4. Instrumentation

With the passage of time, our instrumentation has evolved significantly from that employed in the original FPA measurements.[18] The present light source is still a Spectra-Physics synch-pumped dye laser with Rhodamine 6G as the lasing medium. It delivers pulses of 575–585-nm light with a full width at half maximum (fwhm) of 15 ps at 800 kHz. The original Ar-ion pump laser has been replaced by a stabilized Spectra-Physics frequency-doubled, mode-locked YAG laser, which produces pulses of 532 nm light with a fwhm of 60–80 ps at 82 MHz. This has substantially increased the output power of the dye laser and improved the overall ease of operation, although the long-term power stability is not quite as good. The original RCA 31034C photomultiplier tube is now usually replaced by a Hamamatsu R2809U microchannel plate tube, which has considerably faster response. The original electronics for time-correlated single-photon counting have been replaced by a Tennelec TC-454 amplifier–discriminator, modified according to manufacturer's recommendations, a locally constructed coincidence unit, and an Ortec 567 time-to-amplitude converter (TAC). The timing pulse is now supplied by a Motorola MRD 500 photodiode that is illuminated with a portion of the incident beam. The coincidence unit enables us to run in the TAC forward configuration, wherein the timing pulse starts the TAC, but only when the sample fluorescence triggers a photoelectron pulse on the same shot. The TAC is connected to a Nucleus Spectrum 88 multichannel analyzer (MCA) that is interfaced to a Terak/LSI-11 microcomputer. That in turn is linked to a cluster of departmental Microvaxes, which carry out the extensive data deconvolutions. With the new electronics, the fwhm of the instrument response function is about 500 ps for the old photomultiplier tube and 50–60 ps for the double microchannel plate tube. In addition, by attaching wedge prisms or absorption filters to the back and side faces of the cuvette to eliminate reflected light in the plane of the sample volume, and by using a subtractive double monochromator to eliminate stray light, the fwhm of the instrument response function is further reduced to 30–40 ps. The polarization of the exciting pulse is now changed by a stepper motor, which is controlled by the Terak computer. At present, the duty cycle consists of 50-s data collection for vertical (V) polarization, 4 s to switch polarization from V to horizontal polarization (H), 50 s of data collection for H polarization, and 4 s to switch back from H to V. In this way, the fluorescence intensity curves, $i_V(t)$ and $i_H(t)$, are alternately accumulated for several minutes to acquire one data set. The duty cycle time is far less than the time scale of the power drift. In this way, several such data sets are accumulated for each time span and fitted separately.[18] In other respects, the apparatus is about the same as before.[18] With these improvements in instrumentation, the error bars on the measured torsion constants have declined from about $\pm 8\%$ in the original study[18] to $\pm 4\%$ or less in favorable cases at present. Perhaps more important, the equipment is now much more robust, and the measurements are becoming more routine. Substantially increased throughput has enabled the investigation of certain problems requiring intensive data acquisition.

4.2.5. Protocol and Data Analysis

Our experiments are typically carried out at DNA concentrations of 20–50 μg/ml with 1 ethidium per 300 bp, so that depolarization by excitation transfer is negligible.[18] The sample is excited with 575-nm light, and the fluorescence is detected at 630, 640, or 645 nm. Less than one fluorescent photon is detected for every 100 laser shots. The instrument response function $e(t)$ is determined using 575-nm incident light scattered from a suspension of polystyrene latex spheres.

The measured vertical and horizontal components of the fluorescence intensity, $i_V(t)$ and $i_H(t)$, respectively, are combined to form the two decay curves

$$s(t) = i_V(t) + 2i_H(t), \qquad d(t) = i_V(t) - i_H(t) \tag{4.55}$$

When emitted light from several species (j) and some scattered light are superimposed, the true sum response to a delta-function exciting pulse is

$$S(t) = A_S\, \delta(t) + \sum_j S_j(t) \tag{4.56}$$

and the true difference response is

$$D(t) = r_S A_S\, \delta(t) + \sum_j r_j(t)\, S_j(t) \tag{4.57}$$

where A_S and r_S are, respectively, the integrated intensity and anisotropy of the scattered light, $\delta(t)$ is a delta function, and $S_j(t)$ and $r_j(t)$ are, respectively, the fluorescence intensity response and anisotropy of the jth species. Normally, $S_j(t) = S_j^\circ \exp(-t/\tau_j)$, where τ_j is the relaxation time of the fluorescent excited state. The experimental curves are related to the true curves by the usual convolution relations, which can be written as

$$s(t) = A_S e(t) + \int_0^t dt' e(t') \sum_j S_j(t-t') \tag{4.58}$$

$$d(t) = A_S r_S e(t) + \int_0^t dt' e(t') \sum_j r_j(t-t')\, S_j(t-t') \tag{4.59}$$

For ethidium/DNA complexes, we usually represent $S(t)$ by two exponentials, $S_1(t)$ and $S_2(t)$, plus a delta function to account for a small amount of Raman shifted light from the solvent. $S_1(t)$ represents intercalated dye with a lifetime of 21.8 ± 1.0 ns and is associated with the anisotropy function $r_1(t)$ of the dye/DNA complex. $S_2(t)$ accounts for a (usually) small amplitude of short-lived component ($\tau_2 = 1.5$–2.0 ns) due to nonintercalated dye. If this short-lived nonintercalated dye is completely free, then its $r_2(t)$ will relax with the time constant of free ethidium ($\tau_{rot} = 88 \pm 4$ ps in dilute aqueous solution at 20 °C, as measured using the microchannel plate tube). However, if some of the short-lived nonintercalated dye is (outside) bound to the DNA, then $r_2(t)$ for that species will relax much more slowly, perhaps even as $r_1(t)$ for intercalated dye. In the majority of our experimental conditions, the prevailing amplitude ratio is large, that is, $S_1(0)/S_2(0) \gtrsim 15$, and there is negligible indication of outside bound dye. In such cases, the same best-fit parameters are obtained from Eq. (4.59) over time spans of 20 ns or more by any of the following procedures: (1) use $r_1(t)\,S_1(t) + \exp[-t/(85 \times 10^{-12})]\,S_2(t)$; (2) set $r_2(t) = 0$ and use just $r_1(t)\,S_1(t)$; or (3) use $r_1(t)[S_1(t) + S_2(t)]$ but delete the first 4 ns from the fit. Procedure 2 agrees with procedure 1, because $r_2(t)$ relaxes in a much shorter time than $S_2(t)$, and hence the free dye contributes negligibly to the production of polarized photons. Procedure 3 agrees with procedure 1 for that same reason and for the additional reason that $S_2(t)$ is negligible compared to $S_1(t)$ after 4 ns. In our earlier work, $S(t)$ was often represented by a single exponential, $S_1(t)$, which gives essentially the same fitted parameters as procedures 1–3. If procedure 3 is employed *without* deleting the first 4 ns, a somewhat (15–20%) higher value of the best-fit torsion constant is obtained, which we believe is incorrect.

Ethidium binding is significantly diminished when (i) the added ethidium per base pair is less than 1/200, (ii) the DNA concentration is low ($\lesssim 0.025$ mg/ml), (iii) the salt concentration is high ($\gtrsim 1\,M$), or (iv) another intercalator is present in large excess. Under these conditions, $S_1(0)/S_2(0) \lesssim 10$, and there appears increasing evidence that some of the nonintercalated dye is rotating much more slowly than free dye. In such cases, procedures 1 and 2 yield somewhat ($\sim 15\%$) lower best-fit torsion constants than procedure 3. This implies that some portion of the $S_2(t)$ photons are not depolarized as rapidly as expected for free dye, so they contribute significantly to the anisotropy up to 2–4 ns. If the quantity $r_1(t)[S_1(t) + 0.15 S_2(t)]$ is employed in Eq. (4.59) with *no* times deleted from the fit, the best-fit torsion constants agree with those from procedure 3. This coefficient (0.15) of $S_2(t)$ is a very rough estimate. Nevetheless, it is probably safe to say that a significant fraction, though much less than half, of the rapidly emitting (presumably nonintercalated) ethidium is most likely bound to the DNA and reorients at a comparably slow rate over the first 2–4 ns. In such cases, we employ procedure 3 above, because it omits most of the short-lived emission of nonintercalated dye from the fit. It also yields values in good agreement with those obtained with higher DNA concentration (0.05 mg/ml) or dye/bp ratio (1/150), where the amplitude ratio is higher ($S_1(0)/S_2(0) \gtrsim 15$) and this small component of rapidly emitting, slowly rotating dye is negligible.

Curve fitting is carried out using a nonlinear least-squares convolute and compare approach, which is based on the Marquardt algorithm.[120] For a sum of exponentials, such as $\sum_j S_j(t)$, recursion relations for the function and its derivatives[121] are employed to reduce computing time.[18] After $S(t)$ is determined from $s(t)$, then $r(t)$ is determined from $d(t)$ and $S(t)$ in a second deconvolution step.[18] In that step, the convolution of the product [e.g., $r_1(t)\,S_1(t)$] with $e(t)$ is now carried out by first multiplying their Fourier transforms together, and then taking the inverse transform. With the fast Fourier transform (FFT) routine from the IMSL library, this results in about a fivefold reduction in the overall running time of the fitting program for ~950 data channels, as compared to that when the convolution is calculated by direct summation. In fitting $S(t)$ to $s(t)$, the adjustable parameters are A_s, S_j°, and τ_j. In fitting $D(t)$ to $d(t)$, the adjustable parameters are r_S (typically 0.3 in our experiments) and various parameters in the model function $r_1(t)$, which is *always* given by Eq. (4.24). In most cases, the factor $\frac{2}{5}$ is replaced by an adjustable initial anisotropy r_0, and Eqs. (4.28a–c) are employed for I_n, as discussed in connection with Eqs. (4.30a–d).

Collection of multiple data sets for each time span, with frequent alternation of the polarization, is an essential feature of our protocol. This provides some protection against the effects of drifts in laser power, photomultiplier quantum yield, and absolute calibration of the TAC, photochemical decomposition of the dye, and any other long-term processes that may alter the measured fluorescence response curves. Separate analysis of each data set is necessary to provide an indication of the uncertainty in run-to-run reproducibility and to detect and delete the rare spurious data set.

An especially important aspect of our protocol is the collection and analysis of data over different time spans.[18] This is done using the same number of channels to collect about the same number of photons, but with different channel delays. Typically, data on four time spans, 0 to 16–20 ns, 0 to 35–40 ns, 0 to 65–80 ns, and 0 to 120–130 ns, are collected and analyzed. In our experience, the constancy of the best-fit model parameters with variation in time span is the most sensitive and reliable test for the shape of the model function.

4.2.6. Experimental Results

4.2.6.1. Orientation of the Transition Dipoles

The intramolecular orientations of the transition dipoles for the S_1 and S_2 bands of ethidium were determined from polarized fluorescence studies of the dye in extremely viscous solutions and stretched films.[122, 123] Both dipoles lie in the plane of the phenanthridinium ring and are separated by an angle of about 60°.[123]

For intercalated ethidium, the polar angles of the S_1 and S_2 transition dipoles with respect to the symmetry axis of the DNA were originally estimated to be 70.5 and 67°, respectively.[88] These values were obtained by extrapolating the electric field-induced linear dichroism of restriction fragment/ethidium complexes to infinite electric field (perfect alignment).[88] The reduced linear dichroism (LD^r) for each absorption band of the intercalated dye was determined from both absorbance- and fluorescence-detected absorption anisotropies with respect to the direction of the applied electric field.[88] The validity of the long extrapolation of LD^r as a function of $1/E$ (E = applied electric field) to obtain the limiting LD^r at perfect alignment has been questioned.[124] However, two observations inspire some confidence in these limiting LD^r values. First, for some nonintercalating dyes, the LD^r actually extrapolates to significantly more negative values, corresponding to polar angles near 90°.[88] This argues strongly that the transition dipoles of ethidium are not perpendicular to the rod axis, but are inclined at a significantly lower angle. Second, both electric[25, 124] and flow[125–128] linear dichroism studies of naked DNAs yield limiting LD^r values in nearly the same range. Especially important in this regard is the work of Johnson and co-workers,[126–128] who measured the linear dichroism for more than one UV band and were thereby able to eliminate the orientation factor and avoid the long extrapolation to perfect alignment. They concluded that the transition dipoles, which lie in the planes of the bases, have equilibrium polar angles less than 73°. We discount the possibility that ethidium is perpendicular, while the base planes are tilted (perhaps due to propeller twist within a base pair). Such a circumstance would almost certainly involve significant disruption of the local structure upon intercalation and would be expected to significantly alter the torsional rigidity and dynamics, which is not observed, even up to very high levels of intercalator binding.[53]

Granted the validity of the measured limiting LD^r, there remains still another difficulty. A large amplitude of internal motion could significantly decrease the magnitude of the (negative) LD^r, so the apparent polar angle calculated from LD^r, assuming a rigidly attached dye, would be significantly smaller than the actual polar angle of its equilibrium, or average, orientation. The original investigators did not consider this possibility, which was recently noted by Hård.[86] Hård proposed combining two measurements pertaining to the S_1 band of ethidium, namely, the limiting LD^r and the reduction in FPA amplitude due to rapid libration of the dye, to obtain both the equilibrium polar angle ε_0 and the rms amplitude of angular libration of the dye.[86] Theoretical errors in Hård's treatment invalidate most of his quantitative conclusions.[83] A correct analysis based on Hård's proposal has been presented.[83] The limiting reduced linear dichroism can be written in the form

$$\mathrm{LD}^r = 3I_0(\infty)^{1/2} \tag{4.60}$$

where $I_0(\infty)$ is given by Eq. (4.29a) with $t=\infty$. For any given value of LD^r, Eq. (4.60) provides a parametric relation between ε_0 and $\langle\delta\varepsilon^2\rangle$. Evidently, the LD^r is affected by polar wobble ($\langle\delta\varepsilon^2\rangle$) of the transition dipole, but not by azimuthal wobble ($\langle\delta\zeta^2\rangle$), as expected for completely aligned DNAs. The experimental estimate is $\mathrm{LD}^r = -0.97 \pm 0.05$.[88] The loci of pairs of values $(\varepsilon_0, \langle\delta\varepsilon^2\rangle^{1/2})$ that satisfy Eq. (4.60) for $\mathrm{LD}^r = -0.90$, -1.00, and -1.10 are presented in Figure 4.8.[83] Interpolation yields the curves for $\mathrm{LD}^r = -0.92$, -0.97, and -1.02.

When the internal motion is so rapid that it relaxes before significant relaxation by twisting and bending takes place, there exists a time domain in which $I_n(t) \cong I_n(\infty)$ yet $C_n(t) \cong 1.0$, $F_n(t) \cong 1.0$, and the residual anisotropy is given by

$$A = \sum_{n=0}^{2} I_n(\infty) \tag{4.61}$$

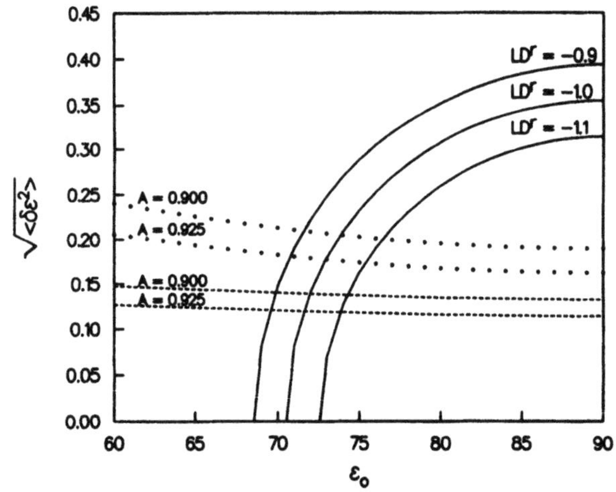

Figure 4.8. Calculated value of the rms amplitude of local polar libration ($\langle\delta\varepsilon^2\rangle^{1/2}$) that satisfies Eq. (4.60) or Eq. (4.61) versus the assumed equilibrium polar angle (ε_0). The solid lines are the solutions of Eq. (4.60) for the indicated values of the reduced linear dichroism (LD^r). The dashed lines are the solutions of Eq. (4.61) for the indicated values of A when the local angular motion of the transition dipole is assumed to be isotropic. The dotted lines are the solutions of Eq. (4.61) for the indicated values of A when the local angular motion of the transition dipole is assumed to be purely polar. The intersection of pairs of curves defines the region "allowed" by a particular pair of LD^r and A values and a particular assumption about the degree of anisotropy of the local angular motion of the transition dipole. If the LD^r lies between -0.92 and -1.02, as indicated by experiment, then for isotropic internal motion, $\varepsilon_0 = 70.5°$, and $\langle\delta\varepsilon^2\rangle^{1/2} = 0.122$ (7°) fall in the allowed region.

where the $I_n(\infty)$ are given by Eqs. (4.29) with $t=\infty$.[83] Using a streak camera, Magde *et al.*[87] detected a small-amplitude relaxation of the anisotropy with a time constant $\tau \approx 100$ ps, which was attributed to local internal motion, or dye wobble.[39] At $t=500$ ps [where $I_n(t) \cong I_n(\infty)$] their data indicate that $A=0.90$–0.925. Comparable values are obtained from the initial anisotropies ($r_0=0.36$–0.37) observed in single-photon counting experiments with a time resolution of 500 ps. For any given value of A, Eq. (4.61) provides a parametric relation among ε_0, $\langle\delta\varepsilon^2\rangle$, and $\langle\delta\zeta^2\rangle$. One strategy is to adopt particular choices for $\langle\delta\zeta^2\rangle$ and determine the parametric relation between ε_0 and $\langle\delta\varepsilon^2\rangle$ for each.[83] At one extreme, it is assumed that the internal motion is isotropic so $\langle\delta\zeta^2\rangle = \langle\delta\varepsilon^2\rangle/\sin^2\varepsilon_0$. At the other extreme, the internal motion is assumed to be purely polar, so $\langle\delta\zeta^2\rangle=0$. For each choice of $\langle\delta\zeta^2\rangle$, the loci of pairs of values (ε_0, $\langle\delta\varepsilon^2\rangle^{1/2}$) that satisfy Eq. (4.61) for $A=0.90$ and 0.925 are determined and plotted in Figure 4.8.[83]

For either choice of $\langle\delta\zeta^2\rangle$, the experimentally "allowed" values of ε_0 and $\langle\delta\varepsilon^2\rangle^{1/2}$ lie in the intersection between the two curves for $A=0.90$ and 0.925 and the two (interpolated) curves for $\mathrm{LD}'=-0.92$ and -1.02. If the internal motion is isotropic, the values $\varepsilon_0=70$–$72°$ and $\langle\delta\varepsilon^2\rangle^{1/2}=6.9$–$8.0°$ satisfy nicely both the linear dichroism and residual anisotropy constraints. Thus, the original estimate of Hogan *et al.*[88] for ε_0 is essentially sustained in this case. If the internal motion is purely polar, then somewhat larger values, $\varepsilon_0=73$–$77°$ and $\langle\delta\varepsilon^2\rangle^{1/2}=10.0$–$11.7°$, are required to satisfy the two constraints. Arguments against the choice of an anisotropic purely polar motion are given elsewhere.[83] Of course, if there is no polar internal motion whatsoever, then $\langle\delta\varepsilon^2\rangle=0$, and Eq. (4.60) gives precisely the ε_0 reported by Hogan *et al.*[88] In any case, ε_0 *must* be less than 77°, and most likely ε_0 is close to 70.5°, as assumed in previous work from this laboratory.

4.2.6.2. *Internal Motion of the Dye*

Magde *et al.*[87] showed that the small-amplitude initial relaxation was independent of solution viscosity over a very wide range. This rules out rotation of free dye as its origin. Their data were analyzed using Eqs. (4.29).[39] It is assumed that $\varepsilon_0=70.5°$ and that the internal motion is isotropic, so the adjustable parameters are $\langle\delta\zeta^2\rangle \cong \langle\delta\varepsilon^2\rangle$ and $\tau_\varepsilon=\tau_\zeta$. The values $\langle\delta\varepsilon^2\rangle^{1/2}=7°$ and $\tau_\varepsilon=100$ ps give a good fit to the data.[39] This value of $\langle\delta\varepsilon^2\rangle^{1/2}$ falls in the allowed range for isotropic motion in Figure 4.8,[83] as expected. Recent experiments in our laboratory yield the same rms amplitude $\langle\delta\varepsilon^2\rangle^{1/2}=7°$, but a somewhat smaller relaxation time $\tau_\varepsilon \leqslant 60$ ps, which has still not been completely resolved.

Although it is assumed in this and the preceding section that the reduction in residual amplitude is due entirely to internal Brownian motion of the dye in its binding site, other mechanisms of amplitude reduction are also possible.

It is conceivable that the equilibrium orientation of the intercalated dye in its S_1 excited state differs from that in its S_0 ground state and that this is what is responsible for the rapid initial relaxation. If so, the rms amplitudes of internal Brownian motion estimated above would be upper limits to the actual values.

The reported initial anisotropy[87] was $r(0) = 0.39 \pm 0.01$, which is within experimental uncertainty of the theoretical value 0.40. Evidently, there is no significant amplitude of faster motions. However, internal motions of the dye on a time scale much longer than 120 ns would not be detected in FPA experiments, as they would provide only an effectively static equilibrium distribution of dye orientations over the observation period.

4.2.6.3. Friction Factor and Hydrodynamic Radius for Rotation around the Symmetry Axis

The friction factor per base pair γ for rotation of DNA around its symmetry axis was determined from FPA studies of restriction fragments containing $N+1=43$ and 69 bp.[109] Both fragments are sufficiently short that a substantial amplitude of $C_n(t)$, and also $F_n(t)$, resides in their Uniform Mode Zones. Particular values of certain parameters were assumed, namely, the rise per base pair $h = 3.4$ Å, the hydrodynamic radius $b = 12$ Å *for transverse motion* in Eqs. (4.43)–(4.47) (which are quite insensitive to b), and $D_\perp = 1.8 \times 10^6\ \text{s}^{-1}$ for 43 bp and $D_\perp = 4.8 \times 10^5\ \text{s}^{-1}$ for 69 bp. The latter values were extrapolated or interpolated from the data of Elias and Eden[26] using an inverse cubic relation between $D_\perp$ and L. They are close to the values calculated using the theory of Tirado and Garcia de la Torre.[129]

The adjustable parameters were α, γ, and r_0. The best-fit friction factors γ for the 43-bp fragment are the same on all four time spans, as shown in Figure 4.9. The best-fit γ values are similarly independent of time span for the 69-bp fragment.[109] The hydrodynamic radius a for azimuthal rotation was calculated from the measured friction factor for uniform azimuthal rotation of the entire filament, $f_\parallel = (N+1)\gamma$, using the formula of Tirado and Garcia de la Torre,[129] $f_\parallel = 3.841\eta\pi a^2 L(1+\delta_\parallel)$, where η is the solvent viscosity, and $\delta_\parallel$ is an end-plate correction, which they tabulate. The same value

$$a = 12.0 \pm 0.6 \text{ Å} \tag{4.62}$$

was obtained for both fragments. However, such good agreement could be achieved only when (1) the twisting correlation functions appropriate for such short filaments were used in the proper time zones, (2) the correct amplitudes $D_n(\infty)$ were employed for the uniform tumbling mode decays in Eq. (4.49), and (3) the data analysis was restricted to times after the bending deformation modes have all died away, leaving just a reduced amplitude of the uniform

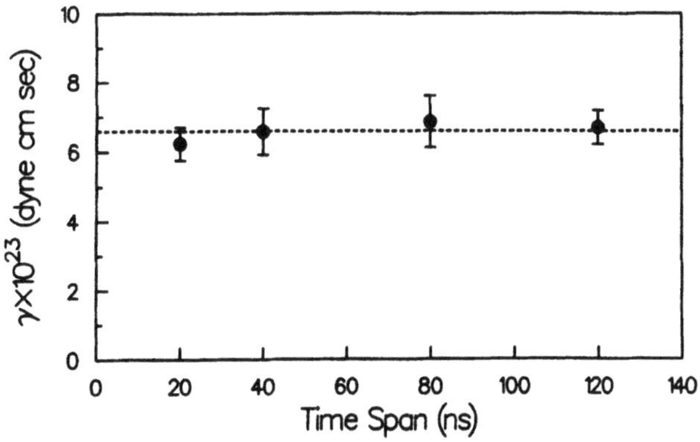

Figure 4.9. Best-fit friction factor γ versus experimental time span for 43-bp fragment. The sample is in 0.1 *M* NaCl, 10 m*M* cacodylate, 1 m*M* EDTA, at pH 8.6 and T = 20 °C. Twelve data sets were averaged to obtain the best-fit friction factors for each time span. The dashed line is the average of the best-fit values for all four time spans.

tumbling mode. When the BZ tumbling correlation functions were used in the data analysis, the hydrodynamic radii came out slightly smaller, and the agreement between the values for the two fragments was not as good, barely within the joint experimental errors.[109]

The friction factor per base pair for azimuthal rotation of a long DNA, for which end-plate corrections are negligible, is calculated from a to be

$$\gamma = 4\eta\pi a^2 h = (6.15 \pm 0.62) \times 10^{-23} \text{ dyn-cm s} \tag{4.63}$$

in water at 20 °C. Because the real DNA is far from a smooth cylinder, the hydrodynamic radii a and b for azimuthal rotation and end-over-end tumbling, respectively, are not required to be identical. Thus, one cannot infer b from a, which is now considerably more precisely determined.

In view of the fact that the DNA cross section perpendicular to the helix axis is roughly an (eccentric) ellipse with a semi-major axis of 10 Å and a semi-minor axis of 5 Å, the effective hydrodynamic radius could conceivably have been as low as $[(10^2 + 5^2)/2]^{1/2} = 8$ Å, yielding a γ only half as large as observed.[109] The measured 12-Å hydrodynamic radius implies that a significant fraction of water in the major and minor grooves must be moving as if more or less rigidly attached to the DNA on this time scale. This interpretation is consistent with the appearance of structurally ordered O atoms of H_2O molecules near the DNA surface in electron density maps of crystalline duplexes.[51] It is also consistent with the formation of an extensive network of H_2O molecules hydrogen-bonded to hydrophilic atoms in the major and minor grooves during a 106-ps molecular dynamics simulation of a 5-bp duplex in 830 water molecules.[52] A decrease in solvent mobility, or increased effective viscosity, near the DNA surface in very cold ethylene glycol:H_2O mixtures has been proposed to account for the temperature dependence of the phosphorescence spectra of a dye associated with DNA.[54]

With respect to the *torsion constants* of these 43- and 69-bp restriction fragments, the fitting procedure was not robust, and acceptable precision could not be attained.

Significantly smaller values of the hydrodynamic radius in the range $a = 10.0 \pm 0.3$ Å were recently obtained for 8-, 12-, and 20-bp synthetic DNAs by depolarized dynamic light scattering[225] (DDLS) and for 12- and 36-bp synthetic DNAs by FPA.[226] The origin of the difference in hydrodynamic radius between these short synthetic DNAs, which contain 83–100% GC, and the restriction fragments studied previously is not yet known, but is currently under investigation.

4.2.6.4. *Torsional Dynamics of Long Linear DNAs*

To assign the motion responsible for relaxation of $r(t)$ from 0.5 to 120 ns, it is necessary to ascertain the functional form of the decay. On theoretical grounds, we believe that bending makes a comparatively small contribution over this time range, so in the first approximation it is neglected, though subsequently it is taken into account. Several functional forms of the decay have been considered:

1. The possibility that internal motion of the dye dominates the relaxation in this time range was tested[39] by assuming $C_n(t) = 1.0$ and $F_n(t) = 1.0$ and fitting Eqs. (4.29a–c) to the data. Adjustable parameters are r_0, $\langle \delta\varepsilon^2 \rangle = \langle \delta\zeta^2 \rangle$, and $\tau_\varepsilon = \tau_\zeta$ for isotropic internal motion. Anisotropic motions with $\langle \delta\varepsilon^2 \rangle = 0$ or $\langle \delta\zeta^2 \rangle = 0$ were also examined. Agreement with the data on any given time span is poor, as judged by reduced chi-square (χ_r^2) values and differences between the best-fit curves and data. Moreover, the best-fit parameters vary strongly with variation in the time span of the data from 0–18 ns to 0–120 ns.[39] Clearly, internal motion of the dye is *not* the primary cause of relaxation from 0.5 to 120 ns.[39]
2. The possibility that DNA twists predominantly at sites of isolated rigidity weaknesses was similarly tested. It is assumed that $F_n(t) = 1.0$ and $C_n(t)$ is given by Eq. (4.36) for the Initial Exponential Decay Zone. If the elementary rod length were 100 bp in a DNA with typical long-range torsional rigidity, then this zone should prevail from 0 to 160 ns. Adjustable parameters are r_0, α, and γ. On any given time span, agreement with the data is satisfactory, though not as good as with the Intermediate Zone formula. This satisfactory agreement is attributable in part to the use of three disposable parameters

(only two are available for the Intermediate Zone formula). However, α and γ both vary strongly, but in opposite directions, with variation of the time span of the data from 0–19 to 0–120 ns.[18] Precisely the same variations of α and γ with time span are obtained when simulated Intermediate Zone data are fitted to the Initial Exponential Decay Zone formula. Evidently, DNA does *not* undergo torsional deformations according to this "linked sausage" model.[18]

3. The possibility that DNA exhibits a uniform torsional rigidity and follows Intermediate Zone dynamics has been quite thoroughly tested.[18, 39] It is assumed that $F_n(t) = 1.0$ and $C_n(t)$ is given by Eq. (4.37) for the Intermediate Zone. In this case, γ is regarded as a known constant, but r_0 and α are taken to be adjustable. The fits are generally very good, with $\chi_r^2 \leqslant 1.1$ in half of all cases, and $\chi_r^2 \leqslant 1.05$ in a third of all cases. The higher χ_r^2 value reported in the original work[18] was due to a spurious factor in the statistical weights, which had no other consequences. Except under special conditions, the best-fit torsion constant α is always independent of time span from 0–20 ns to 0–120 ns, as shown in Figure 4.10. Typically, r_0 lies in the range 0.34 to 0.37, with higher values usually associated with the data sets giving the best fits. These observations provide rather strong evidence that $C_n(t)$ exhibits Intermediate Zone decay for sufficiently long DNAs with more than 2000 bp. Major torsional rigidity weaknesses in the range 1/20 bp to 1/1000 bp are effectively ruled out by these observations.[18]

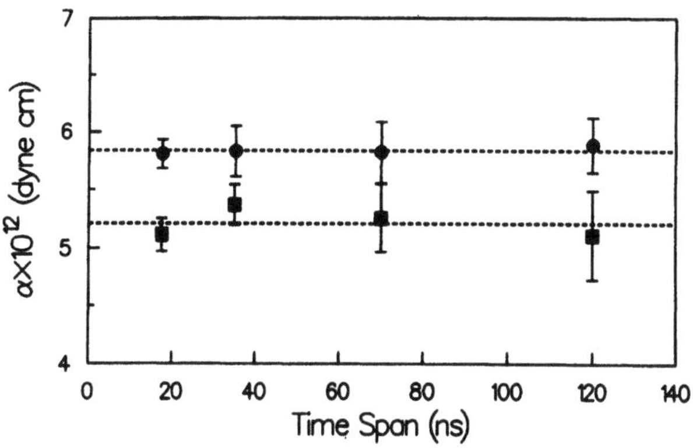

Figure 4.10. Best-fit torsion constant α versus experimental time span for two different samples of supercoiled pUC8 dimer. The two samples are in 10 m*M* NaCl, 10 m*M* Tris, 1 m*M* EDTA, at pH 8 and T = 20 °C, and differ only in their mean superhelix density (σ): ●, $\sigma = -0.048$ (native form); ■, $\sigma = -0.031$. The ethidium concentration was 1 dye for every 300 base pairs in the sample. The results indicate that both samples are adequately described by the Intermediate Zone formula and that the FPA is able to resolve significantly their torsion constants, which differ by only 10%.

Millar *et al.*[20] originally assumed that $\varepsilon_0 = 90°$ and that internal motion of the dye does not affect the amplitude of the $n = 0$ term. Subsequent work partially remedied these deficiencies, but the resulting torsion constants are invalid due to use of the incorrect anisotropy formula of Barkley and Zimm for nonvanishing $\langle \Delta_x(t)^2 \rangle$.[19] Ashikawa and co-workers also assumed that $\varepsilon_0 = 90°$ and used a simple anisotropy expression that is an approximation to the incorrect formula of Barkley and Zimm.[21, 58, 61, 108] Consequently, absolute values of the torsion constants reported by both groups need to be corrected before any detailed comparison with results from other laboratories is possible. In principle, the relative changes in torsion constant are more reliable.

4.2.6.5. *Tumbling Dynamics of Long Linear DNAs*

It is difficult to distinguish $\langle \Delta_x(t)^2 \rangle$ by FPA because it contributes comparatively little to the total relaxation up to 120 ns and the BZ form of $F_n(t)$ implied by Eqs. (4.26) and (4.45) does not differ sufficiently from the Intermediate Zone form of $C_n(t)$.[39] For data analysis, $\langle \Delta_x(t)^2 \rangle$ is calculated first using Eq. (4.45) with a persistence length $P = 500$ Å and hydrodynamic radius $b = 12.5$ Å, and then $F_n(t)$ is calculated using Eq. (4.26). Otherwise, the fitting is carried out precisely as for $F_n(t) = 1.0$. The χ_r^2 values are unchanged, and the best-fit α values are still independent of time span, but they are now 1.9 times larger than found using $F_n(t) = 1.0$.[39] Despite the relatively small ($\lesssim 20\%$) contribution of BZ bending to the overall relaxation at 120 ns, the best-fit torsion constant is strongly affected, in part because the experiment measures directly $\alpha^{1/2}$ instead of α. Even with noise-free simulated data, the fitting program cannot tell whether the simulated data were constructed using $F_n(t) = 1.0$ or the BZ form of $F_n(t)$ from Eqs. (4.26) and (4.45).[39] To cope with this impasse, we note that the best-fit α obtained by assuming $F_n(t) = 1.0$ is a lower bound, because all of the depolarization is assigned to torsion. On the other hand, the best-fit α obtained by assuming the BZ form of $F_n(t)$ must be an upper bound, because the BZ theory provides an upper bound for $\langle \Delta_x(t)^2 \rangle$, and hence a lower bound for $F_n(t)$. Generally, we have reported the lower bound α values and duly noted that the actual values could be as much as 1.9 times larger.

To reduce this ambiguity in α, a reliable estimate of $F_n(t)$ is necessary. The extraordinarily good agreement between the BZ form of $F_0(t)$ and simulations of Allison and McCammon[113, 114] for filaments containing 0.5 and 2.36 persistence lengths strongly suggests that the BZ form of $F_n(t)$ may be a good approximation even for considerably longer filaments at times well beyond 200 ns. Thus, Eq. (4.45) is probably a fairly accurate reflection of the underlying model up to 400 ns or more for filaments of sufficient length. However, one still requires an accurate estimate of the dynamic bending rigidity, or dynamic persistence length, P in order to determine α precisely from FPA data.

A fast component was detected in the off-field decay of the transient electric dichroism of restriction fragments containing 100 to 250 base pairs[25, 130] and assigned to the longest bending mode. Its relaxation times scale in the predicted way with length, approximately as $(N+1)^4$, as indicated in Figure 4.11. However, the experimental values are nearly four times smaller than those calculated by the recent normal-mode theory[216] for a persistence length of 500 Å and correspond almost exactly to those predicted for a dynamic persistence length $P = 2100$ Å.[219] Relaxation times of the fast component detected in the off-field decay of the transient electric birefringence of several restriction fragments of comparable size[220] are found to lie on that same curve for $P = 2100$ Å. When these T_2 values are instead calculated by Eq. (4.44), agreement with experiment is achieved for $P = 1500$ Å. The most straightforward interpretation would be that DNA *on this time scale* is three to four times stiffer than inferred from static persistence length measurements at the same ionic strength (1 m*M*). Recent EPR studies of site-specifically spin-labeled DNAs containing 12, 24, 48, and 96 bp at 100 m*M* ionic strength yield comparable values of the dynamic persistence length (1800–1900 Å), which are again three to four times longer than static persistence lengths measured at that ionic strength.[227] These observations imply that DNA exhibits long-lived bent states, either transient or permanent. This could occur if either stable or thermally accessible but slowly relaxing bent structures contribute significantly to the apparent static flexibility. Evidence for both position-dependent variations in the long-range static curvature and a fairly large bending persistence length ($P \approx 900$ Å) comes from analysis of electron micrographs of DNAs with different markers at either end.[210] With greater spatial resolution, the variations in static curvature and the magnitude of P might be even greater. Three-dimensional structures inferred for several small DNAs (~ 12 bp) from two-dimensional NMR studies in solution show

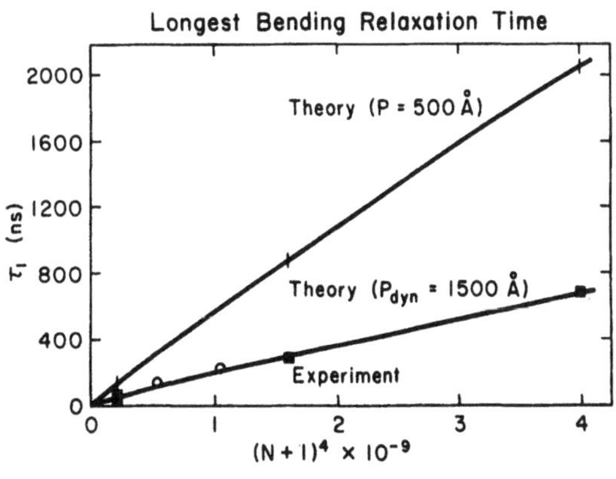

Figure 4.11. Longest bending relaxation time τ_1 as a function of $(N+1)^4$. The symbols are the fast components detected in the off-field transient electric dichroism data of Refs. 25 and 130. The lines were calculated from a recently developed exact normal-mode theory for finite filaments.[216, 219]

surprisingly large variations in curvature.[211–213] Although these curved structures are unconfirmed and could conceivably change with subsequent refinement, and may in any case be straightened somewhat by the greater electrostatic tension prevailing in longer DNAs, they certainly underscore the possibility[228] that sequence-dependent static curvature may contribute to significantly reduce the apparent persistence length obtained from static measurements. Slowly relaxing bent structures may also occur, in view of the accumulated evidence for long-lived conformational isomers.[55, 214, 222, 229] Thus, both permanent and slow transient bends may contribute to the equilibrium curvature, but not to dynamic bending in times less than 10 μs. If our FPA data are deconvoluted using the BZ form of $F_n(t)$ with $P = 1500$ Å, then the best-fit α is 1.35 times larger than that obtained using $F_n(t) = 1.0$.[227]

Another comparison between theory and experiment is provided by the TPD data of Hogan *et al.*[28] for 600-bp DNA/Methylene Blue complexes, shown in Figure 4.12. The theoretical anisotropy is calculated using the BZ form of $F_n(t)$ with $P = 500$ Å, and using the appropriate expressions for $C_n(t)$ with the upper bound $\alpha = (1.90)(3.8 \times 10^{-12}) = 7.22 \times 10^{-12}$ dyn-cm obtained

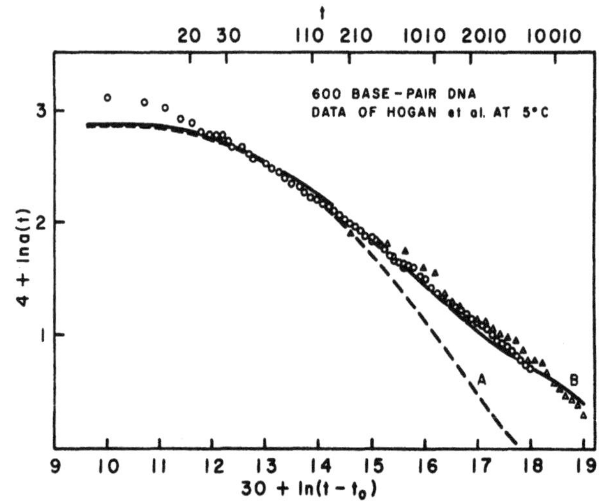

Figure 4.12. Plot of ln $a(t)$ versus $\ln(t - t_0)$, where $a(t)$ is the absorption polarization anisotropy and t_0 is the time of maximum intensity of the polarized photobleaching pulse. The circles and triangles are experimental data of Hogan *et al.*[28] for Methylene Blue intercalated in 600-bp fractionated calf thymus DNA. The curves were calculated by convoluting theoretical $r(t)$s with the pulse profile. Curve A was computed using $\alpha = (1.90)(3.8 \times 10^{-12}) = 7.22 \times 10^{-12}$ dyn-cm, $\gamma = 9.34 \times 10^{-23}$ dyn-cm s, and the Barkley–Zimm tumbling correlation functions from Eqs. (4.26) and (4.45). Curve B was computed using $N + 1 = 600$, $\alpha = 3.8 \times 10^{-12}$ dyn-cm, $\gamma = 9.34 \times 10^{-23}$ dyn-cm s, and the empirical double-exponential correlation function of Elias and Eden given in Eq. (4.52)–(4.54). Both curves use Eqs. (4.37)–(4.41) for the twisting correlation function $C_n(t)$. Both curves are scaled by the same common factor, which is chosen to match curve B to the data at the end of the exciting pulse at $t = 26$ ns.

for a much longer DNA.[39] When the theoretical curve is scaled by r_0 to match the experimental data at the end of the exciting pulse at 26 ns, it agrees closely out to 200 ns, but then begins to fall significantly below the data by about 300 ns,[39] which is close to the relaxation time of the uniform mode of $C_2(t)$ [i.e., $(4k_BT/600\gamma)^{-1} = 346$ ns]. More recent data for 209-bp DNA/Methylene Blue complexes are similarly compared with simulations in which both twisting and bending are admitted and the optical anisotropy is averaged over all subunits.[209] The torsion constant is taken to be $\alpha = 7.6 \times 10^{-12}$ dyn-cm and $P = 500$ Å. The theoretical curve is again scaled to match the experimental values at 26 ns. Agreement is good for a while, but the theory begins to fall significantly below the data at about 80 ns, which is close to the relaxation time for the uniform mode of $C_2(t)$ for this DNA [i.e., $(4k_BT/209\gamma)^{-1} = 80$ ns]. The BZ form of $F_n(t)$ from Eqs. (4.26) and (4.45) is expected to lie *above* the true $F_n(t)$ at sufficiently long times, when the actual $\langle \Delta_x(t)^2 \rangle$ is proportional to $2D_\perp t$ instead of $D_\perp t + \frac{1}{2}B(t)t^{1/4}$. These observations indicate that either (i) the *predicted* amplitude of the uniform mode of $C_2(t)$ is too large, or (ii) the *experimental* amplitude of the $n = 2$ term in $r(t)$ (Eq. 4.24) is somewhat smaller than expected, so its relaxation does not produce the expected relative decrease in $r(t)$ in the appropriate time range. That is, in the experimental system, $I_2(t)$ is too small compared to $I_0(t)$ and $I_1(t)$. These two possibilities are discussed separately below.

(i) The predicted amplitude of the uniform mode of $C_2(t)$ would be too large if the assumed torsion constant were too large. In fact, excellent agreement with the TPD data for 600-bp DNA is achieved from 26 ns to 20 μs by reducing the torsion constant to its lower bound, $\alpha = 3.8 \times 10^{-12}$ dyn-cm, and using Eqs. (4.52)–(4.54) for $F_n(t)$.[39] If the dynamic bending rigidity of DNA were actually four times higher than the static rigidity corresponding to $P = 500$ Å, then the corresponding BZ form of $F_n(t)$ would lie somewhat closer to that in Eqs. (4.52)–(4.54). However, a significant discrepancy between this theory and the TPD data would still remain (unpublished calculations). For the 209-bp fragment, rather good agreement is obtained using $\alpha = 7.6 \times 10^{-12}$ dyn-cm and $P = 1250$ Å. It appears that this agreement, especially in the range from 150 to 500 ns, could be significantly improved by decreasing α and increasing P still further. Although any final conclusion would be premature, *all* of the pertinent observations suggest that the dynamic bending rigidity of DNA greatly exceeds that inferred from its static persistence length and that its torsion constant is closer to the lower bound, as noted above.

(ii) The circumstance wherein $I_2(t)$ is too small compared to $I_0(t)$ and $I_1(t)$ could arise in either of two ways, which are *not* mutually exclusive: (1) A large amplitude of *anisotropic*, *azimuthal* internal motion ($\langle \delta\zeta^2 \rangle$) would substantially decrease $I_2(t)$ relative to $I_1(t)$ and $I_0(t)$; and (2) if the transition dipoles of some of the bound Methylene Blue molecules exhibit a polar angle ε_0 much less than the assumed 72°, then for those dyes $I_2(0)$ would be substantially decreased relative to $I_1(0)$ and $I_0(0)$. Austin and co-workers have made two additional TPD observations on DNA/Methylene Blue complexes that also require explanation. First, the initial anisotropies ($r_0 \lesssim 0.25$) are anomalously low. Second, the TPD dynamics exhibit a strong temperature dependence that would imply a rapid decrease in rigidity with increasing T (R. H. Austin personal communication). This latter result is quite contrary to those obtained by FPA using ethidium dye,[40] by EPR spin-label relaxation,[131, 227] and by dynamic light scattering,[22, 23] which show no appreciable temperature dependence of the twisting and bending rigidities or dynamics apart from that attributable to the factors T and η. Numerous measurements indicate that the static bending rigidity is largely insensitive to T.[22, 23] If the underlying rigidities and $F_n(t)$ and $C_n(t)$ actually are *not* changing significantly with increasing T, then one must conclude that one or more of the amplitude ratios $I_2(t)/I_1(t)$, $I_2(t)/I_0(t)$, and $I_1(t)/I_0(t)$ are strongly increasing functions of T, so that *relative* amplitude is shifted into faster relaxing terms. This could be achieved in any of three ways: (i) the amplitude of polar internal motion $\langle \delta\varepsilon^2 \rangle$ is a strongly increasing function of T; (ii) a large amplitude of azimuthal motion $\langle \delta\zeta^2 \rangle$ is a strongly decreasing function of T; or (iii) the polar angle ε_0 increases significantly with increasing T from a value much less than 72°. These possibilities may require the existence of different binding sites with different $\langle \delta\varepsilon^2 \rangle$ and/or $\langle \delta\zeta^2 \rangle$ and/or ε_0 and that the population shifts with increasing T so as to occupy sites with larger $\langle \delta\varepsilon^2 \rangle$ and/or smaller $\langle \delta\zeta^2 \rangle$ and/or larger ε_0.

In regard to possible explanations of the observations, one can say with certainty only that a substantial amplitude of rapid internal motion is required to account for the low initial anisotropy. The possibility that Methylene Blue is distributed among different binding sites, perhaps even nonintercalated sites, some of which have polar angles ε_0 substantially less than 72° and/or rather different amplitudes of internal motion, is certainly raised by these observations, but is not proven. More direct evidence for nonintercalative binding of some of the TPD probe dye (Methylene Blue) comes from recent fluorescence studies. Under conditions where the Methylene Blue is practically all bound, a substantial fraction (~40%) of the dye fluoresces with lifetimes in the range 300–800 ps, instead of the 25-ps lifetime of the strongly quenched intercalated dye.[230] The long-lived species undergoes a rapid (~100 ps) rotational relaxation with a rather large amplitude that is more easily reconciled with outside binding than with intercalation. Additional evidence for two or more kinds of Methylene Blue binding sites comes from studies of the induced circular dichroism of Methylene Blue[231] and its helix unwinding angle[232] as a function of salt concentration. Variations in the relative populations of different sites with temperature and salt concentration are also observed.[230–232] In view of these complications, any interpretation of the TPD data should be regarded cautiously.

Whether α is closer to its upper or lower bound hinges on whether the dynamic bending rigidity is the same as the static rigidity, corresponding to $P = 500$ Å, or instead is three to four times greater, corresponding to $P = 1500$–2000 Å. The importance of additional measurements of the dynamic bending rigidity of DNA and reliable determinations of the anisotropy of internal motions of ethidium and Methylene Blue in oriented DNA samples is now clear.

Ashikawa and co-workers attempted to determine both the bending and torsional rigidities simultaneously from their FPA data by fitting a simple approximate form that was then compared with the incorrect anisotropy formula of Barkley and Zimm.[21, 58, 61, 108] Neither their claim to distinguish the bending contribution nor their reported bending rigidities can be taken seriously.

4.2.6.6. Values of the Torsion Constant

Upper and lower bounds for the torsion constants between base pairs of several *linear* DNAs at 20 °C are indicated in Table 4.1. Our current best estimate for the torsion constant is obtained by using the BZ theory for $F_n(t)$ with a dynamic persistence length $P = 1500$ Å, which yields α values of $(1.35) \times (3.8 \times 10^{-12}) = 5.1 \times 10^{-12}$ dyn-cm for canonical linear DNAs, 5.8×10^{-12} dyn-cm for linearized pUC8, and 6.5×10^{-12} dyn-cm for linearized pBR322 DNA. The torsional rigidities, often discussed by other investigators, are given by

$$C = h\alpha \tag{4.64}$$

where $h = 3.4$ Å is the rise per base pair. Especially for ϕ29, pUC8, and pBR322 DNAs, measurements were made on many samples under various conditions. Perhaps the outstanding feature of these data is the significantly higher values exhibited by the linearized plasmids pUC8 and pBR322. These might reflect a somewhat different secondary structure. After extended heating at 70–80 °C, the torsion constant for pBR322 at 20 °C fell to 3.8×10^{-12} dyn-cm, but increased steadily over the next eight weeks back to its normal

Table 4.1. Lower-Bound and Upper-Bound Torsion Constants between Base Pairs for Various Linear DNAs at 20°C[a]

DNA	$N+1$ (bp)	% GC	μ^b (M)	$\alpha \times 10^{12c}$ (dyn-cm)	$\alpha' \times 10^{12d}$ (dyn-cm)
ϕ29	17,400	34	0.01	3.8	7.2
ϕ29	17,400	34	0.1	3.5	6.7
Calf thymus	$\sim 10^4$	39	0.01	3.8	7.2
Calf thymus	$\sim 10^4$	39	0.1	3.5	6.7
Lambda	48,000	49	0.01	3.8	7.2
Micrococcus luteus	$\sim 10^4$	77	0.02	4.2	8.0
$(dG\text{-}dC)_n \cdot (dG\text{-}dC)_n$	230	100	0.02	4.0	7.6
$(dG\text{-}dC)_n \cdot (dG\text{-}dC)_n$	590	100	0.02	4.0	7.6
M13mp7	7,238	41	0.01	3.8	7.2
pUC8	2,717	51	0.1	4.3 (5.5)[e]	8.2 (10.4)[e]
pBR322	4,363	54	0.1	4.8 (6.1)[e]	9.1 (11.6)[e]

[a] All torsional rigidities are uniform in the sense that α is independent of the time span of the data.
[b] μ = Ionic strength.
[c] Lower-bound torsion constants α. Estimated uncertainties are about $\pm 8\%$ for ϕ29, calf thymus, lambda, *M. luteus*, and M13mp7, $\pm 10\%$ for $(dG\text{-}dC)_n \cdot (dG\text{-}dC)_n$, and $\pm 6\%$ for pUC8 and pBR322.
[d] Upper-bound torsion constants $\alpha' = (1.9)\alpha$. Percent errors are given in footnote *c*.
[e] Occasionally, high values in this range are observed during the weeks after linearization.

value, where it remained. It is also mentionable that both lower and substantially higher values, indicated in parentheses, were frequently observed for some period during the weeks after linearization of the parent superhelical forms, although after eight weeks the limiting values indicated were reached.

The torsion constant of ϕ29 DNA increased by a factor of 1.25 after 6 months and by a factor of 2.0 after 18 months of storage in solution at 5 °C.[(23)] Its apparent DLS diffusion coefficient at large scattering vector ($K = 4.5 \times 10^5\ \text{cm}^{-1}$), subsequently referred to as D_{plat}, increased by the corresponding factors 1.35 and 1.70, respectively, and the T_m of the 18-month sample went up 9 °C. We have not observed such large long-term *relative* changes in properties of plasmid DNAs, once the plateau is reached at 8–10 weeks after linearization. However, there are some indications that these DNAs may also undergo slow changes at 5 °C. Such observations raise the possibility that the stable secondary structures of DNA at 5 °C might not be identical to that of the native DNA at higher temperatures.

The FPA experiment can also be performed by exciting into the S_2 band of ethidium using 315-nm radiation and detecting at 630 nm (J. H. Shibata, J. C. Thomas, and J. M. Schurr, unpublished results). Substantial cancellation of the two main relaxing terms ($n = 1, 2$) in $r(t)$ considerably diminishes the statistical accuracy of the best-fit α, but nevertheless it remains the same as found for excitation into S_1, which provides a reassuring check. Using intercalated quinacrine as the probe dye, Fan *et al.*[(132)] recently obtained a torsion constant for calf thymus DNA that matches the corresponding value in Table 4.1. This indicates that the values in Table 4.1 are not peculiar to ethidium. The attempt of Fan *et al.* to gain additional precision by fitting simultaneously the FPA data for excitation into the S_2 and S_1 bands was frustrated by the requirement for a second amplitude reduction factor for the S_2 excitation, presumably because of excitation transfer to the dye from DNA bases, which are weakly excited at the shorter excitation wavelength.[132]

If the torsion potential between base pairs actually is Hookean, as assumed for simplicity, then the equilibrium rms twisting displacement *between adjacent base pairs* lies in the range $\langle\phi^2\rangle^{1/2} = (k_B T/\alpha)^{1/2} = 4.5$ to $6.2°$ for ϕ29 DNA in 0.1 *M* NaCl at 20 °C.[23] The equilibrium rms polar angular displacement between base pairs is $\langle\beta^2\rangle^{1/2} = [2(h/P)]^{1/2} = 6.7°$ for a persistence length $P = 500$ Å. The rms angular displacement for twist is evidently quite similar to that for bend. In this sense, DNA is rather isotropic in its deformations. There is no indication that DNA is much more easily twisted than bent, as might have been expected for such a layered structure. For the plasmid DNAs, the rms twist angles are smaller by a factor of 1.4.

The wavelength of the torsion normal mode with relaxation time $\tau = 1$ ns is $\lambda \geqslant 50$ bp for $\alpha \geqslant 3.8 \times 10^{-12}$ dyn-cm [from Eq. (4.34)]. Thus, the shortest torsion normal modes resolved in the FPA have wavelengths extending over about five full turns of the helix. The rms angular displacement of a base pair around its helix axis is about 18° at $t = 1$ ns and increases without bound as t goes to infinity.

4.2.6.7. *Comparison with Results of the Ligation Method*

Linear DNAs with dangling four-base self-complementary ("sticky") single-strand ends form at equilibrium a small population of circular species, which can be enzymatically ligated. Each circular DNA is formed with a particular integral number of turns of one strand around the other, which is called the linking number l. Thermal fluctuations produce different molecules, called topoisomers, with different linking numbers, which are then "locked" in the ligated sample. The ratio of two populations of topoisomers with different linking numbers represents an equilibrium constant from which the free energy change associated with the difference in linking numbers is obtained.[66, 87] In general, the linking number is distributed among twist T and writhe (of the helix axis) W, which obey the constraint[133]

$$l = T + W \tag{4.65}$$

The free energy difference between any two topoisomers of sufficient length can be written as[53]

$$\Delta A = (E_T/N)(\Delta l^2 - \Delta m^2) \tag{4.66}$$

wherein the linking number difference of the first topoisomer is $\Delta l \equiv l - l_0$, $l_0 = N\delta_0/2\pi$ is the (nonintegral) equilibrium net twist in an *unstrained* (nicked) circular molecule containing N base pairs, and $\delta_0 \cong 34.6°$ is the angle between successive base pairs. A corresponding definition applies to the linking number difference Δm of the second topoisomer. The *twist energy parameter* E_T is related to microscopic quantities by[53]

$$E_T = [(2\pi)^2/2k_B T][\alpha\kappa_W/(\alpha + \kappa_W)] \tag{4.67}$$

wherein α is the torsion constant, and κ_W is the *effective* force constant for fluctuations in writhe in a population of nicked circular DNAs; κ_W is proportional to the bending constant κ_β between base pairs, but also contains a length-dependent factor that is known from analytical theory for small circles containing up to eight persistence lengths[134–136] and via Monte Carlo simulations for longer circles.[137–142] An interpolation formula to cover the whole range has also been constructed.[136] From these results, it can be shown for small DNAs that $\kappa_W = [k_B T/\pi^2(0.00385)](P/h)^2\,(1/N) = 2.30 \times 10^{-8}/N$ for $h = 3.4$ Å and $P = 500$ Å.[136] For $N = 200$ bp, $\kappa_W/\alpha \gtrsim (1.15 \times 10^{-10})/(9.1 \times 10^{-12}) = 12$, so α is negligible in the denominator of Eq. (4.67), and κ_W essentially cancels out to leave $E_T \cong [(2\pi)^2/2k_B T]\alpha$. Thus, the static torsion constant α can be obtained from measurements of E_T. For N in the range 200 to 250 bp, two groups have measured $E_T = 4300$,[68, 69] whence $\alpha = 8.8 \times 10^{-12}$ dyn-cm.[136] Curiously, their E_T values differ substantially in a systematic way for larger N.[39]

An alternative method is to examine the relative rates of formation of circles and linear dimers by the ligase as a function of DNA length. Early experiments[143] using this method yielded $\alpha = 7.1 \times 10^{-12}$ dyn-cm.[134] A more recent study using what should be a superior protocol yields a static torsion constant in the range $\alpha = (5.9 \pm 0.6) \times 10^{-12}$ dyn-cm.[221]

For DNAs with $N \gtrsim 10^4$ bp, Monte Carlo results yield the average limiting values[136] $\kappa_W = k_B TP/[(0.095)(2\pi^2)h] = 3.171 \times 10^{-12}$ dyn-cm for $P = 500$ Å. For such long DNAs, it is found that $E_T = 995$.[66, 67] Using these values in Eq. (4.67) yields the static value $\alpha = 5.70 \times 10^{-12}$ dyn-cm. This also lies in the range of the current best estimates of the dynamic torsion constants for ϕ29, pUC8, and pBR322 DNAs.

All estimates of the static torsion constant lie in the expected range between the upper- and lower-bound values of the dynamic torsion constants measured by FPA for pUC8 and pBR322 DNAs. Indeed, the most recent static estimate[221] agrees well with our current best estimates for the dynamic torsion constants of ϕ29, pUC8, and pBR322 DNAs, namely, 5.1×10^{-12}, 5.8×10^{-12}, and 6.5×10^{-12} dyn-cm, respectively.[233] Evidently, the dynamic and static torsion constants are comparable. The origin of the wide variation in the earlier static estimates is not yet known with certainty, but this variation might arise from the temporal changes noted in the preceding section.

Although the most recent estimate of the static torsion constant[221] agrees well with the current best estimates of the dynamic torsion constant,[233] some ambiguity in the latter values will remain until questions about the friction factor for azimuthal rotation and the dynamic bending rigidity are completely resolved. In any case, this ambiguity does not alter the conclusions in subsequent sections regarding relative changes in torsion constant. In the sequel we generally report only the lower-bound value of α, which is proportional to the actual value and faithfully reflects any changes in that.

4.2.6.8. Effect of Salt Concentration

The torsion constant of calf thymus DNA decreases somewhat with increasing NaCl concentration from 0.001 to 1.0 *M*.[19] The torsion constants of the DNAs in Table 4.1 typically remain uniform, but decrease by about 8% between 0.01 and 0.1 *M* NaCl. At much higher NaCl concentrations (>1.0 *M*), the torsion constant of linear pBR322 DNA is still uniform, but it undergoes a substantial decrease, as shown in Figure 4.13. This is accompanied by almost complete disappearance of the positive CD band at 275 nm (U.-S. Kim and J. M. Schurr, unpublished data). The 3300-bp HincII restriction fragment of pBR322 exhibits a substantial (20%) decrease in torsion constant between 0.1 and 1.0 *M* NaCl and then a slower decline at higher salt concentrations, as shown in Figure 4.13. These and other differences in behavior between linearized pBR322 and its restriction fragments are currently under investigation. We find that calf thymus DNA undergoes a 40% decrease in α between 0.1 and 4.0 *M* NaCl. It was formerly believed that DNA adopts the C conformation[144] at high NaCl concentration,[145] but that belief has apparently waned.[208] These decreases in torsion constant are very

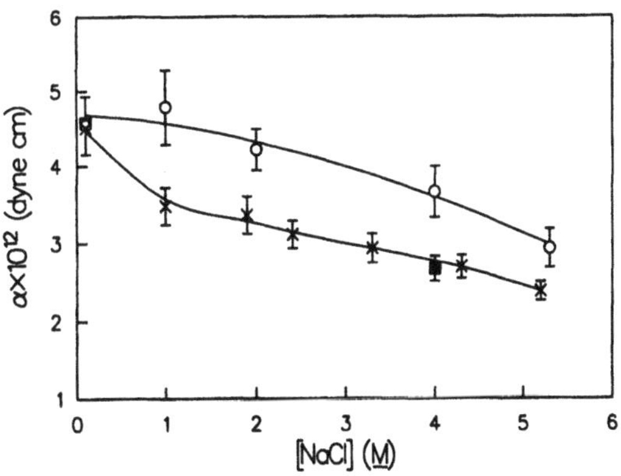

Figure 4.13. Torsion constant α versus NaCl concentration for linear DNAs. ○, Linear pBR322; ■, calf thymus DNA; ×, 3300-bp HincII fragment of pBR322.

likely associated with either a novel, but unknown, secondary structure prevailing at high salt or with junctions between that structure and the normal B-helix. In contrast to the results above, Ashikawa *et al.* reported a decrease in α of calf thymus DNA by only 4% between 0.1 and 3.8 *M* NaCl.[108] The origins of this discrepancy are unknown. In any case, it will be necessary to measure the friction factor γ at high NaCl concentration to establish with certainty whether the torsion constant actually softens at high NaCl concentration or whether instead the hydrodynamic radius for azimuthal rotation decreases.

Mg^{2+} concentrations up to 40 m*M* in 0.1 *M* NaCl have no significant effect on α for linear pBR322 DNA, but cause a modest decrease ($\sim$15%) in α for supercoiled pBR322 DNA.

4.2.6.9. *Effect of Base Composition*

Any effect of base composition on the magnitude or uniformity of the torsion constant is negligible over the range 34–100% GC, as is clear from Table 4.1.[233] As stability against melting increases rapidly with % GC, this argues strongly that locally melted states of the DNA make no significant contribution to the torsional flexibility and Brownian dynamics.

It is notable that the GC samples in Table 4.1 are much too short for the Intermediate Zone formula to apply out to 120 ns. The formulas for subsequent zones of $C_n(t)$ (Eqs. 4.38–4.41) are employed as needed and yield the same value of α for both 230- and 590-bp samples.[146] The 590-bp sample initially exhibited a threefold higher value, which relaxed over several months, during which time many very small fragments dissociated from, or annealed out of, the predominant 590-bp species. This was tentatively attributed to the presence of branched structures, which exhibit high affinity sites for ethidium, in the original material. Both gel electrophoretic and electron microscopic[147] evidence for branched structures in poly(dG-dC) were noted.[146] The 500-bp length from gel electrophoresis was confirmed by sedimentation.[146]

The results of Millar *et al.*[19] for commercial synthetic DNAs with different base compositions must be viewed with caution. Their unfractionated and uncharacterized samples undoubtedly consisted entirely, or in large part, of DNAs so short as to preclude validity of the Intermediate Zone formula, which they used throughout. They also employed the incorrect anisotropy formula of Barkley and Zimm. Similar remarks apply to the results of Ashikawa and co-workers.[108]

The persistence length of poly(dG-dC) was determined to be 800 Å,[148] about 1.6 times the value characteristic of native DNAs. Possible long-term changes in these light scattering samples were not investigated.

FPA measurements on poly(dA-dT) were also undertaken,[146] but the excited-state decay function $S(t)$ contained additional intermediate components besides those of intercalated and free dye. This almost certainly indicates multiple modes of binding that very likely exhibit different values of ε_0 and different amplitudes of internal motion, which would confound the analysis. The possibility of forming branched structures, including cruciforms, is also a concern. No credible values for the torsion constant of poly(dA-dT) have been reported yet.

The absence of any appreciable influence of macroscopic base composition from 34 to 10% GC on the global torsion constant suggests that torsional rigidity is unlikely to be a determining factor in binding specificity. However, the rather large difference in torsional rigidity between the plasmid DNAs (pUC8 and pBR322) and the others might arise from a different sequence-dependent, but more or less global, secondary structure.

4.2.6.10. Effect of Temperature

The torsion constant of linear ϕ29 DNA is independent of temperature from 0 °C up to $T_m = 76$ °C.[40] That both the torsional rigidity and bending rigidity are largely independent of T over this same range was inferred from DLS studies on ϕ29 DNA.[23] EPR spin-label studies likewise indicate that α is independent of temperature.[131] These observations strongly imply that torsional deformations do not occur primarily at sites of high-enthalpy perturbed structures, such as open base pairs. A more quantitative analysis of the results is given elsewhere.[40]

The absence of segmental motion in the FPA relaxation and the invariance of α with respect to changes in base composition and temperature provide very strong arguments that DNA undergoes torsional deformations in a smooth rather than segmental manner (cf. Figure 4.2).

Any anharmonicity of the torsion potential is sufficiently small that the torsion constant is unaffected by changing T from 273 to 351 K.[40]

At $T = 70$ °C, α reproducibly exhibits a value that is somewhat higher (4.4×10^{-12} dyn-cm) than the values (3.51×10^{-12} dyn-cm) prevailing at lower T, though marginally within the joint experimental errors.[40] This may be a manifestation of aggregation in the melting region, which was also detected by DLS[23] and predicted on theoretical grounds.[149]

Ashikawa *et al.* reported 1.3-fold decrease in the torsional rigidity and a 2.2-fold decrease in bending rigidity of calf thymus DNA from 6 to 36 °C.[108] For reasons noted above, their claim to distinguish bending from twisting cannot be taken seriously. In any case, these findings disagree with those for α cited above, as well with as many other works showing that the static bending rigidity varies at most only weakly with T. However, the presence of small amounts of protein contaminants can cause α to decrease significantly with increasing T (J. C. Thomas, unpublished results). The preparation procedure used for this calf thymus DNA[108] does not include a proteinase K digestion step and will very likely leave significant amounts of protein contaminants. Neither fluorescamine nor A_{280}/A_{260} tests were reported to indicate the purity of this sample. In contrast, both tests were reported for the ϕ29 DNA discussed above.(40)

For poly(dG-dC) in 0.1 *M* NaCl, Ashikawa *et al.* reported a 1.4-fold decrease in torsional rigidity and a 1.8-fold decrease in bending rigidity from 6 to 36 °C.(108) Almost certainly, a substantial fraction of this sample is too small for validity of the Intermediate Zone formula used. Hence, the actual decay will vary considerably more strongly with solvent viscosity, which decreases by twofold from 6 to 36 °C. Whether this could account completely for the apparent temperature dependence of the torsional and bending rigidities is not known. A firm conclusion regarding the temperature dependence of the torsional rigidity of calf thymus and poly(dG-dC) DNAs must await additional observation on samples that are much better characterized, and for which the fitting formulas are known to be valid.

4.2.6.11. Z-DNA

Ashikawa *et al.* reported a six- to eightfold *decrease* in apparent torsional rigidity and a sevenfold *decrease* in bending rigidity for poly(dG-dC) in 3.8 *M* NaCl, for which the CD spectrum indicates that the left-handed Z-form predominates.(108) However, a critical assumption of the data analysis, namely, that ethidium is bound to Z-helix, is very likely invalid. It is well known that ethidium binding can induce a cooperative reversion of the secondary structure of poly(dG-dC) from Z to B.(241) Moreover, studies using fluorescence-detected circular dichroism (FDCD) have recently shown that the environment of intercalated ethidium at very low levels in poly(dG-dC) under Z-forming conditions is essentially the same as in B-DNA.(150) There is no evidence in the FDCD spectra for a left-handed binding site. In addition, the magnitude of the 310–330-nm band, which is sensitive to binding ratio in the B-form, is invariant to binding ratio under Z-forming conditions. This indicates that the dye is clustered at constant dye/base pair ratio rather than uniformly dispersed over the sample.(150) Under such conditions, depolarization by excitation transfer may greatly decrease the apparent torsion constant. It is also likely that in a polydisperse sample under Z-forming conditions the dye will cluster primarily on *short* DNAs via all-or-none transitions, so as to minimize the number of (free-)energetically unfavorable B–Z junctions formed. Those shorter species will, of course, depolarize the FPA signal most rapidly. Whether these effects are sufficient to account for the reported relative decrease in torsion constant and bending rigidity is not known.

Electric dichroism data indicate that, when the Z-form is induced by $[Co(NH_3)_6]^{3+}$, the B–Z junctions are not highly flexible.(130) Static light scattering evidence indicates that the persistence length and static bending rigidity are three times larger in the Z-form of poly(dG-dC) than in the B-form.[148]

4.2.6.12. *Long-Range Effects of a $(GC)_8$ Sequence*

DLS, FPA, CD, sedimentation, optical melting, and enzymatic digestion studies were performed on a 1098-bp restriction fragment containing 16 bp of alternating GC inserted near its center and on 1089-bp and 1382-bp control fragments with the identical sequence except for about 30 bp near the center.[222] In 0.1 *M* NaCl, the 1098-bp insert fragment differs from the 1089- and 1382-bp controls by a factor of 0.75 in torsion constant and by a factor of 1.35 in circular dichroism at 273 nm. It also exhibits significantly greater susceptibility to S1 nuclease. While the 1089-bp control melts predominantly in a single main transition with $T_M = 84.1\,°C$, the insert fragment melts in a biphasic manner with a lower transition at $T_M = 84.6\,°C$ and an upper transition at $T_M > 90\,°C$. The latter amounts to 35% or more of the total transition. These and other data constitute almost unequivocal evidence that the $(GC)_8$ insert alters the secondary structure of a substantial fraction (several hundred base pairs, or more) of the total sequence in 0.1 *M* NaCl. With increasing NaCl concentration near 2.5 *M*, the insert fragment undergoes a sigmoidal transition to a stiffer state that must extend over hundreds of base pairs. In 4.3 *M* NaCl (but not in 0.1 *M* NaCl), adding one ethidium per 300 bp induces substantial changes in the DLS and CD of the insert fragment. The control fragments show no sign of either the salt-induced or ethidium-induced transitions. Whether enhancer sequences, which are GC-rich, exert a similar long-range influence on secondary structure is a question that now merits investigation.

4.2.6.13. *Effect of Spermidine in 0.01 M NaCl*

The effect of the trivalent cation spermidine on the torsion constant of ϕ29 DNA in 0.01 *M* NaCl is shown in Figure 4.14.[64] Increasing spermidine concentration induces a small CD change that saturates at 10 μM.[151] Between 10 and 20 μM spermidine, the radius of gyration decreases by a factor of 1.4, indicating partial intramolecular compaction.[151] Beyond 40 μM spermidine, intermolecular aggregation sets in.[151] Even in the partially compacted state at 40 μM spermidine, the apparent torsion constant is not greatly affected, although there is some softening on the longer time spans, which probably reflects zones of softer torsional rigidity. The dynamic light scattering D_{plat} shows a similar relative decrease.[151] There is no sign of any stiffening of α, or hindered torsion, as might have been expected.

When the pH is raised to 10.2, the torsion constant is unaltered on the

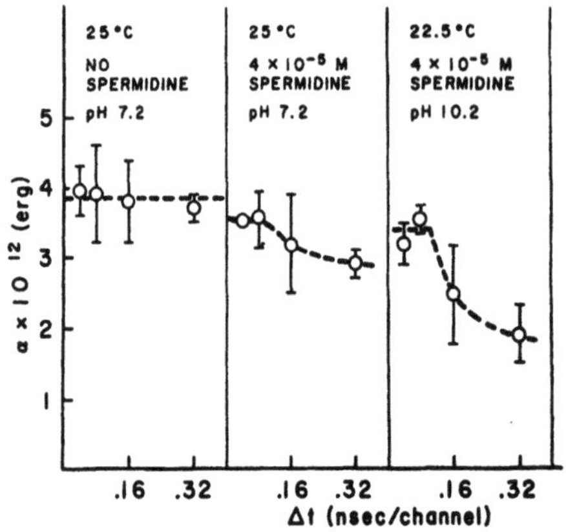

Figure 4.14. Torsion constant α versus channel width for ϕ29 DNA in 0.01 *M* NaCl. In each of the three panels the data points correspond to the experimental time spans 0–20 ns, 0–40 ns, 0–80 ns, and 0–120 ns. The three samples differ in either spermidine concentration or pH. In the absence of spermidine, α is uniform. With the addition of spermidine, α decreases slightly on the longer time spans. The decrease probably represents zones of lower torsional rigidity. The large decrease in α on the longest two time spans when the pH is raised to 10.2 is consistent with the occurrence of occasional (isolated) major rigidity weaknesses.

two shortest time spans, but falls dramatically on the two longer time spans, as would be expected for occasional (isolated) major rigidity weaknesses. These so-called titratable joints are manifested in DLS by comparable relative decreases in D_{plat}.[64, 78, 152] They are attributed to ammonium groups of the bound polyamine, which stabilize the premature opening of a base pair and titration of its imino proton, presumably by serving as the proton donor for the resulting negatively charged imino nitrogen.[152] The distance between rigidity weaknesses can be estimated approximately by noting that they are not yet resolved at 40 ns, but are already well resolved at 80 ns. The wavelengths of the torsion normal modes with such relaxation times are found from Eq. (4.34) to be 310 and 440 bp, respectively. An average gives 375 bp as the estimated distance between major rigidity weaknesses.

When the spermidine concentration is increased to 50 μM at neutral pH, the *apparent* torsion constant undergoes a colossal decrease to $\alpha \cong 0.5 \times 10^{-12}$, but the dynamic light scattering D_{plat} does not exhibit a corresponding decrease, and in fact increases slightly. The dye is still intercalated in the DNA, as judged by its long fluorescence lifetime. However, it is very likely clustered, perhaps on the surface of the aggregates, so that depolarization by excitation transfer might be making a dominant contribution. In the absence of a parallel change in D_{plat}, we are reluctant to interpret our results as indicating an abnormally low torsion constant.

FPA studies on ethidium intercalated into short chicken red-cell DNA, which was condensed by 800 μM spermidine in 5 m*M* Tris, were also reported.[57] Up to 30 ns, the dynamics is the same as for the corresponding free DNA, but at longer times the amplitude of angular motion, or depolarization, is significantly less, presumably due to retardation of tumbling in the aggregate. It appears that the torsional dynamics is largely unaltered, despite the 3.0-Å spacing between helices, as measured by X-ray diffraction.[57]

4.2.6.14. *Effect of Antitumor Drugs*

Covalent complexes of the antitumor compound *cis*-diamminedichloroplatinum(II) and its clinically ineffective *trans* isomer with calf thymus DNA were studied via the FPA of intercalated ethidium.[224] The data are analyzed using Eqs. (4.24), (4.28a–c), (4.37), and (4.45)–(4.47), without deconvoluting the instrument response function. The hydrodynamic radius is assumed to be $a = 13.4$ Å, instead of the measured value of 12 Å, and the dynamic persistence length is assumed to be $P = 500$ Å, independent of bound ligand. Consequently, reported values of the torsion constant are about $(12/13.4)^2\,(1.9) = 1.5$ times larger than the lower bound for that DNA. No assessment was made of the uniformity of the torsion constant over different time spans. Up to 0.2 bound ligand per base pair, the *trans* isomer exerts little or no effect on the product $\alpha\gamma$. However, the *cis* isomer induces a 1.25-fold increase in $\alpha\gamma$ between 0 and 0.01 bound ligand per base pair, and a subsequent 1.8-fold decrease in $\alpha\gamma$ between 0.10 and 0.154 bound ligand per base pair. The increase in $\alpha\gamma$ at low binding levels is tentatively attributed to kink formation and a concomitant increase in γ. The decrease in $\alpha\gamma$ at higher binding levels is attributed to disruption of the DNA secondary structure. Other explanations, such as ligand-induced tilting of both bases and ethidium transition dipoles toward the helix axis at low coverage and a sharp decline in bending rigidity at high coverage, are also possible.

4.2.6.15. *Interaction of Intercalators with Linear and Supercoiled DNAs*

An intercalating dye unwinds the normal B-helix by an angle ϕ, thus decreasing the *equilibrium* net twist l_0 in an *open* (nicked) circular molecule.[70, 71, 153] Intercalation of a dye into a *closed* circular DNA with fixed linking number l generally increases its linking number difference, $\Delta l = l - l_0$, and superhelix density $\sigma = \Delta l/l_0$. Native supercoiled DNAs are normally underwound, so their linking number difference and superhelix density are negative; typically, $\sigma \cong -0.05$. With increasing bound intercalator, Δl and σ rise up to zero and beyond to positive values. The deformational free energy A [cf. Eq. (4.66)] decreases to zero as Δl approaches zero and then increases again for positive Δl. This change in deformational free energy contributes to the effective binding constant of the dye, so in principle the twist energy parameter E_T can be obtained from dye-binding studies. All such determinations to date are based on the assumption that the intercalating dye induces no change in (a) the intrinsic binding constant K, (b) the twisting or bending rigidity, (c) the secondary structure (apart from local unwinding), or (d) the type of tertiary structure (e.g., from interwound to toroidal).[70, 71, 153] A general theory for the binding of one and two different intercalators to supercoiled DNAs under these assumptions is now available.[53] For a dye, such as ethidium or chloroquine, which obeys the nearest-neighbor exclusion model, the predicted binding isotherm is[53]

$$\frac{r}{C} = K\exp[a(r^* - r)](1-2r)^2/(1-r) \tag{4.68}$$

where r is the binding ratio (bound dye/bp), C is the concentration of free intercalator, K is the intrinsic binding constant, $r^* = -\Delta l(360)/\phi N$ is the binding ratio at which $\sigma = 0$, and N is the number of base pairs. The exponential coefficient is $a = 2E_T(\phi/360)^2$. The unwinding angles for ethidium ($\phi_E = 26°$),[154] chloroquine ($\phi = 17°$),[155] and several other dyes are known, so E_T can be determined from a measurement of a.

Literature values for E_T determined by the ligation and ethidium-binding methods were collected and compared.[53] The consensus ligation values, $E_T = 1000 \pm 100$, are twice as large as the consensus dye-binding values, $E_T = 500 \pm 50$! For either method, one can also find one or two exceptional values that lie above the corresponding consensus values by 500–600, but the majority of the data fall in the ranges indicated. This major discrepancy in E_T values most probably reflects a failure of one or more of the underlying assumptions in the dye-binding method, which typically spans a far wider range of superhelix densities.

Equation (4.68) with $a = 0$ applies for linear or nicked circular DNAs. When $r = r^*$, the (initially) supercoiled DNA is predicted to experience no deformational strain, as it is fully relaxed, and to bind the same amount of dye as its linear counterpart with the same concentration of free dye. Under these conditions, the supercoiled and linear DNA/chloroquine complexes are expected to exhibit identical local structures, rigidities, and deformational dynamics. This important corollary to the standard model was untested till recently.[53]

To avoid depolarization by excitation transfer, the DNA is unwound using a second intercalator, for example, chloroquine, that does not engage in excitation transfer to or from the extremely dilute FPA probe, ethidium. Equation (4.68) applies to chloroquine when it is in excess, but the simultaneous binding of trace ethidium obeys a somewhat different relation, which is expressed in terms of the ratio of amplitudes (A_b/A_f) of the bound (slow) and free (fast) components in its fluorescence decay as follows[53]:

$$\frac{A_b}{A_f} = K'_E C^\circ_{\mathrm{bp}} \exp[a_{EC}(r^* - r)](1-2r)^2/(1-r) \tag{4.69}$$

Here r and r^* refer to the predominant chloroquine, K'_E is the product of the intrinsic equilibrium constant for ethidium binding and an experimental efficiency factor,[53] C°_{bp} is the total concentration of base pairs, and

$a_{EC} = 2E_T\phi\phi_E/(360)^2$, where ϕ_E and ϕ are the unwinding angles for ethidium and chloroquine, respectively. Equations (4.68), (4.69), and the conservation equation for chloroquine [$rC^{\circ}_{\text{bp}} + C = xC^{\circ}_{\text{bp}}$, where x is *added* chloroquine per base pair (chl/bp)] are three equations in the three unknowns: r, K'_E, and K for a linear DNA. Once K is determined for the linear DNA, then the same three equations contain the three unknowns r, K'_E, and a (or E_T) for the supercoiled form. Least-squares fits of A_b/A_f data from ethidium fluorescence over a wide range of added chl/bp ratios for both linear and supercoiled pBR322 DNA allow the estimation of K, E_T, and the r values corresponding to each added chl/bp ratio. The values $E_T = 360$ and 460 were obtained for samples with, respectively, normal ($\sigma = -0.048$) and high ($\sigma = -0.083$) native superhelix densities.(53) These E_T values for chloroquine binding lie in or slightly below the consensus ethidium-binding range. Evidently, the discrepancy in E_T values between ligation and dye-binding methods holds for chloroquine as well as ethidium.

4.2.6.16. Effect of Intercalated Chloroquine on the Torsion Constants of Linear and Supercoiled pBR322 DNA

The torsion constant of *linear* pBR322 DNA remains uniform and independent of added chloroquine up to chl/bp = 25, corresponding to $r = 0.19$, as shown in Figure 4.15.(53) The experimental A_b/A_f ratios and the theoretical curve calculated for the best-fit values of K and K'_E are also shown. At the higher chl/bp ratios, ethidium is driven off the DNA by competition for intercalation sites. These observations argue strongly that the FPA is not significantly relaxed by any mobile kinks or solitons, which should be strongly affected by such high levels of intercalation. They also indicate that the discrepancy between ligation and dye-binding values for E_T cannot be ascribed to any significant reduction in the long-range torsional rigidity by intercalated chloroquine, at least in the unstrained *linear* DNA.

If α_d denotes the torsion constant between a dye and a base pair, and α_0 that between two base pairs, then the effective long-range torsion constant is given by(23, 53)

$$\alpha_{\text{eff}} = \alpha_0/[(1-f) + f(\alpha_0/\alpha_d)] \tag{4.70}$$

where $f = 2r/(1+r)$ is the fraction of torsion springs between a dye and a base pair. The observation that $\alpha_{\text{eff}}/\alpha_0 = 1.0 \pm 0.15$ implies that α_0/α_d lies in the range 0.65 to 1.64 with a most probable value near 1.0.(53) Thus, the torsion constant between intercalated chloroquine and a base pair does not differ from that between two base pairs by more than a factor of about 1.5 either way. We conclude that either (a) the chloroquine intercalation site does not correspond to a β-kink,(35) or (b) the torsion constant of the β-kink is not

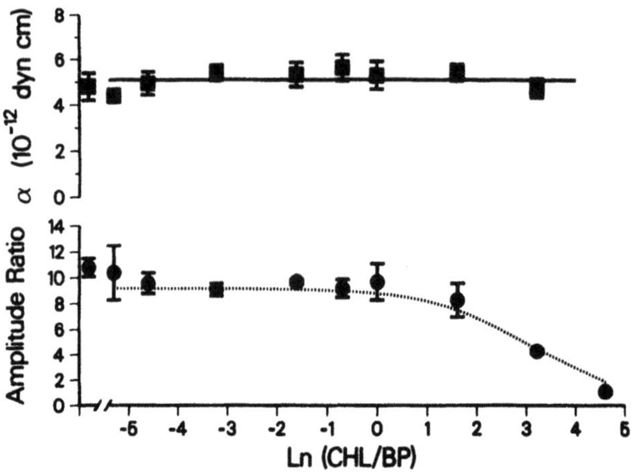

Figure 4.15. Effect of chloroquine on *linear* pBR322 DNA. Torsion constant α (*top*) and fluorescence amplitude ratio of the slow (intercalated) and fast (free) components in the decay of the ethidium fluorescence intensity (*bottom*) versus ln(added chl/bp). The DNA is present in 0.1 *M* NaCl, 10 m*M* Tris, 10 m*M* EDTA, pH 8.5, at 20 °C, and ethidium is present at 1 dye per 300 base pairs. The points at the left of the figure apply for zero added chloroquine. The best-fit theoretical curve (·········) is calculated as described in the text using the values $K = 381\ M^{-1}$ and $K'_E = 2.28 \times 10^5\ M^{-1}$. Linear pBR322 remains uniform and independent of added chloroquine up to chl/bp = 25. The decline in the amplitude ratio at large chl/bp is due to competition between ethidium and chloroquine for binding sites on the DNA and is used to determine the chloroquine binding constant.

smaller than the normal value by as much as a factor of two. Likewise, these results contradict any proposal[94] that the adjacent base pairs are rigidly clamped to the intercalated dye.

In the absence of chloroquine, the apparent torsion constants for supercoiled pBR322 DNAs with normal ($\sigma = -0.048$) and high ($\sigma = -0.083$) twist are uniform and nearly identical to that for the linear DNA, $\alpha = (5.0 \pm 0.4) \times 10^{-12}$ dyn-cm. As described below, superhelical stress apparently induces allosteric transitions in secondary structure, so the secondary structures of supercoiled and linear DNAs might not be identical, despite their similar torsion constants.

Though still uniform, these torsion constants decrease by about 15% with increasing chl/bp ratio up to 10 and 30 for, respectively, the normal-twist and high-twist supercoiled DNAs, as shown in Figure 4.16. For the normal-twist sample, chl/bp = 10 corresponds to $r = 0.141$, or $\sigma = -(r - r^*)(17/360)(10.4) = +0.022$. For the high-twist supercoiled sample, chl/bp = 30 corresponds to $r = 0.249$, or $\sigma = +0.039$. The normal-twist sample passes through $\sigma = 0$ at $r = r^* = 0.097$ and chl/bp = 4.9. The high-twist sample passes through $\sigma = 0$ at $r = r^* = 0.170$ and chl/bp = 10.2. The apparent torsion constants for these relaxed supercoiled DNAs at $\sigma = 0$ clearly differ from those of the corre-

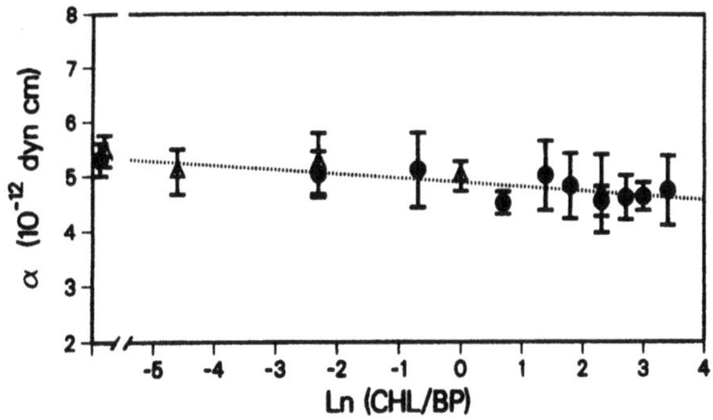

Figure 4.16. Experimental torsion constants α for *supercoiled* pBR322 DNA versus ln(added chl/bp). Data for normal-twist ($\sigma = -0.048 \pm 0.005$) supercoiled DNA are denoted by triangles. Data for high-twist ($\sigma = -0.083 \pm 0.007$) supercoiled DNA are denoted by solid circles. Solution conditions are the same as in Figure 4.15. The points at the left of the figure apply for zero added chloroquine.

sponding linear DNA with the same binding ratios. Even larger differences are observed using proflavine and 9-aminoacridine in place of chloroquine (P.-G. Wu, B. S. Fujimoto, and J. M. Schurr, unpublished results). This 15% reduction in apparent α could arise from (1) a 15% decrease in (actual) torsion constant; (2) a much larger relative decrease in bending rigidity; (3) a change in tilt angle of the ethidium from 70 to 90°; or (4) a clustering of the ethidium, so that excitation transfer contributes significantly to the depolarization process. In any case, the local structure and/or rigidity and dynamics of these relaxed supercoiled DNA/chloroquine complexes at $\sigma = 0$ must differ in some fundamental way from those of their corresponding linear complexes with the same binding ratios. This strongly contradicts the prevailing belief that local properties of linear and relaxed supercoiled DNA/chloroquine complexes with the same binding ratio should be identical.

At binding ratios $r \geqslant 0.27$, both linear and supercoiled DNAs show evidence of a marked structural change. A component with intermediate lifetime ($\tau \simeq 5$ ns) appears in the ethidium fluorescence decay, which may represent a partially intercalated species. The apparent torsion constants become highly nonuniform and exhibit considerably altered values. The long-range torsion constant increases appreciably for the linear DNA, but decreases for the supercoiled DNAs, which are substantially positively supercoiled at that point.[53]

4.2.6.17. Excitation Transfer between Intercalated Ethidium Dyes

Fluorescence depolarization by excitation transfer between intercalated ethidium dyes was originally studied in attempts to determine their unwinding angle.[65, 156–158] The total anisotropy was assumed to be given by a simple product of factors for the contributions of Brownian motion [$r_B(t)$] and excitation transfer [$r_E(t)$], namely, $r(t) = \frac{2}{5} r_B(t) r_E(t)$. Though plausible at the time, this assumption is seen to be incorrect in light of subsequent theoretical developments. Unsatisfactory agreement with experiment prompted a reexamination of the validity of Förster theory for identical fluorophores(65) and of the magnitude of the refractive index along the helix axis.(158) Evidence was adduced that Förster theory greatly overestimates the incoherent transfer rate for identical chromophores, either in DNA or in glycerol, over very short distances, although some observations are compatible with the theory.(65)

In order to assess the contribution of torsional deformations to the FPA of intercalated ethidium at relatively high binding levels, it is necessary to take account of the effect of excitation transfer.(223) Monte Carlo procedures described by Genest and co-workers(157, 158) are employed to simulate the excitation migration along a stationary straight DNA. An ensemble of DNA/dye complexes, each containing 4000 bp, is created by randomly placing dyes in the intercalation sites, subject to nearest-neighbor exclusion and the specified binding ratio. A single excitation is initially placed on the central dye of every DNA in the ensemble, and the subsequent hopping of each is simulated by Monte Carlo techniques. The rate of excitation transfer between two ethidiums *separated by m base pairs* is given by the Förster formula,

$$k_m = A(\cos\eta_m - 3\cos^2\varepsilon_0)^2/n^4 R_m{}^6 \tag{4.71}$$

where $R_m = |m+1|h$ is the distance between dyes, ε_0 is the polar angle of the transition dipoles with respect to the helix axis, $\cos\eta_m = \cos^2\varepsilon_0 + \sin^2\varepsilon_0 \cos\delta_m$ is the cosine of the angle between the transition dipoles, $\delta_m = 10 + 36(m-1)$ is the azimuthal angle of rotation of the second transition dipole (around the helix axis) with respect to the first, and n is the refractive index. The coefficient $A = 1.77 \times 10^7\ \text{ns}^{-1}\ \text{Å}^6$ contains the usual factors for overlap of the absorption and emission spectra and the radiative decay rate.(65, 157) Ultimately, this Monte Carlo procedure yields the probability $p(m, t)$ that at time t the excitation is displaced by m intervening base pairs. The $R_m{}^{-6}$ dependence of the transfer rate strongly favors short hops, over which the DNA is comparatively straight. However, after many such hops, the net displacement of the excitation may be quite large, and the intervening DNA normally is significantly curved. Depolarization due to excitation transfer along a stationary curved DNA then arises from two causes, namely, rotation of the transition dipole around the helix axis and rotation of that helix axis from the initial site to the final site. These two contributions of excitation transfer are incorporated into $F_n(t)$ and $C_n(t)$ [cf. Eqs. (4.25) and (4.26)] by identifying their equivalent Gaussian mean squared angular displacements $\langle\delta_x(t)^2\rangle$ and $\langle\delta_z(t)^2\rangle$ in the frame of the initial site and superimposing them on $\langle\Delta_x(t)^2\rangle$ and $\langle\Delta_z(t)^2\rangle$, respectively.(223)

The chain of subunit symmetry axis vectors (bond vectors) is projected onto a plane containing the first vector. In this plane, the mean squared angular displacement of the $(m+1)$th vector with respect to the first is[109] $\langle \delta_{xm}{}^2 \rangle = |m+1| k_B T/\kappa_\beta = R_m/P$. Thus, we set

$$\langle \delta_x(t)^2 \rangle = \sum_{m \neq 0} p(m, t)\, R_m/P \tag{4.72}$$

Due to its large discrete jumps, rotation around the helix axis is not a continuous Gaussian random process, so its actual mean squared angular displacement cannot be used directly in $C_n(t)$. However, $\langle \delta_z(t)^2 \rangle$ for the equivalent Gaussian random process that makes the same contribution to the FPA is obtained from the relation

$$I_0 + I_1 E(t) + I_2 E(t)^4 = \sum_{m \neq 0} p(m, t)\, P_2(\cos \eta_m) \tag{4.73}$$

wherein $E(t) = \exp[-\langle \delta_z(t)^2 \rangle/2]$, and the I_n are given by Eqs. (4.28a–c). The right side of Eq. (4.73) corresponds to Eq. (4.16) for excitation hopping along a stationary straight chain. Pertinent parameters in the simulation are $\varepsilon_0 = 70.5°$, $h = 3.4$ Å, and $P = 500$ Å. The refractive index $n = 1.7$ is estimated from the refractive index increment and the partial specific volume of DNA and is close to that $(n = 1.75)$ proposed by Harrington.[159]

Experimental data are deconvoluted using Eq. (4.24) with[223]

$$F_n(t) = \exp[-(6 - n^2)\langle \delta_x(t)^2 \rangle/2] \tag{4.74a}$$

$$C_n(t) = \exp[-n^2\{k_B T[t/(\pi \alpha \gamma)]^{1/2} + \langle \delta_z(t)^2 \rangle/2\}] \tag{4.74b}$$

The simulated values of $\langle \delta_x(t)^2 \rangle$ and $\langle \delta_z(t)^2 \rangle$ from Eqs. (4.72) and (4.73) at each experimental time are inserted directly, and the best-fit initial anisotropy r_0 and lower-bound torsion constant α are determined in the usual way. Constraints on computer time have so far limited most of these analyses to the 0- to 40-ns time span. Typical results for linear DNAs are shown in Figure 4.17. In these experiments, the added ethidium is essentially all bound, so the abscissa is effectively the binding ratio r. If no account is taken of excitation transfer, the apparent torsion constant exhibits a very large decrease with increasing binding ratio. However, if excitation transfer is taken into account in the manner described, the apparent torsion constant is essentially independent of binding ratio up to $r = 0.10$. This conclusion is amply confirmed by the fact that D_{plat} from DLS is also independent of r up to $r = 0.20$. Thus, one may conclude that the Förster theory with $n = 1.7$ gives a good account of the excitation transfer dynamics up to a binding ratio $r = 0.10$. For $r \geqslant 0.20$, r_0

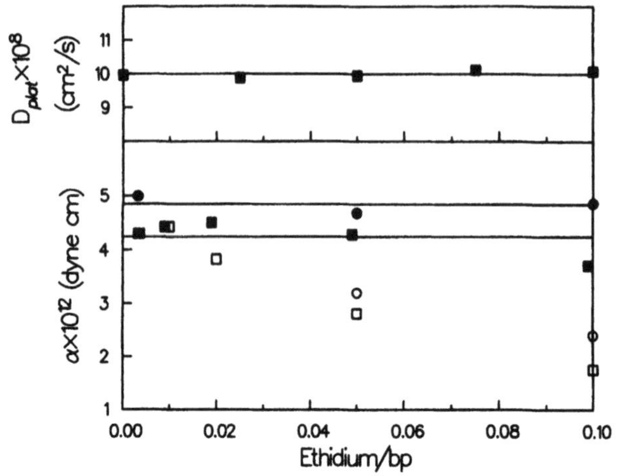

Figure 4.17. Effect of intercalated ethidium on *linear* pUC8 DNA (■, □) and *linear* pBR322 DNA (●, ○). *Top*: D_{plat} of *linear* pUC8 versus ethidium/bp. Dynamic light scattering shows no effect of intercalated ethidium on the internal dynamics. *Bottom*: Torsion constant α versus ethidium/bp for *linear* pUC8 and pBR322 DNA. The filled symbols are torsion constants obtained after correcting for the effects of energy transfer. The open symbols are the apparant torsion constants obtained from FPA data without taking excitation transfer into account. Under the conditions of these experiments, practically all of the added ethidium is intercalated.

decreases and the apparent torsion constant (taking no account of excitation transfer) rises rapidly. This could arise from several possible causes:

1. Rapid unresolved excitation transfers would decrease r_0. Their predominantly azimuthal "motion" would preferentially decrease the amplitude of the $n=2$ term relative to the $n=1$ term and the $n=1$ term relative to the $n=0$ term, thereby shifting *relative* amplitude into the more slowly relaxing terms. That in turn would increase the apparent torsion constant.
2. The DNA might actually stiffen, as found also for chloroquine at somewhat higher binding ratio ($r \gtrsim 0.30$).
3. The dye might be distributed nonrandomly.

Excitation transfer is predicted to cause a 10% decrease in apparent torsion constant at $r=1/50$, a 2–3% decrease at $r=1/100$, and no detectable effect at $r=1/300$.

4.2.6.18. *Effect of Intercalated Ethidium on the Torsion Constants of Linear and Supercoiled DNAs*

Up to $r=0.10$, intercalated ethidium has no significant effect on the torsion constant of *linear* DNAs, as indicated in Figure 4.17.[223] Together with the invariance of D_{plat}, this argues strongly that the torsional rigidity of the filament is not significantly altered at the site of intercalated ethidium.

For supercoiled DNAs, the apparent torsion constant decreases substantially with increasing binding ratio, even after taking account of excitation transfer, as shown in Figure 4.18.[223] The torsion constant of this supercoiled pJMS2 DNA (a slight modification of pBR322) is unaccountably higher than that of supercoiled pBR322 DNA. In any case, pBR322 undergoes a nearly identical decrease from $\alpha = 5.1 \times 10^{-12}$ dyn-cm at $r = 0.0025$ to $\alpha = 3.8 \times 10^{-12}$ dyn-cm at $r = 0.05$, after correction for excitation transfer. For this normal-twist pBR322, the superhelix density is estimated to vanish at $r^* = (\text{EB/BP})^* = 0.06$, where EB/BP is ethidium/base pair. Clearly, the local structure and/or rigidity and dynamics of this relaxed supercoiled DNA/ethidium complex must differ in some fundamental way from that of the corresponding linear complex with the same binding ratio. This is a profound contradiction of the prevailing belief that local properties of linear and relaxed supercoiled DNA/ethidium complexes with the same binding ratio should be identical. Though larger than in the case of chloroquine, this decrease in torsion constant is not sufficient to account for the discrepancy in E_T values between ligation and dye-binding methods.

Another indication that linear and relaxed supercoiled DNA/ethidium

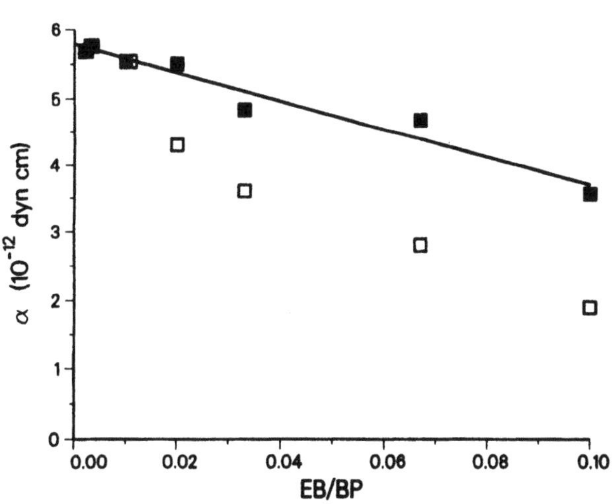

Figure 4.18. Torsion constant α of *supercoiled* pJMS2 DNA versus ethidium/base pair, denoted by EB/BP. The filled symbols are torsion constants obtained after correcting for the effects of energy transfer. The open symbols are the apparent torsion constants obtained from FPA data without taking excitation transfer into account. Even after correcting for excitation transfer, the torsion constant of this supercoiled DNA decreases appreciably with increasing intercalated ethidium, in contrast to linear DNAs.

complexes are not identical at the same binding ratio comes from flow dichroism measurements at constant shear as a function of increasing ethidium/bp.[160] For linear ϕX-174, SV40, and pBR322 DNAs, the dichroism of the ethidium detected at 520 nm closely parallels the dichroism of (mainly) the bases at 260 nm. However, for the corresponding supercoiled DNAs, the dichroism observed at 520 nm definitely does not parallel that at 260 nm, although the latter varies in the expected manner as the DNA is unwound. This suggests that the intercalated dye may be clustered onto a small subset of the molecules, or into a small domain of each, with rather different properties.

A related observation is that fully relaxed supercoiled DNA/dye complexes are somehow different from nicked circular DNA/dye complexes in the presence of the same concentration of free dye, where the binding ratios should be the same. This is readily seen in gel electrophoresis in the presence of sufficient dye concentration so that at least one, but not all, of the topoisomers is positively supercoiled. The slowest moving, presumably fully relaxed, topoisomer migrates significantly faster than the nicked circle, and this difference increases with the amount of dye present. This is not observed with chloroquine, perhaps because the effect is too small. However, it is readily apparent in the original gels of Keller[161] in which ethidium was used to unwind the topoisomers. We have confirmed this effect for ethidium and have observed similar behavior for proflavine, 9-aminoacridine, and quinacrine.

4.2.6.19. Contradictions of the Standard Model of Supercoiled DNA/Dye Complexes

According to the standard model of supercoiled DNAs, the global secondary structure is a simply strained B-helix, whose twisting and bending rigidities are unaffected by changes in superhelical strain. Moreover, interactions with intercalating dyes are supposed to be the same in supercoiled as in linear or nicked circular DNAs, except for the change in superhelical strain due to unwinding of the helix upon intercalation. This standard model of supercoiled DNA/dye complexes and predictions based upon it are profoundly contradicted by the following five observations.[214] (1) The torsion constants of supercoiled DNAs complexed with chloroquine and ethidium decline with increasing r to values that lie significantly below those of the corresponding linear DNA/dye complexes with the same r. This holds even when $r = r^*$ and $\sigma = 0$, so the supercoiled DNAs are completely relaxed. Considerably larger differences in torsion constant between relaxed supercoiled and linear DNA/chloroquine complexes with the same $r \simeq r^*$ are observed in millimolar salt concentration.[215] Substantial differences in torsion constant between relaxed supercoiled and linear DNA/dye complexes with the same $r \simeq r^*$ are also observed for 9-aminoacridine and proflavine. It is extremely unlikely that ligating the linear DNA/dye complexes with $r = r^*$ to form circular DNAs with $\sigma = 0$ would significantly alter their torsion constants. Hence, we are inclined to ascribe the observed differences between relaxed supercoiled and linear DNA/dye complexes to a hypothetical long-lived metastable state with a lower torsion constant that prevails in the relaxed supercoiled DNA. (2) The E_T values obtained from dye-binding methods are about twofold less than those from ligation methods. The observed decreases in torsion constant of supercoiled DNA/dye complexes are not sufficient to account for this discrepancy. However, if the DNA is trapped in a metastable state, so that complete reversion to unstrained B-helix is prevented, a substantial reduction in E_T might result. (3) In gels containing ethidium,[161] 9-aminoacridine, proflavine, or quinacrine, the relaxed (by dye) topoisomer with $\sigma = 0$ migrates significantly faster than the corresponding nicked circle,[214] which should exhibit the same r. However, this discrepancy is not observed in gels containing chloroquine, or in gels containing no dye when the $\sigma = 0$ topoisomer is produced by the action of topoisomerase I.[214] The observed discrepancies indicate that the structures and rigidities of those relaxed supercoiled DNA/dye complexes are not everywhere identical to those of their linear or nicked circular counterparts. If these differences are ascribed to a metastable state, then it is clear that chloroquine is less effective in trapping that state (in gels) than are the other dyes. (4) The flow dichroism anomaly noted in the preceding section provides a further indication that the structure and rigidity of the relaxed supercoiled DNA/dye complex must differ from those of its linear counterpart, at least near the binding site of the dye. This and the preceding observations could be understood if bound dye were to stabilize a particular state in a supercoiled DNA that becomes kinetically trapped, or metastable, upon relaxation of the superhelical strain. (5) Depending upon the time of exposure of native supercoiled DNAs to topoisomerase I action, topoisomers *with the same linking number* (and the same more or less uniform pattern of susceptibility to S1 and P1 nucleases) exhibit significantly different gel mobilities in the presence of ethidium.[217] This difference in response of putatively identical DNAs to added ethidium is most pronounced for partially relaxed topoisomers with $-0.02 \leqslant \sigma \leqslant 0$ in the absence of dye. In the presence of sufficient ethidium that topoisomers in this range are positively supercoiled, those exposed for the longer time to topoisomerase I action migrate faster. Such differences vanish when the exposure times of the DNAs to topoisomerase I action exceed several hours. This implies that there exists a residual difference in presumably metastable secondary structure between these otherwise identical partially relaxed topoisomers with the same linking number and that equilibration of the secondary structure in these topoisomers is catalyzed by topoisomerase I, albeit at a much slower rate than that of the initial relaxation of superhelix density.

4.2.6.20. *Alternate Secondary Structures in Supercoiled DNAs*

Evidence that the *secondary* structure of a supercoiled DNA may exhibit two different alternate forms, or allosteres, was first encountered in studies of the supercoiled replicative form of M13mp7 DNA.(55) In low concentrations of Tris buffer, these DNAs exhibit a torsion constant in the high range, $\alpha = (4.8 \pm 0.6) \times 10^{-12}$ dyn-cm, whereas in low concentrations of citrate (or cacodylate) they exhibit a torsion constant in the low range, $\alpha = (2.8 \pm 0.2) \times 10^{-12}$ dyn-cm, as shown in Figure 4.19. These torsion constants are in all cases uniform. Sufficient NaCl is present so that every sample contains 10 to 12 m*M* univalent cations.

A curious feature of these M13mp7 DNAs is that the preparation procedure yields either of two very different metastable *tertiary* conformers, labeled 1 and 4, but not both simultaneously. Conformer 4 migrates (electrophoretically) 0.4 times as fast as conformer 1 in low-resolution (0.3% agarose) gels, yet its center-of-mass translational diffusion coefficient D_0 is 1.7 times larger. Complete conversion of conformer 4 to conformer 1 takes place over 1.5 to 2.5 months at 5 °C. These observations are consistent with a straight interwound tertiary structure for conformer 1, which comprises two-thirds of the preparations, and a toroidal tertiary structure for conformer 4, which comprises the rest.(55) The equilibrium tertiary conformer, labeled 2, migrates about 0.7 times as fast as conformer 1 in gel electrophoresis, but exhibits a similar D_0. Possibly it has a *Y* or rosette type of tertiary structure. Conversion of conformer 1 to conformer 2 is facilitated by high NaCl concentration, passage over NACS 37 resin, and contact with dialysis tubing and proceeds in a completely homogeneous fashion over about two weeks in 1.0 *M* NaCl at 5 °C.(55) These transitions between tertiary conformations

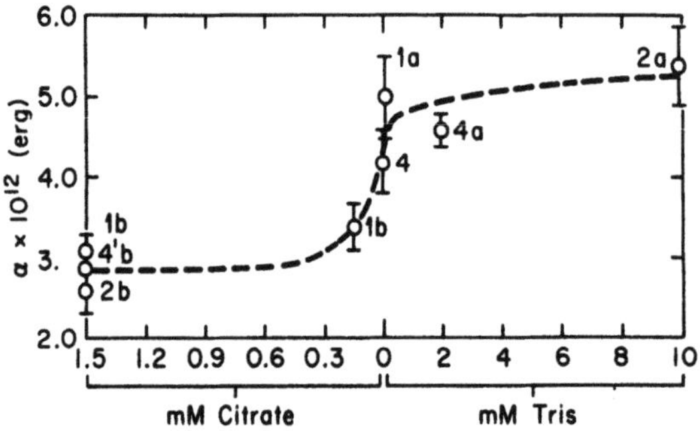

Figure 4.19. Torsion constant α versus buffer concentration for supercoiled M13mp7 DNA in different buffers. All samples except that in 10 m*M* Tris contain 10 m*M* NaCl, so all have between 10 and 12 m*M* univalent positive ions. The sample in the middle contains only 10 m*M* NaCl. The numbers (1, 2, and 4) in the sample label refer to the gel electrophoretic mobilities, which reflect different tertiary structures, as described in the text. The *a* samples all contain varying amounts of Tris, while the *b* samples all contain citrate.

conceivably involve either extrusion or intrusion of cruciform hairpins at a 48-bp palindrome. Small-angle X-ray scattering evidence for similar large-scale variations in tertiary conformation from one preparation to another, even from the same growth batch, has been presented for other supercoiled DNAs.[162, 163]

Amazingly, for all three tertiary conformations of M13mp7 DNA, the torsion constant, which reflects the secondary structure, is switched from low to high upon changing from citrate to Tris (or vice versa), as indicated in Figure 4.19. The change in torsion constant from low to high within tertiary conformers 1 and 2 is accompanied by a substantial change in the CD spectrum, namely, a marked decrease in intensity and red shift of the band at 270 nm.[55] This change is also accompanied by a significant increase in D_{plat}. (Comparable data are not available for conformer 4.) The most straightforward interpretation of these observations is that the change from citrate to Tris induces a more or less global change in secondary structure of this supercoiled DNA from a state (a) with low torsion constant to a state (b) with high torsion constant, irrespective of the prevailing tertiary conformation.

Upon linearization by the restriction enzyme Bgl I, the torsion constants in both buffers decreased by about half. These values were observed 5 to 7 days after linearization. Subsequently, the torsion constant of the sample in Tris evolved through a slight maximum at 6–8 weeks and settled to its equilibrium value, 3.8×10^{-12} dyn-cm, at 10–14 weeks.[214] (The sample in citrate was not similarly tracked.) These observations indicate that neither state a nor state b is a simply deformed B-helix. Instead, these must be allosteres that encounter significant free energy barriers in converting to B-helix. Similar evolution of the torsion constant after linearization was also observed for pBR322 and pUC8 DNAs.[214] For these DNAs, the torsion constant in the equilibrium linear form is nearly the same as in the supercoiled parent, but the initial decrease to anomalously low values during the first week and the evolution through a maximum at 6–8 weeks are clearly evident. Low-resolution gel electrophoretic mobilities give no indication of these temporal changes in torsion constant.[214] However, a very similar evolution of the CD at 273 nm was observed for circular p30δ DNA subsequent to photochemical nicking (unpublished result).

These observations, together with those on supercoiled DNAs relaxed by intercalating dyes and by topoisomerase I, indicate that complete conversion from the prevalent secondary structures in supercoiled DNAs to the normal B-helix must be severely hindered kinetically. It is also clear that the free energies per base pair of the secondary structure states a and b must be nearly identical in order for these states to be interconverted by such a small environmental perturbation.

A similar transition induced by changing the buffer from 10 mM cacodylate to 10 mM Tris is observed in some, but not all, samples of pBR322 DNA

(A. S. Benight, J. Langowski, B. S. Fujimoto, and J. M. Schurr, unpublished results). It appears that the transition is not induced in samples with higher than normal superhelix densities. This suggests that the equilibrium between the secondary structure states a and b might be rather sensitive to superhelical stress. This question is addressed immediately below.

4.2.6.21. Induction of Allosteric Transitions in Secondary Structure by Superhelical Stress

Samples of pUC8 dimer (5434 bp) with different median superhelix densities were prepared by relaxing the native plasmid with topoisomerase I

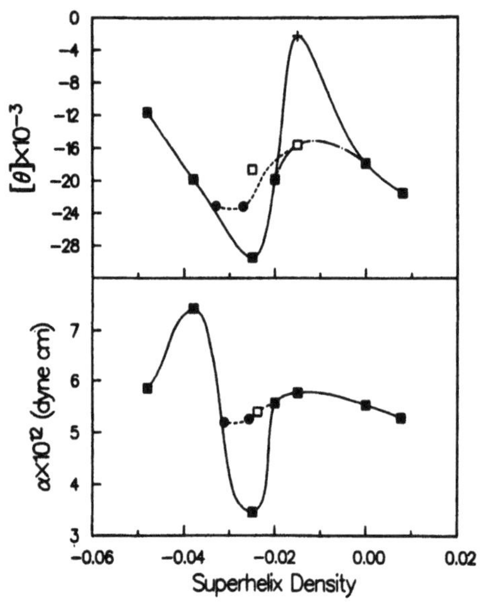

Figure 4.20. Variation of the molar ellipticity [θ] and torsion constant α of pUC8 dimer with superhelix density. All samples were prepared from the same stock solution as described in the text. Solution conditions were 10 m*M* NaCl, 10 m*M* Tris, 1 m*M* EDTA, and pH 8, and the DNA concentrations were between 40 and 50 μg/ml. ■, Five days after preparation; +, 15 days after preparation; □, 2 months after preparation; ●, samples prepared from the same initial stock solution 4 months after the other samples and measured within 5 days. *Top*: [θ] versus superhelix density σ. Only the $\sigma = -0.015$ sample changed significantly over the first few weeks. *Bottom*: α versus superhelix density. Torsion constants are averages for 70- and 120-ns time spans. For the FPA measurements only, ethidium was added to a concentration of 1 dye per 300 base pairs. With the exception of $\sigma = -0.025$, the samples denoted by ■ did not change significantly over a period of 2 months, and it is the final values that are plotted. The final measurement for the $\sigma = -0.025$ sample is denoted by □. The error bars are less than or equal to the size of the symbols in the figure. Complete data for $\sigma = -0.048$ and $\sigma = -0.031$ are given in Figure 4.10, which demonstrates the ability of FPA measurements to distinguish between samples whose torsion constants differ by only 10%.

in the presence of various amounts of ethidium, which was then removed by dialysis.[72, 214] Some results of FPA, DLS, and CD measurements on these samples are shown in Figures 4.20 and 4.21. With increasing negative superhelix density, this DNA undergoes a transition near $\sigma = -0.017$ to an intermediate state with significantly lower torsion constant, D_{plat}, and molar ellipticity $[\theta]$ at 240 nm and a significantly higher D_0.[72, 214] A second transition that begins near $\sigma = -0.035$ eventually restores more normal values of these quantities. All samples were measured at about 5–7 days after topoisomerase I treatment, again at 2–3 weeks, and again at about 50 days, but only the one at $\sigma = -0.015$ exhibits strong temporal evolution during the first few weeks, which may involve significant coupling between secondary and tertiary structure. These samples were maintained and (occasionally) studied at room temperature for 5 months without significant loss of supercoils. Over a time span of 2–3 months, both the molar ellipticity and torsion constants observed for $\sigma = -0.025$, but not D_{plat} and D_0, *increased* appreciably, although they remained significantly lower than those of the native and relaxed samples, as indicated in Figures 4.20 and 4.21. In contrast, the molar ellipticity of the sample $\sigma = -0.015$ *decreased* appreciably from 15 days to 50 days, which implies that it lies on the opposite side of a structural transition. The curve of D_0 versus σ is very similar to curves of sedimentation coefficient (S_0) versus σ for SV40[164] and PM2[165] DNAs (after correcting their superhelix densities).[72, 214] The rather abrupt rise in D_0 at $\sigma = -0.015$ is consistent with a substantially lower bending, as well as twisting, rigidity of the intermediate state. The anomalously high torsion

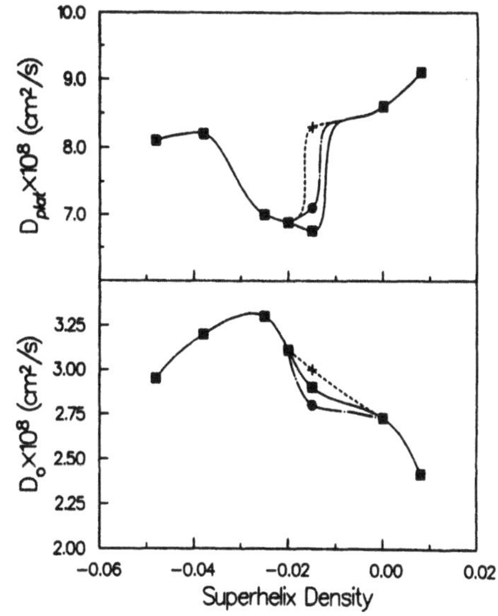

Figure 4.21. D_0 and D_{plat} for pUC8 dimer versus superhelix density. Conditions are the same as in Figure 4.20. All samples were measured at 5–7 days after preparation, again at 2–3 weeks, and finally at $1\frac{1}{2}$–2 months. ●, Five days after preparation; +, 15 days after preparation; ■, 50 days after preparation. Only the $\sigma = -0.015$ sample showed any significant change over time. *Top*: D_{plat} versus superhelix density. D_{plat} is the apparent DLS diffusion coefficient at $K^2 = 20 \times 10^{10}\ \text{cm}^{-2}$. The lower values at $\sigma = -0.015$, -0.020, and -0.025 suggest that, in this intermediate range, DNA exhibits a different secondary structure with decreased torsional and/or bending rigidity. *Bottom*: D_0 versus superhelix density. D_0 is the translational diffusion coefficient of the center of mass.

constant at $\sigma = -0.038$ was nonuniform, but did not evolve over time. The torsion constants of the other samples were fairly uniform.

The most straightforward interpretation of these data is that increasing superhelical stress induces an allosteric transition near $\sigma = -0.017$ to an alternate secondary structure with lower twisting and bending rigidities and induces a subsequent allosteric transition near $\sigma = -0.037$ to yet another secondary structure with (ultimately) more normal rigidities. The observed changes in α, D_{plat}, and $[\theta]$ cannot be ascribed entirely to a progressive change in tertiary structure of a given type (e.g., straight interwound) for the following reasons. D_{plat} and α reflect mainly short-range dynamics over distances of a few hundred base pairs and should be very insensitive to the superhelix density, provided it does not alter the twisting and bending rigidities. $[\theta]$ reflects mainly nearest-neighbor electronic interactions, which are perturbed only slightly by increasing superhelix density. Any effect of a progressive change in tertiary structure on α, D_{plat}, or $[\theta]$ should vary *smoothly* and *monotonically* with increasing $|\sigma|$. However, these quantities all decrease abruptly near $|\sigma| = 0.020$ and rise back up near $|\sigma| = 0.035$. Furthermore, α greatly overshoots at $|\sigma| = 0.038$ and then subsides at $|\sigma| = 0.048$. Each reversal in the direction of change with increasing $|\sigma|$ would require a change in the type of tertiary structure (e.g., from interwound to toroidal). However, gel electrophoresis provides no evidence for different tertiary structures. Finally, it is difficult (or impossible) to see how *any* change in tertiary structure *alone* could cause a *decrease* in apparent torsion constant, as observed in the intermediate region of superhelix density. Highly bent or compact structures can only increase the resistance to torsional motion. Tilting of the transition dipole toward the effective superhelix axis would also increase the apparent α. Thus, it is most unlikely that these experimental observations could be rationalized without invoking significant changes in secondary structure. The possibility that the observed changes arise from transitions to radical secondary structure at only one or a few sites of very small extent is discounted for reasons given elsewhere.[214] The hypothesis that the observed changes are due to allosteric transitions in global secondary structure induced by superhelical stress provides the simplest interpretation of the most data.[214]

The low torsion constant at $\sigma = -0.025$ is very similar to that observed in a supercoiled pBR322 that was partially relaxed by saturation binding of *Escherichia coli* single-strand binding (ssb) protein, and which persisted for over a month.[56] It is also similar to that recently inferred from an *in vivo* assay based on variation in repression efficiency with size of a putative DNA loop.[234] Indeed, it appears that anomalously low torsion constants may be universally encountered in the course of either partial or complete relaxation of supercoiled DNAs, regardless of whether the superhelix density is reduced by action of topoisomerase I, binding of ssb protein, binding of intercalated dyes, or outright linearization. Certainly in the case of linearization, and probably also in the case of complete relaxation by intercalating dyes, the state with low torsion constant is not stable, but instead is a long-lived metastable intermediate. We suspect that as the native superhelix density is progressively relaxed by intercalating dyes, the DNA first converts to the intermediate state with low torsional rigidity, but the subsequent transition to normal B-helix at lower superhelix density is kinetically hindered or blocked. If so, the secondary structure of the "relaxed" DNA/dye complex would correspond, at least in part, to that of the metastable intermediate state. In such a circumstance, one would expect to obtain a significantly lower twist energy parameter from dye-binding experiments than from ligation experiments, which sample superhelix densities only in a very narrow range around $\sigma = 0$, where the secondary structure is simply deformed B-helix. Such metastable "relaxed" DNA/dye complexes should also exhibit lower torsion constants than the corresponding linear complexes and different gel electrophoretic mobilities from nicked circles, as observed.[214]

An important question is whether the secondary structure of pUC8 dimer at native superhelix density is the same as that of the relaxed ($\sigma = 0$) species, as the similarities in α, D_{plat}, and $[\theta]$ would suggest. We suspect not, because their $[\theta]$ values would probably be rather different after correction for the difference in superhelix density.

The abruptness and extremely slow kinetics of these (undriven) allosteric transitions indicate that they may be highly cooperative. This implies a large free energy to form junctions between domains of different secondary structure. Highly cooperative allosteric transitions in DNA secondary structure would enable long-range communication between specifically bound proteins, thus facilitating remote control of polymerase activity by distantly bound regulatory proteins.[55] Indeed, evidence that catabolite activator protein (CAP) binding may induce a long-range change in secondary structure of supercoiled DNAs already has been reported.[47]

4.3. Rotational Dynamics of DNA in Nucleosomes, Chromatin, Viruses, and Sperm

4.3.1. Nucleosomes

The rotational dynamics of nucleosomes containing $\sim(146 \pm 2)$-bp DNA bound to core particles was studied by FPA[21, 59, 60, 235] of intercalated ethidium and by TPD[62] of intercalated Methylene Blue. These studies yield strong evidence that DNA wrapped around the histone core particle exhibits considerable torsional mobility over at least some of its length. The TPD data of Wang *et al.*,[62] which extend to 1.6 μs, show clearly that the anisotropy decays of the nucleosome and free nucleosomal DNA are identical up to 80–100 ns, but that the former decays much more rapidly at longer times due to its more compact shape. By using D_2O to extend the ethidium lifetime to 39 ns, Winzeler and Small were able to access times as long as 200 ns.[235] Their FPA data indicate that the DNA on the core particle undergoes considerably smaller amplitudes of angular motion in any given time than is inferred from the TPD data.[62] Both TPD and FPA data were analyzed using models in which the axis of the torsionally mobile DNA is constrained to girdle the equator of a sphere, and the correct $C_n(t)$ for short DNAs are employed.[63, 235] The model in which the nucleosomal DNA is everywhere free to twist is ruled out, because acceptable fits over the complete time course of the decay could not be achieved.[63] However, excellent fits with a nucleosome hydrodynamic radius $R_H = 63$ Å are obtained using a model in which the nucleosomal DNA is rigidly clamped to the sphere at both ends. Equally good fits could be obtained for a very wide range of lengths $(N-1)$ of the torsionally mobile domain, provided that a particular relation between $N+1$, α, and γ was maintained.[63, 235] A model of overdamped harmonic libration with respect to DNA immobilized at the surface of a sphere, which corresponds roughly to the wagging of free ends of the nucleosomal DNA in solution, also gives a very good fit.[63] These analyses demonstrate unequivocally that, even though much of the DNA is free to twist, it must be rigidly clamped at one or more points for times less than 1 μs. This conclusion has subsequently been confirmed by UV photodamage experiments on reconstituted nucleosomes, in which a regular 10.3-bp phasing of the photodamage with respect to the ends of the nucleosomal DNA is observed.[166] If the 146-bp DNA were clamped at one end, the equilibrium rms amplitude of twist at the other end would be about $\langle \Delta_z^2 \rangle^{1/2} = 75°$, which would yield a standard deviation of only about 2 bp (36° each) of "dephasing" of the photodamage site. Thus, a single rigid attachment anywhere is probably consistent with the TPD, FPA, and photodamage results. Certainly, two or more points of rigid attachment would yield excellent agreement with the TPD, FPA, and photodamage data. Neither TPD nor FPA nor photodamage data provide firm information about the length of the torsionally mobile domain. Contrary to what has been suggested,[59] it is not possible to determine uniquely the length of the mobile region without invoking an additional assumption about α or γ.[63]

Best-fit parameters for the earlier FPA and TPD data[59, 62] are in reasonable accord, but differ significantly from those for the most recent FPA experiments.[235] For the TPD data,[62] α must be at least 2.5 times smaller than that of the free DNA, provided γ is not smaller than for free DNA in solution.[63] Assuming only that one or both ends of the mobile region are clamped to the core particle, Genest *et al.*[59] concluded that α is reduced by about threefold for DNA in core particles compared to the value reported by Thomas *et al.*,[18] for free DNA (cf. Table 4.1). Ashikawa *et al.* similarly assumed that both ends of the DNA are clamped in the nucleosome and would have concluded that α for that DNA was significantly smaller than for the free nucleosomal DNA if they had used the correct $C_n(t)$ to analyze their short free DNA.[60] However, the most recent experiments yield considerably larger α and/or γ values that do not require a reduced value of α for DNA in the nucleosome.[235] The reasons for this discrepancy are presently unknown. If DNA in the nucleosome were actually clamped at only one point, it is not yet determined how the best-fit α would compare with that for free DNA.

Genest *et al.*[59] reported a significant decline in initial anisotropy of nucleosomes from $r_0 = 0.35$ to a plateau value at 0.29 with increasing ethidium binding ratio from 1/1000 to 1/170 bp, which they attributed to dye clustering in a small region (15 bp) and consequent excitation transfer. However, in our studies on free DNA/ethidium complexes, essentially no decline in r_0 is observed up to dye/bp = 1/5, although excitation transfer decreases the apparent α substantially (cf. Figure 4.17). At much higher binding ratio (dye/bp $\geqslant$ 2), a large ($\sim$50%) reduction in r_0 is observed. Evidently, excitation transfer does not reduce r_0 except for the very shortest distances ($\sim$2 bp). It is also conceivable that nucleosomal DNA undergoes some kind of structural change that admits greater libration of its bound dye as progressively more dye is added. Genest *et al.*[59] suggested that the DNA begins to progressively detach from the nucleosome about when the third dye is bound, possibly due to lengthening of the filament.

4.3.2. Chromatin

The FPA of ethidium intercalated in the high-affinity sites of chicken red-cell chromatin was studied as a function of ethidium concentration.[167] A rapid increase in the FPA decay rate with increasing ethidium was attributed to dye clustering and consequent excitation transfer. These data were analyzed using the (invalid) anisotropy expression, $r(t) = \frac{2}{5} r_B(t)\, r_E(t)$, wherein $r_E(t)$ is calculated via the (valid) Monte Carlo procedure described above.[157, 158] Satisfactory agreement with the observations was obtained by assuming that ethidium can bind only to a 28-bp stretch of DNA, which is believed to comprise about half the linker DNA.

FPA studies at extremely low binding ratios (1/400–1/700 bp) to assess the DNA motion were carried out on ethidium intercalated in calf thymus[21, 60] and chicken red-cell[61] chromatin. Under the conditions of these experiments, ethidium is believed to be intercalated only in the linker DNA, and excitation transfer is believed to be negligible. The amplitude of angular motion, or depolarization, at any given time is much lower than in free DNA and is further reduced under conditions that induce the condensed "solenoidal" state of chromatin, such as 50–200 m*M* NaCl or 1 m*M* Ca^{2+} or Mg^{2+}.[60] Ashikawa *et al.*[60] suggested that the torsion constant of the linker DNA stiffens by a factor of 1.4 upon condensation, but pointed out that changes in length of the mobile region between fixed attachments or in the friction coefficient γ are also possible. In chicken red-cell chromatin, the amplitude of angular motion, or depolarization, on any given time span is diminished to a greater extent than in calf thymus chromatin, and the transition to the condensed form is observed at lower NaCl or Mg^{2+} concentrations. Histone H5 stabilizes the condensed form in Mg^{2+} but not in NaCl alone.[61]

4.3.3. Viruses

The rotational dynamics of ethidium intercalated in double-strand DNAs of intact bacteriophages, namely, λ, the deletion mutant $\lambda\Delta$, T4D (wild type), and T4dC (with normal cytosine instead of glucosylated hydroxymethylcytosine) were studied by FPA.[57] The interhelix spacings of their DNAs *in situ* were also studied by X-ray diffraction.[57] Except in the case of T4D, the amplitude of angular motion, or depolarization, at any given time is much less than in the corresponding free DNA, and the relative degree of hindrance, or restriction, increases with decreasing interhelix distance of the DNA in the virus, as expected. For T4D, however, the FPA of DNA in the phage is about the same as that of the free DNA up to 25 ns, and even at longer times it relaxes considerably more rapidly than observed for the other viruses. It is proposed that glucosylation of this DNA restricts ethidium binding to more flexible or mobile regions of the DNA,[57] and ^{31}P-NMR relaxation evidence in support of that is discussed.

An ethanol-condensed aggregate manifests FPA relaxation and X-ray diffraction results similar to those obtained for the viruses with nonglucosylated DNAs, albeit with even more restricted motion.[57] The interhelix spacing in the ethanol-induced aggregate was similar to that of the λ, $\lambda\Delta$, and T4dC DNAs (2.7 nm), but shorter than that in spermidine-induced aggregates (3.0 nm), which showed relatively normal torsional dynamics.[57]

4.3.4. Sperm

FPA studies of ethidium dye intercalated in whole sperm nuclei at different stages in spermatogenesis reveal changes that are ascribed to changes in the mode of DNA packaging.[58] During the period when replacement of histones by protamine is in progress, the apparent amplitude of angular motion, or depolarization, is substantial, but this relaxation is greatly diminished during later periods in development. This is ascribed to the availability of more flexible DNA regions in the former nuclei, although dye clustering, and consequent depolarization by excitation transfer, cannot be ruled out.

4.4. Steady-State Studies of DNA Dynamics

Steady-state FPA experiments on DNA/ethidium complexes were undertaken by Genest *et al.*[168] and Hård and Kearns[169–172] to address a number of questions. Reliability of the results and conclusions is reduced somewhat not only by the lack of time resolution, but also by the following factors. (1) No measure of the relative amounts of intercalated and free dye, or the existence of partially intercalated dye, is available from the fluorescence decay. (2) The partially fractionated samples are all quite polydisperse in length and heterogeneous in composition except in the case of synthetic DNAs and may have dangling single-strand ends. For short DNAs, this introduces a large polydispersity into the uniform mode tumbling relaxation times, which vary as $\sim(N+1)^3$ for $N+1 \lesssim 200$ bp. (3) There is no way to assess the initial anisotropy r_0, or ARF in Eqs. (4.30a–c), which is not the same for all DNAs. In most analyses, it is simply assumed that ARF = 1.0, which is not valid. (4) Quantitative data analyses are performed using a rigid-rod model for anisotropic rotational diffusion. Even for short DNAs that exhibit a substantial amplitude of uniform (rigid-rod) mode, this procedure takes no account of the significant amplitude reductions due to twisting and bending that appear in $B_n(\infty)$ and $D_n(\infty)$, respectively. Dye wobble is incorporated using a model function that corresponds to no known qualitatively correct anisotropy expression. Despite these problems, certain of the qualitative conclusions still should be fairly reliable.

An increase in the steady-state FPA of short DNAs (in the range ~30–90 bp) with increasing concentration was ascribed to the formation of aggregates in which the tumbling motion is restricted.[168, 169] The threshold for aggregation is ~5 mg of DNA/ml, independent of DNA length. This aggregation is favored by low T and high concentrations of Mg^{2+} (>10 mM)[169] and is similar to a sol–gel transition of longer DNA.[178]

The steady-state FPA of large (~500 bp) calf thymus DNA/ethidium complexes is unaffected by addition of proflavine up to one per two base pairs. From this, it is concluded that the torsion constant is unaltered[168] by intercalation of proflavine. However, in our time-resolved FPA studies of linear pBR322 DNA/ethidium complexes, the torsion constant is reduced by the factor 0.60 as proflavine is added from zero to one per two base pairs.[173] Whether this discrepancy is due to a real difference between these DNAs or to some compensating change in the steady-state FPA is not known. This example illustrates the difficulty in extracting *unambiguous* conclusions from steady-state FPA measurements on so complicated a system.

The steady-state FPA from excitation into the S_2 band, as well as that from the S_1 band, was employed in an attempt to assess the relative contributions of azimuthal and end-over-end rotations.[170, 171] In a subsequent analysis of such experiments,[175] steady-state FPA expressions corresponding to the Intermediate Zone for $C_n(t)$ (Eq. 4.37) and $F_n(t) = \exp[-(6-n^2) D_\perp t]$ were derived, and the data of Hård and Kearns[170] were reanalyzed. A sensitivity analysis of the results was also performed to estimate uncertainties in the $D_\perp$ and $C = k_B T/(\pi\alpha\gamma)^{1/2}$ extracted from the two FPA data. For DNAs with $N+1 \gtrsim 2000$ bp, one can obtain quantitative information about C, albeit with larger uncertainties than found for time-resolved FPA measurements. However, no useful information about $D_\perp$ can be obtained, because the uncertainty in $D_\perp$ greatly exceeds its magnitude.[175] The values of $D_\perp$ and C extracted from the data are most sensitive to the assumed ARF and angle $\phi \simeq 60°$ between the transition dipoles. Because no goodness-of-fit criterion can be applied to the determination of two numbers from two data points, this two-wavelength method cannot provide an indication of a preference for one model of the motion, or one binding site geometry, over another.[175] In view of the theoretical problems noted above, the conclusion of Hård and Kearns[171] that their data are consistent with a model of substantial dye wobble within the intercalation site is very likely not warranted.[175] In any case, substantial dye wobble could in no sense be inferred from the data. In their analysis, Fujimoto and Schurr treated both tilted ($\varepsilon_0 = 70.5°$) and perpendicular ($\varepsilon_0 = 90°$) binding site geometries, contrary to the comment of Hård and Kearns,[172] and obtained unphysical negative values of $D_\perp$ in both cases.[175] This probably arises from the use of an inadequate model for the tumbling motion. In any case, no useful information can be obtained regarding motions on time scales much longer than the ethidium lifetime for any model, because the uncertainties in the slow motion rate constants would substantially exceed their magnitudes.

4.5. DNA Dynamics by Fluorescence Microscopy

Direct observation of single DNA molecules with fluorescent bound dyes, mainly 4′,6-diamino-2-phenylindole dihydrochloride (DAPI), was achieved by enhanced video microscopy using a silicon intensified target camera.[176] Not only is the translational diffusion coefficient readily measured, but long-wavelength changes in shape of the random coil, corresponding to the longer Rouse–Zimm modes, are clearly resolved for T4 (160 kbp), BF23 (11.1 kbp), λ (48.3 kbp), and T3 (19.6 kbp) DNAs. When the instantaneous shapes are modeled as prolate ellipsoids, the time-averaged axial ratio is about 2.2 to 2.5, which demonstrates the expected nonspherical shape of finite random coils. The persistence lengths estimated from the observed radii of gyration lie in the expected range from 470 to 700 Å. The relaxation time of the slowest extensional mode of the random coil form of T4 DNA is 0.2 to 0.7 s.(176) This compares favorably with the Langevin relaxation time, $\tau = 0.7$ s, for the longest Rouse–Zimm mode, which is estimated from the empirical relation(177, 178) $\tau = (10^{-13})M^{1.6}$, where M is the molecular weight. When one end of the DNA is attached to the glass, the DNA can be highly extended by shear flow to a very thin filament, whose average length is about 80% of the contour length, and whose observed maximum length is the expected contour length. The rate of retraction of the thin filament upon breakage can also be measured. This rate agrees fairly well with that calculated from the estimated tension and friction factor per Kuhn length (twice the persistence length).(176)

These studies also reveal some unexpected phenomena. At reduced shear flow, the thin filaments contract to yield a thick filament that is still much more extended than the normal random coil forms, typically, one-seventh to one-half of the contour length. After the flow is arrested, free thick filaments contract to random coils, but those with one end attached to the glass are metastable. They also show visible substructure that resembles domains connected by thin filaments. There are also indications of waviness, or superhelicity, in the structures of these attached thick filaments.(176) The role of the glass surface in stabilizing these thick-filament (non-random coil) structures remains to be elucidated. Conceivably, this is one manifestation of a more general surface phenomenon and is related to the catalysis of aggregation and change in tertiary structure of M13mp7 DNA by contract with NACS 37 resin and dialysis tubing.(55)

The electrophoretic migration of large DNAs (labeled with fluorescent intercalators) through thin layers of agarose gel were directly visualized by video microscopy.(236, 237) The average migration rate measured microscopically agrees with that observed by conventional means in macroscopic gels of the same agarose concentration. The retarding forces exerted on the DNA by the gel network appear to be extremely nonuniform, so that only a few points of high friction dominate the drag. The DNA strands downfield (toward the positive electrode) from the high friction points can be highly extended even in rather weak electric fields, and the head ends are visibly brighter than their trailing stems or tails, in agreement with theory.(238) The motion is more or less episodic, in the sense that the forward-extending head eventually encounters an obstacle, which allows the tail to catch up as the molecule contracts to a more compact coil, from which eventually a new head emerges to extend forward, and so on.

The extensions of individual topologically snared (by the gel) circular

DNAs were measured as a function of electric field strength by video microscopy.[239] The data for extension versus electric field strength are consistent with the theoretically predicted form and yield a best-fit effective charge of about 0.1 electrons per base pair. The effective charge is that which, when multiplied by the macroscopic electric field, yields the total force per base pair on the DNA. The dynamics of individual U-shaped molecules sliding over a single friction point in a gel were also investigated by video microscopy.[240] Quantitative analysis of the data yields an estimate of the friction coefficient for DNA passing over the pivot. These estimates of the effective charge and friction coefficient will hopefully improve the quantitative accuracy of future simulations of gel electrophoresis.

Fluorescence microscopy techniques were also applied to study chromatin and chromosomes,[179] but those studies lie outside the scope of this chapter.

4.6. Dynamics of tRNAs

An excellent review of tRNA structures and dynamics was presented in 1983.[180] Only subsequent fluorescence decay and FPA studies are reviewed here. The use of excitation transfer to measure intramolecular distances[181, 182] and the use of fluorescence as a probe of protein/tRNA interactions[182–185] lie outside the scope of this chapter.

4.6.1. Ethidium Fluorescence

Since tRNA is more varied structurally than DNA, ethidium could reside in pockets as well as intercalate into double-strand regions. The fluorescence decay provides information about the type, or types, of binding sites occupied by ethidium. It is currently believed that the excited state of ethidium is quenched by proton transfer to the solvent[186] and that its lifetime is reduced with increasing solvent exposure. If ethidium occupies two or more kinds of sites with different degrees of exposure to solvent, then its fluorescence decay is expected to be multiexponential.

Satisfactory fits of the fluorescence decays for ethidium bound to yeast tRNA^{Phe} and *E. coli* $\text{tRNA}_1^{\text{Val}}$ require (at least) two exponentials in the sum response $S(t)$ [cf. Eq. (4.56)] under all conditions studied.[187, 188] The normalized amplitudes and lifetimes for $\text{tRNA}_1^{\text{Val}}$ (extrapolated to zero concentration) are $S_1^\circ = 0.917$, $\tau_1 = 25.6$ ns and $S_2^\circ = 0.082$, $\tau_2 = 5.6$ ns.[187] The results for tRNA^{Phe} are similar.[188] This requirement for two (or more) exponentials is unequivocal evidence for at least two ethidium binding sites. The dominant component has a lifetime similar to, but slightly longer than, that of ethidium intercalated in DNA and is taken to represent ethidium intercalated into a double-strand region of tRNA, very likely in the acceptor stem.[189] The minor component has a lifetime intermediate between those of the free and the normally intercalated dye. It presumably represents ethidium that is either bound in a pocket or only partially intercalated. Crystallographic studies show that ethidium can bind in the P_{10} cavity in the knee of the tRNA, where it would indeed be less shielded from the solvent than in a normal intercalation site.[190] It is suggested, but not proven, that this is the binding site of the minor component. If these assignments are correct, then it appears that crystal packing forces enormously enhance binding to the P_{10} pocket relative to intercalative binding in the acceptor stem or elsewhere.

If the tRNA can be treated as a rigid ellipsoid of revolution, then its anisotropy is expressed by Eq. (4.24) with I_n given by Eqs. (4.28a–c), $C_n(t) = \exp(-n^2 D_{\|} t)$, and $F_n(t) = \exp[-(6-n^2) D_{\perp}]$, where $D_{\|}$ and $D_{\perp}$ are the rotational diffusion coefficients around the symmetry and transverse axes, respectively. In general, a multiexponential decay is expected. However, the data for $tRNA^{Phe}$ and $tRNA_1^{Val}$ in the presence of endogenous Mg^{2+} are well fitted by the single-exponential function $r(t) = r_0 \exp(-t/\tau_R) + r_\infty$.[187, 188] The nonzero baseline must be due to incomplete rotational relaxation at longer times and is presumed to arise from occasional large (slowly rotating) aggregates or other objects containing tRNA or additional nucleic acids. The decay does not exhibit three (or more) components, as expected for a non-spherical object, yet the tRNA is definitely nonspherical. It was noted earlier that when ε_0 is approximately equal to 0°, 40°, or 90°, $r(0)$ is dominated by, respectively, I_0, I_1, and I_2.[191] Although there is insufficient information to distinguish among these possibilities, one or the other probably prevails.

The rotational relaxation time of $tRNA_1^{Val}$ remains constant at $\tau_R = 24.0 \pm 0.8$ ns, independent of tRNA concentration from 0.3 to 54 mg/ml at 100 m*M* ionic strength, and is also largely independent of ionic strength from 0 to 130 m*M* at 54 mg/ml,[187] without any correction for solution viscosity. At the higher tRNA concentrations and lower ionic strengths, tRNA exhibits significant liquidlike ordering in its intermolecular structure factor, as well as other evidence of strong spatial correlations.[192] It is remarkable that, despite such strong interactions and spatial correlations, the rate of rotational diffusion is largely unaffected. This justifies an assumption commonly invoked in NMR studies of tRNAs, namely, that the rotational relaxation time is the same in the concentrated NMR samples as in dilute solution. Theories for the effect of hydrodynamic interactions predict that the rotational diffusion coefficient should decrease with increasing concentration, whereas an increase of very small magnitude is actually observed.[187] The inapplicability of these theories is perhaps due to the long-range repulsive forces that greatly reduce the number of close approaches of neighboring tRNAs.[187]

The rotational relaxation time τ_R can be combined with time-dependent nuclear Overhauser effect (NOE) measurements to determine interproton distances. If the molecule is (roughly) spherical, the cross-relaxation rate (σ_{ij}) between two protons is related to their interproton distance r_{ij} by[193]

$$\sigma_{ij} = \frac{\hbar^2 \gamma^4}{10 r_{ij}^6} \tau_R \tag{4.75}$$

where γ is the proton gyromagnetic ratio, $\hbar = h/2\pi$, and h is Planck's constant. This formula applies also for a cylindrically symmetric molecule, provided the internuclear vector is oriented along the symmetry axis (i.e., $\varepsilon_0 = 0$). The cross saturation of N3-H of U-64 by N1-H of G-50 in $tRNA_1^{Val}$ is measured as a function of preirradiation time, and the initial cross-relaxation rate determined.[187] The distance between these two protons of the G · U wobble pair is calculated from τ_R and σ_{12} to be $r_{12} = 0.272$ nm, with a relative error significantly less than 5%.[187] This is the most accurate measurement to date for the distance between imino protons of a G · U wobble pair.

Removal of endogeneous Mg^{2+} by rigorous treatment (heating to 80 °C in the presence of 10 m*M* EDTA) introduces a fundamental difference between yeast $tRNA^{Phe}$ and *E. coli* $tRNA_1^{Val}$, although their τ_R values remain similar and are only slightly diminished to 21.3 and 20.1 ns, respectively.[188] *As* Mg^{2+} *is gradually restored*, τ_R for treated $tRNA^{Phe}$ *increases* to its original value, $\tau_R = 25.9$ ns, at 40 added Mg^{2+} ions per tRNA. In contrast, τ_R for treated $tRNA_1^{Val}$ *declines* steadily to 16.2 ns at 40 added Mg^{2+} ions per tRNA.[188] Moreover, the FPA data for treated $tRNA_1^{Val}$ are fitted significantly better by two exponentials with relaxation times near 29 and 5.5 ns, and an amplitude ratio that decreases from 2.8 to 2.1, as the number of added Mg^{2+} ions per tRNA is increased from 5 to 40. Neither excess Mg^{2+} nor transient heating restores $tRNA_1^{Val}$ to its original state prior to the Mg^{2+} removal treatment.[188] The second rotational relaxation time (5.5 ns) in treated $tRNA_1^{Val}$ is incompatible with a simple change in tertiary structure, and the possibility that it arises from intramolecular torsion or flexure is suggested. This possibility is consistent with the significantly sharper NMR lines observed for treated $tRNA_1^{Val}$ than for treated $tRNA^{Phe}$,[188] which presumably exhibits a much smaller amplitude of such torsional or flexural motion.

4.6.2. Wyebutine Fluorescence

The wyebutine base at position 37 in the anticodon loop of $tRNA^{Phe}$ has also been used as a fluorescence probe.[194–200] The wyebutine base, like ethidium, exhibits a reduced fluorescence lifetime when exposed to aqueous solvents.[195–197] The number and relative size of exponential terms needed to fit the fluorescence decay provides information about the conformation of the anticodon loop.

The fluorescence decay is multiexponential.[199, 200] This is unequivocal evidence that the wyebutine base can be bound in at least two different conformations with different solvent shielding. Wells and Lakowicz[200] resolved two exponential components. They measured the normalized amplitudes and lifetimes for the wyebutine fluorescence at two different concentrations of added Mg^{2+}: $S_1^\circ = 0.50$, $\tau_1 = 1.7$ ns, $S_2^\circ = 0.50$, and $\tau_2 = 5.9$ ns with no added Mg^{2+} present, and $S_1^\circ = 0.16$, $\tau_1 = 0.6$ ns, $S_2^\circ = 0.84$, and $\tau_2 = 6.0$ ns with 10 mM Mg^{2+}. Since the 6ns component is the longest lifetime present, it must represent the conformation that shields the wyebutine to the greatest extent and is generally believed to involve a 3′ stack of bases 34–38.[180, 199–201] The fraction of the tRNA in this conformation increases when Mg^{2+} is added to the solution. This structure is also observed in crystal structures which include Mg^{2+}.[202–204] In the other conformation(s), the wyebutine is more exposed to the solvent. A 5′ stack, which does not include bases 37 and 38, is one possibility. The wyebutine base would be more exposed to the solvent and have a shorter fluorescence lifetime as a result. However, both NMR data[205, 206] and chemical modification studies[207] are inconsistent with a 5′ stack. For the moment, this matter is unresolved.

The results from fitting the anisotropy decay support the above conclusions. Wells and Lakowicz[200] resolved two exponential components in the anisotropy decay. They obtained $r_0' = 0.11$, $\tau_R' = 0.3$ ns, $r_0'' = 0.15$, and $\tau_R'' = 18.5$ ns for the sample with no added Mg^{2+}, and $r_0' = 0.05$, $\tau_R' = 0.4$ ns, $r_0'' = 0.17$, and $\tau_R'' = 17.4$ ns for a sample with 10 mM Mg^{2+}. Here r_0' and r_0'' are the amplitudes of the fast and slow components. The longer rotational relaxation time corresponds to overall tumbling of the tRNA, although its amplitude is reduced by much more rapid local motions. The shorter relaxation time corresponds directly to a rapid local motion. Upon addition of Mg^{2+}, the *relative* amplitude of the rapid local motion decreases, while that of the overall tumbling increases. This implies that the wyebutine base is held in a more rigid or constrained state, such as a 3′ stack, in the presence of Mg^{2+}. In that state, the amplitude of local angular motion is substantially diminished in comparison with that in the alternate state that prevails in the absence of Mg^{2+}. As noted before, the exact nature of these conformation(s) is unresolved.

Claesens and Rigler[199] also studied the wyebutine fluorescence and obtained similar results for the effect of Mg^{2+} on the conformation of the anticodon loop. In addition, they also studied the effect of codon–anticodon interactions by binding the $tRNA^{Phe}$ to $tRNA^{Glu}$. They concluded that the anticodon loop of $tRNA^{Phe}$ undergoes a conformational change after binding and suggested that the 3′ stack has shifted to a 5′ stack.

4.7. Summary and Outlook

The application of time-resolved fluorescence techniques and new theory have provided the first major insights into the rapid Brownian dynamics of elastically deformable filaments, specifically DNA. Substantial knowledge about the torsional deformation process, the long-range torsion potential, and the influence of various environmental factors has been acquired almost entirely from time-resolved FPA experiments. The remarkable sensitivity with which fluorescence techniques have detected novel structural changes in DNAs and tRNAs in solution is also amply documented. It is highly likely that some of these changes play an important role in the processing of genetic information.

Future studies are expected to exploit current technical advances, such as the improved time resolution of microannel plate detectors, molecular biology techniques to prepare a variety of novel samples, and the use of electric and shear fields to partially orient samples. Particularly important problems concern the anisotropy and dynamics of local motions of intercalated dyes, the torsional dynamics and rigidity of short circular DNAs with 200–250 bp (as were used for ligation measurements of the static torsional rigidity), and the possible occurrence of rigidity weaknesses associated with special sequences (e.g., B–Z junctions) or with sequence-specific protein or drug binding. Additional measurements of the dynamic bending rigidity of restriction fragments by a variety of optical methods are required to reduce our present uncertainty regarding the tumbling correlation functions. Parallel Brownian dynamics simulations will be an essential component of such experiments, especially for molecules that exhibit permanent bends and/or anisotropic bending. In the case of long-range allosteric changes in DNA secondary structure induced by superhelical stress or by proteins bound to supercoiled DNAs, conventional diffraction and 2-D NMR structure methods are not likely to be applicable in the foreseeable future. In such cases, fluorescence methods in combination with other physical techniques, such as TPD, DLS, spin-label EPR, and CD, and various chemical modification and cutting methods are likely to provide the majority of new structural information.

Acknowledgment

Support of our work by the National Science Foundation and the National Institutes of Health is gratefully acknowledged.

References

1. P. R. Callis, *Annu. Rev. Phys. Chem. 34*, 329–357 (1983).
2. E. L. Elson, *Annu. Rev. Phys. Chem. 36*, 379–406 (1985).

3. R. F. Steiner and Y. Kubota, in: *Excited States of Biopolymers* (R. F. Steiner, ed.), pp. 203–254, Plenum Press, New York (1983).
4. A. R. Morgan, J. S. Lee, D. E. Pulleyblank, N. L. Murray, and D. H. Evans, *Nucleic Acids Res. 7*, 547–569 (1983).
5. A. R. Morgan, D. H. Evans, J. S. Lee, and D. E. Pulleyblank, *Nucleic Acids Res. 7*, 571–594 (1983).
6. M. D. Barkley, A. A. Kowalzyk, and L. Brand, *Biomolecular Stereodynamics Proceedings of Symposium 1*, 391–403, Academic Press, Schenectady, New York (1981).
7. R. Rigler and W. Wintermeyer, *Annu. Rev. Biophys. Bioeng. 12*, 475–505 (1983).
8. J. M. Schurr, *CRC Crit. Rev. Biochem. 4*, 371–431 (1977).
9. P. R. Callis and N. Davidson, *Biopolymers 8*, 379–390 (1969).
10. A. Anders, *Chem. Phys. Lett. 81*, 270–272 (1981).
11. T. Aoki and P. R. Callis, *Chem. Phys. Lett. 92*, 327–332 (1982).
12. S. Georghiou, T. M. Nordlund, and A. M. Saim, *Photochem. Photobiol. 41*, 209–212 (1985).
13. D. Magde, E. Elson, and W. W. Webb, *Biopolymers 13*, 29–61 (1974).
14. J. Bresloff and D. M. Crothers, *J. Mol. Biol. 95*, 103–123 (1975).
15. P. Wahl, J. Paoletti, and J. B. LePecq, *Proc. Natl. Acad. Sci. U.S.A. 65*, 417–421 (1970).
16. M. D. Barkley and B. H. Zimm, *J. Chem. Phys. 70*, 2991–3007 (1979).
17. S. A. Allison and J. M. Schurr, *Chem. Phys. 41*, 35–59 (1979).
18. J. C. Thomas, S. A. Allison, C. J. Appellof, and J. M. Schurr, *Biophys. Chem. 12*, 177–188 (1980).
19. D. P. Millar, R. J. Robbins, and A. H. Zewail, *J. Chem. Phys. 76*, 2080–2094 (1982).
20. D. P. Millar, R. J. Robbins, and A. H. Zewail, *Proc. Natl. Acad. Sci. U.S.A. 77*, 5593–5597 (1980).
21. I. Ashikawa, K. Kinosita, Jr., A. Ikegami, Y. Nishimura, M. Tsuboi, K. Watanabe, and K. Iso, *J. Biochem. (Tokyo) 93*, 665–668 (1983).
22. J. M. Schurr and K. S. Schmitz, *Annu. Rev. Phys. Chem. 37*, 271–305 (1986).
23. J. Wilcoxon and J. M. Schurr, *Biopolymers 22*, 2273–2321 (1983).
24. J. Wilcoxon, J. H. Shibata, J. C. Thomas, and J. M. Schurr, in: *Biomedical Applications of Laser Light Scattering* (D. B. Sattelle, W. I. Lee, and B. R. Ware, eds.), pp. 21–36, Elsevier Biomedical Press, Amsterdam (1982).
25. S. Diekmann, W. Hillen, B. Morgenmeyer, R. D. Wells, and D. Pörschke, *Biophys. Chem. 15*, 263–270 (1982).
26. J. G. Elias and D. Eden, *Macromolecules 14*, 410–19 (1981).
27. P. J. Hagerman, *Biopolymers 20*, 1503–1535 (1981).
28. M. Hogan, J. Wang, R. M. Austin, C. Monitto, and S. Hershkowitz, *Proc. Natl. Acad. Sci. U.S.A. 79*, 3518–3522 (1982).
29. J. M. Schurr, *Chem. Phys. 84*, 71–96 (1984).
30. P. N. Pusey and R. J. A. Tough, Particle interactions, in: *Dynamic Light Scattering: Applications of Photon Correlation Spectroscopy* (R. Pecora, ed.), pp. 85–179, Plenum Press, New York (1985).
31. J. L. Skinner and P. G. Wolynes, *J. Chem. Phys. 73*, 4015–4021 (1980).
32. J. L. Skinner, *J. Chem. Phys. 79*, 1955–1964 (1983).
33. G. S. Manning, *Biopolymers 22*, 689–729 (1983).
34. F. H. Crick and A. Klug, *Nature 255*, 530–533 (1975).
35. H. M. Sobell, T. D. Sakore, J. A. Banerjee, K. K. Bhanclary, B. S. Reddy, and E. D. Lozansky, *Cold Spring Harbor Symp. Quant. Biol. 47*, 273–314 (1983).
36. M. Guéron, M. Kochoyan, and J.-L. Leroy, *Nature 328*, 89–92 (1987).
37. M. Kochoyan, J.-L. Leroy, and M. Guéron, *J. Mol. Biol. 196*, 599–609 (1987).
38. A. S. Benight, J. M. Schurr, P. F. Flynn, B. R. Reid, and D. E. Wemmer, *J. Mol. Biol. 200*, 377–399 (1988).

39. J. H. Shibata, B. S. Fujimoto, and J. M. Schurr, *Biopolymers 24*, 1909–1930 (1985).
40. J. C. Thomas and J. M. Schurr, *Biochemistry 22*, 6194–6198 (1983).
41. W. R. McClure, *Annu. Rev. Biochem. 54*, 171–204 (1985).
42. M. G. Fried and D. M. Crothers, *Nucleic Acids Res. 11*, 141–158 (1983).
43. A. Kolb, A. Spassky, C. Chapon, B. Blazy, and H. Buc, *Nucleic Acids Res. 11*, 7833–7852 (1983).
44. H. M. Wu and D. H. Crothers, *Nature 308*, 509–513 (1984).
45. M. G. Fried, H. M. Wu, and D. M. Crothers, *Nucleic Acids Res. 11*, 2479–2494 (1983).
46. H.-N. Liu-Johnson, M. Gartenberg, and D. M. Crothers, *Cell 47*, 995–1005 (1986).
47. A. S. Benight, J. Langowski, P.-G. Wu, J. Wilcoxon, J. H. Shibata, B. S. Fujimoto, N. S. Ribeiro, and J. M. Schurr, in: *Laser Scattering Spectroscopy of Biological Objects* (J. Stepanek, P. Anzenbacher, and B. Sedlacek, eds.), pp. 407–422, Elsevier, Amsterdam (1987).
48. M. G. Fried, *Biophys. J. 47*, 390a (1985).
49. J. E. Anderson, M. Ptashne, and S. C. Harrison, *Nature 326*, 846–852 (1987).
50. M. E. Hogan and R. H. Austin, *Nature 329*, 263–266 (1987).
51. M. L. Kopka, A. V. Fratini, H. R. Drew, and R. E. Dickerson, *J. Mol. Biol. 163*, 129–146 (1985).
52. G. L. Seibel, U. Singh, and P. A. Kollman, *Proc. Natl. Acad. Sci. U.S.A. 82*, 6537–6540 (1985).
53. P.-G. Wu, L. Song, J. B. Clendenning, B. S. Fujimoto, A. S. Benight, and J. M. Schurr, Interaction of chloroquine with linear and supercoiled DNAs. Effect on the torsional dynamics, rigidity, and twist energy parameter, *Biochemistry 27*, 8128–8144 (1988).
54. J. C. Milton and W. C. Galley, *Biopolymers 25*, 1673–1684 (1986).
55. J. H. Shibata, J. Wilcoxon, J. M. Schurr, and V. Knauf, *Biochemistry 23*, 1188–1194 (1984).
56. J. Langowski, A. S. Benight, B. S. Fujimoto, and J. M. Schurr, *Biochemistry 24*, 4022–4028 (1985).
57. I. Ashikawa, T. Furuno, K. Kinosita, Jr., A. Ikegami, H. Takahashi, and H. Akutsu, *J. Biol. Chem. 259*, 8338–8344 (1984).
58. I. Ashikawa, K. Kinosita, Jr., A. Ikegami, and T. Tobita, *Biochim. Biophys. Acta 908*, 263–267 (1987).
59. D. Genest, P. Wahl, M. Erard, M. Champagne, and M. Daune, *Biochimie 64*, 419–427 (1982).
60. I. Ashikawa, K. Kinosita, Jr., A. Ikegami, Y. Nishimura, M. Tsuboi, K. Watanabe, K. Iso, and T. Nakano, *Biochemistry 22*, 6018–6026 (1983).
61. I. Ashikawa, K. Kinosita, Jr., A. Ikegami, Y. Nishimura, and M. Tsuboi, *Biochemistry 24*, 1291–1297 (1985).
62. J. Wang, M. Hogan, and R. H. Austin, *Proc. Natl. Acad. Sci. U.S.A. 79*, 5896–5900 (1982).
63. J. M. Schurr and R. L. Schurr, *Biopolymers 24*, 1931–1940 (1985).
64. J. Wilcoxon, J. H. Shibata, J. C. Thomas, and J. M. Schurr, in: *Biomedical Applications of Laser Light Scattering* (W. I. Lee, D. B. Sattelle, and B. R. Ware, eds.), pp. 27–36, Elsevier Biomedical Press, Amsterdam (1982).
65. M. LeBret, J.-B. LePecq, J. Barbet, and B. P. Roques, *Nucleic Acids Res. 4*, 1361–1379 (1977).
66. R. E. Depew and J. C. Wang, *Proc. Natl. Acad. Sci. U.S.A. 72*, 4275–4279 (1975).
67. D. E. Pulleyblank, M. Shure, D. Tang, J. Vinograd, and H. P. Vosberg, *Proc. Natl. Acad. Sci. U.S.A. 72*, 4280–4284 (1975).
68. D. Shore and R. L. Baldwin, *J. Mol. Biol. 170*, 957–1007 (1983).
69. D. Horowitz and J. C. Wang, *J. Mol. Biol. 173*, 75–91 (1984).
70. W. Bauer and J. Vinograd, *J. Mol. Biol. 47*, 419–435 (1970).
71. T. S. Hsieh and J. C. Wang, *Biochemistry 14*, 527–535 (1975).

72. L. Song, B. S. Fujimoto, and J. M. Schurr, *Biophys. J. 53*, 306a (1988).
73. J. M. Schurr, *Chem. Phys. 65*, 417–424 (1982).
74. A. Szabo, *J. Chem. Phys. 81*, 150–167 (1984).
75. E. P. Wigner, *Group Theory*, Academic Press, New York (1959).
76. A. R. Edmonds, *Angular Momentum in Quantum Mechanics*, Princeton University Press, Princeton, New Jersey (1974).
77. J. M. Schurr, *Mol. Phys. 40*, 1025–1051 (1980).
78. S. C. Lin, J. C. Thomas, S. A. Allison, and J. M. Schurr, *Biopolymers 20*, 209–230 (1981).
79. T. Yoshizaki, M. Fujii, and H. Yamakawa, *J. Chem. Phys. 82*, 1003–1013 (1985).
80. D. Wallach, *J. Chem. Phys. 47*, 5258–5268 (1967).
81. S. A. Allison, J. H. Shibata, J. Wilcoxon, and J. M. Schurr, *Biopolymers 21*, 729–762 (1982).
82. J. Langowski, B. S. Fujimoto, D. E. Wemmer, A. S. Benight, G. Drobny, J. H. Shibata, and J. M. Schurr, *Biopolymers 24*, 1023–1056 (1985).
83. J. M. Schurr and B. S. Fujimoto, *Biopolymers 27*, 1543–1569 (1988).
84. J. M. Schurr, *Chem. Phys. 65*, 417–424 (1982).
85. J. M. Schurr, *Biopolymers 24*, 1233–1246 (1985).
86. T. Hård, *Biopolymers 26*, 613–618 (1987).
87. D. Magde, M. Zappala, W. H. Knox, and T. M. Nordlund, *J. Phys. Chem. 87*, 3286–3288 (1983).
88. M. Hogan, N. Dattagupta, and D. M. Crothers, *Biochemistry 18*, 280–288 (1979).
89. R. J. Wittebort and A. Szabo, *J. Chem. Phys. 69*, 1722–1736 (1979).
90. G. Lipari and A. Szabo, *Biochemistry 20*, 6250–6256 (1981).
91. R. King and O. Jardetzky, *Chem. Phys. Lett. 55*, 15–23 (1978).
92. D. E. Woessner, *J. Chem. Phys. 36*, 1–4 (1962).
93. D. E. Woessner, *J. Chem. Phys. 37*, 647–654 (1962).
94. M. E. Hogan and O. Jardetzky, *Proc. Natl. Acad. Sci. U.S.A. 76*, 6341–6345 (1979).
95. M. E. Hogan and O. Jardetzky, *Biochemistry 19*, 2079–2085 (1980).
96. M. E. Hogan and O. Jardetzky, *Biochemistry 19*, 3460–3468 (1980).
97. P. H. Bolton and T. L. James, *Biochemistry 19*, 1388–1392 (1980).
98. P. H. Bolton and T. L. James, *J. Am. Chem. Soc. 102*, 25–31 (1980).
99. P. H. Bolton and T. L. James, *J. Phys. Chem. 83*, 3359–3366 (1979).
100. J. W. Keepers and T. L. James, *J. Am. Chem. Soc. 104*, 929–939 (1982).
101. G. C. Levy, D. J. Craik, A. Kumar, and E. London, *Biopolymers 22*, 2703–2726 (1983).
102. N. Assa-Munt, J. Granot, R. W. Behling, and D. R. Kearns, *Biochemistry 23*, 944–955 (1984).
103. P. A. Mirau, R. W. Behling, and D. R. Kearns, *Biochemistry 24*, 6200–6211 (1985).
104. B. A. Scalettar, M. P. Klein, and J. E. Hearst, *Biopolymers 26*, 1287–1299 (1987).
105. S. A. Allison, *Macromolecules 16*, 421–425 (1983).
106. S. A. Allison, *Macromolecules 15*, 1544–1548 (1982).
107. S. C. Lin and J. M. Schurr, *Biopolymers 17*, 425–461 (1978).
108. I. Ashikawa, K. Kinosita, Jr., and A. Ikegami, *Biochim. Biophys. Acta 789*, 87–93 (1984).
109. P.-G. Wu, B. S. Fujimoto, and J. M. Schurr, *Biopolymers 26*, 1463–1488 (1987).
110. H. Yoshimura, N. Takuhiro, K. Mihashi, K. Kinosita, Jr., and A. Ikegami, *J. Mol. Biol. 179*, 453–467 (1984).
111. H. Yamakawa and T. Yoshizaki, *J. Chem. Phys. 75*, 1016–1030 (1981).
112. L. D. Landau and E. M. Lifschitz, *Theory of Elasticity*, Addison-Wesley, Reading, Massachusetts (1959).
113. S. A. Allison and J. A. McCammon, *Biopolymers 23*, 363–375 (1984).
114. S. A. Allison, *Macromolecules 19*, 118–124 (1986).
115. D. W. Ding, R. Rill, and K. E. Van Holde, *Biopolymers 11*, 2109–2124 (1972).
116. K. S. Schmitz and J. M. Schurr, *Biopolymers 12*, 1543–1564 (1973).

117. H. Yamakawa, *Annu. Rev. Phys. Chem. 35*, 23–47 (1984).
118. T. Yoshizaki and H. Yamakawa, *J. Chem. Phys. 84*, 4684–4692 (1986).
119. H. Yamakawa, T. Yoshizaki, and M. Fujii, *J. Chem. Phys. 84*, 4693–4707 (1986).
120. P. R. Bevington, *Data Reduction and Error Analysis for the Physical Sciences*, McGraw-Hill, New York (1969).
121. A. Grinvald and I. Z. Steinberg, *Anal. Biochem. 59*, 583–598 (1974).
122. C. Houssier, B. Hardy, and E. Fredericq, *Biopolymers 13*, 1141–1160 (1974).
123. I. Zimmermann and H. W. Zimmermann, *Ber. Bunsen-Ges. Phys. Chem. 81*, 81–88 (1977).
124. E. Charney, H. H. Chen, E. R. Henry, and D. C. Rau, *Biopolymers 25*, 885–904 (1986).
125. Y. Matsuoka and B. Nordén, *Biopolymers 22*, 1731–1746 (1983).
126. G. C. Causley and W. C. Johnson, Jr., *Biopolymers 21*, 1763–1780 (1982).
127. S. P. Edmondson and W. C. Johnson, Jr., *Biopolymers 24*, 825–841 (1985).
128. S. P. Edmondson and W. C. Johnson, Jr., *Biopolymers 25*, 2335–2348 (1986).
129. M. M. Tirado and J. Garcia de la Torre, *J. Chem. Phys. 73*, 1986–1993 (1980).
130. D. Pörschke, W. Zacharias, and R. D. Wells, *Biopolymers 26*, 1971–1974 (1987).
131. B. H. Robinson, L. S. Lerman, A. Beth, H. L. Frisch, L. R. Dalton, and C. Auer, *J. Mol. Biol. 139*, 19–44 (1980).
132. P. Fan, T. Hård, D. Magde, and D. R. Kearns, *Biophys. J. 53*, 305a (1988).
133. F. B. Fuller, *Proc. Natl. Acad. Sci. U.S.A. 68*, 815–819 (1971).
134. J. Shimada and H. Yamakawa, *Macromolecules 17*, 689–698 (1984).
135. J. Shimada and H. Yamakawa, *Biopolymers 23*, 853–857 (1984).
136. J. Shimada and H. Yamakawa, *J. Mol. Biol. 184*, 319–329 (1985).
137. M. LeBret, *Biopolymers 18*, 1709–1725 (1979).
138. M. LeBret, *Biopolymers 19*, 619–637 (1980).
139. M. LeBret, *Biopolymers 23*, 1835–1867 (1984).
140. Y. Chen, *J. Chem. Phys. 75*, 2447–2453 (1981).
141. A. V. Vologodskii, V. V. Anshelevich, A. V. Lukashin, and M. D. Frank-Kamenetskii, *Nature (London) 280*, 294–298 (1979).
142. S. D. Levene and D. M. Crothers, *J. Mol. Biol. 189*, 73–83 (1986).
143. D. Shore and R. L. Baldwin, *J. Mol. Biol. 170*, 957–1007 (1983).
144. M. J. B. Tunis-Schneider and M. F. Maestre, *J. Mol. Biol. 52*, 521–541 (1970).
145. S. Hanlon, S. Brudno, T. T. Wu, and B. Wolf, *Biochemistry 14*, 1648–1660 (1975).
146. B. S. Fujimoto, J. H. Shibata, R. L. Schurr, and J. M. Schurr, *Biopolymers 24*, 1009–1022 (1985).
147. H. Castleman and B. F. Erlanger, *Cold Spring Harbor Symp. Quant. Biol. 47*, 133–142 (1983).
148. T. J. Thomas and V. A. Bloomfield, *Nucleic Acids Res. 11*, 1919–1929 (1983).
149. J. H. Shibata and J. M. Schurr, *Biopolymers 20*, 525–549 (1981).
150. M. L. Lamos, G. T. Walker, T. R. Krugh, and D. H. Turner, *Biochemistry 25*, 687–691 (1986).
151. S. A. Allison, J. C. Herr, and J. M. Schurr, *Biopolymers 20*, 469–488 (1981).
152. J. Wilcoxon, J. M. Schurr, and R. A. J. Warren, *Biopolymers 23*, 1188–1194 (1984).
153. W. Bauer, *Annu. Rev. Biophys. Bioeng. 7*, 287–313 (1979).
154. J. C. Wang, *J. Mol. Biol. 89*, 783–801 (1974).
155. R. L. Jones, A. C. Lanier, R. A. Keel, and W. D. Wilson, *Nucleic Acids Res. 8*, 1613–1624 (1980).
156. D. Genest and P. Wahl, in: *Dynamic Aspects of Conformational Change in Biological Macromolecules* (C. Sadron, ed.), pp. 367–379, D. Reidel, Dordrecht, Holland (1973).
157. D. Genest, P. Wahl, and J. C. Auchet, *Biophys. Chem. 1*, 266–278 (1974).
158. D. Genest and P. Wahl, *Biophys. Chem. 7*, 317–323 (1978).
159. R. E. Harrington, *J. Am. Chem. Soc. 92*, 6957–6964 (1970).

160. C. E. Swenberg, S. E. Carberry, and N. E. Geacintov, *Biopolymers 29*, 1735–1744 (1990).
161. W. Keller, *Proc. Natl. Acad. Sci. U.S.A. 72*, 4876–4880 (1975).
162. G. W. Brady, C. J. Benham, and D. Foos, *Biopolymers 23*, 2963–2966 (1984).
163. S. W. Chen, S. Rothenberg, D. Schaak, D. B. Fein, and G. W. Brady, *Biophys. J. 53*, 306a (1988).
164. W. B. Upholt, H. B. Gray, Jr., and J. Vinograd, *J. Mol. Biol. 61*, 21–38 (1971).
165. J. C. Wang, *J. Mol. Biol. 87*, 797–816 (1974).
166. J. M. Gale and M. J. Smerdon, *Biophys. J. 53*, 103a (1988).
167. D. Genest, G. Sabeur, P. Wahl, and J. C. Auchet, *Biophys. Chem. 13*, 77–87 (1981).
168. D. Genest, P. A. Mirau, and D. R. Kearns, *Nucleic Acids Res. 13*, 2603–2615 (1985).
169. T. Hård and D. R. Kearns, *Biopolymers 25*, 1519–1529 (1986).
170. T. Hård and D. R. Kearns, *J. Phys. Chem. 90*, 3437–3444 (1986).
171. T. Hård and D. R. Kearns, *Nucleic Acid Res. 14*, 3945–3956 (1986).
172. T. Hård and D. R. Kearns, *J. Phys. Chem. 91*, 2004 (1987).
173. P. Wu, A. S. Benight, and J. M. Schurr, *Biophys. J. 53*, 307a (1988).
174. M. Fried and V. A. Bloomfield, *Biopolymers 23*, 2141–2155 (1984).
175. B. S. Fujimoto and J. M. Schurr, *J. Phys. Chem. 91*, 1947–1951 (1987).
176. M. Yanagida, Y. Hiraoka, and I. Katsura, *Cold Spring Harbor Symp. Quant. Biol. 47*, 177–187 (1983).
177. J. C. Thomas, S. A. Allison, J. M. Schurr, and R. D. Holder, *Biopolymers 19*, 1451–1474 (1980).
178. P. R. Callis and N. Davidson, *Biopolymers 8*, 379–390 (1969).
179. M. Yanagida, K. Morikawa, Y. Hiraoka, S. Matsumoto, T. Uemura, and S. Ukada, in: *Applications of Fluorescence in the Biomedical Sciences*, pp. 321–345, Alan R. Liss, New York (1980).
180. R. Rigler and W. Wintermeyer, *Annu. Rev. Biophys. Bioeng. 12*, 475–505 (1983).
181. L. Stryer, *Annu. Rev. Biochem. 47*, 819–846 (1978).
182. A. Favre and G. Thomas, *Annu. Rev. Biophys. Bioeng. 10*, 175–195 (1981).
183. W. Wintermeyer, J. M. Robertson, H. Weidman, and H. G. Zauchau, in: *Transfer RNA: Structure, Properties and Recognition* (P. R. Schimmel, D. Söll, and J. N. Abelson, eds.), 445–457, Cold Spring Harbor Laboratory, Cold Spring Harbor, New York (1979).
184. R. Ehrlich, J. F. Lefevre, and P. Remy, *Eur. J. Biochem. 103*, 145–153 (1980).
185. J. F. Lefevre, R. Ehrlich, M. C. Kilhoffer, and P. Remy, *FEBS Lett. 114*, 219–224 (1980).
186. J. Olmsted III and D. R. Kearns, *Biochemistry 16*, 3647–3654 (1977).
187. J. C. Thomas, J. M. Schurr, and D. R. Hare, *Biochemistry 23*, 5407–5413 (1984).
188. J. C. Thomas, J. M. Schurr, B. R. Reid, N. S. Ribeiro, and D. R. Hare, *Biochemistry 23*, 5414–5420 (1984).
189. B. D. Wells and C. R. Cantor, *Nucleic Acids Res. 8*, 3229–3246 (1980).
190. M. Liebmann, J. Rubin, and M. Sundaralingham, *Proc. Natl. Acad. Sci. U.S.A. 74*, 4821–4825 (1977).
191. T. Tao, *Biopolymers 8*, 609–632 (1969).
192. A. Patkowski, E. Gulari, and B. Chu, *J. Chem. Phys. 73*, 4178–4184 (1980).
193. A. A. Bothner-By, in: *Biological Applications of Magnetic Resonance* (R. G. Shulman, ed.), pp. 177–219, Academic Press, New York (1979).
194. K. Beardsley, T. Tao, and C. R. Cantor, *Biochemistry 9*, 3524–3532 (1970).
195. D. Labuda, T. Haertle, and J. Augustyniak, *Eur. J. Biochem. 79*, 293–301 (1977).
196. D. Labuda and J. Augustyniak, *Eur. J. Biochem. 79*, 303–307 (1977).
197. N. Okabe and F. Cramer, *J. Biochem. 89*, 1439–1443 (1981).
198. B. D. Wells, *Nucleic Acids Res. 12*, 2157–2170 (1984).
199. F. Claesens and R. Rigler, *Eur. Biophys. J. 13*, 331–342 (1986).
200. B. D. Wells and J. R. Lakowicz, *Biophys. Chem. 26*, 39–43 (1987).

201. C. Urbanke and G. Maass, *Nucleic Acids Res. 5*, 1551–1560 (1978).
202. S. R. Holbrook, J. L. Sussman, R. W. Warrant, and S.-H. Kim, *J. Mol. Biol. 123*, 631–660 (1978).
203. M. M. Teeter, G. J. Quigley, and A. Rich, in: *Nucleic Acid–Metal Ion Interactions* (T. Spiro, ed.), p. 145, Wiley, New York (1980).
204. D. Labuda and D. Pörschke, *Biochemistry 21*, 49–53 (1982).
205. L. S. Kan, P. O. P. Ts'o, F. von der Haar, M. Sprinzl, and F. Cramer, *Biochemistry 14*, 3278–3291 (1975).
206. L. S. Kan, P. O. P. Ts'o, M. Sprinzl, F. von der Haar, and F. Cramer, *Biochemistry 16*, 3143–3154 (1977).
207. D. C. Fritzinger and M. J. Fournier, *Biochemistry 21*, 2118–2127 (1982).
208. J. T. Bokma, C. W. Johnson, Jr., and J. Blok, *Biopolymers 26*, 893–909 (1987).
209. S. A. Allison, R. H. Austin, and M. E. Hogan, *J. Chem. Phys. 90*, 3845–3854 (1989).
210. B. Théveny, D. Coulaud, M. LeBret, and B. Révet, in: *Structure and Expression*, Vol. 3, *DNA Bending and Curvature* (W. K. Olson, M. H. Sarma, R. H. Sarma, and M. Sundaralingam, eds.), pp. 039–055, Adenine Press, Schenectady, New York (1988).
211. W. Nerdal, D. H. Hare, and B. R. Reid, *J. Mol. Biol. 201*, 717–739 (1988).
212. K. M. Banks, D. R. Hare, and B. R. Reid, *Biochemistry 28*, 6996–7010 (1989).
213. P. F. Flynn, Ph.D. thesis, University of Washington (1989).
214. L. Song, B. S. Fujimoto, P.-G. Wu, J. C. Thomas, J. H. Shibata, and J. M. Schurr, *J. Mol. Biol. 214*, 307–326 (1990).
215. P.-G. Wu and J. M. Schurr, *Biopolymers 28*, 1695–1703 (1989).
216. L. Song, S. A. Allison, and J. M. Schurr, *Biopolymers 29*, 1773–1791 (1990).
217. R. Negri, F. Della Seta, E. di Mauro, and G. Camilloni, Topological evidence for allosteric transitions in DNA secondary structure, *Biophys. J.*, submitted.
218. G. B. Koudelka, P. Harbury, S. C. Harrison, and M. Ptashne, *Proc. Natl. Acad. Sci. U.S.A. 85*, 4633–4637 (1988).
219. L. Song and J. M. Schurr, *Biopolymers 30*, 229–237 (1990).
220. D. Eden and C. Sunshine, in: *Dynamic Behavior of Macromolecules, Colloids, Liquid Crystals and Biological Systems by Optical and Electro-Optical Methods* (H. Watanabe, ed.), pp. 000–000, Hirokawa, Tokyo (1989).
221. W. H. Taylor and P. J. Hagerman, *J. Mol. Biol. 212*, 351–362 (1990).
222. U. S. Kim, B. S. Fujimoto, and J. M. Schurr, *Biophys. J. 55*, 364a (1989).
223. J. M. Schurr, P. Wu, and B. S. Fujimoto, in: *Time-Resolved Laser Spectroscopy in Biochemistry II* (J. R. Lakowicz, ed.), *Proc. SPIE*, 368–379 (1990).
224. D. P. Millar, K. M. Ho, and A. J. Aroney, *Biochemistry 27*, 8859–8606 (1988).
225. W. Eimer, J. R. Williamson, S. G. Boxer, and R. Pecora, *Biochemistry 29*, 799–811 (1990).
226. B. S. Fujimoto and J. M. Schurr, *Biophys. J. 59*, 303a (1991).
227. E. J. Hustedt, A. Spaltenstein, J. E. Kirchner, C. Mailer, P. B. Hopkins, and B. H. Robinson, *Biophys. J. 59*, 303a (1991).
228. E. N. Trifonov, R. K.-Z. Tan, and S. C. Harvey, *Structure and Expression*, Vol. 3, *DNA Bending and Curvature* (W. K. Olson, M. H. Sarma, R. H. Sarma, and M. Sundaralingam, eds.), pp. 243–253, Adenine Press, Schenectady, New York (1987).
229. A. Stellwagen and N. C. Stellwagen, *Biopolymers 30*, 309–324 (1990).
230. B. S. Fujimoto and J. M. Schurr, Abstracts of the 10th International Biophysics Congress, Vancouver, British Columbia, July 29–August 3, 1990, p. 58.
231. B. Norden and F. Tjernfeld, *Biopolymers 21*, 1713–1734 (1982).
232. C. OhUigin, D. J. McConnell, J. M. Kelley, and W. J. M. van der Putten, *Nucleic Acids Res. 15*, 7411–7427 (1987).
233. B. S. Fujimoto and J. M. Schurr, *Nature 344*, 175–178 (1990).

234. G. R. Bellomy and M. T. Record, Jr., Helical repeat of a region of supercoiled plasmid DNA *in vivo*, *J. Mol. Biol.*, submitted.
235. E. A. Winzeler and E. W. Small, in: *Time-Resolved Laser Spectroscopy in Biochemistry II* (J. R. Lakowicz, ed.), *Proc. SPIE 1204*, 297–302 (1990).
236. S. B. Smith, P. A. Aldridge, and J. B. Callis, *Science 243*, 203–206 (1989).
237. D. C. Schwartz and M. Koval, *Nature 338*, 520–522 (1989).
238. J. M. Schurr and S. B. Smith, *Biopolymers 29*, 1161–1165 (1990).
239. S. B. Smith and A. J. Bendich, *Biopolymers 29*, 1167–1173 (1990).
240. L. Song and M. F. Maestre, *Biophys J. 59*, 308a (1991).
241. F. M. Pohl, T. M. Jovin, W. Baehr, and J. J. Holbrook, *Proc. Natl. Acad. Sci. U.S.A. 69*, 3805–3809 (1972).

5

Fluorescence in Membranes

Christopher D. Stubbs and Brian Wesley Williams

5.1. Introduction

The properties of membranes commonly studied by fluorescence techniques include motional, structural, and organizational aspects. Motional aspects include the rate of motion of fatty acyl chains, the head-group region of the phospholipids, and other lipid components and membrane proteins. The structural aspects of membranes would cover the orientational aspects of the lipid components. Organizational aspects include the distribution of lipids both laterally, in the plane of the membrane (e.g., phase separations), and across the membrane bilayer (phospholipid asymmetry) and distances from the surface or depth in the bilayer. Finally, there are properties of membranes pertaining to the surface such as the surface charge and dielectric properties. Fluorescence techniques have been widely used in the studies of membranes mainly since the time scale of the fluorescence lifetime coincides with the time scale of interest for lipid motion and since there are a wide number of fluorescence probes available which can be used to yield very specific information on membrane properties.

In Table 5.1 some of the main areas of interest concerning membranes which are amenable to investigation using fluorescence techniques associated with major fluorescence properties are listed. In this chapter, sections are organized under the major fluorescence attributes of lifetime, anisotropy, and quenching. Membrane surface-related properties are dealt with under solvent relaxation and surface charge properties. Since the subject of fluorescence in membranes is very large, details of many interesting techniques and theoretical treatments could not be included. Instead, we attempt to briefly introduce each area and try to concentrate on more recent developments of interest. Also, we have confined coverage to intramembrane properties and have not covered aspects such as membrane fusion studies.

Christopher D. Stubbs • Department of Pathology and Cell Biology, Thomas Jefferson University, Philadelphia, Pennsylvania 19107. *Brian Wesley Williams* • Department of Chemistry, Bucknell University, Lewisburg, Pennsylvania 17837.

Topics in Fluorescence Spectroscopy, Volume 3: Biochemical Applications, edited by Joseph R. Lakowicz. Plenum Press, New York, 1992.

Table 5.1. Applicability of Fluorescence Techniques to the Study of Membrane Properties

Membrane property	Fluorescence technique(s)[a]
Motion	
Rotational diffusion and orientation	B
Lateral diffusion	A, C
Organization	
Lateral (phase separations)	A
Longitudinal (distance from surface)	B, C, D
Surface density	C, D
Lipid–lipid and protein–protein association	C, D
Surface properties	
Surface charge	A
Dielectric properties	E

[a] A, Intensity and lifetime; B, anisotropy; C, quenching; D, energy transfer; E, solvent effects.

5.2. Fluorescence Lifetimes

The measurement of fluorescence lifetimes is an integral part of the anisotropy, energy transfer, and quenching experiment. Also, the fluorescence lifetime provides potentially useful information on the fluorophore environment and therefore provides useful information on membrane properties. An example is the investigation of lateral phase separations. Recently, interest in the fluorescence lifetime itself has increased due to the introduction of the lifetime distribution model as an alternative to the discrete multiexponential approach which has been prevalent in the past.

5.2.1. The Use of Fluorescence Lifetimes for Membrane Organizational Studies

The fluorescence lifetime is sensitive to the environment of the fluorophore, and in membranes this usually means the surrounding fatty acyl chains or the membrane protein interfacial region (see summary in Table 5.3). Generally, the lifetime of membrane-bound fluorophores is rather less sensitive to the types of subtle alterations which are encountered in membranes as compared to the fluorescence anisotropy parameters. The gel-to-liquid crystalline phase transition is a notable exception where most fluorophores show an alteration in lifetime properties. Although, again, the anisotropy (see below) is the most sensitive parameter in this regard, the fluorescence lifetime has been used with considerable success in the study of phase transitions and lateral phase separations. Fluorophores used to yield information on the fatty acyl chain environment in this way include, for example, 1,6-diphenyl-1,3,5-hexatriene (DPH)[1] and parinaric acids (see reviews in Refs. 2 and 3). The parinaric acids gained considerable popularity when it was found that *trans*-parinaric acid preferentially partitioned into gel-phase lipids while *cis*-parinaric acid showed a more equal distribution with some preference for liquid-crystalline phase lipids. These types of studies utilized the finding that the lifetime of *trans*-parinaric acid was much longer in gel-phase lipids as compared to liquid-crystalline phase lipids. A problem with the use of the fluorescence lifetimes is that there is a heterogeneity sometimes even in a simple organic solvent.[4] Heterogeneity refers to the number of discrete lifetimes which can be ascribed to the fluorophore. One cause of a fluorophore having a heterogeneous lifetime would be multiple environments, a case likely to occur in a membrane, but its occurrence in a simple homogeneous system complicates interpretation of data from a membrane system. However, the basic observation of phase partitioning behavior of the parinaric acids continues to provide useful information, particularly on model systems, and has recently been confirmed using NMR.[5]

The use of DPH lifetimes for the analysis of phase separations and membrane properties has been described for model systems.[1,6] In the case of both parinaric acids and DPH, one of the motivations for examining phase separation in a model lipid bilayer is the possibility that phase separations might be detectable in natural membranes. However, this technique has not been able to satisfactorily resolve lateral phase separations in natural membranes, either because they do not exist or because they are much more complex and even possibly transient in nature. Alternatively, it could be argued that if a probe could be found with the characteristics of *trans*-parinaric acid but perhaps with an even greater phase partitioning ability, then this approach might be reevaluated.

Another factor affecting the lifetime of a membrane fluorophore probe is its proximity to the surface. The lifetimes of the DPH, DPH-phosphatidylcholine (DPH-PC), and trimethylammonium-DPH (TMA-DPH) probes decrease in the order DPH > DPH-PC > TMA-DPH, as the probe locates nearer to the surface of the lipid bilayer.[7] The same is found for the anthroylstearate probes.[8] More recently, it has been shown that with TMA-DPH, the lifetime appears to be fairly sensitive to the differences in lipid bilayer packing induced by differing degrees of unsaturation in the phospholipid fatty acyl chains.[9] This aspect of the use of TMA-DPH and possibly other probes remains to be further exploited.

5.2.2. Fluorescence Lifetime Distributions

The conventional analysis of fluorescence decay is described in terms of a sum of one or more exponential terms, each with a characteristic lifetime and weight. Recently, several workers have expanded upon this analysis and suggested that for some systems, usually involving proteins, a better description of fluorescence decay might be afforded by a distributional approach.[10–14] Since this is a relatively new and interesting development, we discuss this in some detail. In this approach the fluorescence decay is modeled as arising from some continuous distribution of fluorophore states, in contrast to the multiexponential model, in which the fluorescence decay is interpreted as arising from a few discrete states. Several situations involving fluorescence in membranes appear to be amenable to this new treatment, including processes involving anisotropy, time-dependent spectral shifts, quenching, and energy transfer.[15–17] Chapter 2 gives complementary information regarding distributional analysis.

The mathematical basis of the distributional approach can be understood by reference to the equations used for multiexponential decay. Integrals replace finite sums of terms, while the discrete parameters inside the sums are replaced by continuous functions of these parameters. The equation

$$I(t) = \sum_{i=1}^{N} a_i \exp(-t/\tau_i) \tag{5.1}$$

representing the decay intensity $I(t)$ as a sum of N terms of lifetime τ_i and preexponential coefficients or weights a_i is replaced by

$$I(t) = \int_0^\infty a(\tau) \exp(-t/\tau)\, d\tau \tag{5.2}$$

where $a(\tau)$ now represents a distribution function of the continuous lifetime variable. The domain of this variable is all positive values, as shown by the limits on the integral. The distribution function replaces the weight terms a_i in the sum, but still serves the same function of expressing the relative contribution of each particular lifetime to the total decay. The discrete multiexponential lifetime analysis now becomes a subcase of the distributional analysis in which the distribution function is represented as a weighted sum of Dirac delta functions:

$$a(\tau) = \sum_{i=1}^{N} c_i\, \delta(\tau - \tau_i) \tag{5.3}$$

Note also that Eq. (5.2) is equivalent to the common Laplace transform. A comparison of double-exponential and distributional analyses is represented in Figure 5.1. The distribution function shows width about central values which the double-exponential fit cannot express because of its mathematical form. Here the appearance of central values may partially be a consequence of the model functions assumed in the solution. Nevertheless, width directly

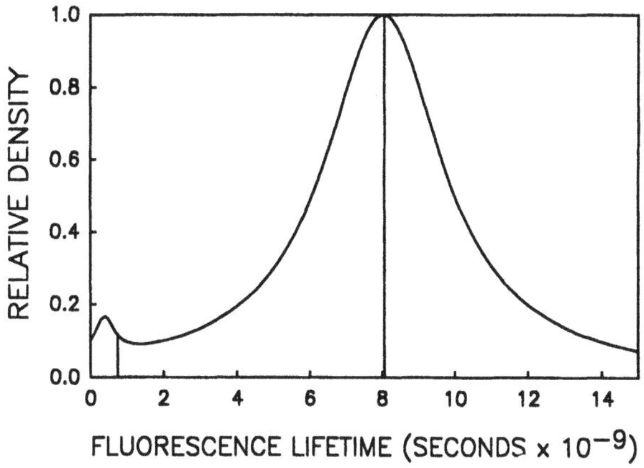

Figure 5.1. Representations of double-exponential and bimodal Lorentzian distribution analyses of DPH fluorescent decay lifetimes in liver microsomal membranes. Results (see Table 5.2) are normalized to the major component. The double-exponential analysis, represented by the vertical lines, recovers lifetimes near the centers of the Lorentzian distributions. The width of the distributions represents contributions from a variety of lifetimes. (From Ref. 17.)

represents the existence of a continuum of fluorescent states. An intuitive physical interpretation of this continuum is that it reflects heterogeneity in the fluorophore environment.

Equation (5.2) is often modified by introducing terms which relate to the normalization of the distribution function or by its reexpression in terms of a rate variable which is the reciprocal of the lifetime. Normalization can be understood by reference to Eq. (5.1), where the preexponential a_i are usually subject to the normalization condition that their sum equal 1. The analogous condition for distribution functions once again replaces this sum with an integral over all positive values. Also by analogy to the preexponential a_i, the distribution functions are usually positively valued. Negative preexponential terms or distribution function values can arise, however, in cases such as excited-state reactions.[17]

Although satisfactory criteria for deciding whether data are better analyzed by distributions or multiexponential sums have yet to established, several methods for determining distributions have been developed. For pulse fluorometry, James and Ware[11] have introduced an "exponential series" method. Here, data are first analyzed as a sum of up to four exponential terms with variable lifetimes and preexponential weights. This analysis serves to establish estimates for the range of the preexponential and lifetime parameters used in the next step. Next, a "probe" function is developed with fixed lifetime values and equal preexponential factors. An iterative Marquardt[18] least-squares analysis is undertaken with the lifetimes remaining fixed and the preexponential constrained to remain positive. When the preexponential factor for a particular lifetime drops below a cutoff, defined as a fraction of the largest preexponential value, that term in the exponential series is dropped. The eventual stopping point of the analysis is determined when a parameter based on the square of the difference of the calculated fit and the experimental data (χ^2) can no longer be minimized. An exponential series of between 50 and 120 terms is taken to favor the distributional representation. A process of removing exponential terms whose preexponential factors fall below some cutoff was introduced since the search procedure often failed to converge when these terms were included.

The mathematical basis for the exponential series method is Eq. (5.3), the use of which has recently been criticized by Phillips and Lyke.[19] Based on their analysis of the one-sided Laplace transform of model excited-state distribution functions, it is concluded that a small, finite series of decay constants cannot be used to represent a continuous distribution. Livesey and Brouchon[20] described a method of analysis using pulse fluorometry which determines a distribution using a "maximum entropy method." Similarly to Phillips and Lyke, they viewed the determination of the distribution function as a problem related to the inversion of the Laplace transform of the distribution function convoluted with the excitation pulse. Since Laplace transform inversion is very sensitive to errors in experimental data,[21] physically and nonphysically realistic distributions can result from the same data. The latter technique provides for the exclusion of nonrealistic trial solutions and the determination of a physically realistic solution. These authors noted that this technique should be easily extendable to data from phase–modulation fluorometry.

Data from phase–modulation fluorometry have been analyzed using an alternative approach to those described above, as expounded by Gratton and co-workers[14,12,13,22] and Lakowicz *et al.*[16] Here, Lorentzian or Gaussian distribution functions with widths and centers determined by least-squares analysis are used to model the unknown distribution function. While this approach may introduce assumptions about the shape of the ultimate distribution function since these trial functions are symmetric, it has the advantage of minimizing the number of parameters involved in the fit. Here, a minimum χ^2 is sought, where

$$\chi^2 = 1/(2n - f - 1) \sum_{\omega_i} (\{[P_m(\omega_i) - P_c(\omega_i)]/dP_m\}^2 + \{[M_m(\omega_i) - M_c(\omega_i)]/dM_m\}^2) \tag{5.4}$$

Here n equals the number of modulation frequencies used, f is the number of free parameters, $P_{m,c}(\omega_i)$ are the measured and calculated phase data at frequency ω_i, $M_{m,c}(\omega_i)$ are the measured and calculated modulation data, and dP_m, dM_m are the uncertainties in the measured phase and modulation values. Detailed algorithms using this approach have not yet been published. However, from the form of the distribution functions published by Alcala *et al.*,[12–14] it appears that in this case the Lorentzian or Gaussian distribution is represented by some number of Dirac delta functions which are determined by the width and center of the model distribution function. Changing the centers and widths of the trial functions varies the position and weights of the delta function set.

The computerized least-squares analyses used in these methods and in the exponential series method are not without difficulties. Trial values for the fitted parameters are required before initiation of analysis, and these may not be obvious. For inappropriate values, convergence to some final satisfactory set might not be possible. When convergence is achieved, another problem lies in determining whether or not the solution is the best possible, or merely represents some "local" minimum value of the figure of merit used. Beyond this, the decision needs to be made as to whether a distributional analysis is warranted in contrast to a multiexponential analysis. In addressing this last question, both Alcala *et al.*[12] and Lakowicz *et al.*[16] have modeled the effects of multiexponential and model distribution function approaches on the χ^2 value. Lakowicz *et al.* have examined several cases and concluded that a unimodal distribution is difficult to distinguish from a double-exponential decay and that a bimodal distribution is difficult to distinguish from a triple-exponential decay on the basis of these effects. Nevertheless, these authors appear to consider the distributional approach a viable alternative to multi-exponential fits, particularly in cases where a distribution of lifetimes might be expected from the physical properties of the system under examination.

Studies carried out so far with lipid vesicle systems and natural membranes have concentrated on whether distributions can be discerned in these systems and whether or not this offers a more informative approach. Fiorini *et al.*[22] showed that for DPH in dipalmitoyl phosphatidylcholine (DPPC) and dimyristoyl phosphatidylcholine (DMPC) vesicles, analyzed using a Lorentzian bimodal function, broad widths are found below the phase transition (in the major component). Above the transition, narrow widths essentially indistinguishable from exponential terms were found. The authors suggested that this result stems from the distribution of DPH in different environments characterized by differences in dielectric constant along the membrane normal. At higher temperatures, increased probe mobility allows the averaging of these environments, resulting in the observed decrease in width. James *et al.*[23] have also investigated the effects of the phase transition on DMPC unilamellar vesicles labeled with the parinaric acids. Unimodal distributions determined using the exponential series method were suggested to better describe the decay than exponential fits with a few discrete terms.

Natural membranes have also been examined using a distributional approach.[17,24,25] In erythrocyte membranes, Fiorini *et al.*[25] found that a bimodal Lorentzian fit to the data gave a greater width in the intact membrane as compared to vesicles of extracted lipids, and it was concluded that intact membranes show greater heterogeneity by comparison with the vesicles made from extracted lipids. However, the small differences observed in the χ^2 values between the multiexponential and distributional fits serve to illustrate the difficulties faced in choosing the appropriate model in a real situation lacking any clear-cut physical basis for distinction.

In an attempt to determine the physical basis for lifetime distributions in natural membranes, we have been examining a variety of natural membranes and model vesicle systems.[17] Typical data for a selection of phospholipid bilayers and a natural membrane are illustrated in Table 5.2 and Figure 5.1. The general conclusion would be that DPH would appear to inhabit a large number of distinct environments in natural membranes during the excited state as reflected by the broad distribution of the fluorescence lifetime. However, even in phospholipid vesicles containing a mixture of different unsaturated fatty acyl constituents, there is still a reasonable case for a multi-environment interpretation. With palmitoyloleoylphosphatidylcholine vesicles (which lack the heterogeneity of fatty acyl chain degree of unsaturation), however, the width of the major component of a bimodal Lorentzian analysis becomes very narrow so that for this component the result obtained from the Lorentzian analysis is indistinguishable from that given by the double-exponential model. We conclude that there may be a number of different conditions which could underlie fluorescence lifetime distributions, with fatty acyl unsaturation being just one. More experimentation will be required before

Table 5.2. Double-Exponential and Bimodal Lorentzian Analyses of DPH Fluorescence Decay Lifetimes in Liver Microsomes and Vesicles of Extracted Phospholipids[a,b]

	Double exponential					Bimodal Lorentzian						
	f_1	τ_1	f_2	τ_2	χ^2	f_1	c_1	w_1	f_2	c_2	w_2	χ^2
Microsomes	0.90	8.06	0.10	0.74	3.88	0.91	8.04	3.96	0.09	0.39	0.67	1.03
Phospholipids	0.95	8.18	0.05	1.12	3.04	0.96	8.12	3.54	0.04	0.27	0.80	1.00
Phosphatidylcholine	0.95	7.68	0.05	1.11	1.23	0.94	7.72	1.35	0.06	1.14	2.13	0.95
POPC[c,d]	0.92	7.90	0.08	1.59	1.10	0.90	7.95	0.00	0.10	1.93	1.61	1.03

[a] From Ref. 17, with permission.
[b] Membranes and vesicles were labeled at a DPH/lipid ratio of 1/400 and measured using phase–modulation fluorometry at ten frequencies between 5 and 90 MHz at 37°C. χ^2 values were calculated assuming errors of 0.2 and 0.002 in the phase and modulation, respectively, except where otherwise noted. $f_{1,2}$, Fraction of Exponential term or Lorentzian; $\tau_{1,2}$, lifetime (ns); $c_{1,2}$, center of Lorentzians (ns); $w_{1,2}$, half-width of Lorentzians (ns).
[c] Synthetic 1-palmitoyl-2-oleoyl phosphatidylcholine.
[d] χ^2 values calculated assuming errors of 0.15 and 0.0015 in the phase and modulation, respectively.

the physical basis for the behavior occurring in a complex natural membrane can be understood. Once this understanding has been gained, there may be considerable potential in this approach for the understanding of membrane structure and dynamics.

5.2.3. Excimer Probes

The excimer-forming fluorophores are a special class of probes which have been used in the study of membranes for some time and continue to provide useful information. Their placement in this section is arbitrary although the commonly used excimer probe pyrene is of interest because of its long fluorescence lifetime. Nevertheless, this aspect is of less importance than its ability to form an excited-state complex. In the next section, we consider fluorescence anisotropy, which provides information on membrane lipid dynamics due to the rotational properties of the fluorophore. Excimer formation, however, can be used to give information on the lateral diffusion properties of a membrane, since the formation of an excimer is governed by the lateral diffusion of the surrounding lipids. This subject has been reviewed,[26–28] and methods for obtaining the lateral diffusion constant have been described. Pyrene is the most common excimer probe used in membrane studies. It has several emission maxima centered around 400 nm and a broad emission at 475 nm from the excimer state. In practice, obtaining the ratio of the excimer to the monomer emission intensity is often sufficient to study effects on the membrane of interest. Other methods for obtaining the lateral diffusion properties of a membrane fluorophore probe include the use of fluorescence quenching (see below) and the fluorescence recovery after photobleaching technique.[29]

5.3. Fluorescence Anisotropy

The polarization properties of light in combination with fluorescence can be used as a powerful tool for determining motional properties of membranes. This is possible due to the fact that the time scale of interest for membrane lipids falls within the time frame of the fluorescence decay phenomena (0–100+ ns). This, coupled with high sensitivity, low perturbing properties of fluorescent probes, and the large number of available probes, makes the fluorescence approach the method of choice for membrane motional studies.

The motional characteristics of interest are typically those governed by the phospholipid fatty acyl chains and head-group region and the neutral lipid or protein components of membranes. Rotational motion can be subdivided into a structural component, the *order* or degree of orientational constraint, and a dynamic component, the *rate* of motion. In membranes the lateral rate of motion is also often considered. The most common types of questions asked concern the motional properties of specific types or mixtures of lipids, effects of proteins on lipid motion, and the response to perturbation by various agents such as drugs and anesthetics.

The first decision to be made in designing an experiment to measure the motional properties of membrane lipids concerns the type of probe molecule. Too often, this choice is made from the point of view of convenience or tradition rather than suitability, although there is now a considerable range of suitable fluorophores from which to choose. The second consideration is the type of measurement to be made. The most detailed and complete motional information is obtained from a time-resolved fluorescence anisotropy measurement which is able to separate the structural or orientational aspects from the dynamic aspects of fluorophore motion. Steady-state anisotropy measurements, which are much easier to perform, provide a more limited physical parameter relating to both of these aspects.

5.3.1. Anisotropy Parameters

The easiest parameter to measure, which is related to motion, is the steady-state anisotropy (r^s). However, there are a number of pitfalls to be aware of in the determination of r^s. It is also the most misunderstood parameter. The relationship used to determine r^s is

$$r_s = (I_{VV} - I_{VH}G)/(I_{VV} + 2I_{VH}G) \tag{5.5}$$

where I is the fluorescence intensity, and the first and second subscripts indicate the orientation, vertical (V) or horizontal (H), of the excitation polarizer and the emission polarizer, respectively. G is an optical correction factor (I_{HH}/I_{HV}). The intensity values for the unlabeled sample have to be subtracted and corrections made for scattering artifacts which can be particularly troublesome with membrane suspensions or intact cells.[30,31] A common method for dealing with this problem is to measure the r^s and optical density (at the emission wavelength) for different dilutions of the sample. The corrected r^s is obtained from a plot of r^s versus optical density extrapolated to zero optical density.

The r^s is often equated with the term "membrane fluidity," which itself is a vague term relating to the motional condition of membrane lipids. Nevertheless, membrane fluidity continues to be a useful concept in studies with natural cell membranes. This subject has been rigorously reviewed elsewhere[32–34] and will therefore not be dealt with in detail here. In spite of the problem that r^s contains both rate and orientational contributions (see below), it retains a central importance as a part of many fluorescence experiments and can be used with caution to assess the value of performing a time-resolved fluorescence measurement.

5.3.2. Time-Resolved Anisotropy

In time-resolved anisotropy measurements, the static or orientational components of motion and the rate of motion are derived. The time-resolved derivation of r^s is revealed as

$$r^s = \int_0^\infty r(t)\, s(t)\, dt \Big/ \int_0^\infty s(t)\, dt \tag{5.6}$$

where $r(t) = I_{\parallel}(t) - I_{\perp}(t)$ and $s(t) = I_{\parallel}(t) - 2I_{\perp}(t) = \sum_i \alpha_i \exp(-t/\tau_i)$, and the τ_i are the fluorescence lifetimes. Experimentally, it is found that the anisotropy $r(t)$ does not decay to zero in lipid bilayers but to a finite value r_∞ (see Figure 5.2).

The anisotropy decay is given by

$$r(t) = \sum_i \beta_i \exp(-t/\phi_i) \tag{5.7}$$

where the ϕ_i are the rotational correlation times. Equation (5.7) can be simplified either to

$$r(t) = (r_0 - r_\infty) \exp(-t/\phi) + r_\infty \tag{5.8}$$

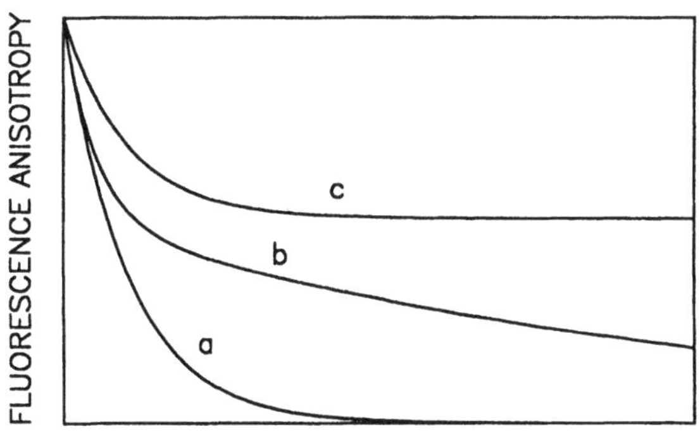

Figure 5.2. Simulated fluorescence anisotropy decays for (a) an isotropic system, (b) a system such as a lipid bilayer with short and long rotational correlation times, and (c) a system in which one of the rotational correlation times is infinite, and there is therefore a residual anisotropy or r_∞.

or

$$r(t) = \beta_1 \exp(-t/\phi_1) + \beta_2 \exp(-t/\phi_2) \tag{5.9}$$

The r_∞ term is the anisotropy at times long compared to the fluorescence lifetime, whereas in Eq. (5.9) ϕ_2 will be long. If there is no r_∞, then Eq. (5.8) reduces to the familiar Perrin equation for an isotropic rotator. Earlier, some confusion existed in this field since it was not recognized that an r_∞ term was required for the case of membrane lipid bilayers. For the most part, time-resolved anisotropy measurements have a short rotational correlation time and an r_∞ term. However, it has been recognized that a more adequate description may be to use two rotational correlation times, where the second may be quite long but not infinite as the r_∞ implies.(35, 36)

The steady-state anisotropy r^s is related to r_∞ [from Eq. (5.8)] according to

$$r_s = \frac{r_0 - r_s}{1 + \tau/\phi} + r_\infty \tag{5.10}$$

Thus, the r^s is a complex term which embodies the fluorescence lifetime, rotational correlation time (ϕ), and also r_0 (r^s in the absence of depolarizing motion). The most common type of experiment involves a comparison of r^s for two experimental conditions; however, such a comparison of r^s ignores possible changes in τ, ϕ, and r_0. Nevertheless, for many cases a comparison of r^s values alone may be satisfactory although a more rigorous analysis requires a time-resolved measurement. A comparison of the effects of changes in common membrane properties on time-resolved fluorescence parameters is shown in Table 5.3.

The observation that $r(t)$ decays to the finite value r_∞ soon led to the recognition that the fluorophore DPH has an orientational motion which is restricted due to the surrounding lipid chains.(35–37) From this the "wobbling-in-cone" model(37, 43, 44) was developed. In this model, DPH was assumed to wobble within a cone of half angle θ_c (which relates to the degree of orientational constraint, order, or r_∞) with a wobbling diffusion constant D_w (which relates to the rate of motion) with

$$r_\infty / r_0 = [\tfrac{1}{2} \cos \theta_c (1 + \theta_c)]^2 \tag{5.11}$$

An alternative "model-free" form would be

$$r_\infty / r_0 = \langle (3 \cos \theta - 1)/2 \rangle^2 \tag{5.12}$$

This formulation led to the recognition that an order parameter, S, could be derived analogous to that obtained from ^{2}H-NMR and EPR studies,(37, 45–49) where $S = (r_\infty / r_0)^{1/2}$. Thus, for time-resolved studies at

Table 5.3. Alteration of Membrane Lipid Properties and the Resultant Approximate Directions of Changes of Fluorescence Parameters of a Typical Membrane Fluorophore Probe of the Fatty Acyl Chain Region[a]

Increase of:	r_∞ (order)	ϕ (rate)	τ	τ width[b]
Temperature through T_c[c]	↓[d]	↑[d]	↓[d]	↓[e]
Cholesterol : phospholipid	↑[f]	(↓)[f,g]	(↑)[f,g]	?
Phospholipid unsaturation	↓[h,i]	↓[h]	~[h]	↑ (?)[j]
Protein : lipid	↑[k]	↓[k]	~[k]	↑[l,j]

[a] See text for definitions of parameters.
[b] An increase indicates a broadening lifetime distribution.
[c] Gel–liquid-crystalline phase transition.
[d] Refs. 35–38.
[e] Ref. 22.
[f] Ref. 39.
[g] Parentheses denote a small effect.
[h] Ref. 40.
[i] Ref. 41.
[j] Ref. 17.
[k] Refs. 7 and 42.
[l] Ref. 25.

present it is common to use either the rotational correlation time(s) or r_∞ and the calculated order parameter to describe the fluorophore motion in the membrane of interest. It should be emphasized that the order parameter is probe specific and will differ, for example, between TMA-DPH and DPH. Comparisons of the order parameters obtained from fluorescence and NMR or EPR probably result in similar values only in very restricted circumstances.[45,46] Indeed, it is important to realize that the order parameter is only a very partial and incomplete description of acyl chain orientation and that there is no "absolute" or "real" value. The same applies to all of the fluorescence parameters. The parameters extracted are thus "probe specific." This is illustrated in the comparison of time-resolved fluorescence parameters for DPH, TMA-DPH, and DPH-PC in sarcoplasmic reticulum membranes in Table 5.4.

Since steady-state data are much easier to obtain, some effort has been directed to methods for deriving time-resolved anisotropy parameters from the steady-state anisotropy.[2,45–49] A number of relationships have been described, some of which require knowledge of r_0 and the fluorescence lifetime (see, e.g., Ref. 48). An example[50] of such an empirical relationship is

$$r_\infty = r_0 r_s / [r_0 r_s + (r_0 - r_s)^2 / m] \tag{5.13}$$

where m is an adjustable parameter. Different equations which can be used to obtain r_∞ from r_s have been discussed.[50] Generally, the calculation is successful if the r^s value is relatively high. This would apply to lipid bilayers below the phase transition temperature and to natural membranes. Liquid-

Table 5.4. Fluorescence Parameters of DPH, DPH-PC, and TMA-DPH in Intact Sarcoplasmic Reticulum Membranes and Liposomes Made from Total Lipid Extracts of the Membranes at 37°C[a,b]

	r_s	r_∞	θ_c	ϕ	D_w	α_1	τ_1	α_2	τ_2
Intact membranes									
DPH	0.122	0.071	56.7	1.33	0.151	0.26	1.54	0.74	8.67
DPH-PC	0.187	0.126	47.6	1.12	0.145	0.63	1.48	0.37	6.96
TMA-DPH	0.238	0.182	39.8	0.57	0.209	0.69	1.51	0.31	5.14
Membrane lipid liposomes									
DPH	0.071	0.021	70.0	0.99	0.248	0.27	1.18	0.73	8.14
DPH-PC	0.166	0.101	51.6	1.05	0.170	0.58	1.53	0.43	6.02
TMA-DPH	0.183	0.125	47.8	0.79	0.215	0.65	1.30	0.22	4.65

[a] From Ref. 7, with permission.
[b] Definitions of parameters: r^s, Steady-state anisotropy; r_∞, residual equilibrium anisotropy; θ_c, cone angle; ϕ, apparent relaxation time (nanoseconds); D_w, wobbling diffusion constant; $\tau_{1,2}$ and $\alpha_{1,2}$ fluorescence lifetimes (nanoseconds) under double-exponential approximation and respective decay amplitudes.

crystalline phase phospholipid bilayers may exhibit rather low r^s values, and therefore the calculation of r_∞ becomes less successful; however, the formulation of Eq. (5.13) is an attempt to allow some flexibility in the range of calculable values. The conversion of r^s to r_∞ appears to be successful for the DPH and anthroylstearate probes but less so for parinaric acid.[(50)] Knowledge of r_0, the anisotropy in the absence of depolarizing motion, is necessary both for time-resolved anisotropy measurements and for calculations of r_∞ from r^s where the empirical relationships employed may assume a certain value of r_0. For DPH, r_0 values have been measured in glycerin, yielding a wide range of values within the range 0.362–0.395.[(50)] Alternatively, r_0 may be left as a free parameter (Eq. 5.8), although this results in a rather low value. In a study of the behavior of r_0 left as a free parameter,[(50)] nonsystematic effects on other parameters were demonstrated, and it was therefore concluded that a fixed value was more appropriate.

Although the calculation of order parameters from r_∞ has become an area of intense interest, the current position is that it is better to quote steady-steady anisotropy data in terms of r^s as well as possibly calculating an order parameter if this is useful, whereas calculations of microviscosities or rotational correlation times (from the Perrin equation) should be avoided.

Extensions of the analysis of time-resolved fluorescence anisotropy decay data in terms of two order parameters have also been developed (see, e.g., Refs. 51–54). Thus, the corresponding higher order parameter term is $\langle P_4 \rangle$ given by[(53)]

$$\langle P_4 \rangle = \tfrac{1}{8}\langle 35\cos^4\beta - 30\cos^2\beta + 3\rangle \tag{5.14}$$

where the angle brackets denote an ensemble average. This is for molecules possessing cylindrical symmetry about the long axis. Consideration of $\langle P_4 \rangle$ has led to the suggestion[54] that the differences in the ordering of DPH in different systems may be due to different fractions of the probe molecules lying with their long axes parallel to the bilayer surface.[54,9] This "bimodal" distribution has been investigated in a series of unsaturated phosphatidylcholines[9] and the same result obtained. TMA-DPH does not have a bimodal distribution, however, since it is tethered at the head-group region. DPH-PC must also lack a bimodal distribution, as the fluorophore is attached to the *sn*-2 position of the glycerol backbone of the phospholipid.

At the present time, two methods are in common use for the determination of time-resolved anisotropy parameters—the single-photon counting or pulse method[55–56] and the frequency-domain or phase fluorometric methods.[57–59] These are described elsewhere in this series. Recently, both of these techniques have undergone considerable development, and there are a number of commercially available instruments which include analysis software. The question of which technique would be better for the study of membranes is therefore difficult to answer. Certainly, however, the multifrequency phase instruments are now fully comparable with the time-domain instruments, a situation which was not the case only a few years ago. Time-resolved measurements are generally rather more difficult to perform and may take considerably longer than the steady-state anisotropy measurements, and this should be borne in mind when samples are unstable or if information of kinetics is required. It is therefore important to evaluate the need to take such measurements in studies of membranes. Steady-state instruments are of course much less expensive, and considerable information can be extracted, although polarization optics are not usually supplied as standard.

5.3.3. Applications to Membrane Studies

There has been considerable interest in using fluorescence anisotropy to detect multiple environments in membranes as with fluorescence lifetimes (see above). For example, if a fluorophore is located in two environments with long and short lifetimes, then the fluorescence anisotropy decay process at longer times after excitation will be dominated by the long-lived fluorescent species. This occurs with parinaric acids, and this situation has been explored for a number of theoretical cases.[60] A similar situation has been found for DPH in two-phase lipid systems by collecting anisotropy decay-associated spectra at early and late times after excitation.[61] Evidence was found for more than one rotational environment in vesicles of a single lipid of it is at the phase transition temperature. It is important to identify systems showing "associated anisotropy decays" with more than one correlation time, each of which is associated with a distinct physical environment (e.g., lipid phase) as distinct from nonassociated anisotropy decays.[60,62] The latter may be due to populations of conformers, excited-state processes, or transient effects in quenching.[62]

If a collisional quencher of the fluorophore is also incorporated into the membrane, the lifetime will be shortened. The time resolution of the fluorescence anisotropy decay is then increased,[63] providing the collisional quenching itself does not alter the anisotropy decay. If the latter condition does not hold, this will be indicated by an inability to simultaneously fit the data measured at several different quencher concentrations to a single anisotropy decay process. This method has so far been applied to the case of tryptophans in proteins[63] but could potentially be extended to lipid-bound fluorophores in membranes. If the quencher distribution in the membrane differed from that of the fluorophore, it would also be possible to extract information on selected populations of fluorophores possibly locating in different membrane environments.

There have been rather few studies of the location of probes in whole cells. DPH incorporates into most subcellular fractions (see, e.g., Ref. 64), whereas with TMA-DPH, early after introduction only the plasma membranes appear to be labeled.[64,65] There is considerable interest in examining the lipid motional properties of living cells by fluorescence techniques. In this type of study the location of the probe has to be carefully checked before conclusions can be drawn. This is carried out by separate measurements of the recovery of probe from intact labeled cells in isolated subcellular fractions and/or by fluorescence microscopy.

5.3.4. Fluorescent Probes for Lifetime and Anisotropy Studies

The choice of fluorophore for studying membrane properties is governed by a number of requirements. Some of these have been discussed previously.[66] The main points (mainly with reference to anisotropy) are:

1. The fluorophore should be well characterized in terms of absorption and emission transition moments, quantum yield, polarization bands of interest, and behavior at different temperatures. The quantum yield should be high enough so that the level of probe needed for acceptably low signal noise would not be great enough to cause significant perturbation effects.
2. The scale of the fluorescence lifetime should coincide with the time scale of the physical process of interest.
3. For lipid dynamics studies, the probe should be rigid, preferably rod- or disk-shaped with at least biaxial symmetry so that information on order can be obtained.

4. The partitioning into the membrane should be high in combination with a low solubility/fluorescence in water.
5. The location of the fluorophore in the lipid bilayer should be known.
6. The fluorophore should be stable under the conditions of measurement. Some fluorophores (e.g., parinaric acid), for example, may be incorporated into phospholipids in natural membranes.[67] Conversely, phospholipids with the fluorophore attached to one of the fatty acyl chains (e.g., DPH-PC) may be cleaved by the action of phospholipases. Also, DPH is susceptible to photobleaching so that a low excitation intensity has to be used. Parinaric acids are liable to oxidize and therefore have to be kept under argon.

For the investigation of gel–liquid-crystalline phase transitions, DPH type probes are excellent choices. Parinaric acids are also useful due to the preferential partitioning of *trans*-parinaric acid into the gel phase and can be utilized for the investigation of mixed phase systems. There are a number of DPH probes which are commercially available, and the properties of others have also been described.[68] Apart from free DPH, there are also the positively charged and negatively charged versions, trimethylammonium-DPH[69] (TMA-DPH) and DPH-propionic acid. DPH has also been conjugated to phosphatidylcholine (1-palmitoyl-2[[2-[4-(6-phenyl-*trans*-1,3,5-hexatrienyl)-phenyl]ethyl]carboxyl]-3-*m*-phosphatidylcholine or DPH-PC) and a sterol.[68] DPH itself locates in the bilayer central region, while the charged species locate at or near the head-group region, and for DPH-PC the DPH moiety is located where the *sn*-2 fatty acyl chain would normally reside. The properties of DPH have been discussed in detail in earlier reviews,[70, 71, 66] and recently the advantages of using DPH-PC have been reviewed.[72] In the investigation of the lipid dynamics of natural membranes, DPH has been the fluorophore of choice mainly due to its being the first of the DPH type probes available. Recently, TMA-DPH has been increasingly used to complement DPH, since it probes more toward the lipid head-group region and has a known location whereas the location of DPH is less precisely known. Thus, a number of recent studies have shown that DPH can orient to a small extent parallel to the plane of the bilayer as well as in the direction of the fatty acid chains (see Section 5.3.1). This could mean that in some circumstances interpretation of results could be made more difficult, particularly if the order of the fatty acyl chains is being inferred from the extracted fluorescence anisotropy parameters, and for this reason it has been suggested that DPH is not suitable as a membrane probe.[54] However, providing the results are interpreted with caution and perhaps confirmed with a second probe, such as TMA-DPH, DPH can provide much useful information, and in fact its lack of "tethering" may allow it to assume orientations at the protein–lipid interface, which may not be possible for other probes.

The anthroylstearate series of fluorescent probes can be used to give information on lipid dynamics at different depths into the lipid bilayer.[73,74] Compared with DPH, the anthroylstearates are less fluorescent and higher probe: lipid ratios (1:100 compared to <1:400 with DPH) are needed. Although it is possible to extract time-resolved fluorescence anisotropy parameters with the anthroylstearates, so far studies have been largely confined to model lipid bilayers.[8,75,76]

Another series of probes which has attracted much interest in membrane studies has been the 2-*N*-(4-nitrobenzo-2-oxa-1,3-diazole) (NBD)-labeled lipids. These have been used for studies of the lipid trafficking in intact cells using mainly fluorescence microscopic techniques[77], for fluorescence quenching[78,79] and for studies of phospholipase A_2 activities in membranes.[80] To date, they have received less attention in the study of lipid dynamics.

The dansyl (dimethylaminonaphthalenesultonyl) group has been attached to various lipids, most notably at the phospholipid head group, where the probe is sensitive to solvent effects (see below).

There has been a continued interest in examining the properties of intact living cells using fluorescence microscopy. This field has seen considerable advances since the application of digital imaging techniques. In examining whole cells, one has to be especially aware of the location(s) of the probe. This is particularly important when bulk measurements are to be made on intact cells.

The method of introduction of the fluorophore into the membrane is also important. Many probes are introduced into preexisting vesicles, natural membranes, or whole cells by the injection of a small volume of organic solvent containing the fluorophore. For DPH, tetrahydrofuran is commonly used, while methanol is often employed for other probes. The amount of solvent used should be the absolute minimum possible to avoid perturbation of the lipids, since the solvent will also partition into the membrane. With lipid vesicles this potential problem can be avoided by mixing the lipids and fluorophore followed by evaporation of the solvent and codispersing in buffer. For fluorophores attached to phospholipids, this is the only way to get the fluorophore into the bilayer; with natural membranes, phospholipid exchange proteins or other techniques may have to be employed.

5.4. Fluorescence Energy Transfer

Fluorescence energy transfer is the transfer of electronic energy from a molecule in an excited state (donor) to another molecule (acceptor). The efficiency of this process is dependent on the distance between the donor and the acceptor. The fluorescence energy transfer process may or may not lead to emission of fluorescence by the acceptor. The transfer is due to a dipole–dipole interaction which can occur over distances of 0.2–0.5 nm providing there is sufficient overlap of the donor emission and acceptor absorption spectra. The original theory was for donors and acceptors in solution (see Vol. 1 in this series) but has also been developed for use in oriented systems such as membranes.

The fluorescence energy transfer process has been widely used to determine the distance between fluorophores, the surface density of fluorophores in the lipid bilayer, and the orientation of membrane protein or protein segments, often with reference to the membrane surface and protein–protein interactions. Membranes are intrinsically dynamic in nature, so that so far the major applications have been the determination of fixed distances between molecules of interest in the membrane.

In this section we will briefly outline the theory of fluorescence energy transfer as applied to the study of the more simple case of the surface distribution of acceptor and donor in the same plane. A number of theories for interpretation of fluorescence energy transfer data have been developed for more complex situations which cannot be elaborated here due to space limitations; however, these are referred to where appropriate.

5.4.1. Surface Distribution of Fluorophore-Labeled Lipids

An important contribution to the use of fluorescence energy transfer was the work of Fung and Stryer[81] (also see the review in Ref. 82), who considered the dependence of the transfer efficiency on the surface density of unassociated of the transfer efficiency on the surface density of unassociated donors and acceptors. Useful general guidelines for choices of donors and acceptors were listed. These were that the donor and acceptor be good analogous of membrane lipids, there should be a random distribution in the plane of the membrane, the donor and acceptor should be located at the same depth into the membrane, preferentially near the head-group region (since the surface distribution was being determined and contributions from the opposite side of the bilayer were to be excluded), and, lastly, the donor and acceptor should have a wide range of R_0 values (the distance for a 50% efficiency). These considerations obviously preclude some types of investigation, beyond studies of surface distributions of fluorophores. Thus, for example, since this work appeared, a number of theories and applications for situations in which the donor and acceptor may not be in the same plane have been described. The method of Fung and Stryer[81] is a numerical solution of the distance for 50% energy transfer. The basic relationship relating the rate of energy transfer, k_T, of a donor and acceptor separated by a distance r is

$$k_T = (1/\tau_0)(R_0/r)^6 \tag{5.15}$$

where τ_0 is the fluorescence lifetime of the donor in the absence of acceptor. R_0 is given by

$$R_0 = (Jk^2 Q_0 n^{-4})^{1/6} \times 9.79 \times 10^3 \tag{5.16}$$

where J is the overlap integral (in $cm^3\ M^{-1}$), k^2 is the dipole–dipole orientation factor, Q_0 is the quantum yield of the donor in the absence of acceptor, and n is the refractive index. Fung and Stryer considered a special case in which there is no transfer of energy between energy donors, R_0 is the same for all donor and acceptor pairs, the number of acceptors in the excited state is small compared with the number in the ground state, and the acceptor–donor distance does not change during the excited-state lifetime of the donor. One limitation is the last assumption, since it is obvious that many membrane processes of interest would entail motion of the donor and/or acceptor during the lifetime of the donor (see below). The calculated energy transfer efficiency was then plotted against the acceptor surface density for different values of R_0. For this, the following relationships were employed:

$$F(t) = F_0 \exp(-t/\tau) \exp[-\sigma S(t)]$$

and

$$S(t) = \int_{\alpha}^{\infty} \{1 - \exp[-(t/\tau)(R_0/r)^6]\}\, 2\pi r\, dr \tag{5.17}$$

where $\exp[-\sigma S(t)]$ is the energy transfer term, F_0 is the initial fluorescence intensity, σ is the surface density of energy acceptors, and α is the distance of closest approach of donor and acceptor. The efficiency of the energy transfer process is given by

$$E = 1 - (1/\tau_0) \int_0^{\infty} F(t)/F_0\, dt \tag{5.18}$$

Thus, the efficiency of energy transfer between donors and acceptors randomly distributed in a plane depends on R_0, σ, and α, and the transfer efficiency is independent of σ. The important point was made that surface density of the acceptor could be 1 per 500 phospholipids for $R_0 > 30$ Å. Using these equations for different donor and acceptor concentrations, the data were matched against the different theoretical curves to obtain the R_0. An example of the application of the method of Fung and Stryer(81) is the study of energy transfer between the tryptophan of a membrane protein (or peptide models of proteins) and DPH,(83) in which it was shown that efficient energy transfer can occur without any special interaction being required between DPH and the proteins in specific areas of the membrane.

While Fung and Stryer present a numerical solution for the determination of R_0 and the area per lipid molecule in a bilayer, an analytical solution has also been formalized.[84] In this method, Wolber and Hudson extended the treatment to consider the case where acceptors are excluded from a region surrounding each donor or are bound to the donors. More recently, Davenport *et al.*[85] used energy transfer to determine the location of DPH in the bilayer. For this a theory for energy transfer from donors situated outside a random planar distribution of acceptors was developed. The theory also included orientation effects previously considered in detail by Dale *et al.*[86]

Fluorescence energy transfer has been used to examine the distribution of bacteriorhodopsin in lipid vesicles using energy transfer from DPH to the acceptor retinal.[87] It was pointed out that care must be taken in fluorescence anisotropy studies if the fluorescence lifetime of a probe is decreased by energy transfer (in this case to retinal) since a shorter lifetime will lead to an erroneously high anisotropy value.

5.4.2. Location of the Longitudinal and Lateral Position of Membrane Proteins

A number of studies have taken advantage of the fact that membrane proteins contain one or more tryptophans, the fluorescence of which can be used to determine the conformation of the protein or its position in the membrane.[88-91] Of course, the information is limited by the number of tryptophans and the fact that a tryptophan may not be positioned in the region of the protein of interest. While a single tryptophan often simplifies the situation, most often there are a number in the protein so that it is difficult to extract useful information.

With cytochrome b_5, an intrinsic microsomal membrane protein, a number of approaches have been made. Using the energy transfer from tryptophan to trinitrophenyl- or dansyl-labeled lipids,[88] and a theory[89] to evaluate energy transfer between nonassociated, membrane-bound chromophores, the precise location relative to the bilayer surface was determined. The approach taken used a direct calculation of E and took into account possible variation of the dipole–dipole orientation factor (k) from the dynamic average of $\frac{2}{3}$. Friere *et al.*[90] used energy transfer between the intrinsic tryptophan fluorescence and pyrenedecanoic acid to show that cytochrome b_5 has a random distribution in the bilayer. More recently, Kleinfeld and Lukacovic[91] have shown that the positions of more than one tryptophan can be determined or "mapped" if several acceptors are used, in this case, the anthroylstearate series of probes. This was possible assuming that the distribution of the protein and acceptors is uniform (oligomeric association of protein being permitted), the probes remain in the outside half of the bilayer, and the depth of the acceptors is known (which is the case for anthroylstearates). The method was applied to the cytochrome b_5 problem, and using tryptophan-to-heme and anthroylstearate-to-heme energy transfer measurements, it was shown that the heme moiety is about 1.5 nm from the membrane surface.[92]

Kometani *et al.*[93] used a theory for energy transfer from a donor to acceptors in a plane to determine the location of the retinal chromophore relative to the membrane surface. Another similar study on the location of the active site of chloroplast ATPase relative to the membrane surface has also been carried out.[94]

The interaction of an extrinsic membrane protein with a lipid bilayer can also be investigated by energy transfer. The interaction of cytochrome *c* has attracted much attention, and in an early study by Shaklai *et al.*[95] the number of binding sites per red cell was determined. It was shown that an equation analogous to the Stern–Volmer relationship could be derived:

$$1/\tau = 1/\tau_0 + K_q \sigma \tag{5.19}$$

where τ_0 and τ are the fluorescence lifetimes (unquenched and quenched, respectively), σ is the density of quencher molecules, and K_q is the quenching constant. In a more recent, similar study,[96] concerning anthracyclin binding to synthetic and natural membranes (mitochondria), the energy transfer between DPH and adriamycin and between tryptophan and adriamycin was determined. It was demonstrated that the drugs interact with both the phospholipids and proteins.

5.4.3. Protein–Protein Associations

The association of membrane proteins in polymeric forms lends itself to the energy transfer approach. Vanderkooi *et al.*[97] used this approach to show that purified sarcoplasmic reticulum ATPase covalently labeled with *N*-iodoacetyl-N^1-(sulfo-1-naphthyl)ethylenediamine (1,5-IAEDANS; donor) and iodoacetamidofluorescein (IAF; acceptor) self-associated in the lipid phase, demonstrating an oligomeric structure. This type of approach has since been repeated in a more quantitative manner with a number of membrane proteins. Bacteriorhodopsin aggregation was investigated[98] by a theory developed by Dewey and Hammes[99] for energy transfer from multiple donors on a two-dimensional surface to multiple acceptors in a circular patch. It was shown that aggregation decreased with temperature and was dependent on the type of lipids present. The same theory can be used to investigate phase separations, antibody–receptor clustering, and membrane fusion.

5.5. Fluorescence Quenching

There are two mechanisms of quenching, static and dynamic. Static quenching is the nonradiative return of an excited state to the ground state, which occurs with the fluorophore and quencher remaining at a fixed distance during the lifetime of the excited state. With dynamic quenching, the fluorophore–quencher distance changes rapidly, and the quenching occurs when the quencher closely approaches the fluorophore. Static quenching is characterized by a lack of effect on the fluorescence lifetime while dynamic quenching results in a decrease in the fluorescence lifetime. Quenching can occur by Förster dipole–dipole energy transfer, heavy-atom quenching, or quenching by paramagnetic molecules. Examples of commonly used lipophilic quenchers include spin-labeled compounds, especially *n*-DOXYL-stearates ($n = 3$, 6, 9, and 12). Halogenated compounds, especially brominated lipids, are often used as quenchers, and a number of compounds of pharmacological interest such as halothane, chlorpromazine, and tetracaine fall into this class. Cobalt and copper salts, and iodide can be used as quenchers of fluorescence in the study of bilayer penetrative ability.

Quenching of fluorescence can be used to determine the partitioning of a quencher into a membrane or a region of the membrane. The region could be specified with reference to the depth in the membrane or in terms of laterally separated areas. Quenching can also be used to determine the degree of binding of a quencher, or competing nonquencher, to a membrane protein.

5.5.1. Determination of Partitioning and Binding of Fluorophore Quenchers to Membranes

The illustration in Figure 5.3 shows three types of curves which are commonly obtained when fluorescence quenching data are plotted in the manner shown. Curve a would be typical of static quenching, while curve b would be

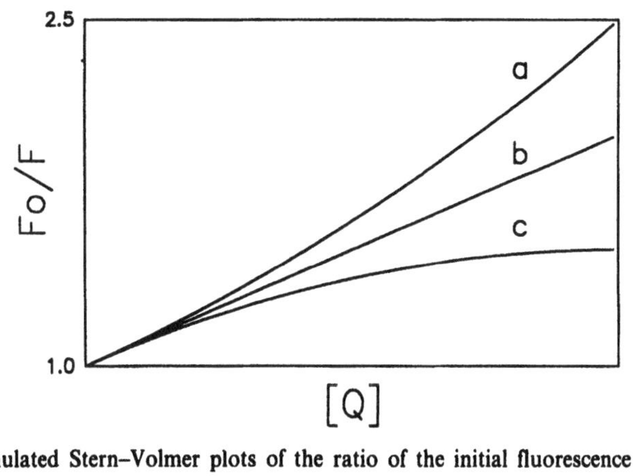

Figure 5.3. Simulated Stern–Volmer plots of the ratio of the initial fluorescence intensity F_0 to the intensity F in the presence of quencher of concentration [Q] showing (a) static quenching, (b) dynamic quenching (linear), and (c) binding and/or inaccessible quenchers.

found for dynamic or collisional quenching. Curve c would be obtained if a proportion of the fluorophores were inaccessible to the quencher[100] or if there is "binding" of the quencher to the region of the fluorophore.[101] The more a membrane-bound fluorophore is quenched, the greater the amount of quencher that is partitioning into the membrane. A method for the determination of the molar partition coefficient of membrane-soluble quenchers was described by Lakowicz *et al.*[102] using the following relationship:

$$\frac{1}{\tau}=\frac{1}{\tau_0}+\frac{k_m K_p[\mathrm{Q}]_\mathrm{T}}{K_p\alpha_m+(1-\alpha_m)}=\frac{1}{\tau_0}+k_{\mathrm{app}}[\mathrm{Q}]_\mathrm{T} \tag{5.20}$$

where k_m is the bimolecular quenching constant for the fluorophore bound to the membrane, K_p is the molar partition coefficient, $[\mathrm{Q}]_T$ is the total quencher concentration, α_m is the volume fraction of the membrane phase, τ_0 is the fluorescence lifetime in the absence of quencher and τ that in its presence, and k_{app} is an apparent quenching constant, which is given by

$$\frac{1}{k_{\mathrm{app}}}=\alpha_m\left[\frac{1}{k_m}-\frac{1}{k_m K_p}\right] \tag{5.21}$$

Plots of τ_0/τ for varying $[\mathrm{Q}]_\mathrm{T}$ are first otained for different concentrations of lipid. Then, form the slope and intercept of a plot of k_{app}^{-1} against α_m, the value of the partition coefficient is obtained. This method has been applied to the partitioning of lindane into lipid bilayers.[102, 103] An example is shown in Figure 5.4 for the quenching of DPH by 5-DOXYL-decane in dimyristoylphosphatidylcholine vesicles[104]; the increase in the partition coefficient as the

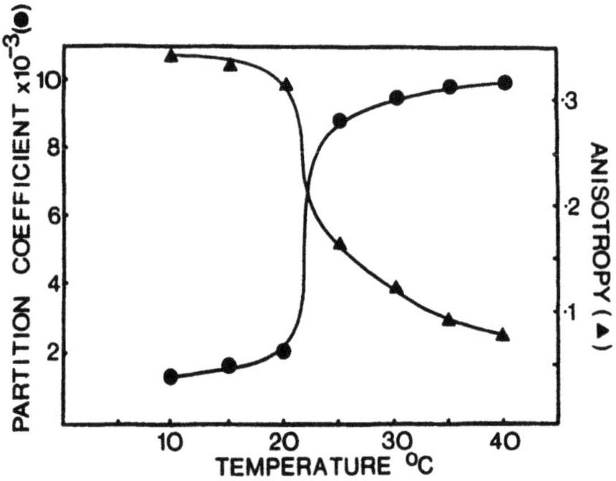

Figure 5.4. Effect of the phase transition on the partition coefficient of 5-DOXYL-decane calculated on the basis of its ability to quench the fluorescence of DPH in dimyristoylphosphatidylcholine vesicles. The fluorescence anisotropy of the DPH is also shown. (From Ref. 104, with permission.)

lipid goes from the gel to the liquid-crystalline phase is clearly shown. In this experiment the fluorescence anisotropy was measured before the addition of quencher so that from one experiment both details of the lipid order and the partitioning properties of the membrane were obtainable. With natural membranes the intrinsic tryptophans can also be used as the fluorophore, as was done in the study of the effect of hexachlorocyclohexanes on Ca^{2+}-ATPase from sarcoplasmic reticulum.[105] Alternatively, it has been shown that carbazoylundecanoic acid will biosynthetically incorporate into membrane phospholipids, following which the partition coefficient of a suitable quencher can then be determined for various membrane fractions.[106]

In systems where only dynamic quenching occurs, then steady-state fluorescence intensities can be measured instead of lifetimes.[101,103,107] In experiments where comparisons are being made (i.e., for a comparison of different experimental conditions or types of membrane), it is important that the lifetime of the fluorophore (τ_0) is not affected by the experimental conditions. Fluorescence intensities can be obtained much more rapidly and without specialized instrumentation. Blatt and Sawyer[101] have employed a relationship essentially the same as Eq. (5.20) in this way. They have pointed out that since the quenching mechanism is collisional, the partition coefficient that is derived is a partition coefficient of the quencher into the immediate environment of the fluorophore and is therefore a "local K_p." It is therefore possible to investigate the partition coefficient gradient across the lipid bilayer by using a series of probes, such as the anthroylstearates,[108] located at different depths. In their method, Eq. (5.20) has the form

$$\frac{F}{F_0 - F}[\mathrm{Q}]_\mathrm{T} = \frac{\alpha_m}{K_q \tau_0} + \frac{1}{K_q K_p \tau_0} \tag{5.22}$$

where F and F_0 are the fluorescence intensities corresponding to the τ and τ_0 above. The partition coefficient may be obtained without recourse to lifetime measurements even if static quenching is found, usually, but not necessarily, indicated by an upward curvature in the Stern–Volmer plot (Figure 5.3). Thus, the slopes and intercepts of plots of $[\mathrm{Q}]_\mathrm{T} F/(F_0 - F)$ versus α_m are each plotted versus $[\mathrm{Q}]_\mathrm{T}$, and from the slope and intercept values obtained from an extrapolated to zero quencher concentration (where the static component does not contribute since the concentration of the equilibrium complex between the quencher and the ground state of the fluorophore becomes infinitely small) K_p and K_q are then obtained.[101] The possibility that quenching can occur due to the combined contribution of both static and dynamic quenching for both partitioning and binding has been considered,[109] and it has been shown that the partition coefficient and binding association constants along with the number of binding sites can be extracted from quenching data. Such a binding may be thought to occur not only to a "site," perhaps on the surface of a protein, but also to other lipid-bound fluorophores although the binding would be more an "association."

In another method, the partition coefficient for the quencher into the entire membrane is determined rather than a local partition coefficient.[101] With membranes, usually only a single population of fluorophores and quenchers has been considered. However, for membrane proteins with multiple tryptophans or for coexisting lipid phases where fluorophores show partitioning behavior,[110] then the Stern–Volmer plot will reflect the different "sites" for interaction with the quencher. It should then be possible to separate the different interactions which occur. A potential problem with this approach for membrane proteins is that there may be binding sites on the protein for the quencher which do not have a tryptophan present,[110] and in this type of situation long-range energy transfer quenching may have to be employed. It should also be possible to gain some information about the interaction of a fluorophore with the surface of a membrane protein using the quenching approach. Thus, if the fluorophore and quencher encounter at the surface was longer than in the bulk lipid phase, one might expect to see static quenching occurring.[111]

The relative binding constants for different phospholipids to a membrane protein can also be determined using a fluorescence quenching technique, as demonstrated for purified Ca^{2+}-ATPase from sarcoplasmic reticulum[112–115] and, more recently, with bacteriorhodopsin,[116] reconstituted with different phospholipids, and phosphatidylcholine with either a spin-labeled[112–114] or brominated fatty acid[115,116] as the quencher of the tryptophan fluorescence. Basically, the method depends on competition between a quencher and the nonquenching lipid of interest. The quenching of the tryptophan fluorescence is first determined. The binding constants of the lipids of interest are then determined relative to that of a chosen "reference" phospholipid. In this way, effects of variations in chain length and fatty unsaturation on the binding were determined.

In addition to the partition coefficient, the bimolecular quenching constant (k_m) is obtained from quenching experiments,[100,117,118] and, in principle, this can be used to obtain the lateral diffusion constant of the quencher by using the Smoluchowski equation:

$$k_m = 4\pi N\gamma(D_Q + D_F)(R_Q + R_F)/1000 \qquad (5.23)$$

where γ is the quenching efficiency of the fraction of effective collisional encounters, $D_{Q,F}$ ($cm^2\,s^{-1}$) and $R_{Q,F}$ are the diffusion coefficients and the molecular radii of the quencher and fluorophore, and N is Avogadro's number.[100,117] The problem is to determine γ and the values for $D_{F,Q}$ and $R_{Q,F}$. This makes this type of calculation difficult in practice. For an example, the reader is referred to the work of Fato *et al.*[117] where lateral diffusion constants of different ubiquinones in phosphatidylcholine vesicles are calculated. The oxygen quenching of pyrene fluorescence has also been used to determine the lateral diffusion as shown by Chong and Thompson,[119] who also pointed out that the diffusion constant can be regarded as a microscopic diffusion constant, in contrast to that provided by the fluorescence recovery after photobleaching measurement, which is a macroscopic diffusion coefficient.

5.5.2. Location of Fluorophores

Quenching can also be used to determine the location of a fluorophore in the membrane (e.g., distance from the surface), and this has most commonly been carried out using spin-labeled fatty acids. This method has been applied to the location of the tryptophans in a number of studies (see, e.g., Refs. 120 and 121), and this application of spin-labeled fatty acids has been reviewed by London.[110] London has pointed out potential problems which must be considered. These include the possibility that not all of the spin-labeled fatty acids are associated with the membrane and the possibility of the influence of electrostatic repulsion or attraction on the interaction of charged spin labels and fluorophores. Also, the possible influence of orbital orientation on nitroxide quenching must be considered, as well as the fact that the motion of the spin labels will vary with the depth in the bilayer and hence affect the amount of quenching. Notwithstanding these points, the nitroxide-labeled fatty acids have been used in the study of the location of tryptophans in gramacidin[120] and the location of the NBD fluorophore as attached to different lipids.[78] Brominated lipids have also been used for this purpose in the study of the position of the tryptophans of cytochrome b_5.[121] Instead of using quenchers which are bound to lipids, it is possible to study the location of membrane-bound fluorophores using potassium iodide or cobalt[49,101] as the quencher, the degree of quenching being an indication of the distance of the fluorophore from the membrane surface. Finally, quenching can be used to increase the time resolution of fluorescence anisotropy decay meaurements (see Section 3).

5.6. Solvent Relaxation

The region at the surface of membranes and the underlying phospholipid head-group region are of particular interest since these regions can vary considerably with variations in the membrane phospholipid composition and under the influence of external molecules such as ions and hydrophobic molecules. The fluorescence anisotropy parameter tends to be less useful for examining this region since it is already intrinsically disordered and the anisotropy of head-group-labeled phospholipids tends to be rather insensitive to environmental changes. Methods available to probe this region of membranes include the use of fluorescence quenching, energy transfer, and techniques for measuring surface charge and dielectric properties. The dielectric properties are accessible through solvent relaxation measurements.

The wavelength of fluorescence is longer than the absorption wavelength due to several processes which cause a loss in energy. The absorption of light results in an increase in, or the formation of, a dipole moment. The surrounding solvent molecules then reorient or relax to accommodate the newly formed dipole. The time scale of the solvent relaxation process reflects directly the properties of the solvent. This is one aspect of a complex process which includes[100] general interactions concerning the electronic polarizability of the solvent, as described by the refractive index, and molecular polarizability of the solvent resulting from the solvent dipole orientation, which is related to the dielectric constant. Other specific factors include hydrogen bonding, proton loss, and charge transfer complex formation. The theoretical treatment of these processes has been described.[123] The excitation of a fluorophore at or near the membrane surface is likely to result in solvent relaxation effects, and this can be used to assess the dynamics of this region of the membrane. Although the relaxation process can be approximated by a single relaxation time τ_R, it is more properly described by a continuum of relaxation times. It is possible to identify solvent relaxation processes by obtaining time-resolved fluorescence spectra (TRES) at a time early after excitation of a fluorophore (e.g., 0–3 ns) and later (e.g., 30–50 ns), when a fully relaxed emission spectrum, which is shifted to a longer wavelength, will be collected. Methods for obaining early and late gated spectra by pulse fluorometric (see, e.g., Refs. 124–128) and phase (e.g., Refs. 129–131) methods have been described. If the relaxation process occurs on a comparable time scale to the lifetime of the excited state, then the fluorescence intensity decay observed after pulsed excitation will appear to increase at first followed by the decay. If the emission at the red edge is collected, then this will be more apparent, and analysis of the collected decay as a multiexponential process will reveal a component with a negative preamplitude:

$$F(t) = \sum_{i=1}^{m} \alpha_i \exp(-t/\tau_{Fi}) - \sum_{j=1}^{n} \beta_j \exp(-t/\tau_{Rj}) \tag{5.24}$$

The τ_{Fi} terms are the fluorescence lifetimes of fractional contributions α_i, and the τ_{Rj} indicate decay constants due to solvent relaxation (or other excited-state processes) of fractional contribution β_j. The negative sign is indicative of a relaxation process (red shift). Usually, the relaxation process is approximated to a single relaxation time τ_R by assuming an initial excited state and a final fully relaxed state (see, e.g., Ref. 128). A steady-state fluorescent measurement can also be used to obtain an indication of solvent relaxation (see, e.g., Refs. 132 and 133). By excitation at the red edge of the absorption maxima, the solvent-relaxed species are isolated since this results in the excitation of the fluorophores which are interacting most strongly with the solvent, and as a result a red-shifted emission spectrum is obtained. Red-shifted spectra are most easily observed when the solvent relaxation process is long compared to the fluorescence lifetime.

Fluorophores which have been used to study solvent relaxation processes in membranes include 2-(*p*-toluidinyl)naphthalene-6-sulfonate (TNS),[130, 132, 133–135] dansylated lipids, and drugs.[128, 136]

5.7. Surface Charge

Membranes possess charge on their surface arising from ionization of component lipids and proteins as well as adsorbed ions. Surface potential is a consequence of this surface charge, representing the electrical potential between the membrane–solution interface and the external (bulk) solution. As biological membranes generally have a net negative surface charge, surface potential affects membrane functions. Membrane enzyme activity and transport[137] and redox reactions[138] are but two recently reviewed examples. A variety of spectroscopic and other methods have been developed to accomplish the measurement of these surface properties.[139] In the following, we focus on reported uses of fluorescent probes for this purpose. Complete understanding of these measurements requires familiarity with the Gouy–Chapman–Stern theory, which mathematically relates surface charge to surface potential and is the subject of several papers.[137, 140–142] Also, no coverage will be given to the related topic of membrane potential (see Ref. 142).

The effect of surface potential on interfacial ionic concentration is given by a Boltzmann distribution relating the total solution concentration to that at the interface. For charged, amphiphilic species, binding constants replace these ionic concentrations, and the expression

$$\Delta\psi = \psi_1 - \psi_2 = -(RT/zF)\ln(K_2/K_1) \tag{5.25}$$

is obtained, relating the change in surface potential ψ to changes in these binding constants. Here R is the gas constant, T is absolute temperature, z is the valence of the probe, F is he Faraday constant, and K_1 and K_2 are binding constants under two different conditions. It may be useful to classify probes as "extrinsic" or "intrinsic," the former pertaining to amphiphile binding constants, and the latter to ionic concentration. Extrinsic probes are charged and change their fluorescent intensity according to their partitioning between the external solution and the interface. With intrinsic probes, the chromophore is positioned such that it is responsive to the ionic concentration at the interface through its attachment to long acyl chains or phospholipids. These latter probes are usually pH indicators, as surface potential affects interfacial proton concentration. The relevant equation replaces the logarithmic term of Eq. (5.25) with differences in pH and a multiplicative factor. For either class of probe, note that Eq. (5.25) gives the absolute surface potential only if the binding constant or response of the probe in the absence of surface potential can be found. Some general approaches used to achieve this include increasing the concentration of external electrolyte in order to mask the effects of surface potential, measuring an uncharged chemical analogue of the original probe, or comparing the response of the probe on a neutral or reference surface with that on a charged one.

Important considerations for surface potential measurement have been enumerated by Eisenberg *et al.*[143] First, a probe must locate in the interfacial region of interest. An extrinsic probe which permeates the membrane may respond to membrane as well as surface potential, while an intrinsic probe undergoing trans-bilayer motion will not report on external surfaces alone. Second, the probe response should be high, allowing its use at low concentrations. High concentrations of charged probes will alter existing surface charge by their presence. Finally, any effects the external solution may have on probe properties must be accounted for. Hydroxycoumarin pH indicators, for example, have been shown to be affected by changes in dielectric constant in addition to changes in pH.[144]

Common extrinsic probes which have been used to determine surface potential include 8-anilino-1-naphthalenesulfonate (ANS), TNS, merocyanine, and Rhodamine 6G dyes. The anions ANS and TNS show large increases in fluorescence upon binding to membranes, while the merocyanine anions and the Rhodamine 6G cation show increased fluorescence in the aqueous phase. ANS has long been applied in studies of membranes, and surface-related properties have been recently summarized by Ehrenberg.[139] Two recent studies suggest problems with the use of ANS. Gibrat *et al.*[145] have questioned the common assumption that high salt concentrations allow accurate determination of the binding constant of ANS. Their results with liposomes suggest that residual surface potential is maintained even under external KCl concentrations as high as 1.5 *M*. These authors have given alternative procedures for binding constant determination on natural and neutral surfaces. In rat liver mitochondria, Robertson and Rottenberg[146] noted that apparently two types of binding sites with different constants exist for ANS, and also that ANS appears to respond to changes in membrane as well as surface potential. In contrast to the above, Tanabe *et al.* used ANS and a neutral analogue of ANS in measurements on the surface potential of a living protozoan[147] and observed good agreement with complementary electrophoretic zeta potential measurements.

TNS has been used by McLaughlin and co-workers in several studies.[143,148] One recent application was an experimental test of theories of discrete charge distribution which predicted significant differences from Gouy–Chapman theory: no such discrepancies were found.[148] These authors took the response of TNS on phosphatidylcholine vesicles as their base reference for surface potential and used the total fluorescence in Eq. (5.25) in place of binding constants. Since quantum yields and emission spectra show little change in the various lipids examined, the authors concluded that this replacement is reasonable. An additional concern of these authors has been probe permeabilization. TNS does not appear to penetrate neutral or negatively charged bilayers, although it may permeate positively charged ones. These data suggest that TNS may be superior to ANS, at least for measurements in model vesicle systems. Merocyanine dyes and Rhodamine 6G do not appear to be as well characterized as ANS and TNS. Masamoto *et al.*[149] in their work on photosynthetic membranes showed that although partitioning of the merocyanine dyes depends on surface potential, agreement with Gouy–Chapman predictions is only qualitative, and careful choice of experimental conditions is necessary. Fluorescent changes of Rhodamine 6G observed in chemotactic responses in a ciliated protozoan appear to be related to membrane potential as well as surface potential.[150]

Following Barber and co-workers,[151–153] other authors have taken a different approach in applying the extrinsic probe 9-aminoacridine to surface property measurement. This cationic dye shows fluorescence quenching upon association with negatively charged surfaces. Addition of cations to the bulk solution decreases surface potential, and increased fluorescence results from dissociation of the probe from the interface. Care is taken to ensure that cation adsorption is minimized. Application of Gouy–Chapman theory in the case of different cation valences allowed the estimation of surface charge densities of thylakoid membranes,[152,153] while the surface charge density and potential of plasmalemma vesicles from plant roots[154] and yeast cells[155] have also been examined. From studies with 9-aminoacridine, Cerbon *et al.* suggest that phosphatidylinositols participate in the maintenance and regulation of surface potential in yeast cells.[156]

The pH indicator 4-heptadecylhydroxycoumarin has seen use as an intrinsic probe in several studies of vesicle systems.[144,157–160] The results of a critical study of its properties by Drummond and Grieser[144] suggest, however, that some care must be taken. Given the influence of dielectric constant, their interpretation of results on phosphatidylcholine vesicles using a micelle reference is that either this chromophore senses a negative electrostatic potential or it is located on average in a region of low dielectric constant. If the local environment of the reference micelle resembles that in the vesicles, these authors believe that the local electrostatic potential in the glycerol backbone region is some −120 mV. Among later studies, Pal *et al.*[160] measured the surface potential of vesicular stomatitis virus membranes and showed that viral glycoproteins contribute to negative surface charge. Cholesterol levels also appeared to affect surface charge. Cholesterol levels also appeared to affect surface potential. Dansyl chromophores have also served as intrinsic probes,[161] and measurement of the quenching of NBD phospholipid derivatives by cobalt has been suggested as a possible method.[162,163] Given the great variety of effects surface potential can induce in membrane systems and the relative simplicity of instrumentation required, extrinsic and intrinsic fluorescence probes should enjoy greater future use in surface charge and potential measurement.

5.8. Future Directions

Most studies which have utilized fluorescence techniques have looked at the membrane as essentially a static structure apart from the motional aspects of anisotropy. Also, physical parameters have usually been extracted on the basis of the assumption that the fluorophore can adopt only one or two states within the lifetime of the excited state. Membranes are dynamic structures, however, and motional changes in lipid organization and in protein–protein interaction and protein position and protein conformational changes are potentially accessible using fluorescent techniques. One possibility with respect to energy transfer, investigated by Haas and Steinberg,[164] is the study of intramolecular dynamics as a function of the distance from donor and acceptor ends of the molecule. This could potentially be applied to separate donor and acceptor molecules in a membrane. It is also possible to conceive of a situation in which, instead of either a fixed donor–acceptor distance or an average distance, there may be a distribution of distances. The distributional approach has already been initiated in fluorescence lifetime measurements, and fluorescence anisotropy and quenching should soon see a similar approach. There is a great deal of information waiting to be discovered about membrane structure and dynamics, and one can expect that much of this will be learned using fluorescent techniques.

Acknowledgments

This work was supported in part by U.S. Public Health grants NIAAA 08022, 07215, 07463, and 07186.

References

1. D. A. Barrow and B. R. Lentz, Membrane structural domains, *Biophys. J. 48*, 221–234 (1985).
2. L. A. Sklar, The partition of *cis*-parinaric acid and *trans*-parinaric acid among aqueous, fluid lipid, and solid lipid phases, *Biochemistry 32*, 169–177 (1980).
3. B. Hudson and S. A. Cavalier, in: *Studies of Membrane Dynamics and Lipid–Protein Interactions with Parinaric Acid in Spectroscopic Membrane Probes Vol. 1* (L. Loew, ed.), CRC Press, Inc., Boca Raton, Florida (1988).
4. T. Parasassi, F. Conti, and E. Gratton, Study of heterogeneous emission of parinaric acid isomers using multifrequency phase fluorometry, *Biochemistry 23*, 5660–5664 (1984).
5. V. B. Yashar, M. Menashe, R. L. Biltonen, M. L. Johnson, and Y. Barenholz, Interaction of *trans*-parinaric acid with phosphatidylcholine bilayers: Comparison with the effect of other fluorophores, *Biochim. Biophys. Acta 904*, 117–124 (1987).
6. T. Parassassi, F. Conti, M. Glaser, and E. Gratton, Detection of phospholipid phase separation. A multifrequency phase fluorimetry study of 1,6-diphenyl-1,3,5-hexatriene fluorescence, *J. Biol. Chem. 259*, 14011–14017 (1984).
7. C. D. Stubbs, K. Kinosita, Jr., F. Munkonge, P. J. Quinn, and A. Ikegami, The dynamics of lipid motion in sarcoplasmic reticulum membranes determined by steady-state and time-resolved fluorescence measurements on 1,6-diphenyl-1,3,5-hexatriene and related molecules, *Biochim. Biophys. Acta 775*, 374–380 (1984).
8. M. Vincent and J. Gallay, Time-resolved fluorescence anisotropy study of effect of a cis double bond on structure of lecithin and cholesterol-lecithin bilayers using *n*-(9-anthroyloxy) fatty acids as probes, *Biochemistry 23*, 6514–6522 (1984).
9. M. Straume and B. J. Litman, Equilibrium and dynamic structure of large, unilamellar, unsaturated acyl chain phosphatidylcholine vesicles. Higher order analysis of 1,6-diphenyl-1,3,5-hexatriene and 1-[4-(trimethylammonio)phenyl]-6-phenyl-1,3,5-hexatriene anisotropy decay, *Biochemistry 26*, 5113–5120 (1987).
10. D. R. James and W. R. Ware, A fallacy in the interpretation of fluorescence decay parameters, *Chem. Phys. Lett. 120*, 455 (1985).
11. D. R. James and W. R. Ware, Recovery of underlying distributions of lifetimes from fluorescence decay data, *Chem. Phys. Lett. 126*, 7–11 (1986).
12. J. R. Alcala, E. Gratton, and G. G. Prendergast, Resolvability of fluorescence lifetime distributions using phase fluorometry, *Biophys. J. 51*, 587–596 (1987).
13. J. R. Alcala, E. Gratton, and F. G. Prendergast, Fluorescence lifetime distributions in proteins, *Biophys. J. 51*, 597–604 (1987).
14. J. R. Alcala, E. Gratton, and F. G. Prendergast, Interpretation of fluorescence decays in proteins using continuous lifetime distributions, *Biophys. J. 51*, 925–936 (1987).
15. J. R. Lakowicz, R. F. Steiner, and I. Gryczynski, The distribution of donor-to-acceptor distances in troponin C from frequency-domain fluorometry, *Biophys. J. 49*, 106a (1986).
16. J. R. Lakowicz, M. L. Johnson, W. Wiczk, A. Bhat, and R. F. Steiner, Resolution of a distribution of distances by fluorescence energy transfer and frequency-domain fluorometry, *Chem. Phys. Lett. 138*, 587–593 (1987).
17. B. W. Williams and C. D. Stubbs, Properties influencing fluorophore lifetime distributions in membranes, *Biochemistry 27*, 7994–7999 (1988).
18. D. W. Marquardt, An algorithm for least squares estimation of non-linear parameters, *J. Soc. Ind. Appl. Math. 11*, 431–441 (1963).
19. S. B. Phillips and R. L. Lyke, Fluorescence kinetics of systems with a quasi-continuum of excited states, *Chem. Phys. Lett. 136*, 247–251 (1987).
20. A. K. Livesey and J. C. Brouchon, Analyzing the distribution of decay constants in pulse-fluorimetry using the maximum entropy method, *Biophys. J. 52*, 693–706 (1987).

21. J. G. McWhirter and E. R. Pike, On the numerical inversion of the Laplace transform and similar Fredholm integral equations of the first kind, *J. Phys. A: Math. Gen. 11*, 1729–1745 (1978).
22. R. Fiorini, M. Valentino, S. Wang, M. Glaser, and E. Gratton, Fluorescence lifetime distributions of 1,6-diphenyl-1,3,5-hexatriene in phospholipid vesicles, *Biochemistry 26*, 3864–3870 (1987).
23. D. R. James, J. R. Turnbull, B. D. Wagner, W. R. Ware, and N. O. Petersen, Distributions of fluorescence decay times for parinaric acids in phospholipid membranes, *Biochemistry 26*, 6272–6277 (1987).
24. T. Parassassi, F. Conti, E. Gratton, and O. Sapora, Membrane modification of differentiating proerythroblasts. Variation of 1,6-diphenyl-1,3,5-hexatriene lifetime distributions by multifrequency phase and modulation fluorimetry, *Biochim. Biophys. Acta 898*, 196–201 (1987).
25. R. M. Fiorini, M. Valentino, E. Gratton, E. Bertoli, and G. Curatola, Erythrocyte membrane heterogeneity studies using 1,6-diphenyl-1,3,5-hexatriene fluorescence lifetime distribution, *Biochem. Biophys. Res. Commun. 147*, 460–466 (1987).
26. J. M. Vanderkooi and J. B. Callis, Pyrene. A probe of lateral diffusion in the hydrophobic region of membranes, *Biochemistry 13*, 4000–4007 (1974).
27. E. Sackmann, On the application of excimers as optical probes in membrane research, *Z. Phys. Chem. 101*, 391–416 (1976).
28. H.-J. Galla and E. Sackmann, Lateral diffusion in the hydrophobic region of membranes: Use of pyrene excimers as optical probes, *Biochim. Biophys. Acta 339*, 103–115 (1974).
29. D. Axelrod, D. E. Koppel, J. Schlessmjer, E. Elson, and W. W. Webb, Mobility measurement by analysis of fluorescence photobleaching recovery kinetics, *Biophys. J. 16*, 1055–1069 (1976).
30. F. W. J. Teale, Fluorescence depolarization by light scattering in turbid solutions, *Photochem. Photobiol. 10*, 363–374 (1969).
31. B. R. Lentz, Light scattering effects in the measurement of membrane microviscosity with DPH, *Biophys. J. 25*, 489–494 (1979).
32. C. D. Stubbs, Membrane fluidity: Structure and Dynamics of membrane lipids, *Essays in Biochemistry 19*, 1–39 (1983).
33. W. van der Meer, Physical aspects of membrane fluidity, in: *Physiology of Membrane Fluidity* (M. Shinitzky, ed.), Vol. 1, pp. 54–71, CRC Press, Boca Raton, Florida (1984).
34. C. D. Stubbs and A. D. Smith, The modification of mammalian membrane polyunsaturated fatty acid composition in relation to membrane fluidity and function, *Biochim. Biophys. Acta 779*, 89–137 (1984).
35. R. E. Dale, L. A. Chen, and L. Brand, Rotational relaxation of the "microviscosity" probe diphenylhexatriene in paraffin oil and egg lecithin vesicles, *J. Biol. Chem. 252*, 7500–7510 (1977).
36. L. A. Chen, R. E. Dale, S. Roth, and L. Brand, Nanosecond time-dependent fluorescence depolarization of diphenylhexatriene in dimyristoyllecithin vesicles and the determination of "microviscosity," *J. Biol. Chem. 252*, 2163–2169 (1977).
37. S. Kawato, K. Kinosita, Jr., and A. Ikegami, Dynamic structure of lipid bilayers studied by nanosecond fluorescence techniques, *Biochemistry 16*, 2319–2324 (1977).
38. J. R. Lakowicz, H. Cherek, and B. P. Maliwal, Time-resolved fluorescence anisotropies of diphenylhexatriene and perylene in solvents and lipid bilayers obtained from multifrequency phase-modulation fluorometry, *Biochemistry 24*, 376–383 (1985).
39. K. Kinosita, Jr. and A. Ikegami, Reevaluation of the wobbling dynamics of diphenylhexatriene in phosphatidylcholine and cholesterol/phosphatidylcholine membranes, *Biochim. Biophys. Acta 769*, 523–527 (1984).
40. C. D. Stubbs, T. Kouyama, K. Kinosita, Jr., and A. Ikegami, Effect of double bonds on

the dynamic properties of the hydrocarbon region of lecithin bilayers, *Biochemistry 20*, 4257–4262 (1981).

41. M. Straume and B. J. Litman, Influence of cholesterol on equilibrium and dynamic bilayer structure of unsaturated acyl chain phosphatidylcholine vesicles as determined from higher order analysis of fluorescence anisotropy decay, *Biochemistry 26*, 5121–5126 (1987).
42. K. Kinosita, Jr., S. Kawato, A. Ikegami, S. Yoshida, and Y. Orii, The effect of cytochrome oxidase on lipid chain dynamics, *Biochim. Biophys. Acta 647*, 7–17 (1981).
43. K. Kinosita, Jr. and A. Ikegami, On the wobbling-in-cone analysis of fluorescence anisotropy decay, *Biophys. J. 37*, 461–464 (1982).
44. K. Kinosita, Jr., S. Kawato, and A. Ikegami, Dynamic structure of biological and model membranes: Analysis by optical anisotropy decay measurement, *Adv. Biophys. 17*, 147–203 (1984).
45. F. Jahnig, Structural order of lipids and proteins in membranes: Evaluation of fluorescence anisotropy data, *Proc. Natl. Acad. Sci. U.S.A. 76*, 6361–6365 (1979).
46. M. P. Heyn, Determination of lipid order parameters and rotational correlation times from fluorescence depolarization experiments, *FEBS Lett. 108*, 359–364 (1979).
47. L. W. Engel and F. G. Prendergast, Values for and significance of order parameters and "cone angles" of fluorophore rotation in lipid bilayers, *Biochemistry 20*, 7338–7345 (1981).
48. F. Hare, Simplified derivation of angular order and dynamics of rodlike fluorophores in models and membranes, *Biophys. J. 42*, 205–218 (1983).
49. H. Pottel, W. van der Meer, and W. Herreman, Correlation between the order parameter and the steady-state fluorescence anisotropy of 1,6-diphenyl-1,3,5-hexatriene and an evaluation of membrane fluidity, *Biochim. Biophys. Acta 730*, 181–186 (1983).
50. W. van der Meer, R. B. van Hoeven, and W. J. van Blitterswijk, Steady-state fluorescence polarization data in membranes. Resolution into physical parameters by an extended Perrin equation for restricted rotation of fluorophores, *Biochim. Biophys. Acta 854*, 38–44 (1986).
51. W. van der Meer, H. Pottel, W. Herreman, M. Amelott, H. Hendrickx, and H. Schroder, Effect of orientational order on the decay of the fluorescence anisotropy in membrane suspensions, *Biophys. J. 46*, 515–523 (1984).
52. M. Amelott, H. Hendrickx, W. Herreman, H. Pottel, F. Van Cauwelaert, and W. van der Meer, Effect of orientational order on the decay of the fluorescence anisotropy in membrane suspensions, *Biophys. J. 46*, 525–539 (1984).
53. J. M. Martin, van De Ven, and Y. K. Levine, Angle-resolved fluorescence depolarization of macroscopically ordered bilayers of unsaturated lipids, *Biochim. Biophys. Acta 777*, 283–296 (1984).
54. F. Mulders, H. van Langen, G. van Ginkel, and Y. K. Levine, The static and dynamic behaviour of fluorescent probe molecules in lipid bilayers, *Biochim. Biophys. Acta 859*, 209–218 (1986).
55. M. G. Badea and L. Brand, Time-resolved fluorescence measurements, *Methods Enzymol. 61*, 378–425 (1979).
56. D. V. O'Connor and D. Phillips, *Time Correlated Single Photon Counting*, Academic Press London (1984).
57. E. Gratton and M. Limkeman, A continuously variable frequency cross-correlation phase fluorometer with picosecond resolution, *Biophys. J. 44*, 315–324 (1983).
58. J. R. Lakowicz and B. P. Maliwal, Construction and performance of a variable-frequency phase-modulation fluorometer, *Biophys. Chem. 21*, 61–78 (1985).
59. J. R. Lakowicz, Fluorescence studies of structural fluctuations in macromolecules as observed by the time, lifetime and frequency domains, *Methods Enzymol. 131*, 518–567 (1986).
60. R. D. Ludescher, L. Peting, S. Hudson, and B. Hudson, Time-resolved fluorescence

anisotropy for systems with lifetime and dynamic heterogeneity, *Biophys. Chem. 28*, 59–75 (1987).

61. L. Davenport, J. R. Knutson, and L. Brand, Anisotropy decay associated fluorescence spectra and analysis of rotational heterogeneity. 2. 1,6-Diphenyl-1,3,5-hexatriene in lipid bilayers, *Biochemistry 25*, 1811–1816 (1986).
62. H. Szmacinski, R. Jayaweera, H. Cherek, and J. R. Lakowicz, Demonstration of an associated anisotropy decay by frequency-domain fluorometry, *Biophys. Chem. 27*, 233–241 (1987).
63. J. R. Lakowicz, H. Cherek, I. Gryczynski, N. Joshi, and M. L. Johnson, Enhanced resolution of fluorescence anisotropy decays by simultaneous analysis of progressively quenched samples, *Biophys. J. 51*, 755–768 (1987).
64. C. D. Stubbs, W. M. Tsang, J. Belin, A. D. Smith, and S. M. Johnson, Incubation of exogenous fatty acids with lymphocytes. Changes in fatty acid composition and effects on the rotational relaxation time of 1,6-diphenyl-1,3,5-hexatriene, *Biochemistry 19*, 2756–2762 (1980).
65. J.-G. Kuhry, G. Kuportail, C. Bronner, and G. Laustriat, Plasma membrane fluidity measurements on whole living cells by fluorescence anisotropy of trimethylammonium-diphenylhexatriene, *Biochim. Biophys. Acta 845*, 60–67 (1985).
66. C. Zannoni, A. Arcioni, and P. Cavatorta, Fluorescence depolarization in liquid crystals and membrane bilayers, *Chem. Phys. Lipids 32*, 179–250 (1983).
67. W. E. Harris and W. L. Stahl, Incorporation of *cis*-parinaric acid, a fluorescent fatty acid, into synaptosomal phospholipids by an acyl-CoA acyltransferase, *Biochim. Biophys. Acta 736*, 79–91 (1983).
68. M. Cranney, R. B. Cundall, G. R. Jones, J. T. Richards, and E. W. Thomas, Fluorescence lifetime and quenching studies on some interesting diphenylhexatriene membrane probes, *Biochim. Biophys. Acta 735*, 418–425 (1983).
69. F. G. Prendergast, R. P. Haugland, and P. J. Callahan, 1-[4-(Trimethylamino)phenyl]-6-phenylhexa-1,3,5-triene: synthesis, fluorescence properties, and use as a fluorescence probe of lipid bilayers, *Biochemistry 20*, 7333–7338 (1983).
70. M. Shinitsky and Y. Barenholtz, Fluidity parameters of lipid regions determined by fluorescence polarization, *Biochim. Biophys. Acta 515*, 367–394 (1976).
71. R. E. Dale, Membrane structure and dynamics by fluorescence probe depolarization kinetics, in: *Time-Resolved Fluorescence Spectroscopy in Biochemistry and Biology* (R. B. Cundall and R. E. Dale, eds.), pp. 555–612, Plenum, New York (1984).
72. R. A. Parente and B. R. Lentz, Advantages and limitations of 1-palmitoyl-1-[[2-[4-(6-phenyl-*trans*-1,3,5-hexatrienyl)phenyl]ethyl]carbonyl]-3-*sn*-phosphatidylcholine as a fluorescent membrane probe, *Biochemistry 24*, 6178–6185 (1985).
73. K. R. Thulborn, L. M. Tilley, W. H. Sawyer, and E. Treloar, The use of *n*-(9-anthroyloxy) fatty acids to determine fluidity and polarity gradients in phospholipid bilayers, *Biochim. Biophys. Acta 558*, 166–178 (1979).
74. D. Schachter, U. Cogan, and R. E. Abbot, Asymmetry of lipid dynamics in human erythrocyte membranes studied by permeant fluorophores, *Biochemistry 21*, 2146–2150 (1982).
75. M. Vincent, B. de Foresta, J. Gallay, and A. Alfsen, Nanosecond fluorescence anisotropy decays of *n*-(9-anthroyloxy) fatty acids in dipalmitoylphosphatidylcholine vesicles with regard to isotropic solvents, *Biochemistry 21*, 708–716 (1982).
76. M. Vincent, J. Gallay, J. de Bony, and J.-F. Tocanne, Steady-state and time-resolved fluorescence anisotropy study of phospholipid molecular motion in the gel phase using 1-palmitoyl-2-[9-(2-anthryl)-nonanoyl]-*sn*-glycero-3-phosphocholine as probe, *Eur. J. Biochem. 250*, 341–347 (1985).
77. R. E. Pagano and R. G. Sleight, Defining rapid transport in animal cells, *Science 229*, 1051–1057 (1985).

78. A. Chattopadhyay and E. London, Parallax method for direct measurement of membrane penetration depth utilizing fluorescence quenching by spin-labeled phospholipids, *Biochemistry 26*, 39–45 (1987).
79. A. Chattopadhyay and E. London, Spectroscopic and ionization properties of *N*-(7-nitro-2,1,3-benzoxadiazol-4-yl)-labeled lipids in model membranes, *Biochim. Biophys. Acta 938*, 24–34 (1988).
80. C. D. Stubbs, B. W. Williams, C. L. Pryor, and E. Rubin, Ethanol-induced modifications to membrane lipid structure—Effect on phospholipase A_2–membrane interactions, *Arch. Biochem. 262*, 560–573 (1988).
81. B. K.-K. Fung and L. Stryer, Surface density determination in membranes by fluorescence energy transfer, *Biochemistry 17*, 5241–5248 (1978).
82. L. Stryer, Fluorescence energy transfer as a spectroscopic ruler, *Annu. Rev. Biochem. 47*, 819–846 (1978).
83. T. Le Doan, M. Takasugi, I. Aragon, G. Boudet, T. Montenay-Garestier, and C. Helene, Excitation energy transfer from tryptophan residues of peptides and intrinsic proteins to diphenylhexatriene in phospholipid vesicles and biological membranes, *Biochim. Biophys. Acta 735*, 259–270 (1983).
84. P. K. Wolber and B. S. Hudson, An analytic solution to the Forster energy transfer problem in two dimensions, *Biophys. J. 28*, 197–210 (1979).
85. L. Davenport, R. E. Dale, R. H. Bisby, and R. B. Cundall, Transverse location of the fluorescent probe 1,6-diphenyl-1,3,5-hexatriene in model lipid bilayer membrane systems by resonance excitation energy transfer, *Biochemistry 24*, 4097–4108 (1985).
86. R. E. Dale, The orientational freedom of molecular probes, *Biophys. J. 26*, 161–194 (1979).
87. M. Rehorek, N. A. Dencher, and M. P. Heyn, Fluorescence energy transfer from diphenylhexatriene to bacteriorhodopsin in lipid vesicles, *Biophys. J. 43*, 39–45 (1983).
88. P. J. Fleming, D. E. Koppel, A. L. Y. Lau, and P. Strittmatter, Intramembrane position of the fluorescent tryptophanyl residue in membrane-bound cytochrome b_5, *Biochemistry 18*, 5458–5464 (1979).
89. D. E. Koppel, P. J. Fleming, and P. Strittmatter, Intramembrane positions of membrane-bound chromophores determined by excitation energy transfer, *Biochemistry 18*, 5450–5457 (1979).
90. E. Friere, T. Markello, C. Rigell, and P. W. Holloway, Calorimetric and fluorescence characterization of interactions between cytochrome b_5 and phosphatidylcholine bilayers, *Biochemistry 22*, 1675–1680 (1983).
91. A. M. Kleinfeld and M. F. Lukacovic, Energy-transfer study of cytochrome b_5 using the anthroyloxy fatty acid membrane probes, *Biochemistry 24*, 1883–1890 (1985).
92. A. M. Kleinfeld, Tryptophan imaging of membrane proteins, *Biochemistry 24*, 1874–1882 (1985).
93. T. Kometani, K. Kinosita, Jr., T. Furuno, T. Kouyama, and A. Ikegami, Transmembrane location of retinal in purple membrane, *Biophys. J. 52*, 509–517 (1987).
94. B. A. Baird, U. Pick, and G. G. Hammes, Structural investigation of reconstituted chloroplast ATPase with fluorescence measurements, *J. Biol. Chem. 254*, 3818–3825 (1979).
95. N. Shaklai, J. Yguerabide, and H. M. Ranney, Interaction of hemoglobin with red blood cell membranes as shown by a fluorescent chromophore, *Biochemistry 16*, 5585–5592 (1977).
96. E. A. Griffin, J. M. Vanderkooi, G. Maniara, and M. Erecinska, Anthracycline binding to synthetic and natural membranes. A study using resonance energy transfer, *Biochemistry 25*, 7875–7880 (1986).
97. J. M. Vanderkooi, A. Ierokomas, H. Nakamura, and A. Martonosi, Fluorescence energy transfer between Ca^{2+} transport ATPase molecule in artifical membranes, *Biochemistry 16*, 1262–1267 (1977).
98. C. A. Hasselbacher, T. L. Street, and T. G. Dewey, Resonance enery transfer as a monitor

of membrane protein domain segregation: Application to the aggregation of bacteriorhodopsin reconstituted into phospholipid vesicles, *Biochemistry 23*, 6445–6452 (1984).
99. T. G. Dewey and G. G. Hammes, Calculation of fluorescence resonance energy transfer on surfaces, *Biophys. J. 32*, 1023–1036 (1980).
100. J. R. Lakowicz, *Principles of Fluorescence Spectroscopy*, Plenum, New York (1983).
101. E. Blatt and W. H. Sawyer, Depth-dependent fluorescent quenching in micelles and membranes, *Biochim. Biophys. Acta 822*, 43–62 (1985).
102. J. R. Lakowicz, D. Hogen, and G. Omann, Diffusion and partitioning of a pesticide, lindane, into phosphatidylcholine bilayers, *Biochim. Biophys. Acta 471*, 401–411 (1977).
103. O. T. Jones and A. G. Lee, Interactions of hexachlorocyclohexanes with lipid bilayers, *Biochim. Biophys. Acta 812*, 731–739 (1985).
104. Y. Nie, C. D. Stubbs, B. W. Williams, and E. Rubin, Ethanol causes decreased partitioning into biological membranes without changes in lipid order, *Arch. Biochem. Biophys. 268*, 349–359 (1989).
105. O. T. Jones, R. J. Froud, and A. G. Lee, Interactions of hexachlorocyclohexanes with the $(Ca^{2+} + Mg^{2})$-ATPase from sarcoplasmic reticulum, *Biochim. Biophys. Acta 812*, 740–751 (1985).
106. G. M. Omann and M. Glaser, Biosynthetic incorporation of fluorescent carbazolylundecanoic acid into membrane phospholipids of LM cells and determination of quenching constants and partition coefficients of hydrophobic quenchers, *Biochemistry 23*, 4962–4969 (1984).
107. R. Fato, M. Battino, G. P. Castelli, and G. Lenaz, Measurement of the lateral diffusion coefficients of ubiquinones in lipid vesicles by fluorescence quenching of 12-(9-anthroyl) stearate, *FEBS Lett. 179*, 238–242 (1985).
108. K. A. Sikaris, K. R. Thulborn, and W. H. Sawyer, Resolution of partition coefficients in the transverse plane of the lipid bilayer, *Chem. Phys. Lipids 29*, 23–36 (1981).
109. E. Blatt, R. C. Chatelier, and W. H. Sawyer, Effects of quenching mechanism and type of quencher association on Stern–Volmer plots in compartmentalized systems, *Biophys. J. 50*, 349–356 (1986).
110. E. London, Investigation of membrane structure using fluorescence quenching by spin-labels, *Mol. Cell. Biochem. 45*, 181–188 (1982).
111. A. C. Simmonds, J. M. East, O. T. Jones, E. K. Ronney, J. McWhirter, and A. G. Lee, Annular and non-annular binding sites on the $(Ca^{2+} + Mg^{2+})$-ATase, *Biochim. Biophys. Acta 693*, 398–406 (1982).
112. E. London and G. W. Feigenson, Fluorescence quenching in model membranes. 1. Characterization of quenching caused by a spin-labeled phospholipid, *Biochemistry 20*, 1932–1938 (1981).
113. E. London and G. W. Feigenson, Fluorescence quenching in model membranes. 2. Determination of the local lipid environment of the calcium adenosinetriphosphatase from sarcoplasmic reticulum, *Biochemistry 20*, 1939–1948 (1981).
114. M. Caffrey and G. W. Feigenson, Fluorescence quenching in model membranes. 3. Relationship between calcium adenosinetriphosphatase enzyme activity and the affinity of the protein for phosphatidylcholines with different acyl chain characteristics, *Biochemistry 20*, 1949–1961 (1981).
115. J. M. East and A. G. Lee, Lipid selectivity of the calcium and magnesium ion dependent adenosinetriphosphatase, studied with fluorescence quenching by a brominated phospholipid, *Biochemistry 21*, 4144–4151 (1982).
116. E. K. Rooney, M. G. Gore, and A. G. Lee, Two classes of binding site for hydrophobic molecules on bacterioopsin, *Biochemistry 26*, 3688–3697 (1987).
117. R. Fato, M. Battino, M. D. Esposti, G. P. Castelli, and G. Lenaz, Determination of partition and lateral diffusion coefficients of ubiquinones by fluorescence quenching of

n-(9-anthroyloxy)stearic acids in phospholipid vesicles and mitochondrial membranes, *Biochemistry 25*, 3378–3390 (1986).

118. M. F. Blackwell, K. Gounaris, S. J. Zara, and J. Barber, A method for estimating lateral diffusion coefficients in membranes from steady-state fluorescence quenching studies, *Biophys. J. 51*, 735–744 (1987).
119. P. L.-G. Chong and T. E. Thompson, Oxygen quenching of pyrene-lipid fluorescence in phosphatidylcholine vesicles, *Biophys. J. 47*, 613–621 (1985).
120. E. A. Haigh, K. R. Thulborn, and W. H. Sawyer, Comparison of fluorescence energy transfer and quenching methods to establish the position and orientation of components within the transverse plane of the lipid bilayer. Application to the gramicidin A–bilayer interaction, *Biochemistry 18*, 3525–3532 (1979).
121. T. Markello, A. Zlotnick, J. Everett, J. Tennyson, and P. W. Holloway, Determination of the topography of cytochrome b_5 in lipid vesicles by fluorescence quenching, *Biochemistry 24*, 2895–2901 (1985).
122. D. B. Chalpin and A. M. Kleinfeld, Interaction of fluorescence quenchers with the *n*-(9-anthroyloxy) fatty acid membrane probes, *Biochim. Biophys. Acta 731*, 465–474 (1983).
123. N. G. Bakhshiev and I. V. Piterskaya, Universal molecular interactions and their effect on the electronic spectra of molecules in two-component solutions, *Opt. Spectrosk. 19*, 390–395 (1965).
124. J. H. Easter, R. P. DeToma, and L. Brand, Nanosecond time-resolved emission spectroscopy of a fluorescence probe adsorbed to L-α-egg lecithin vesicles, *Biophys. J. 16*, 571–583 (1976).
125. M. G. Badea, R. P. DeToma, and L. Brand, Nanosecond relaxation processes in liposomes, *Biophys. J. 43*, 197–209 (1978).
126. S. R. Meech, D. V. O'Connor, A. J. Roberts, and D. Phillips, On the construction of nanosecond time-resolved emission spectra, *Phochem. Photobiol. 33*, 159–172 (1980).
127. R. P. DeToma, Solvent relaxation, in: *Time-Resolved Fluorescence Spectroscopy in Biochemistry and Biology* (R. B. Cundall and R. E. Dale, eds.), pp. 393–410, Plenum, New York (1984).
128. C. D. Stubbs, S. R. Meech, A. G. Lee, and D. Phillips, Solvent relaxation in lipid bilayers with dansyl probes, *Biochim. Biophys. Acta 815*, 351–360 (1985).
129. J. R. Lakowicz and A. Balter, Analysis of excited-state processes by phase-modulation fluorescence spectroscopy, *Biophys. Chem. 16*, 117–132 (1982).
130. J. R. Lakowicz, R. B. Thompson, and H. Cherek, Phase fluorometric studies of spectral relaxation at the lipid–water interface of phospholipid vesicles, *Biochim. Biophys. Acta 734*, 295–308 (1983).
131. J. R. Lakowicz, D. R. Bevan, B. P. Maliwal, H. Cherek, and A. Balter, Synthesis and characterization of a fluorescence probe of the phase transition and dynamic properties of membranes, *Biochemistry 22*, 5714–5722 (1983).
132. J. R. Lakowicz and S. Keating-Nakamoto, Red-edge excitation of fluorescence and dynamic properties of proteins and membranes, *Biochemistry 23*, 3013–3021 (1984).
133. A. Gafni, R. P. DeToma, R. E. Manrow, and L. Brand, Nanosecond decay studies of a fluorescence probe bound to apomyoglobin, *Biophys. J. 17*, 155–168 (1977).
134. A. P. Demchenko and N. V. Shcherbatska, Nanosecond dynamics of charged fluorescent probes at the polar interface of a membrane phospholipid bilayer, *Biophys. Chem. 22*, 131–143 (1985).
135. J. R. Lakowicz and D. Hogen, Dynamic properties of the lipid–water interface of model membranes as revealed by lifetime-resolved fluorescence emission spectra, *Biochemistry 20*, 1366–1373 (1981).
136. K. P. Ghiggino, A. G. Lee, S. R. Meech, D. V. O'Connor, and D. Phillips, Time-resolved emission spectroscopy of the dansyl fluorescence probe, *Biochemistry 20*, 5381–5389 (1981).

137. L. Wojtczak and M. J. Natecz, The surface potential of membranes: Its effect on membrane-bound enzymes and transport processes, in: *Structure and Properties of Cell Membranes* (G. Benga, ed.), Vol. II, pp. 215–242, CRC Press, Boca Raton, Florida (1985).
138. S. Itoh and M. Nishimura, Rate of redox reactions related to surface potential and other surface-related parameters in biological membranes, *Methods Enzymol. 125*, 58–86 (1986).
139. B. Ehrenberg, Spectroscopic methods for the determination of membrane surface charge density, *Methods Enzymol. 127*, 678–696 (1986).
140. S. McLaughlin, Electrostatic potentials at membrane–solution interfaces, *Curr. Top. Membr. Transp. 9*, 71–144 (1977).
141. J. Barber, Membrane surface charges and potentials in relation to photosynthesis, *Biochim. Biophys. Acta 594*, 253–308 (1980).
142. N. Kamo and Y. Kobatake, Changes of surface and membrane potentials in biomembranes, *Methods Enzymol. 125*, 46–58 (1986).
143. M. Eisenberg, T. Gresalfi, T. Riccio, and S. McLaughlin, Adsorption of monovalent cations to bilayer membranes containing negative phospholipids, *Biochemistry 18*, 5213–5223 (1979).
144. C. J. Drummond and F. Grieser, Absorption spectra and acid–base dissociation of the 4-alkyl derivatives of 7-hydroxycoumarin in self-assembled surfactant solution: Comments on their use as electrostatic surface potential probes, *Photochem. Photobiol. 45*, 19–34 (1987).
145. R. Gibrat, C. Romieu, and C. Grignon, A procedure for estimating the surface potential of charged or neutral membranes with 8-anilino-1-naphthalenesulphonate probe, *Biochim. Biophys. Acta 736*, 196–202 (1983).
146. D. E. Robertson and H. Rottenberg, Membrane potential and surface potential in mitochondria, *J. Biol. Chem. 258*, 11039–11048 (1983).
147. H. Tanabe, N. Kamo, and Y. Kobatake, Fluorometric estimation of surface potential change associated with chemotactic stimulation in *Tetrahymena pyriformis*, *Biochim. Biophys. Acta 805*, 345–353 (1984).
148. A. P. Winiski, A. C. McLaughlin, R. V. McDaniel, M. Eisenberg, and S. McLaughlin, An experimental test of the discreteness-of-charge effect in positive and negative lipid bilayers, *Biochemistry 25*, 8206–8214 (1986).
149. K. Masamoto, K. Matsuura, S. Itoh, and M. Nishimura, Surface potential dependence of the distribution of charged dye molecules onto photosynthetic membranes, *J. Biochem. 89*, 397–405 (1981).
150. T. Aiuchi, H. Tanabe, K. Kurihara, and Y. Kobatake, Fluorescence changes of rhodamine 6G associated with chemotactic responses in *Tetrahymena pyriformis*, *Biochim. Biophys. Acta 628*, 355–364 (1980).
151. W. S. Chow and J. Barber, Salt dependent changes of 9-aminoacridine as a measure of charge-densities of membrane surfaces, *J. Biochem. Biophys. Methods 3*, 173–185 (1980).
152. W. S. Chow and J. Barber, 9-Aminoacridine fluorescence changes as a measure of surface charge density of the thylakoid membrane, *Biochim. Biophys. Acta 589*, 346–352 (1980).
153. G. F. W. Searle, J. Barber, and J. D. Mills, 9-Amino-acridine as a probe of the electrical double layer associated with the chloroplast thylakoid membranes, *Biochim. Biophys. Acta 461*, 413–425 (1977).
154. I. M. Moller, T. Lundborg, and A. Berczi, The negative surface charge density of plasmalemma vesicles from wheat and oat roots, *FEBS Lett. 167*, 181–185 (1984).
155. A. P. R. Theuvenet, W. H. H. Van De Wijngaard, J. W. van De Pijke, and G. W. F. H. Borst-Pauwels, Application of 9-aminoacridine as probe of the surface potential. *Biochim. Biophys. Acta 775*, 161–168 (1984).
156. J. Cerbon, C. Ontiveros, and A. Janovitz, Phosphoinositides provide a regulatory mechanism of surface charge and active transport, *Biochim. Biophys. Acta 887*, 275–282 (1986).

157. R. Kramer, Interaction of membrane surface charges with the reconstituted ADP/ATP-carrier from mitochondria, *Biochim. Biophys. Acta 735*, 145–159 (1983).
158. M. S. Fernandez, Determination of surface potential in liposomes, *Biochim. Biophys. Acta 646*, 23–26 (1981).
159. S. Lukac, Surface potential at surfactant and phospholipid vesicles as determined by amphiphilic pH indicators, *J. Phys. Chem. 89*, 5045–5050 (1984).
160. R. Pal, W. A. Petri, Jr., Y. Barenholz, and R. R. Wagner, Lipid and protein contributions to the membrane surface potential of vesicular stomatitits virus probed by a fluorescent pH indicator, 4-heptadecyl-7-hydroxycoumarin, *Biochim. Biophys. Acta 729*, 185–192 (1983).
161. W. L. C. Vaz, A. Nicksch, and F. Jahnig, Electrostatic interactions at charged lipid membranes, *Eur. J. Biochem. 83*, 299–305 (1978).
162. R. Homan and M. Eisenberg, A fluorescence quenching technique for the measurement of paramagnetic ion concentrations at the membrane/water interface. Intrinsic and X537A-mediated cobalt fluxes across lipid bilayer membranes, *Biochim. Biophys. Acta 812*, 485–492 (1985).
163. S. J. Morris, D. Bradley, and R. Blumenthal, The use of cobalt ions as a collisional quencher to probe surface charge and stability of fluorescently labeled bilayer vesicles, *Biochim. Biophys. Acta 818*, 365–372 (1985).
164. E. Haas and I. Z. Steinberg, Intramolecular dynamics of chain molecules monitored by fluctuations in efficiency of excitation energy transfer, *Biophys. J. 46*, 429–437 (1984).

6

Fluorescence and Immunodiagnostic Methods

Thomas M. Li and Richard F. Parrish

6.1. Introduction

Competitive protein-binding methods permit specific, sensitive, and relatively rapid assays for a variety of substances. Radioimmunoassay (RIA), which is a competitive protein-binding technique, has been widely utilized for determinations of analytes present in minute quantities in biological fluids. The RIA technique, however, is plagued by several major disadvantages. It utilizes radioactive material, which is a potential health hazard for those performing the assay. It generates low-level radioactive waste, the disposal of which has become a complex environmental and political problem. The short half-life of some of the isotopes utilized leads to limited shelf life for commercial products. The RIA technique is always heterogeneous: it requires a separation step to separate free and bound analytes. It utilizes relatively expensive and sophisticated instrumentation. During the last ten years, a great deal of research effort has been expended to replace RIAs with assays that have similar sensitivity and specificity but do not require the use of radioactive materials or a separation step to discriminate between free and bound analytes. Fluorescence immunoassay can be homogeneous, sensitive, and specific. It utilizes stable, safe, and inexpensive reagents. It has a large dynamic range and can be performed quickly on relatively simple, inexpensive instruments. During the past decade many novel fluorescence immunoassay protocols have been developed. In this chapter we will discuss some of these methodologies and the strategies used to implement them. We will concentrate essentially on the therapeutic drug monitoring role of fluorescence immunoassays and in this context will limit the discussion to assays for two of the common and important analytes, theophylline and digoxin, which have

---

Thomas M. Li and Richard F. Parrish • Development Department, Syva, Palo Alto, California 94304.

Topics in Fluorescence Spectroscopy, Volume 3: Biochemical Applications, edited by Joseph R. Lakowicz. Plenum Press, New York, 1992.

therapeutic windows in the microgram per milliliter and nanogram per milliliter range, respectively.

6.2. Assay Formats

Fluorescence immunoassays, as is the case with all immunoassays, can be either homogeneous or heterogeneous. Homogeneous fluorescence immunoassays all rely on the observation that the fluorescence obtained from the free fluorophore is sufficiently different from the fluorescence obtained after binding of the fluorophore to the antibody. Thus, the concentration of one of the free or antibody-bound fluorescent species can be measured in the presence of the other fluorescent species without the necessity of a separation step. In practice, this translates into the competitive ligand binding assay format. In this assay configuration, the analyte and an analyte–fluorophore conjugate compete for the available antibody binding sites. If the fluorescent parameter to be measured increases as the concentration of the analyte increases, the assay is referred to as a positive reading assay. If, on the other hand, the fluorescence parameter to be measured decreases as the concentration of the analyte to be measured increases, the assay is referred to as a negative reading assay. In general, positive reading assays are to be preferred since at low analyte concentrations fluorescence increases against a low or minimal background. In negative reading assays, however, the change in observed fluorescence at low analyte concentration is a small difference between two large values, and hence the method has lower precision and accuracy.

Heterogeneous fluorescence immunoassays have many different assay formats, but all possess one unifying feature: the free fluorophore remaining in solution after binding of some of the fluorophore to antibody must be removed before quantification can be achieved. Again, both negative and positive reading assay formats have been employed.

6.3. Fluorescence Polarization Immunoassay

Polarization of fluorescence measurements has been utilized to monitor homogeneous competitive ligand binding reactions. A low-molecular-weight fluorescent molecule (such as theophylline–umbelliferone conjugate) is free to rotate rapidly in solution and as such has a low fluorescence polarization. However, when antibody to theophylline is added to theophylline–umbelliferone conjugate, the resulting theophylline–umbelliferone conjugate–antibody complex is a much larger kinetic unit than the initial fluorophore, and the rotation of the conjugate–antibody complex is reduced, resulting in an increase in polarization of the emitted fluorescence. Increasing the amount of antibody eventually results in a leveling off of the fluorescence polarization (Figure 6.1A). When variable amounts of theophylline are added to a constant amount of theophylline antibody and theophylline–umbelliferone conjugate, a theophylline concentration-dependent reduction in fluorescence polarization will result (Figure 6.1B). This competition between theophylline and theophylline–umbelliferone conjugate forms the basis for a quantitative, homogeneous, fluorescence polarization immunoassay for theophylline.[1]

Although the potential to develop a fluorescence polarization immunoassay has been available for more than 20 years, its use in the clinical laboratory was not initially widespread, presumably for several reasons. First, fluorescence polarization immunoassays are applicable only to low-molecular-weight analytes. High-molecular-weight conjugates (fluorophore-labeled proteins) have a much slower rotational time than low-molecular-weight conjugates. The high initial fluorescence polarization results in an unacceptably small difference between the fluorescence polarization of the free and the bound fluorophore. Second, fluorescence polarization measurements were technically more challenging than fluorescence intensity measurements and required more sophisticated instrumentation. The introduction of the Abbott TDx automatic clinical analyzer in 1981,[2] however, has shown that fluorescence polarization has the potential to be utilized in routine clinical analyses and has opened the door for further development work in this area.

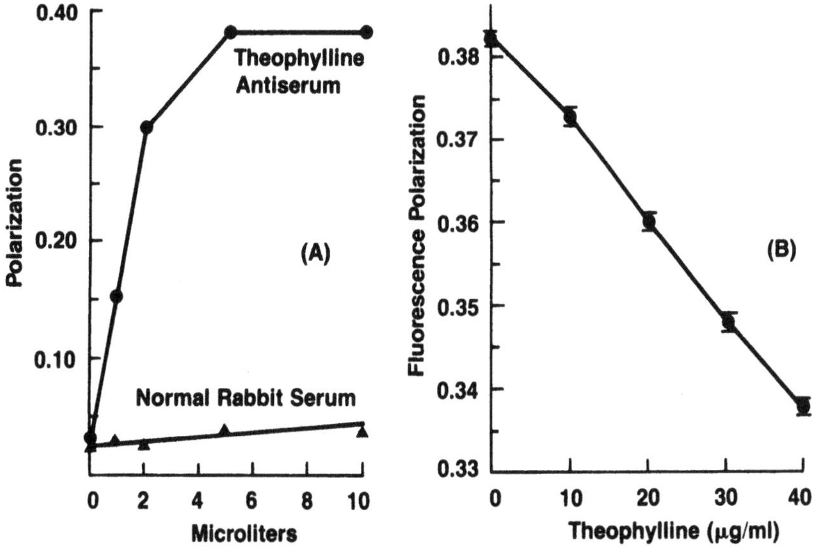

Figure 6.1. Fluorescence polarization immunoassay for theophylline. (A) Effect of theophylline rabbit antiserum (●) and normal rabbit serum (▲) on the fluorescence polarization of the theophylline–umbelliferone conjugate. (B) Fluorescence polarization of the theophylline–umbelliferone conjugate in the presence of varying concentrations of theophylline. (Reprinted from Ref. 1, with permission from Academic Press.)

6.4. Substrate-Labeled Fluorescent Immunoassay (SLFIA)

The hydrolysis of a nonfluorescent enzyme substrate to a fluorescent product is widely utilized to measure the activity of a large number of enzymes. Binding of enzyme substrates by antibodies often protects the enzymatically labile bond from hydrolysis. By the combination of these two formats, the substrate-labeled fluorescent immunoassay (SLFIA) was developed.[3]

The β-Galactosylumbelliferone–theophylline conjugate (Figure 6.2) is essentially nonfluorescent. In the presence of β-galactosidase, the galactosyl–umbelliferone bond is cleaved, releasing the highly fluorescent umbelliferone–theophylline conjugate. However, when the β-galactosylumbelliferone–theophylline conjugate is incubated with antibody to theophylline and β-galactosidase, the hydrolysis of the nonfluorescent β-galactosidase substrate to fluorescent product is prevented by the antibody in a concentration-dependent manner (Figure 6.3A). If free theophylline is present in the sample, competition will result between the analyte (theophylline) and conjugate (β-galactosylumbelliferone–theophylline) for the antibody binding sites. The higher the concentration of free theophylline, the more conjugate will be available to be hydrolyzed by the β-galactosidase. Fluorescence intensity, therefore, is directly proportional to the concentration of theophylline present (Figure 6.3B). The substrate-labeled fluorescent immunoassay technique has

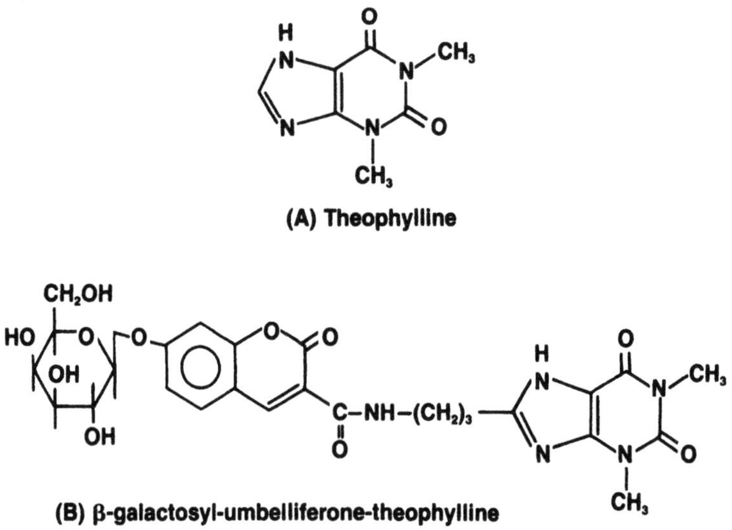

Figure 6.2. Chemical structure of 8-[3-(7-β-galactosylcoumarin-3-carboxyamido)propyl] theophylline (B) used in the homogeneous substrate-labeled fluorescent immunoassay for theophylline (A). (Reprinted from Ref. 3, with permission from the American Association for Clinical Chemistry.)

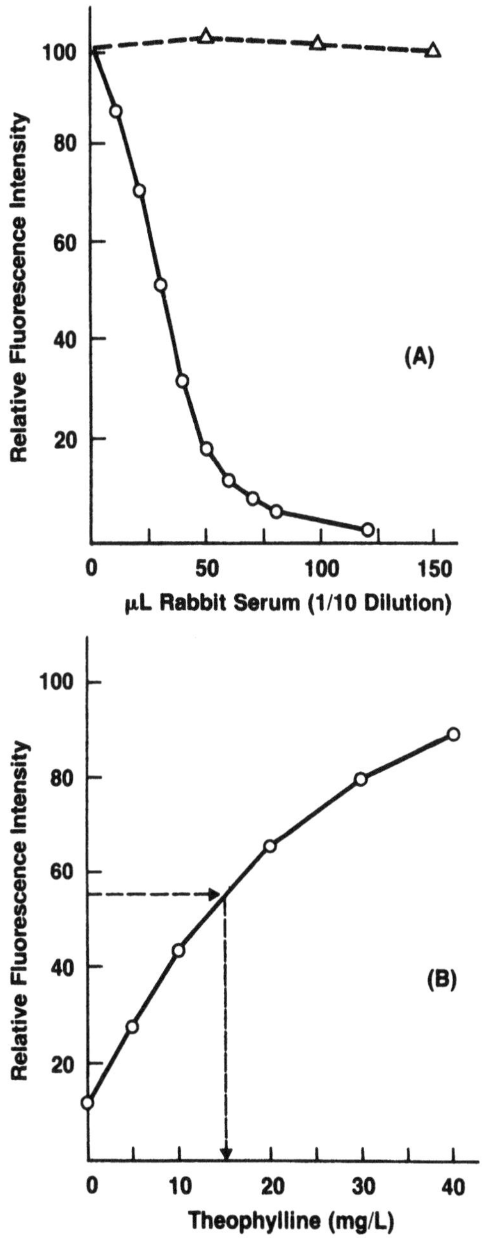

Figure 6.3. Substrate-labeled fluorescent immunoassay for theophylline. (A) Effect of theophylline rabbit antiserum (○) and normal rabbit serum (△) on the enzymatic hydrolysis of 8-[3-(7-β-galactosylcoumarin-3-carboxyamido)propyl]theophylline. (B) Effect of various concentrations of theophylline on the fluorescence resulting from enzymatic hydrolysis of 8-[3-(7-β-galactosylcoumarin-3-carboxyamido)propyl]theophylline. (Reprinted from Ref. 3, with permission from the American Association for Clinical Chemistry.)

been utilized for many therapeutic drug assays and can be run on many inexpensive commercial filter fluorometers. In addition, the technique has been adapted for automated analysis.[4]

6.5. Intramolecularly Quenched Fluorescent Immunoassay

A potential problem associated with the SLFIA technique can occur if the analyte coupled to the fluorophore is a fluorescence quencher. Clearly, if the analyte quenched all of the fluorescence, no assay would be possible. However, the more likely event is that the analyte quenches some of the fluorescence and thereby decreases the sensitivity of the assay. Separating the fluorescent portion from the analyte portion in the conjugate with a fluorescence quencher that is connected to the fluorophore by an enzyme-labile bond has been proposed as a method to circumvent potential analyte quenching of an SLFIA. A theophylline assay[5] that performs according to this rationale is shown schematically in Figure 6.4. Flavin adenine dinucleotide (FAD) was coupled to theophylline via the adenine portion of the molecule to produce an FAD–theophylline conjugate. The adenine in the adenosine monophosphate (AMP) portion of the molecule is a very efficient quencher of the fluorescence of the isoalloxazine ring of flavin mononucleotide (FMN). Enzymatic hydrolysis by nucleotide pyrophosphatase separates the highly fluorescent FMN portion of the conjugate from the quencher AMP–theophylline portion of the molecule and allows full expression of the FMN fluorescence. In the presence of antibody to theophylline, the enzymatically labile bond in the conjugate is protected from hydrolysis. Increasing levels of antibody result in decreased fluorescence. In the presence of theophylline, competition exists between analyte and conjugate for the available antibody binding sites. The higher the analyte concentration is, the higher the concentration of conjugate available for hydrolysis and the higher the fluorescence due to enzymatic hydrolysis of the bond between fluorophore and quencher. The intramolecularly quenched substrate-labeled fluorescent immunoassay has several advantages: (a) The enzymatically labile bond can be far removed from the fluorophore; (b) because of the presence of the quencher molecule between the fluorophore and the analyte, conjugation between analyte and fluorophore is not limited to a specific site of the fluorophore where conjugation might diminish its fluorescence properties; and (c) full expression of the fluorescence potential of the fluorophore is realized since hydrolysis frees the fluorophore from any potential quenching that might occur if the analyte was directly linked to the fluorophore.

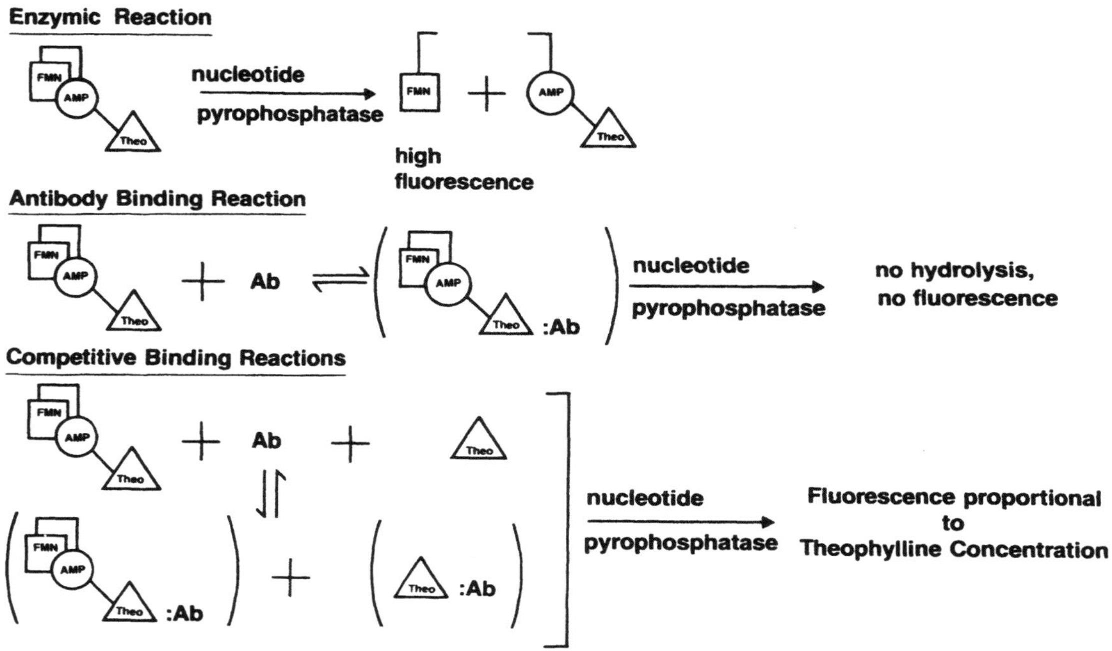

Figure 6.4. Schematic representation of a fluorescent immunoassay for theophylline utilizing enzymatic hydrolysis of an intramolecularly quenched theophylline conjugate of flavin adenine dinucleotide. (Reprinted from Ref. 5, with permission from Academic Press.)

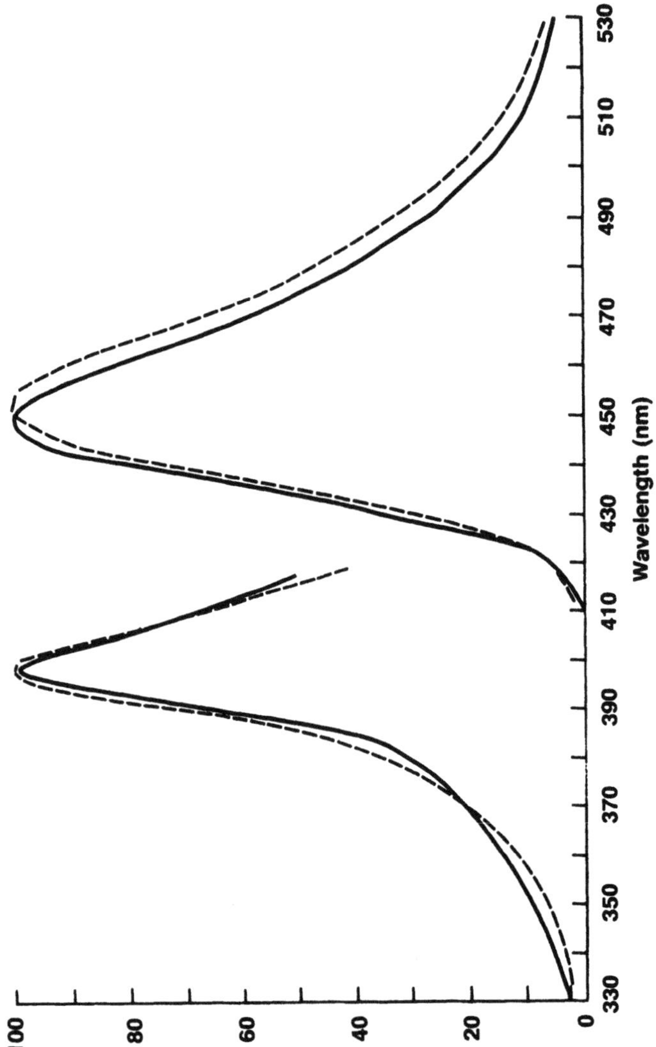

Figure 6.5. Comparison of excitation and emission spectra of the theophylline–umbelliferone conjugate on paper (——) and in solution (– – –). (Reprinted from Ref. 6, with permission from the American Association for Clinical Chemistry.)

6.6. Homogeneous Fluorescent Immunoassay in a Dry Reagent Format

Another advance in the utilization of the substrate-labeled fluorescent immunoassay technique has been the demonstration of a dry reagent format using front-face fluorescence measurments.[6] Figure 6.5 compares the excitation and emission spectra of the umbelliferone–theophylline conjugate on paper and in solution. Excitation and emission maxima are almost identical for the solution- and the solid-phase system. As analyzed in a double reciprocal plot,[6] enzymatic hydrolysis of the conjugate on the paper pad follows Michaelis–Menten kinetics with K_m on the pad (0.33 mM) almost identical to K_m in solution (0.29 mM). In one format, β-galactosidase and antibody to theophylline are impregnated onto a paper pad. Addition of a solution of β-galactosylumbelliferone–theophylline conjugate to the paper pad results in substrate hydrolysis and the appearance of fluorescence. As the amount of antibody to theophylline is increased on the paper pad, the amount of fluorescence generated decreases, indicating binding of conjugate to the antibody and prevention of enzymatic hydrolysis of the fluorogenic β-galactosidase substrate. When known amounts of theophylline are added to a fixed amount of β-galactosylumbelliferone–theophylline conjugate, and this material is added to the paper pad, the free analyte competes with conjugate for antibody binding sites on the paper. As the concentration of analyte increases, the fluorescence also increases.

The dry reagent format can be simplified even further by impregnating the enzyme, antibody, and conjugate onto the paper. This is accomplished by two dip procedures. In the first dip, theophylline antibody and β-galactosidase dissolved in Bicine buffer are impregnated. After drying, the paper is dipped into another solution (β-galactosylumbelliferone–theophylline conjugate dissolved in acetone) and dried. Dissolving the conjugate in acetone prevents hydrolysis of the conjugate by the enzyme already present on the paper. With the test pad that is impregnated with enzyme, antibody, and conjugate, the assay is performed by simply adding a theophylline-containing sample solution to the test pad and measuring the resulting fluorescence from the surface of the pad. Using this format, the assay is rapid and exceptionally easy to perform.

6.7. Fluorescence Excitation Transfer Immunoassay

The theory of resonance transfer of electronic excitation energy between donor and acceptor molecules of suitable spectroscopic properties was first presented by Förster.[7] According to this theory, the rate constant for singlet energy transfer from an excited donor to a chromophore acceptor which may or may not be fluorescent is proportional to r^{-6}, where r is the distance between the molecule involved. Förster proposed that singlet-singlet transfer occurs by a resonance interaction of the dipole pair between the energy donor and the acceptor chromophores. In this quantitive treatment, the distance R_0, the critical distance at which transfer efficiency is 50%, is related to the quantum yield of energy donor, spectral overlap between the emission spectrum of the donor and the absorption spectrum of the acceptor, and the orientation factor K. Under appropriate experimental conditions, this energy transfer can occur over substantial distances, up to 84 Å. Using the energy transfer between the donor-acceptor pair in the analyte-antibody complex, a fluorescent excitation transfer immunoassay has been developed.[8] In this excitation transfer immunoassay, the analyte is labeled with a fluorescent molecule. This conjugate maintains its fluorescence after binding to the antibody directed against the analyte. The antibody directed against the analyte is labeled with a quencher molecule that is not fluorescent at the wavelength(s) of emission of the conjugate molecule. When the conjugate binds to the antibody, the fluorescent energy from the labeled antigen is transferred to the antibody-bound acceptor and the fluorescence is quenched. At a constant antibody concentration, which is sufficient to bind the conjugate, as the concentration of nonlabeled antigen increases and competes with the conjugate, more conjugate will remain in solution and the fluorescence will increase.

6.8. Design of Fluorescent Probes

When choosing a fluorescent compound to conjugate to an analyte of interest, several parameters must be considered. The fluorophore should have a high extinction coefficient, as high a quantum yield as possible, and a large Stokes shift. It should possess a reactive functional group that can be utilized to link the fluorophore to the analyte of interest. It should also demonstrate sufficient fluorescence intensity in aqueous medium so that the fluorescence generated by the specific assay can be distinguished from any inherent background fluorescence present in the sample matrix. Only a few fluorescent compounds possess enough of these properties to be useful in fluorescence immunoassays.

Many fluorescence immunoassays have utilized umbelliferone, a 7-hydroxycoumarin (excitation and emission maxima of 360 nm and 447 nm, respectively), as the fluorescence reporter group. Normal serum, however, exhibits emission maximum at 520 nm and excitation maxima of 330 nm and 440 nm. Even with this overlap between the spectra of umbelliferone and normal serum, umbelliferone has one property that makes it suitable for fluorescence immunoassays: conversion of the free hydroxyl group to an ester or a glucoside effectively quenches the fluorescence of umbelliferone, and hydrolysis of the ester or glycoside allows for recovery of fluorescence. It is this property that makes umbelliferone useful in substrate-labeled fluorescent immunoassays. Nonetheless, while umbelliferone is satisfactory for analyses where the analyte of interest is present in reasonably high concentrations (μg/ml) and high serum dilutions can be utilized, it is not the reagent of choice when high-sensitivity assays (ng/ml) demand minimal serum dilutions.

Fluorescein (excitation and emission maxima of 492 nm and 520 nm, respectively) has also been utilized in fluorescence assays. Although its excitation maximum is higher than that of umbelliferone, it suffers from a problem similar to that of umbelliferone in that albumin-bound bilirubin has excitation and emission maxima of 460 nm and 515 nm, respectively. In addition, commercial preparations have been reported to contain two isomers, which may cause heterogeneity during conjugate preparation.

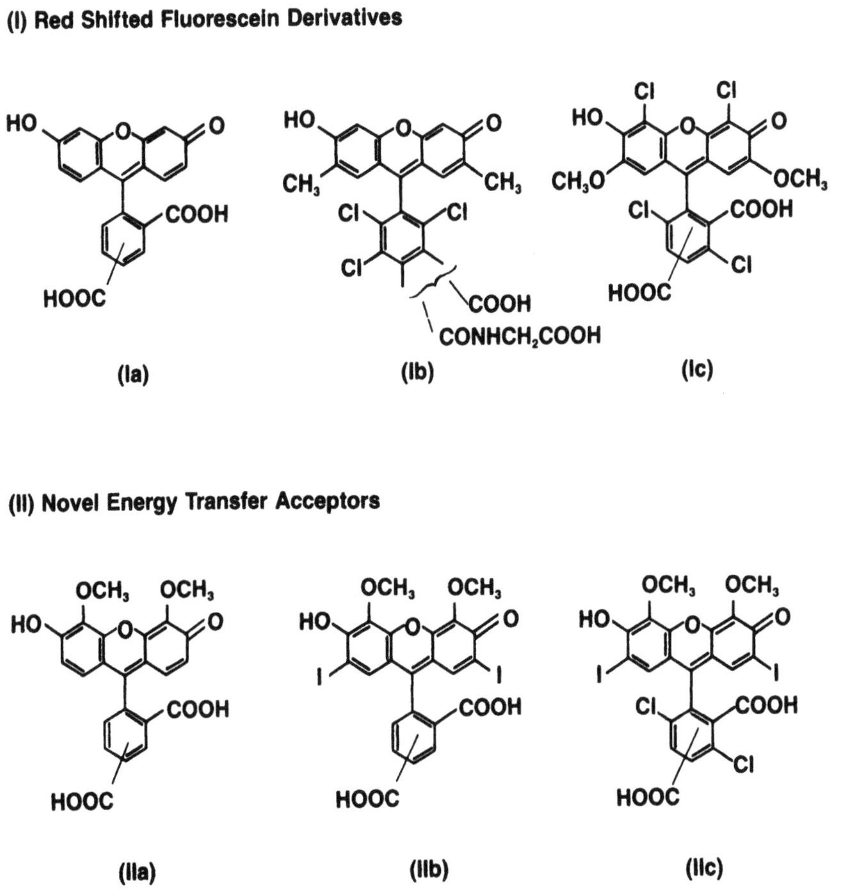

Figure 6.6. Structures of chemically modified fluorescein derivatives with long-wavelength emission (I) and with novel energy transfer acceptor characteristics (II).

Table 6.1. Spectral Properties of Excitation Energy Transfer Donors and Acceptors

Compound	Absorption maximum (nm)	Stokes shift (nm)	Quantum yield
Red-shifted fluorescein derivatives			
Ia	492	22	0.87
Ib	517	18	0.86
Ic	536	22	0.82
Novel energy transfer acceptors			
IIa	512		4×10^{-4}
IIb	533		3×10^{-4}
IIc	549		3.6×10^{-4}

With these potential interference problems associated with both umbelliferone and fluorescein, it appeared that new fluorophores would be required to realize the full sensitivity of fluorescence immunoassays. These new compounds would preferably have excitation maxima at wavelengths long enough (greater than 575 nm) to minimize the background interference of serum. Figure 6.6 and Table 6.1 show the results of one such study[9] on the chemical modification of fluorescein compounds. Compounds Ia, Ib, and Ic indicate the upward shift in absorbance maxima that can be achieved. Compound Ic with its absorbance maximum at 536 nm and a 22-nm Stokes shift is approaching the target absorbance maximum of 575 nm. Figure 6.6 and Table 6.1 also give the structures and spectral properties of some novel energy transfer acceptors.[10] Because of excellent emission and absorption overlap, use of these novel energy donor and acceptor pairs offers efficient energy transfer in the excitation transfer immunoassay. R_0 values of 57, 61, and 62 Å are obtained for fluorescer–quencher pair Ia–IIa, Ib–IIb, and Ic–IIc, respectively.

6.9. Phycobiliproteins

In addition to chemical modification as one way of obtaining fluorophores with long-wavelength emission, there exist in nature algae phycobiliproteins that absorb light energy and transfer the energy to chlorophyll. The properties of the four most important forms are shown in Table 6.2. The very large extinction coefficient ($2.4 \times 10^6\ M^{-1}\ cm^{-1}$) and emission maximum at 575 nm, coupled with the large quantum yield (0.85) and absorption maximum at

Table 6.2. Spectral Properties of Four Phycobiliproteins[a]

	Absorption		Emission	
Phycobiliprotein	λ_{max} (nm)	ε_{max} (M^{-1} cm^{-1})	λ_{max} (nm)	Quantum yield
B-Phycoerythrin	546	2.4×10^6	575	0.59
R-Phycoerythrin	566	2.0×10^6	574	0.85
C-Phycocyanin	614	5.8×10^5	643	0.51
Allophycocyanin	650	7.0×10^5	660	0.68

[a] Selected values from Kronick.(11)

566 nm, make phycoerythrin an ideal fluorophore for use in immunoassays.(11) Kronick and Grossman(12) first used phycoerythrin in energy transfer immunoassay for human IgG in a model system. Subsequently, a practical commercial clinical assay for digoxin was developed and described by Khanna(13) and Plebani and Burlina.(14) The commercial assay showed a fluorophore sensitivity of better than 10^{-10} M, which represents an order of magnitude improvement over performance possible with conventional dyes used in the fluorescence excitation transfer immunoassay format.

6.10. Phase-Resolved Fluorescence Immunoassay

In addition to fluorescence intensity and polarization, fluorescence spectroscopy also includes measurement of the lifetime of the excited state. Recent improvements in the design of fluorescence instrumentation for measuring fluorescence lifetime have permitted additional applications of fluorescence techniques to immunoassays. Fluorescence lifetime measurement can be performed by either phase-resolved or time-resolved fluorescence spectroscopy.

It has been shown that phase-resolved fluorescence spectroscopy can be used for simultaneous determinations of a single species in two different biological microenvironments on the basis of differences in fluorescence lifetime. A homogeneous fluorescent immunoassay for phenobarbital based on fluorescence lifetime selectivity has recently been demonstrated.(15) The analyte, phenobarbital, is labeled with fluorescein. By phase-resolved fluorescence spectroscopy, the free labeled fluorescein has a lifetime of 4.04 ns. Antibody-bound labeled fluorescein, however, has a lifetime of 3.94 ns. This difference in lifetime has been utilized to construct an immunoassay in which phase-sensitive fluorescence intensity for the free and the antibody-bound fraction was measured. Bright and McGown have shown that the combination of five detector phase angles provides good accuracy and the smallest average error for the phenobarbital assay.

6.11. Time-Resolved Fluorescence Immunoassay

The use of differences in fluorescence lifetime to achieve selectivity in heterogeneous fluorescent immunoassay by means of single-photon time-resolved fluorescence has also been recently described.[16] The basis of time-resolved fluorescence immunoassays is quite simple. Rare earth metal chelates have fluorescence lifetimes in the microsecond range. However, the fluorescence lifetimes of most materials that cause background fluorescence in immunoassays have fluorescence lifetimes in the nanosecond range. This approximately three orders of magnitude difference in lifetimes can be exploited by allowing the background fluorescence to decay before the fluorescence of the metal chelate (in many cases, an europium chelate) is measured. The primary gain from this technology is that background fluorescence is reduced to almost zero in a typical immunoassay. To perform a solid-phase heterogeneous immunoassay for digoxin,[16] digoxin conjugated to rabbit serum albumin is immobilized on the polystyrene surface of microtitration strip wells. Antibody against digoxin is labeled with an europium chelate (diazophenyl-EDTA-Eu). In the competitive immunometric assay, the Eu-labeled antibody is competitively distributed between the solid-phase and the sample digoxin. After bound and free antibodies are separated by washing the strips, the europium is dissociated from antibody at low pH and measured by time-resolved fluorescence in a micellar solution containing Triton X-100, β-diketone, and a Lewis base. The detergent solubilizes the chelating compounds and excludes water from the fluorescent ligand–europium complex. Using a 1-s counting time, europium as low as 5×10^{-14} M can be detected.

6.12. Conclusion

In the past ten years, numerous applications of fluorescence methods for monitoring homogeneous and heterogeneous immunoassays have been reported. Advances in the design of fluorescent labels have prompted the development of various fluorescent immunoassay schemes such as the substrate-labeled fluorescent immunoassay and the fluorescence excitation transfer immunoassay. As sophisticated fluorescence instrumentation for lifetime measurement became available, the phase-resolved and time-resolved fluorescent immunoassays have also developed. With the current emphasis on satellite and physician's office testing, future innovations in fluorescence immunoassay development will be expected to center on the simplification of assay protocol and the development of solid-state miniaturized fluorescence readers for on-site testing.

References

1. T. M. Li, J. L. Benovic, and J. F. Burd, Serum theophylline determination by fluorescence polarization immunoassay utilizing an umbelliferone derivative as a fluorescent label, *Anal. Biochem. 118*, 102–107 (1981).
2. M. E. Jolley, S. D. Stroupe, K. S. Schwenzer, C. J. Wang, M. Lu-Steffes, H. D. Hill, S. R. Popelka, J. T. Holen, and D. M. Kelso, fluorescence polarization immunoassay III. An automated system for therapeutic drug determination, *Clin. Chem. 27*, 1575–1579 (1981).
3. T. M. Li, J. L. Benovic, R. T. Buckler, and J. F. Burd, Homogeneous substrate-labeled fluorescent immunoassay for theophylline in serum, *Clin. Chem. 27*, 22–26 (1981).
4. T. M. Li, S. P. Robertson, T. H. Crouch, E. E. Pahuski, G. A. Bush, and S. J. Hydro, Automated fluorometer/photometer system for homogeneous immunoassays, *Clin. Chem. 29*, 1628–1634 (1983).
5. T. M. Li and J. F. Burd, Enzymic hydrolysis of intramolecular complexes for monitoring theophylline in homogeneous competitive protein-binding reactions, *Biochem. Biophys. Res. Commun. 103*, 1157–1165 (1981).
6. A. C. Greenquist, B. Walter, and T. M. Li, Homogeneous fluorescent immunoassay with dry reagents, *Clin. Chem. 27*, 1614–1617 (1981).
7. T. Förster, *Ann. Physik (Leipzig) 2*, 55–75 (1948).
8. E. F. Ullman, M. Schwarzberg, and K. E. Rubenstein, Fluorescent excitation transfer immunoassay, a general method for determination of antigens. *J. Biol. Chem. 251*, 4172–4178 (1976).
9. E. F. Ullman and P. L. Khanna, Fluorescence excitation transfer immunoassay (FETI), *Methods Enzymol. 74*, 28–60 (1981).
10. P. L. Khanna and E. F. Ullman, 4′, 5′-Dimethoxy-6-carboxyfluorescein: A novel dipole–dipole coupled fluorescence energy transfer acceptor useful for fluorescence immunoassays, *Anal. Biochem. 108*, 156–161 (1980).
11. M. N. Kronick, The use of phycobiliproteins as fluorescent labels in immunoassay, *J. Immunol. Methods 92*, 1–13 (1986).
12. M. N. Kronick and P. D. Grossman, Immunoassay techniques with fluorescent phycobiliprotein conjugates, *Clin. Chem. 29*, 1582–1586 (1983).
13. P. Khanna, Energy transfer immunoassays using phycobiliproteins. Presentation at Conference of Phycobiliprotein in Biology and Medicine, Seattle, Washington, September 9–10, 1985.
14. M. Plebani and A. Burlina, Fluorescence energy transfer immunoassay of digoxin in serum, *Clin. Chem. 31*, 1879–1881 (1985).
15. F. V. Bright and L. B. McGown, Homogeneous immunoassay of phenobarbital by phase-resolved fluorescence spectroscopy, *Talanta 32*, 15–18 (1985).
16. P. Helsingius, I. Hemmilä, and T. Lövgren, Solid-phase immunoassay of digoxin by measuring time-resolved fluorescence, *Clin. Chem. 32*, 1767–1769 (1986).

Total Internal Reflection Fluorescence

Daniel Axelrod, Edward H. Hellen, and Robert M. Fulbright

7.1. Introduction

The distribution and dynamics of molecules at or near surfaces are central to numerous phenomena in biology: for example, binding to and triggering of cells by hormones, neurotransmitters, and antigens; the deposition of plasma proteins upon foreign surfaces, leading to thrombogenesis; electron transport in the mitochondrial membrane; cell adhesion to surfaces; enhancement of the reaction rate with membrane receptors by nonspecific adsorption and surface diffusion of ligand; and the dynamical arrangement of submembrane cytoskeletal structures involved in cell shape, motility, and mechanoelastic properties.

In most of these examples, certain functionally relevant molecules coexist in both a surface-associated and a nonassociated state. If such molecules are detected by a conventional fluorescence technique (such as epi-illumination in a microscope), the fluorescence from surface-associated molecules may be dwarfed by the fluorescence from nonassociated molecules in the adjacent detection volume. As an optical technique designed to overcome this problem, total internal reflection fluorescence (TIRF) allows selective excitation of just those fluorescent molecules in close (~ 100 nm) proximity to the surface. TIRF can be used quantitatively to measure concentrations of fluorophores as a function of distance from the substrate or to measure binding/unbinding equilibria and kinetic rates at a biological surface. As applied to biological cell cultures, TIRF allows selective visualization of cell/substrate contact regions, even in samples in which fluorescence elsewhere would otherwise obscure the fluorescent pattern in contact regions. TIRF can be used qualitatively to observe the position, extent, composition, and motion of these contact regions.

Daniel Axelrod, Edward H. Hellen, and Robert M. Fulbright • Department of Physics and Biophysics Research Division, University of Michigan, Ann Arbor, Michigan 48109.

Topics in Fluorescence Spectroscopy, Volume 3: Biochemical Applications, edited by Joseph R. Lakowicz. Plenum Press, New York, 1992.

TIRF is conceptually simple. An excitation light beam traveling in a solid (e.g., a glass coverslip or tissue culture plastic) is incident at a high angle θ upon the solid/liquid interface to which the cells adhere. That angle θ, measured from the normal, must be large enough for the beam to totally internally reflect rather than refract through the interface, a condition that occurs above some "critical angle." TIR generates a very thin (generally less than 200 nm) electromagnetic field in the liquid with the same frequency as the incident light, exponentially decaying in intensity with distance from the surface. The field is called the "evanescent wave" and is capable of exciting fluorophores near the surface while avoiding excitation of a possibly much larger number of fluorophores farther out in the liquid. In Section 7.2, the electromagnetic field which excites TIR fluorescence is discussed.

As an excitation system, TIRF does not specifically refer to the pattern, intensity, or lifetime of the fluorescence emitted from the near-surface molecules which become excited. However, these emission characteristics are somewhat different from those far from a surface, and some of these differences may become experimentally useful. In Section 7.3, the emission pattern of a fluorophore near a dielectric surface (particularly the interface of water with either bare glass or metal-coated glass) is discussed.

TIRF is easy to set up on a conventional upright or inverted microscope with a laser light source or, in a special configuration, with a conventional arc source. TIRF is completely compatible with standard epi-fluorescence, bright-field, dark-field, or phase contrast illumination so that these methods of illumination can be switched back and forth readily. Some practical optical arrangements for observing TIRF through a microscope are described in Section 7.4.

As a technique for selective surface illumination at liquid/solid interfaces, TIRF was first introduced by Hirschfeld[1] in 1965. Other important early applications were pioneered by Harrick and Loeb[2] in 1973 for detecting fluorescence from a surface coated with dansyl-labeled bovine serum allbumin, by Kronick and Little[3] in 1975 for measuring the equilibrium constant between soluble fluorescent-labeled antibodies and surface-immobilized antigens, and by Watkins and Robertson[4] in 1977 for measuring kinetics of protein adsorption following a concentration jump. Previous reviews[5–7] contain additional references to some important early work. Section 7.5 presents a literature review of recent work.

7.2. Theory of TIR Excitation

7.2.1. Single Interface

When a light beam propagating through a transparent medium 3 of high index of refraction (e.g., glass) encounters an interface with medium 1 of lower index of refraction (e.g., water), it undergoes total internal reflection for incidence angles (measured from the normal to the interface) greater than the "critical angle." The critical angle θ_c for TIR is given by

$$\theta_c = \sin^{-1}(n_1/n_3) = \sin^{-1} n \tag{7.1}$$

where n_1 and n_3 are the refractive indices of the liquid and the solid, respectively, and $n = n_1/n_3$ where $n < 1$ for TIR to occur. For incidence angle $\theta < \theta_c$, much of the light propagates through the interface with a refraction angle (also measured from the normal) given by Snell's law. (Some of the incident light internally reflects back into the solid.) For $\theta > \theta_c$, all of the light reflects back into the solid. However, even with TIR, some of the incident energy penetrates through the interface and propagates parallel to the surface in the plane of incidence. The field in the liquid, called the "evanescent field" (or "wave"), is capable of exciting fluorescent molecules that might be present near the surface.

For a finite-width beam, the evanescent wave can be pictured as the beam's partial emergence from the solid into the liquid, travel for some finite distance along the surface, and then reentrance into the solid. The distance of propagation along the surface is measurable for a finite-width beam and is called the Goos–Hanchen shift.

For an infinitely wide beam (i.e., a beam width many times the wavelength of the light, which is a very good approximation for our purposes), the intensity of the evanescent wave (measured in units of energy per unit area per second) exponentially decays with perpendicular distance z from the interface:

$$I(z) = I(0)e^{-z/d} \tag{7.2}$$

where

$$d = \frac{\lambda_0}{4\pi}(n_3^2 \sin^2\theta - n_1^2)^{-1/2} \tag{7.3}$$

with λ_0 the wavelength of the incident light in vacuum. Depth d is independent of the polarization of the incident light and decreases with increasing θ. Except for $\theta \to \theta_c$ (where $d \to \infty$), d is on the order of λ_0 or smaller.

A physical picture of refraction at an interface shows TIR to be part of a continuum, rather than a sudden new phenomenon appearing at $\theta = \theta_c$. For small θ, the light waves in the liquid are sinusoidal, with a certain characteristic period noted as one moves normally away from the surface. As θ approaches θ_c, that period becomes longer as the refracted rays propagate increasingly parallel to the surface. At exactly $\theta = \theta_c$, that period is infinite, as the wave fronts of the refracted light are normal to the surface. This situation corresponds to $d=\infty$. As θ increases beyond θ_c, the period becomes mathematically imaginary; physically, this corresponds to the exponential decay of Eq. (7.2).

The factor $I(0)$ in Eq. (7.2) is a function of θ and the polarization of the incident light; these features are discussed shortly. However, we first examine the remarkable amplitude, polarization, and phase behaviors of the electric fields [from which $I(0)$ is derived] and the magnetic fields of the TIR evanescent wave. The field components are listed below, with incident electric field amplitudes $A_{p,s}$ and phase factors relative to those of the incident **E** field's phase at $z=0$. (The coordinate system is chosen such that the x–z plane is the plane of incidence. Incident polarizations p and s are parallel and perpendicular to the plane of incidence, respectively.)

$$E_x=\frac{(2\cos\theta)(\sin^2\theta-n^2)^{1/2}}{(n^4\cos^2\theta+\sin^2\theta-n^2)}A_p e^{-i(\delta_p+\pi/2)} \tag{7.4}$$

$$E_y=\frac{2\cos\theta}{(1-n^2)^{1/2}}A_s e^{-i\delta_s} \tag{7.5}$$

$$E_z=\frac{2\cos\theta\sin\theta}{(n^4\cos^2\theta+\sin^2\theta-n^2)^{1/2}}A_p e^{-i(\varphi_p+\pi/2)} \tag{7.6}$$

where

$$\delta_p\equiv\tan^{-1}\left[\frac{(\sin^2\theta-n^2)^{1/2}}{n^2\cos\theta}\right] \tag{7.7}$$

$$\delta_s\equiv\tan^{-1}\left[\frac{(\sin^2\theta-n^2)^{1/2}}{\cos\theta}\right] \tag{7.8}$$

Fields polarized parallel (p) or perpendicular (s) to the plane of incidence are given by

$$\mathbf{E}_p=E_x\hat{x}+E_z\hat{z}\qquad \mathbf{E}_s=E_y\hat{y} \tag{7.9}$$

where $\hat{x}$, $\hat{y}$, and $\hat{z}$ are unit vectors. For p-polarized fields,

$$\mathbf{E}_p=E_p(0)\exp(-z/2d)(\alpha_x\hat{x}+\alpha_z\hat{z}) \tag{7.10}$$

where

$$E_p(0)=A_p\frac{(2\cos\theta)(2\sin^2\theta-n^2)^{1/2}}{n^2\cos\theta+i(\sin^2\theta-n^2)^{1/2}} \tag{7.11}$$

and

$$\alpha_x=-i\left(\frac{\sin^2\theta-n^2}{2\sin^2\theta-n^2}\right)^{1/2}\qquad \alpha_z=\frac{\sin\theta}{(2\sin^2\theta-n^2)^{1/2}} \tag{7.12}$$

For s-polarized fields,

$$\mathbf{E}_s = E_s(0)\exp(-z/2d)\,\hat{y} \tag{7.13}$$

where

$$E_s(0) = A_s \frac{2\cos\theta}{\cos\theta + i(\sin^2\theta - n^2)^{1/2}} \tag{7.14}$$

Note that the evanescent field is purely transverse to the propagation direction $(+x)$ only for the s polarization. The p-polarized field $\mathbf{E}_p$ "cartwheels" along the surface with a spatial period of $\lambda_0/(n_3 \sin\theta)$ as shown schematically in Figure 7.1. The nonzero longitudinal component E_x distinguishes an evanescent field from freely propagating subcritical refracted light, which has no longitudinal component. As one might expect, E_x approaches zero as the incidence angle is reduced from the supercritical range back toward the critical angle.

For a finite-width incident beam, the incidence angle dependence of the phase factors δ_p and δ_s gives rise to the measurable longitudinal shift of the beam, known as the Goos–Hanchen shift. This shift ranges from a fraction of a wavelength at $\theta = 90°$ to infinite at $\theta = \theta_c$, which of course corresponds to the refracted beam skimming along the interface. A finite incidence beam can be expressed as a weighted integral over infinite plane waves approaching at a range of incidence angles; each plane wave at each angle gives rise to its own exponentially decaying evanescent field of infinite lateral extent. The x–y intensity profile of the evanescent field for the finite beam can then be calculated by the weighted integral of these plane-wave-generated evanescent fields over the range of incident plane-wave angles. For a TIR Gaussian laser beam focused with a typically narrow angle of convergence, the evanescent

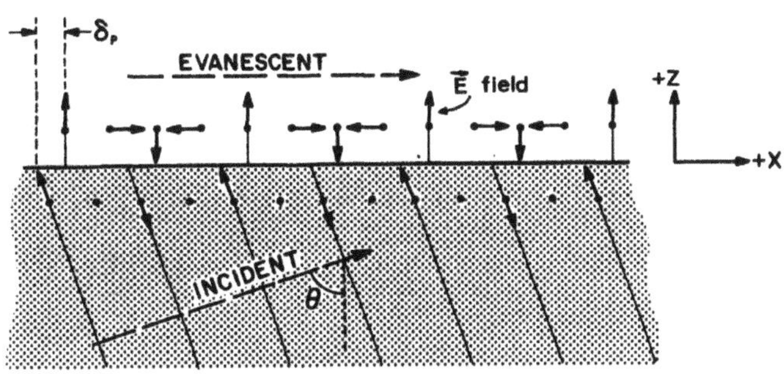

Figure 7.1. Electric field vectors of incident and evanescent light for the p-polarization, showing the phase lag δ_p and the elliptical polarization of the evanescent field in the plane of propagation. Both the incident and evanescent field vectors, shown here below and above the interface for pictorial clarity, refer to the $z = 0$ position.

illumination is approximately an elliptical Gaussian profile, and the polarization and penetration depth are approximately equal to those of a single infinite plane wave.[8]

For absorbers with magnetic dipole transitions, the evanescent magnetic field **H** leads to absorption of electromagnetic energy. Assuming equal magnetic permeabilities at both sides of the interface, the components of the evanescent field **H** at $z=0$ are

$$H_x = \frac{(2\cos\theta)(\sin^2\theta - n^2)^{1/2}}{(1-n^2)^{1/2}} A_s e^{-i(\delta_s - \pi)} \tag{7.15}$$

$$H_y = \frac{2n^2\cos\theta}{(n^4\cos^2\theta + \sin^2\theta - n^2)^{1/2}} A_p e^{-i(\delta_p - \pi/2)} \tag{7.16}$$

$$H_z = \frac{2\cos\theta\sin\theta}{(1-n^2)^{1/2}} A_s e^{-i\delta_s} \tag{7.17}$$

The average energy flux in the evanescent wave is given by the real part of the Poynting vector $\mathbf{S} = (c/4\pi)\mathbf{E}\times\mathbf{H}^*$. However, the probability of absorption of energy per unit time from the evanescent wave by an electric dipole-allowed transition of moment $\boldsymbol{\mu}_a$ in a fluorophore is proportional to $|\boldsymbol{\mu}_a \cdot \mathbf{E}|^2$. Note that $\mathrm{Re}\,\mathbf{S}$ and $|\boldsymbol{\mu}_a \cdot \mathbf{E}|^2$ are not proportional to each other: they have a different dependence on θ.

Given randomly oriented dipoles, the absorption probability rate is proportional to the "intensity" $I_{p,s} \equiv |\mathbf{E}_{p,s}|^2$. At $z=0$, the evanescent intensities are

$$I_x(0) = |A_p|^2 \frac{(4\cos^2\theta)(\sin^2\theta - n^2)}{n^4\cos^2\theta + \sin^2\theta - n^2} \tag{7.18}$$

$$I_z(0) = |A_p|^2 \frac{4\cos^2\theta\sin^2\theta}{n^4\cos^2\theta + \sin^2\theta - n^2} \tag{7.19}$$

which gives, from $I_p = I_x + I_z$,

$$I_p(0) = |A_p|^2 \frac{(4\cos^2\theta)(2\sin^2\theta - n^2)}{n^4\cos^2\theta + \sin^2\theta - n^2} \tag{7.20}$$

Also,

$$I_s(0) = I_y(0) = |A_s|^2 \frac{4\cos^2\theta}{1-n^2} \tag{7.21}$$

Intensities $I_{p,s}(0)$ are plotted versus θ in Figure 7.2a, assuming the incident intensities in the glass, $|A_{s,p}|^2$, are set equal to unity. The plots can be

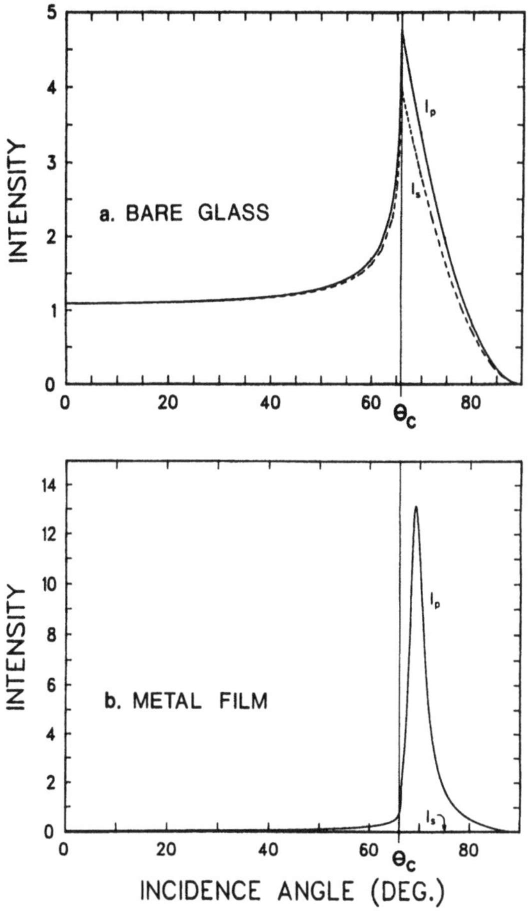

Figure 7.2. Intensity $I_p(0)$ (—) and $I_s(0)$ (- - -) versus incidence angle θ at (a) a bare fused-silica/water interface and (b) a 20-nm aluminum film-coated fused-silica/water interface. Intensity $I_s(0)$ for the aluminum film case is essentially zero. Refractive indices $n_3 = 1.46$ and $n_1 = 1.33$ are assumed, corresponding to $\theta_c = 65.7°$. Note the difference in ordinate scales (always set for $A_{p,s}^2 = 1$) between the bare glass and metal film case.

extended without breaks to the subcritical angle range (based on calculations with Fresnel coefficients), again illustrating the continuity of the transition to TIR. The evanescent intensity approaches zero as $\theta \to 90°$. On the other hand, for supercritical angles within ten degrees of θ_c, the evanescent intensity is as great as or greater than the incident light intensity.

7.2.2. Intermediate Layer

In actual experiments in biophysics, the interface may not be a simple interface between two media, but rather a stratified multilayer system. One example is the case of a biological membrane or lipid bilayer interposed between glass and aqueous media. Another example is a thin metal film coating, which quenches fluorescence within the first ~10 nm of the surface (see Section 7.3). We discuss first qualitatively, then quantitatively, the TIR evanescent wave in a three-layer system in which incident light travels from medium 3 (refractive index n_3) through the intermediate layer (n_2) toward medium 1 (n_1).

Qualitatively, several features can be noted:

1. Insertion of an intermediate layer never thwarts TIR, regardless of the intermediate layer's refractive index, n_2. The only question is whether TIR takes place at the $n_3{:}n_2$ interface or the $n_2{:}n_1$ interface. Since the intermediate layer is likely to be very thin (no deeper than several tens of nanometers) in many applications, precisely which interface supports TIR is not important for qualitative studies.
2. Regardless of n_2 and the thickness of the intermediate layer, the evanescent wave's profile in medium 1 will be exponential with a characteristic decay distance given by Eq. (7.3). However, the overall distance of penetration of the field measured from the surface of medium 3 is affected by the intermediate layer.
3. Irregularities in the intermediate layer can cause scattering of incident light, which then propagates in all directions in medium 1. This subject has been treated theoretically.[9] Experimentally, scattering appears not to be a problem on samples even as inhomogeneous as biological cells. Direct viewing of incident light scattered by a cell surface lying between the glass substrate and an aqueous medium confirms that scattering is many orders of magnitude dimmer than the incident or evanescent intensity and should thereby excite a correspondingly dim contribution to the fluorescence.

To handle the three-layer problem quantitatively, we write the $z=0$ electric field components in terms of complex three-layer Fresnel coefficients:

$$E_x = -A_p \alpha_{13} T_p \tag{7.22}$$

$$E_y = A_s T_s \tag{7.23}$$

$$E_z = A_p (\varepsilon_3/\varepsilon_1)^{1/2} \sin\theta T_p \tag{7.24}$$

Parameters ε_i ($\equiv n_i^2$) are the dielectric constants of the respective media (which may be complex for light-absorbing materials). Parameters $T_{p,s}$ are the Fresnel coefficients for transmission through a stratified three-medium system with the beam incident from the medium 3 side and an intermediate medium 2 of thickness δ[10]:

$$T_{p,s} = \frac{t_{32}^{p,s} t_{21}^{p,s} e^{ik_1 \delta \alpha_{23}}}{1 + r_{32}^{p,s} r_{21}^{p,s} e^{2ik_1 \delta \alpha_{23}}} \tag{7.25}$$

where the t_{ij} and r_{ij} are the transmission and reflection coefficients for p- and s-polarized light for a single interface. They are listed here for convenience:

$$r_{ij}^p = \frac{\varepsilon_j \alpha_{i3} - \varepsilon_i \alpha_{j3}}{\varepsilon_j \alpha_{i3} + \varepsilon_i \alpha_{j3}} \qquad r_{ij}^s = \frac{\alpha_{i3} - \alpha_{j3}}{\alpha_{i3} + \alpha_{j3}} \tag{7.26}$$

$$t_{ij}^p = \frac{2(\varepsilon_i \varepsilon_j)^{1/2}\, \alpha_{i3}}{\varepsilon_j \alpha_{i3} + \varepsilon_i \alpha_{j3}} \qquad t_{ij}^s = \frac{2\alpha_{i3}}{\alpha_{i3} + \alpha_{j3}} \tag{7.27}$$

where

$$\alpha_{ij} \equiv \left(\frac{\varepsilon_i}{\varepsilon_1} - \frac{\varepsilon_j}{\varepsilon_1} \sin^2 \theta \right)^{1/2} \tag{7.28}$$

and $k_1 \equiv (\omega/c)\varepsilon_1^{1/2}$, where ω is the excitation light angular frequency. Note that α_{ij} and $T_{p,s}$ can be complex. As expected, Eqs. (7.22)–(7.24) reduce to Eqs. (7.4)–(7.6) for $\varepsilon_2 = \varepsilon_3$. [Note that Eq. (7.26) corrects misprints in Eqs. (18) and (20) in Reference 6.]

The three-layered interface gives rise to evanescent intensities as follows:

$$I_p = |A_p|^2\, |T_p|^2 \left(2\frac{\varepsilon_3}{\varepsilon_1} \sin^2 \theta - 1 \right) e^{-z/d} \tag{7.29}$$

$$I_s = |A_s|^2\, |T_s|^2\, e^{-z/d} \tag{7.30}$$

I_p and I_s at $z=0$ are affected by n_2. If the intermediate layer is a thin (~20 nm) film of metal, the effect is dramatic (Figure 7.2b). A metal has a dielectric constant consisting of a negative real part and a positive imaginary part (for aluminum, $\varepsilon_2 = -32.5 + 8.4i$ at $\lambda_0 = 520$ nm). The s-polarized evanescent intensity $I_s(0)$ becomes negligibly small. However, the p-polarized behavior is quite interesting. At a certain angle of incidence θ_p, the denominator of t_{21}^p becomes quite small (due to the oppositely signed real parts of ε_2 and ε_1). At that incidence angle, the p-polarized evanescent intensity becomes an order of magnitude brighter than the incident light at the peak. This resonance-like effect is due to excitation of a surface plasmon mode at the metal/water interface. The peak is at the "surface plasmon angle," due to a resonant excitation of electron oscillations at the metal/water interface.[11–13] For an aluminum film at a glass/water interface, θ_p is greater than the critical angle θ_c for TIR. The intensity enhancement is rather remarkable since a 20-nm-thick metal film is almost opaque to the eye.

There are some potentially useful experimental consequences of TIR excitation through a thin metal film coated on glass. As discussed in Section 5.3, fluorescence from molecules less than 10 nm from the metal is strongly quenched. However, TIR can still be used to selectively excite fluorophores in the 10- to 200-nm distance range from metal-film-coated glass. Also, a light beam incident upon a 20-nm-thick Al film from the glass side at a glass/aluminum film/water interface evidently does not have to be collimated to produce TIR. Those rays that are incident at the surface plasmon angle will create a strong evanescent wave; those rays that are too low or high in incidence angle will create a negligible field in the water. This phenomenon may ease the practical requirement for a collimated incident beam in TIR. Lastly, the metal film leads to a highly polarized evanescent wave (provided $A_p \neq 0$), regardless of the purity of the incident polarization.

7.3. Emission by Fluorophores near a Surface

Although the probability of absorption of TIR evanescent energy by a fluorophore of given orientation decreases exponentially with distance z from a dielectric surface, the intensity of the fluorescence actually viewed by a detector varies with z in a much more complicated fashion. Both the angular pattern of the emitted radiation and the fluorescent lifetime are altered as a function of z by the proximity of the surface.

These effects are not limited to fluorophores excited by TIR, although TIR excitation is necessarily near a surface. The discussion in this section is of relevance to any mode of excitation of surface-proximal fluorescence. In many of the experiments involving fluorescence in cell biology, the fluorophores are located near a surface. Usually, this surface is an aqueous buffer/glass or plastic interface upon which cells grow. Occasionally, the interface may have a thin coating on it, such as a synthetic polymer, a metal, or a lipid bilayer.

Various aspects of fluorophore emission at surfaces have been investigated, particularly within the past two decades. For nondissipative surfaces (e.g., bare glass), the lifetime[14] and the inversely related total radiated power[15] for a single emission dipole, modeled as a continuous classical oscillator, have been calculated as functions of orientation and distance from the surface. The radiated intensity emitted from a continuous dipole oscillator has been calculated as a function of observation angle, dipole orientation, and distance.[16–21]

For metal surfaces, most theoretical attention has been devoted to the dramatic decrease in lifetime and concomitant decrease in total radiated power near metal blocks, films, or microscopic islands.[12, 22–26] Several optical phenomena occur, each being important at a particular range of distances of the fluorophore from the surface. At distances comparable with or longer than visible wavelengths ($z > 500$ nm), the interference between the propagating (far-field) emitted light and its reflection dominates.[24, 25] At intermediate ranges ($10\text{ nm} < z < 500$ nm), the evanescent near field of the dipole transfers some energy into propagating surface plasmons and into heat by way of the local resistivity of the metal (i.e., electron scattering).[12, 13, 22, 27–29] At very close ranges ($z < 10$ nm), energy transfer into electron–hole pairs may become significant,[30, 31] and at atomic scale distances ($z < 0.25$ nm), non-homogeneous local field effects become important.[32]

We present here a condensed explanation and summary of the effects. A complete discussion can be found in a paper by Hellen and Axelrod[33] which directly calculates the amount of emission light gathered by a finite-aperture objective from a surface-proximal fluorophore under steady illumination. The effects referred to here are not "quantum-chemical," that is, effects upon the orbitals or states of the fluorophore in the presence of any static fields associated with the surface. Rather, the effects are "classical-optical," that is, effects upon the electromagnetic field generated by a classical oscillating dipole in the presence of an interface between any media with dissimilar refractive indices. Of course, both types of effects may be present simultaneously in a given system. However, the quantum-chemical effects vary with the detailed chemistry of each system, whereas the classical-optical effects are more universal. Occasionally, a change in the emission properties of a fluorophore at a surface may be attributed to the former when in fact the latter are responsible.

This chapter deals only with the classical-optical effects. It emphasizes the emission properties as they might be observed through a microscope, with particular attention to the bare-glass/water and metal-film-coated glass/water interfaces. The results suggest some practical experiments that take advantage of the special optical effects at surfaces. These experiments include deducing the relative concentration of fluorophore as a function of distance from the surface, quenching unwanted "background" fluorescence from fluorophores nonspecifically adsorbed to a substrate, and optimizing collection of fluorescence by a microscope objective.

7.3.1. Description of the Model

Surface optical effects can be calculated at various levels of approximation. The simplest (and least accurate) approach is to model the fluorophore as an oscillating electric dipole of fixed amplitude generating only rays of propagating light (the "far field"). The rays (which are actually symbols for propagating plane waves) interact with the surface according to Snell's law and the law of reflection. Their uniformly spaced wave fronts (within each uniform refractive index medium) extend as semi-infinite planes. This approach considers the interference between such rays of propagating light directly emitted from the dipole and light rays reflecting off the interface. The results are valid only for distances z from the surface of greater than the wavelength of the light (~ 500 nm).

A better approximation must consider the so-called "near field." The mathematical form of a dipole radiation pattern cannot be expressed simply as a superposition of plane waves/rays propagating in different directions with direction-dependent amplitudes. Rather, it is necessary to suppose that some of the wave fronts do not extend infinitely far from the dipole but instead exponentially decay. A whole set of such exponentially decaying fields exists with a continuous range of decay constants. When a fluorophore (say, in water) is near a higher refractive index surface (say, glass), each of these exponentially decaying near-field components can interact with the surface and ultimately some become propagating waves in the glass at their own unique angles θ (with respect to the normal). Angle θ is always greater than the critical angle θ_c for total internal reflection. This conversion of exponentially decaying waves from the near field of the dipole in water into supercritical angle propagating waves in the glass can be significant for fluorophores within about one wavelength of the surface. At a metal surface, consideration of the near field is even more important, because the metal converts the electromagnetic energy into heat.

Another feature of the simplest model that needs modification is the assumption of a fixed dipole amplitude. Because of the efficient capture of nonpropagating near fields by a surface, a fixed-amplitude dipole emits more power, the closer it moves to a surface. However, in steady-state fluorescence, the emitted power can only be as large as the (constant) absorbed power (or less, if the intrinsic quantum yield of the isolated fluorophore is less than 100%). Therefore, the fluorophore must be modeled as a constant-*power* (and variable-*amplitude*) dipole. Many of the earlier theoretical references listed above deal only with constant-amplitude dipoles, so their results must be considered to be an approximation.

The two above features which modify the simplest theory extend the range of distances z between the fluorophore and the surface over which the results remain valid, from a minimum of several hundred nanometers without the modifications to less than ten nanometers with them. Those two features are incorporated into the results displayed here. Other refinements, not included here, involve consideration of energy transfer to electron–hole pairs (for metals only at $z < 10$ nm) and nonhomogeneous atomic field effects ($z < 0.25$ nm). We first assume that the intrinsic quantum yield is 100%; then we will modify that assumption.

7.3.2. Mathematical and Physical Basis

The model here consists of a medium 3 (refractive index n_3; e.g., glass), a sandwiched layer of thickness t (n_2; e.g., polymer, metal, lipid, or more of medium 3), and a medium 1 (n_1; e.g., water). The $z = 0$ origin is at the $n_1 : n_2$ interface, and the dipole resides in medium 1 at $z \geqslant 0$.

In general, an electric field $\mathbf{E}'(\mathbf{r})$ emitted from an isolated, fixed-amplitude dipole (i.e., no surfaces nearby) can be expanded as an integral over plane waves (with sinusoidal time dependence suppressed) as follows:

$$\mathbf{E}'(\mathbf{r}) = \int d\mathbf{k} \exp(i\mathbf{k} \cdot \mathbf{r})\, \mathbf{E}'(\mathbf{k}) \tag{7.31}$$

where wave vector $\mathbf{k} = k_x\hat{x} + k_y\hat{y} + k_z\hat{z}$, and the integration extends over all $\mathbf{k}$, subject only to the restriction that the frequency of the light, ω, remain fixed. Emission field $\mathbf{E}'$ is primed to distinguish it from the excitation field $\mathbf{E}$, which is unprimed and discussed in Section 7.2. Vector $\mathbf{r}$ extends from the origin at the $n_1 : n_2$ interface to the point at which $\mathbf{E}'$ is observed. $\mathbf{E}'(\mathbf{k})$ can be determined directly via Maxwell's equations, or from the known form of dipole radiation $\mathbf{E}'(\mathbf{r})$ via Eq. (7.31). Since the magnitude of $\mathbf{k}$ is

$$|\mathbf{k}| = (k_x^2 + k_y^2 + k_z^2)^{1/2} = \omega n_1/c \tag{7.32}$$

the integration in Eq. (7.31) requires only that $|\mathbf{k}|$ be held fixed. Propagating (sinusoidal) waves are described by k_x^2, k_y^2, and k_z^2, all positive. However, if we allow $(k_x^2 + k_y^2) > |\mathbf{k}|^2$, then $k_z^2 < 0$. Those "plane waves" have an imaginary k_z and correspond to exponentially decaying waves in either direction along the z-axis starting from the position of the dipole. This set of plane waves is the dipole's "near field." The amplitude and phase of each wave $\exp(i\mathbf{k} \cdot \mathbf{r})$, whether propagating or exponentially decaying, is given by $\mathbf{E}'(\mathbf{k})$.

The near-field wave fronts can be chosen to be parallel to the z-axis, exponentially decaying in amplitude in either direction along the z-axis starting from the dipole position and traveling radially outward parallel to the x–y plane. The apparent wavelength of each exponentially decaying wave is shorter than that of the propagating waves, corresponding to the large radial k-vector amplitude given by $k_r = (k_x^2 + k_y^2)^{1/2}$.

The electric field $\mathbf{E}_1'(\mathbf{r})$ observed at point $\mathbf{r}$ in medium 1 is then the superposition of the direct field [calculated from Eq. (7.31) for waves traveling away from the surface in the $+z$ direction] and plane reflected waves (calculated as described above), integrated over all allowed $\mathbf{k}$ vectors. The electric field $\mathbf{E}_3'(\mathbf{r})$ observed in medium 3 is an integral of the plane refracted waves, also integrated over all allowed $\mathbf{k}$. Note that Snell's law demands that the spacing between successive wave fronts as projected on a dielectric boundary must be the same on both sides of the boundary. This requirement is the same as the statement that k_r must be continuous across the boundary. Then the exponentially decaying near-field waves, with their short wave front spacing, will refract only into supercritical angles into medium 3.

For fluorophores close to the surface, exponentially decaying waves with a wide range of decay constants ik_z will extend to the surface, giving rise to a wide band of supercritical refraction angles. However, for fluorophores somewhat farther away, only those exponentially decaying waves with the longest decay distances (i.e., the smallest ik_z) will reach the surface, giving rise to a rather narrow band of supercritical emission angles extending only slightly above the critical angle. Therefore, by viewing only supercritical angle emission at a fixed angle θ, one will detect only fluorophores near the surface. The higher the θ, the closer to the surface are the detected fluorophores.

One *might* conclude that (1) at any particular supercritical observation angle θ in the glass, the observed intensity will decrease exponentially with z as a fluorophore is pulled away from the surface; and (2) the total emission into all other angular ranges is unaffected by moving the fluorophore. However, neither of these conclusions is correct. Recall that a fluorophore must be modeled as a fixed-power, rather than fixed-amplitude, dipole. This means that any increase in emitted power into any one set of directions, for example, supercritical angles into the glass, will be at the expense of emitted power into other directions. Furthermore, the intensity emitted into medium 1 is determined partly by the phase-dependent interference between direct and reflected plane waves, which is also a function of z. The intensity at any angle must be normalized by a function describing the total power $P_T(z)$ released by the dipole (including any lost into heat in a dissipative medium 2). In general, $P_T(z)$ can be calculated from

$$P_T(z) = (\omega/2)\,\mathrm{Im}(\boldsymbol{\mu} \cdot \mathbf{E}') \tag{7.33}$$

Physically, this formula describes the power dissipated by a harmonic oscillator (the emission dipole with moment $\boldsymbol{\mu}$) as it is driven by the force felt at its own location from its own emitted and reflected electric field. P_T is calculable given all the refractive indices and Fresnel coefficients of the layered model.[12, 33]

Incorrect conclusion 1 above is sometimes said to derive from the "reciprocity principle," which states that light waves in any optical system all could be reversed in direction without altering any paths or intensities and remain consistent with physical reality (because Maxwell's equations are invariant under time reversal). Applying this principle here, one notes that an evanescent wave set up by a supercritical ray undergoing total internal reflection can excite a dipole with a power that decays exponentially with z. Then (by the reciprocity principle) an excited dipole should lead to a supercritical emitted beam intensity that also decays exponentially with z. Although this prediction would be true *if* the fluorophore were a fixed-amplitude dipole in both cases, it cannot be modeled as such in the latter case.

The radiated intensity $\hat{S}(\mathbf{r}, z)$ from a fluorophore near a surface per unit of adsorbed power can be derived from the Poynting vector magnitude. In terms of $\mathbf{E}_1'(\mathbf{r})$ and $\mathbf{E}_3'(\mathbf{r})$, it is

$$\hat{S}_i(\mathbf{r}, z) = \frac{cn_i\, |\mathbf{E}_i'(\mathbf{r}, z)|^2}{8\pi P_T(z)} \tag{7.34}$$

where c is the speed of light in vacuum. If we multiply $\hat{S}$ by the input power to the fluorophore's absorption dipole, then (assuming a 100% quantum yield) we get the intensity $I'(\mathbf{r}, z)$ radiated from the fluorophore:

$$I_i'(\mathbf{r}, z) = |\boldsymbol{\mu}_a \cdot \mathbf{E}|^2\, \hat{S}_i(\mathbf{r}, z) \tag{7.35}$$

Two additional feature can be incorporated into Eqs. (7.32)–(7.35): the dipole orientation distribution and the concentration distribution in systems consisting of many dipoles. The orientation of the dipole with respect to the surface, described by angles $\Omega' \equiv (\theta', \phi')$, affects $\mathbf{E}_i'$ and all the other measurables derived from it.[33] Consider a concentration distribution of dipoles in both orientation and distance from the surface specified by $C(\theta', \phi', z)$. Since the dipoles all oscillate incoherently with respect to one another, the integrated intensity $\mathscr{I}$ due to this distribution is simply:

$$\mathscr{I}_i(\mathbf{r}) = K \int d\Omega'\, dz\, C(\Omega', z)\, I_i'(\mathbf{r}, \Omega', z) \tag{7.36}$$

The total fluorescence power $\mathscr{F}$ collected from the fluorophore distribution by a microscope objective centered in the normal line at a distance r is an integral of $\mathscr{I}$ over the objective's aperture which subtends a solid angle Ω:

$$\mathscr{F}_i(r) = \int r^2\, d\Omega\, \mathscr{I}_i(\mathbf{r}) \tag{7.37}$$

Returning to the case of a single dipole, we find another parameter to be useful: the fluorescence collection efficiency Q. Parameter Q is the fraction of the total energy dissipated by a fixed-power dipole that is collected by a microscope objective centered on the normal at distance r:

$$Q(z, \theta') \equiv \frac{\text{collected power}}{\text{total dissipated power}} = \int r^2\, d\Omega\, \hat{S}_i(\mathbf{r}, z, \theta') \tag{7.38}$$

Combining the above equations, we can write a useful expression for the collected fluorescence from a distribution of dipoles in terms of the collection efficiency Q for a single dipole:

$$\mathscr{F} = K \int dz\, d\Omega'\, C(z, \phi', \theta')\, |\boldsymbol{\mu}_a \cdot \mathbf{E}|^2\, Q(z, \theta') \tag{7.39}$$

As shown by Hellen and Axelrod,[33] $Q(z, \theta')$ can be written in terms of the Q fractions for dipoles which are perpendicular $(\perp, \theta'=0)$ and parallel $(\|, \theta'=\pi/2)$ to the surface:

$$Q(z, \theta') = \left(\frac{Q^{\perp}(z)}{1+\eta(z)\tan^2\theta'} + \frac{Q^{\|}(z)}{1+[\eta(z)\tan^2\theta']^{-1}} \right) \tag{7.40}$$

where $\eta(z) \equiv P_T^{\|}/P_T^{\perp}$ is the ratio of the total power dissipated by fixed-amplitude dipoles oriented parallel ($\|$) to the interface to that dissipated by those oriented perpendicular ($\perp$) to the interface. Equation (7.40) shows that at each dipole distance the collected energy can be written as a weighted average of the collection efficiencies for perpendicular- and parallel-oriented dipoles.

7.3.3. Graphical Results

Rather than displaying the rather complicated explicit forms for P_T, η, Q, $\hat{S}$, $\mathscr{I}$, and $\mathscr{F}$ in general terms of η_i, t, Ω', Ω, and C, we show here graphical results for certain specific configurations based on numerical integration. The qualitative features of these results will be relevant for most other configurations. We specialize in two particular interfaces: bare glass/water, where the intermediate layer is just an extension of medium 3 (i.e., $n_2 = n_3$); and glass/aluminum film (22 nm thick)/water. We will assume that all the dipoles are oriented either parallel or perpendicular to the interface; this assumption will be extended to a random orientation distribution later.

Figure 7.3 shows the radiated intensity $\hat{S}$ as a function of the observation angle θ for a dipole 80 nm from the surface. For simplicity, the azimuthal angle of observation ϕ is averaged. (This is equivalent to assuming that the excited dipole distribution is azimuthally symmetric about the surface normal.)

In the bare glass case, note that a rather strong peak of intensity is drawn into the glass, maximal at exactly $\theta = \theta_c$, but with significant intensity into supercritical angles. The effect is especially pronounced for dipoles oriented perpendicular to the surface, but is present for any position or orientation distribution.

In the aluminum film case, a peak of intensity directed into the glass is again present, but here centered in an extremely narrow band at some $\theta = \theta_p > \theta_c$. Angle θ_p is called the "surface plasmon" angle and arises from near-field waves from the dipole whose radial k-vector magnitude k_r is exactly matched to resonant electronic vibrations which can propagate on the metal surface and then reemit light into the hollow cone pattern depicted.[13] Note that dipoles perpendicular to the metal surface can furnish energy into the

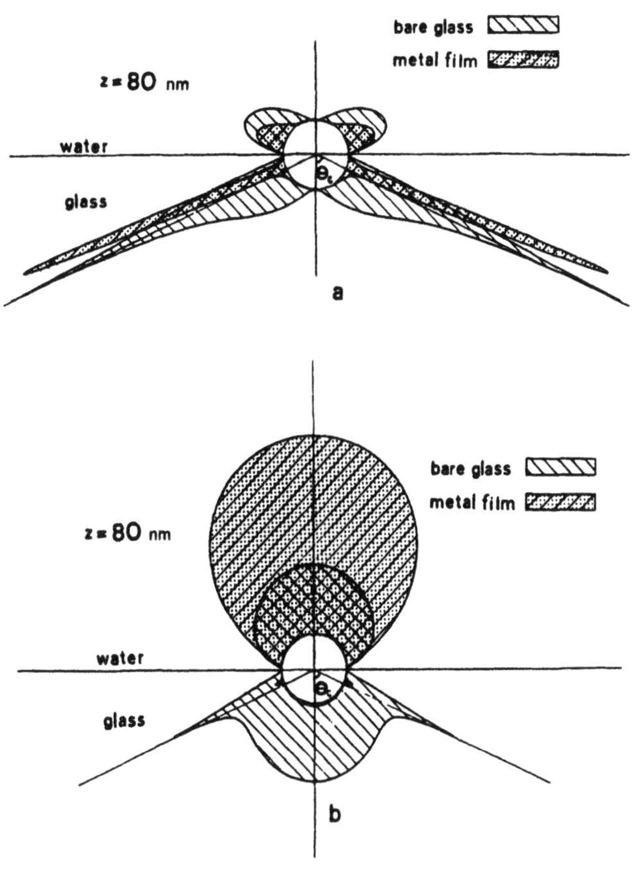

Figure 7.3. Intensity $\langle \hat{S} \rangle^{\parallel,\perp}$ versus polar observation angle θ for dipoles oriented perpendicular (a) and parallel (b) to the surface, normalized for constant total power of emission and averaged over the azimuthal angle of observation. The dipoles are assumed to be at $z = 80$ nm. At each angle, the radial distance from the center circle (surrounding the dipoles' location) is proportional to the radiated power. The bare glass case is a fused-silica ("glass")/water interface; the metal film case is a fused-silica/Al film (22 nm thick)/water interface. Parameters assumed are $\varepsilon_1 = 1.77$ ($n_1 = 1.33$); $\varepsilon_2 = -32.5 + 8.4i$; $\varepsilon_3 = 2.13$ ($n_3 = 1.46$); $\lambda_0 = 520$ nm. Note that at $z = 80$ nm from a metal film, the dipole is well out of range of surface quenching, and constructive interference with the reflected light actually increases the parallel dipole emission intensity into the water.

surface plasmons quite effectively, leading to an apparent transmission of light through a virtually opaque metal film. However, dipoles parallel to the surface are very unsuccessful at coupling with surface plasmons, and almost all the radiated emission appears in the water.

To make any progress in calculating $\mathscr{F}$ from Eqs. (7.39) and (7.40), we must know the total powers $P_T^{\parallel,\perp}(z)$ and their ratio $\eta(z)$; these are shown in Figure 7.4. Note that for the bare glass case (Figure 7.4a), the power P_T

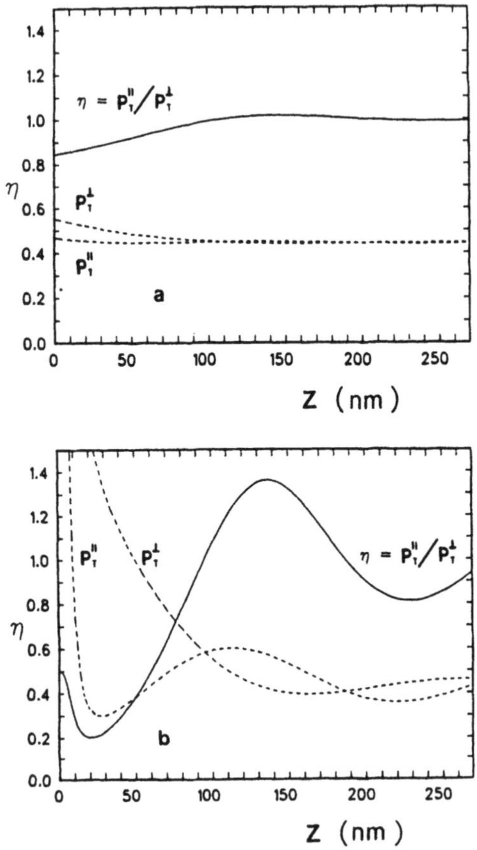

Figure 7.4. Total power P_T (in arbitrary units) dissipated by fixed-amplitude dipoles oriented parallel and perpendicular to the interface (---), and the ratio of these powers η (—) versus distance of the dipole z from the interface with medium 1, assumed to be water. (a) Bare glass; (b) aluminum film on glass. All optical parameters are the same as in Figure 7.3.

increases slightly for perpendicular dipoles very close to the surface. This corresponds to an approximately 10% decrease in the fluorescence lifetime of fluorophores. This effect should be taken into account when measuring fluorescence lifetimes near dielectric surfaces. For parallel dipoles, P_T exhibits only slight undulations.

Figure 7.4b shows P_T for the aluminum film case. There is a dramatic increase in P_T for both dipole orientations at small distances. Virtually all of that energy is converted into heat in the metal, thereby accounting for the strong fluorescence quenching on metal surfaces. For dipoles oriented parallel, an additional factor further promotes quenching: when the dipole is very near the surface ($z \ll \lambda$), its oppositely charged mirror image in the metal virtually cancels out the emitted electric field by simple wave interference.

The remaining variable required for calculation of $\mathscr{F}$ from Eqs. (7.39) and (7.40) is the collection efficiency Q, which measures the fraction of the total power emitted by a fluorophore that can be gathered as light by the microscope objective. Figure 7.5 shows $Q^{\|,\perp}$ for both parallel and per-

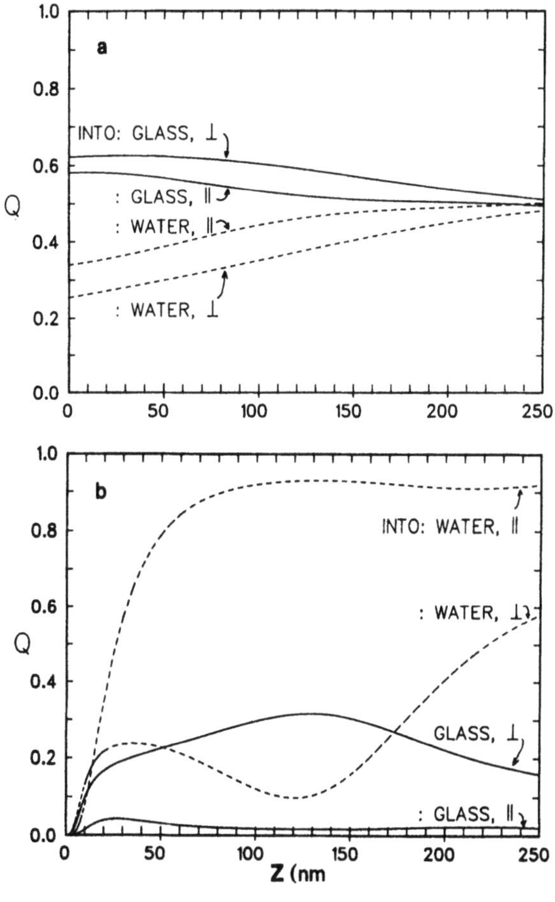

Figure 7.5. Collection efficiencies $Q^{\parallel,\perp}$ versus z, for an objective of numerical aperture 1.4 centered on the normal to the interface. Results are given both for the objective positioned "beneath" the interface to collect light emitted through the glass and "above" the interface to collect light emitted in the water. (a) Bare glass; (b) aluminum film on glass. All optical parameters are the same as in Figure 7.3.

pendicular dipole orientations, for an objective positioned to look either through the water or through the glass substrate.

Figure 7.5a shows the bare glass case. It shows that viewing through the glass substrate is more efficient than viewing through the water, at least for fluorophores very near the substrate surface. Around 60% of the emitted energy can be captured by a numerical aperture 1.4 objective by viewing through the glass substrate; only around 30% can be captured by viewing through the water. Much of this advantage is due to the ability of the high aperture to gather the emitted peak centered at $\theta=\theta_c$ (see Figure 7.3). For smaller aperture objectives, the relative advantage of viewing through the substrate diminishes. Clearly, a 1.4-aperture objective is much better than a 1.3-aperture objective because of its special ability to gather the $\theta = \theta_c$ peak.

Figure 7.5b shows the aluminum film case. In the example shown, the surface plasmon peak is gathered (but just barely) by the objective; if it were not, the collected emission into the glass would be much less. However, even with this high aperture, it is still more efficient to view the fluorophores through the water for most distances outside the strong quenching region of $z < 10$ nm. For large z distances, viewing through the water is very efficient. This is simply because the metal surface acts like a mirror for far-field propagating light emitted by the dipole.

Figure 7.6 shows the intensity I that would be detected in the glass at a particular supercritical angle, given an excitation intensity that is *not* a function of z (e.g., epi-illumination rather than TIR). Only the results for perpendicular dipoles are shown (so that averaging over azimuthal ϕ is unnecessary). The result for parallel dipoles is qualitatively similar except that the metal film case would be very much reduced in overall intensity.

In the bare glass case, note that the decay is not exponential, as otherwise would be expected if the "reciprocity principle" had been misapplied here. Nevertheless, by viewing only supercritical angles, one can selectively observe only those fluorophores within several hundred nanometers of the surface, even if the excitation (rather than the emission) is not surface-selective at all.

In the metal film case, the intensity is virtually zero for distances less than 5 nm. This quenching effect occurs at all angles, not just supercritical ones. The excitation energy is almost entirely converted into heat in the metal film. At larger distances, the dipole near field couples with surface plasmons whose emission into the glass is centered around $\theta = \theta_p$. At even larger distances, the near field is too weak to interact with the surface, and the supercritical intensity drops toward zero.

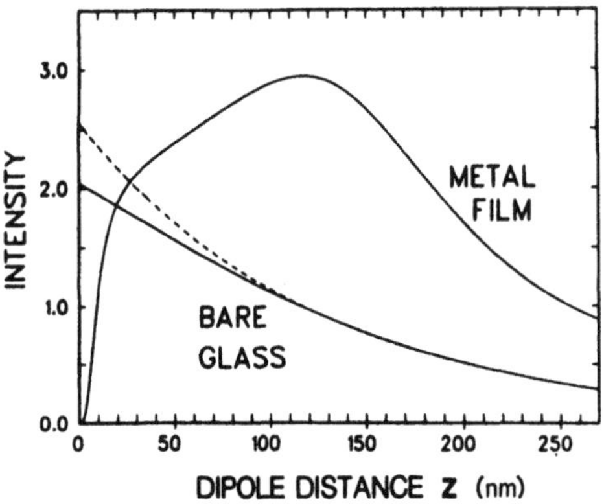

Figure 7.6. Intensity emitted at supercritical angle $\theta = 70°$ into the glass versus dipole distance z for dipoles oriented perpendicular to the interface. The solid lines show the properly normalized (fixed-power dipole) intensities for the bare glass and Al film surfaces. The dashed line shows the exponentially decaying intensity that would be obtained for bare glass by omitting the required normalization P_T in Eq. (7.34). All optical parameters are the same as in Figure 7.3.

7.3.4. Theoretical Results for a Distribution of Dipoles: Random Orientations

The collected fluorescence $\mathscr{F}$ [from Eq. (7.39)] clearly depends on the orientation distribution of the dipoles and the incident polarization through the dependences on θ' and **E**. We will assume a special but common case here: randomly oriented dipoles with a z-dependent concentration near the surface, excited by a p-polarized evanescent wave.

In this case, $C(z, \phi', \theta') = C(z)$. Equation (7.39) can then be written as an integral over the dipole distance z:

$$\mathscr{F} = K(\mu_a)^2 \int dz\, C(z)\, |E_p(z)|^2\, [w^{\perp}(z)\, Q^{\perp}(z) + w^{\parallel}(z)\, Q^{\parallel}(z)] \tag{7.41}$$

where the weighting terms $w^{\parallel,\perp}$ depend in general upon the excitation polarization. For the case here, in which the incident polarization is p and the absorption dipole $\boldsymbol{\mu}_a$ is parallel to the emission dipole $\boldsymbol{\mu}$, these weightings are

$$w^{\parallel,\perp}(z) = |\alpha_x|^2\, w_x^{\parallel,\perp}(z) + |\alpha_z|^2\, w_z^{\parallel,\perp}(z) \tag{7.42}$$

where the $\alpha_{x,z}$ factors determine the amplitude of the $\hat{x}$- and $\hat{z}$-components of the excitation field as given in Eq. (7.12). (The form of $\alpha_{x,z}$ shown in Eq. (7.12) corrects a misprint in Eq. 56 of Ref. 33.) The $w_{x,y,z}^{\parallel,\perp}(z)$ weighting factors, written out explicitly by Hellen and Axelrod[33] for the randomly oriented dipole case, are functions of $\eta(z)$ and are depicted in Figure 7.7.

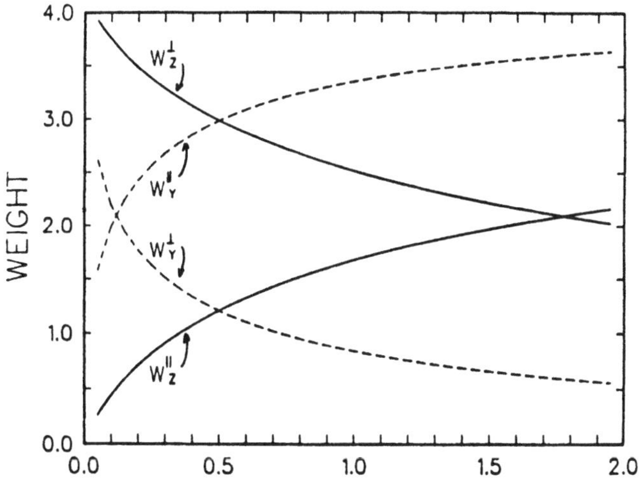

Figure 7.7. The weightings $w_{y,z}^{\parallel,\perp}$ versus $\eta(z)$ for the special case of p-polarized incident light and randomly oriented dipoles. In this case, $w_x^{\parallel,\perp} = w_y^{\parallel,\perp}$.

The remaining functions of z that are needed in order to do the integral in Eq. (7.41) are the excitation intensity $|E_p(z)|^2$ and the concentration profile $C(z)$. The excitation intensity is easily obtained from Eq. (7.10). Note that the $\exp(-z/d)$ implicit in the $|E(z)|^2$ factor can be written as $\exp(-zk_1\beta)$, where

$$\begin{aligned} k_1 &\equiv 2\pi n_1/\lambda_0 \\ \beta &\equiv (2/n)(\sin^2\theta - n^2)^{1/2} \end{aligned} \tag{7.43}$$

so that $(\beta k_1)^{-1} = d$, the evanescent wave depth.

The concentration profile may be known, for example, a delta function at a particular z (say, $z=0$) for closely adsorbed material, or a step function out to a particular $z>0$. However, often, to deduce $C(z)$ may be the purpose of the whole exercise. Equation (7.41) can be cast into a form which is similar, but not equivalent to, a Laplace transform of the concentration profile $C(z)$. Variables $w^{\|,\perp}$ and $E_p(0)$ in the integrand of Eq. (7.41) are functions of β, since they are all functions of the incidence angle θ. Then Eq. (7.41) can be written as

$$\mathscr{F} = \int_0^\infty dz \exp(-\beta k_1 z)\, C(z)\, h(z, \beta) \tag{7.44}$$

where

$$h(z, \beta) \equiv K(\mu_a)^2\, |E_p(0, \beta)|^2\, [w^\perp(z, \beta)\, Q^\perp(z) + w^\|(z, \beta)\, Q^\|(z)] \tag{7.45}$$

The presence of the $h(z, \beta)$ factor makes Eq. (7.44) different from a Laplace transform of $C(z)$. If the z dependence of $h(z, \beta)$ is ignored,[34–36] then calculated concentrations of fluorophore near an interface derived from collected fluorescence are approximations. Also, the β dependence in the $w^{\perp,\|}$ causes the integral in Eq. (7.44) to differ from the form of a Laplace transform even after the excitation term is factored out.

If the excitation electric field is an s-polarized evanescent field instead of the above p-polarized example, then $w^{\perp,\|}$ $[=w_y^{\perp,\|}(z)]$ does not depend upon β. Therefore, an approximate $C(z)$ can be calculated from the observed fluorescence $\mathscr{F}(\beta)$ (obtained experimentally by varying θ) by ignoring the z dependence in the bracketed term in Eq. (7.45) and by inverse Laplace transforming Eq. (7.44) after the $|E_s(0, \beta)|^2$ term has been factored out.[37–39]

7.3.5. Consequences for Experiments

7.3.5.1. Polarization

A complete treatment[33] shows the polarizations of the emitted fields from single dipoles, but the polarization from an orientational and spatial distribution of dipoles, presumably calculated by numerical integration in particular cases, has yet to be done. The graphs in Sections 7.3.3 and 7.3.4 above assume that no polarizing analyzer is used. The polarization field of each emitting dipole in a distribution is always in the same plane as that dipole's orientation at the instant of emission and perpendicular to the direction of emitted light propagation. However, the polarization from excited dipoles spread over an orientational distribution near an interface in general will be different from that observed from an identical distribution far from an interface, because the emission collection efficiency from a dipole at an interface depends upon dipole orientation and distance from the surface (see Figure 7.5). An additional complication is that the polarization will depend upon the angle of observation and therefore will be a function of the numerical aperture of the collecting lens.

7.3.5.2. Fluorescence Efficiency and Lifetime

The preceding treatment assumed, for simplicity, that the quantum yield of the isolated dipole (i.e., at $z = \infty$) was 100%. Here we assign it a more general value of q_0. The following definitions are useful:

$$p \equiv P_T(z)/P_T(\infty)$$
$$f \equiv \frac{\int d\Omega\, S(z, \Omega)}{\int d\Omega\, S(\infty, \Omega)} \tag{7.46}$$

Both p and f are ratios of a power emitted at position z relative to that for an isolated dipole; p refers to total power (light plus heat) whereas f refers to radiated power only, derived by integrating the fixed-amplitude dipole radiated intensity S [given by Eq. (7.34) without the P_T normalization in the denominator] over 4π steradians.

We seek an expression for the actual quantum efficiency observed for a fluorophore at distance z, which we denote as q. One can show that

$$\frac{1}{q} = \frac{1}{f} \cdot \left(\frac{1}{q_0} + p - 1 \right) \tag{7.47}$$

Given a "natural" (i.e., no radiationless decay) fluorescence lifetime τ_0 for an isolated fluorophore, one can show that the actual observed lifetime for a real fluorophore near an interface is

$$\tau = q\tau_0/f \tag{7.48}$$

Note that this expression differs from the more familiar $\tau = q\tau_0$ applicable to systems in which the rate constant of the fluorescence emission path to the ground state is constant and alterations in the fluorescence power arise only from changes in the radiationless decay rate. Near a surface, however, the fluorescence is affected in three ways: (1) the radiated power rate can change due to interference between the dipole's reflected and directly emitted fields; (2) the near field of the dipole, which normally carries away no energy, may be converted into a radiating field in the denser medium by interaction with the surface; and (3) the surface may be a dissipative medium, such as a real metal, thereby converting the dipole near field into heat. In most cases with $z < 10$ nm, effects 2 and 3 combine to increase the total dissipated power P_T, hence decreasing the lifetime for fluorophores close to the surface. Note that the degree of lifetime shortening does depend on the orientation of the dipole.

It is possible in principle for effect 1 to exert an opposing effect by tending to decrease P_T through destructive interference between the reflected and direct fields. However, for most materials likely to be encountered, either dielectrics or metals, the net result will be a lifetime decrease, not an increase, for small z.

For bare glass, the lifetime decrease is only slight ($\sim 10\%$ for $q_0 = 1$ and 5% for $q_0 = 0.5$); for metal-coated glass, the effect is dramatic. Particularly on metals, the expected decrease in lifetime may help protect the fluorophore against photobleaching that arises from excited-state chemical reactions involving diffusional collisions.

7.3.5.3. Collection System Design

The significant anisotropy in the emission pattern from a fluorophore near a surface (Figure 7.3) suggests how to maximize the collection of emitted light. When viewing through the glass, it is clearly desirable to use an objective with a numerical aperture (N.A.) high enough to collect the sharp peaks in the pattern.

For a bare glass/water interface, this criterion is

$$\text{N.A.} > n_1 = 1.33 \tag{7.49}$$

For a glass/metal film/water interface, this criterion is

$$\text{N.A.} > \text{Re}\left[\left(\frac{\varepsilon_1 \varepsilon_2}{\varepsilon_1 + \varepsilon_2}\right)^{1/2}\right] \tag{7.50}$$

This condition predicts that N.A. = 1.4 would capture the surface plasmon peak from an aluminum film surface but not from a silver-film surface. Therefore, since objectives with aperture higher than 1.4 are rather rare, an aluminum film is a better choice.

7.3.5.4. Selective Detection of Adsorbed Fluorophores

On bare glass, the supercritical angle emission that occurs only from fluorophores near the glass suggests a method of selective detection of such fluorophores even in the presence of excited fluorophores farther out in the solution. A microscope objective with a high numerical aperture (e.g., 1.4) can be masked at its back focal plane to exclude any emission at less than the critical angle into the glass, thereby excluding all light from distant fluorophores. This approach avoids the necessity of selective excitation near the surface, as is done by evanescent wave excitation. However, an appropriately masked objective (at an accessible plane in the microscope equivalent to the back focal plane) appears in practice to provide rather poor resolution.

7.3.5.5. Selective Surface Quenching

On a metal-coated surface, the highly effective and highly z-dependent quenching could be used to distinguish between fluorophores close to the surface (e.g., $z < 2$ nm, strongly quenched) and those farther out (e.g., $z > 7$ nm, only weakly quenched). For example, a metal-coated surface could be coated with an artificial, reconstituted, or flattened biological phospholipid bilayer membrane. Fluorescence from the more distal bilayer half would not be quenched nearly so strongly as that from the proximal half. This possibility of selectively detecting fluorophores in only one half of a bilayer may find application to studies of membrane asymmetry, transmembrane transport, and lipid "flip-flop." This selective quenching can be used quite generally, since an aluminum-coated surface can be chemically treated with organosilanes to derivatize it with a wide range of functionalities (see Section 7.4.5).

Metal coatings are subject to heating more than dielectric coatings because, for most incidence angles, a significant portion of the incident light can be absorbed. At high but accessible incident focused laser intensities, microscopic boiling in the water can be seen. As a general rule, the incident laser intensity should be reduced by at least two orders of magnitude from this point.

Semiconductor surfaces also quench nearby fluorescence. This effect has been applied experimentally[40] but not yet treated theoretically.

7.4. TIRF for a Microscope

A wide range of optical arrangements for TIRF have been employed, both with and without a microscope.[5] The arrangements coupled to a microscope[41] are particularly appropriate where small observation and illumination areas are required, for example, for examining biological cells and for measuring local adsorption kinetic and surface diffusion rates.

7.4.1. Inverted Microscope

Figure 7.8 shows a possible arrangement for an inverted microscope. This arrangement is fairly easily switched to phase contrast transmitted illumination or to conventional epi-fluorescence and is also usable with even the shortest working distance objectives. The key element in the optical system is an optical glass or fused-silica cubical prism that permits the incident laser beam to strike the TIR interface (which may be the surface of a microscope slide or coverslip placed in optical contact with the prism via a drop of immersion oil or glycerol) at greater than the critical angle. This prism need not be matched in refractive index to the TIR interface material nor need it be cubical. As an illustration of the optical effect seen with this setup, Figure 7.9 shows a fibroblast in culture labeled with a lipid probe. TIRF clearly illuminates with high contrast only the surface-contacting regions of the cell.

Other TIRF configurations for inverted microscopes have been employed. Figure 7.10 shows an alternative system.[42] Instead of a prism fixed with respect to the beam as above, the prism is fixed with respect to the sample. The glass slide substrate propagates the incident beam toward the

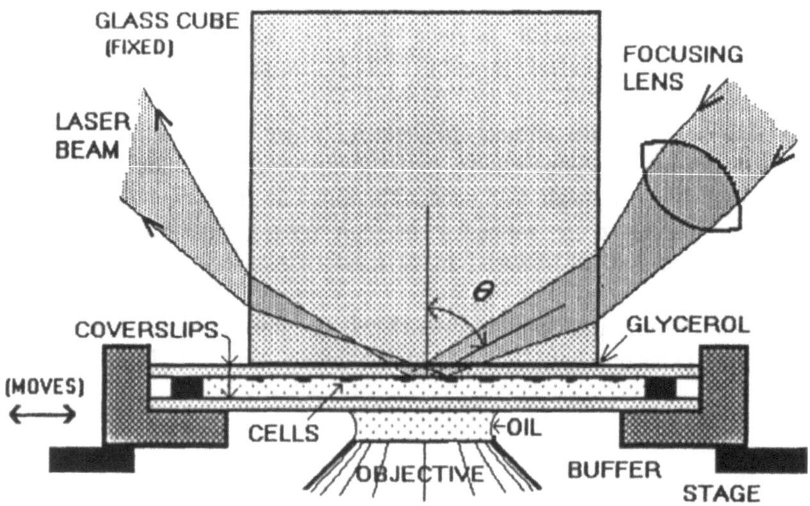

Figure 7.8. TIRF optical configuration for an inverted microscope; detail of the sample chamber region.

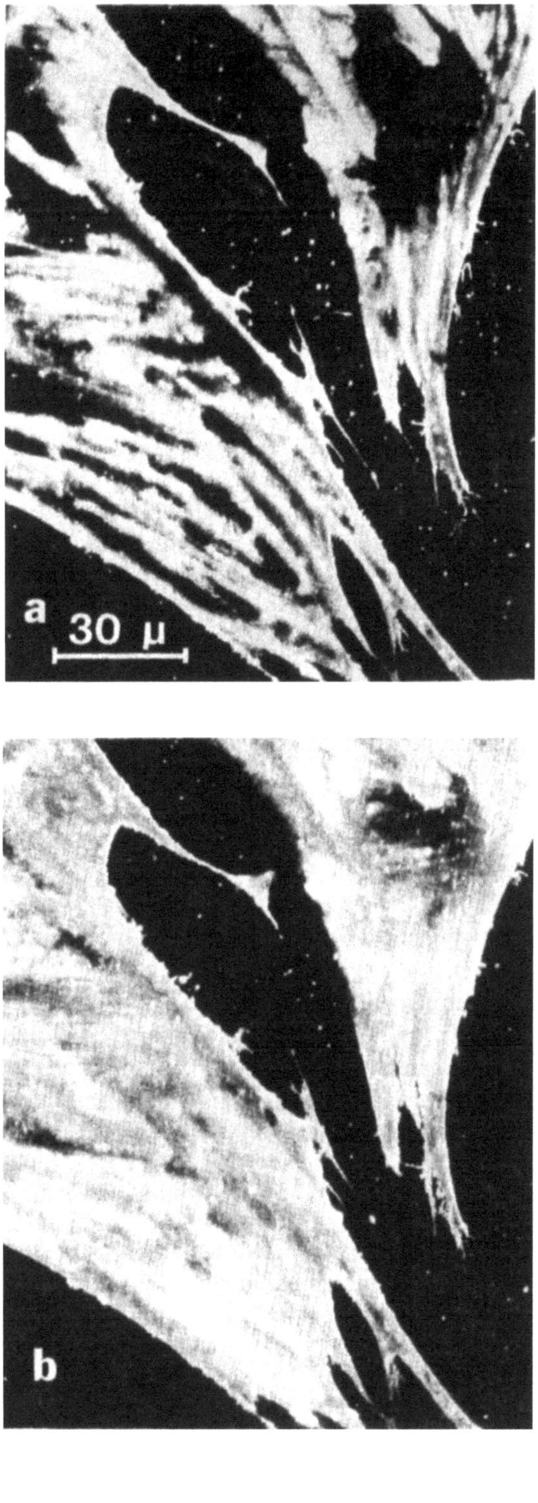

Figure 7.9. Human skin fibroblasts labeled with diI and illuminated by TIRF at two different angles of incidence. (a) $\theta = 74.3°$, $d = 105$ nm; (b) $\theta = 67.9°$, $d = 406$ nm. In this setup, $\theta_c = 67.5°$. Notice how more of the cell surface is immersed in the deeper evanescent wave of (b).

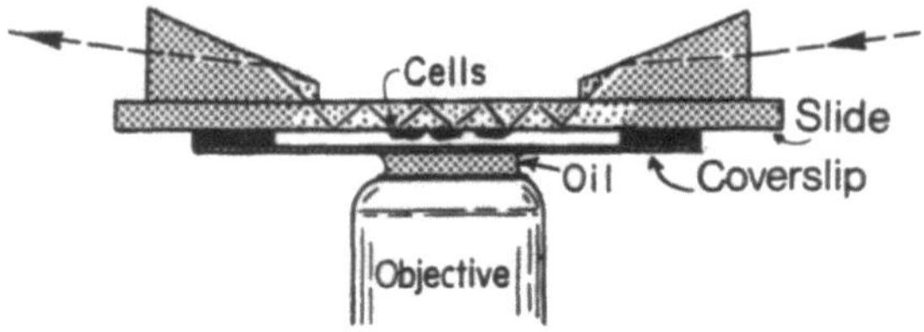

Figure 7.10. TIRF microscopy with prisms fixed to the substrate plate. (Adapted from Weis *et al.*[42])

microscope's optical axis via multiple internal reflections. The illuminated TIRF area will move with translation of the sample in this system.

7.4.2. Upright Microscope

Figure 7.11 shows a TIRF arrangement for an upright microscope. This setup is particularly appropriate for viewing culture cells growing in standard plastic culture dishes. The prism is a trapezoid (actually, a truncated 60° equilateral triangle made of high-refractive index flint glass) brought into optical contact (via a drop of oil or glycerol) with the bottom of the culture dish. This arrangement has the advantage that (1) the culture dish can be inserted and removed easily; and (2) the illuminated region does not move when the microscope is focused. It has the disadvantage that the incidence angle is not variable.

7.4.3. Prismless TIRF

By using an objective with a high numerical aperture (generally, 1.4), supercritical angle incident light can be cast upon the sample by epi-ellumination through the objective.[43] The incident beam must be constrained to pass through the periphery of the objective's pupil and must emerge with only a narrow spread of angles; this can be accomplished by ensuring that the incident beam is focused off-axis at the objective's back focal plane. It emerges into the immersion oil ($n_3 = 1.52$) at a maximum angle θ given by

$$\text{N.A.} = n_3 \sin\theta \tag{7.51}$$

For total internal reflection to take place at the sample surface, θ must be greater than the critical angle θ_c, given by

$$n_1 = n_{\text{water}} = 1.33 = n_3 \sin\theta_c \tag{7.52}$$

From Eqs. (7.51) and (7.52), it is evident that N.A. must be greater than 1.33, preferably by a substantial margin. Several possible arrangements are shown

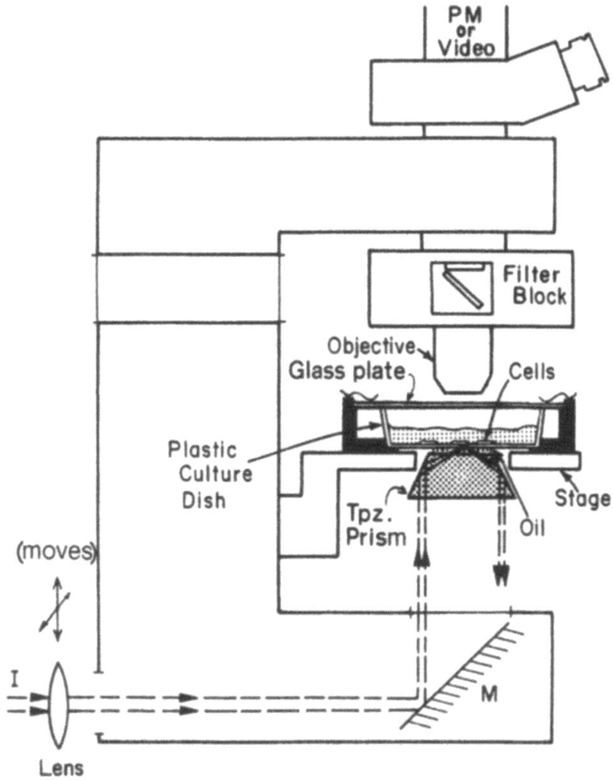

Figure 7.11. TIRF adapted to an upright microscope, shown here with a special sample chamber for long-term viewing of cells in a plastic tissue culture dish. The prism is a truncated equilateral triangle. The region of the sample chamber is shown enlarged relative to the rest of the microscope for pictorial clarity. The sample chamber is provided with a slow flow of humidified 5% CO_2/air, and 37 °C air is blown over the whole region.

in Figure 7.12, one of which even can utilize a conventional arc source rather than a laser beam.

7.4.4. TIRF Interference Fringes

By intersecting two laser beams at the TIR surface, finely spaced interference fringes can be produced. These fringes are useful for studies of surface diffusion rates, as discussed in Section 7.5.6. The interfringe (peak-to-peak) spacing is $\lambda_0[2n_3 \sin\theta \sin(\beta/2)]^{-1}$, where β is the intersection angle between the two beams in the plane of the TIR surface. TIRF fringes were first introduced by Weis *et al.*[42] as an aid to focusing on the TIRF surface.

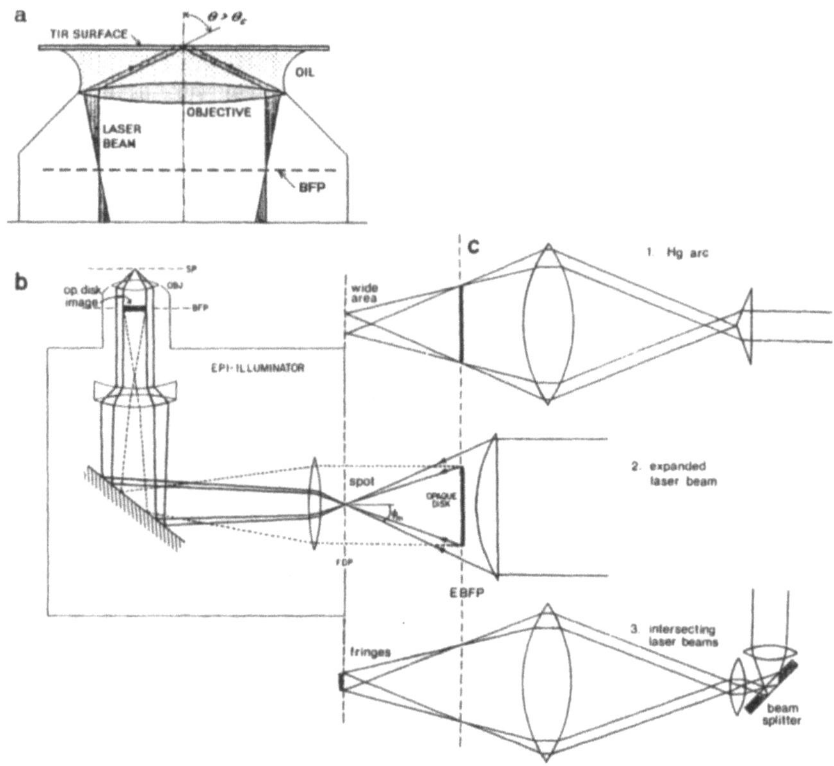

Figure 7.12. Optical arrangements for three prismless TIRF configurations through a 1.4-aperture microscope objective. Abbreviations: SP, Sample plane; OBJ, objective lens; BFP, back focal plane of the objective; EBFP, equivalent back focal plane outside the microscope; FDP, field diaphragm plane. Any sharp feature at the EBFP will form a sharp image at the BFP; likewise for the FDP and the SP (the TIR surface). Panels a (detail of the objective) and b (detail of the epi-illuminator) are common for all three configurations; panel c shows each in the region upbeam from the epi-illuminator: (1) mercury arc source, including the use of a conical lens to increase the power of supercritical light at the expense of subcritical light; (2) laser source focused to a spot with a single focusing lens; (3) laser source split into two mutually coherent beams which intersect at the FDP, forming two evanescent beams intersecting at 180° at the sample plane. This produces a TIR fringe pattern with a submicroscopic 0.18-μm internode spacing. Additional configurations are given in Ref. 43.

Figure 7.12c shows intersecting beam TIRF based around a prismless system, as discussed above, with a high-aperture object. This system is mechanically very stable, enabling one to achieve interfringe spacings of $\sim$0.18 μm without the blurring effects of small vibrations. Another arrangement uses a parabolic mirror to direct the intersecting beams into a hemispherical prism (see Figure 7.15); although somewhat more awkward, this system allows a wider range of incidence angles.

7.4.5. General Experimental Suggestions

Regardless of the optical configuration chosen, the following suggestions may be helpful:

1. The prism used to couple the light into the system and the (usually disposable) slide or coverslip in which TIR takes place need not be matched exactly in refractive index.
2. The prism and slide may be optically coupled with glycerol, cyclohexanol, or microscope immersion oil, among other liquids. Immersion oil has a higher refractive index (thereby avoiding possible TIR at the prism/coupling liquid interface at low incidence angles), but it tends to be more autofluorescent (even the "extremely low" fluorescence types).
3. The prism and slide can both be made of ordinary optical glass for many applications, unless shorter penetration depths arising from higher refractive indices are desired. Optical glass does not transmit light below about 310 nm and also has a dim autoluminescence with a long (several hundred microsecond) decay time, which can be a problem in some photobleaching experiments. The autoluminescence of high-quality fused silica (often called "quartz") is much lower. Tissue culture dish plastic (particularly convenient as a substrate in the upright microscope setup) is also suitable, but tends to have a significant autofluorescence compared to ordinary glass. More exotic high-n_3 materials such as sapphire, titanium dioxide, and strontium titanate can yield exponential decay depths d as low as $\lambda_0/20$.
4. The TIR surface need not be specially polished: the smoothness of a standard commercial microscope slide is adequate.
5. Illumination of surface-adsorbed proteins can lead to apparent photochemically induced cross-linking. This effect is observed as a slow, continual, illumination-dependent increase in the observed fluorescence. It can be inhibited by deoxygenation (aided by the use of an O_2-consuming enzyme/substrate system such as protocatechuic deoxygenase/protocatechuic acid or glucose/glucose oxidase) or by 0.05 *M* cysteamine. Photobleaching, which produces a slow decrease in fluorescence, can be reduced by deoxygenation as above or by 0.01 *M* sodium dithionite, among other substances.[44–49]
6. Virtually any laser with a total visible output in the 0.5 W or greater range should be adequate. The most popular laser for cell biological work with a microscope appears to be a 3-W continuous-wave argon laser.
7. TIRF experiments often involve specially coated substrates. A glass surface can be chemically derivatized to yield special physi- or

chemisorptive properties. Covalent attachment of certain specific chemicals is particularly useful in cell biology and biophysics; such chemicals include poly-L-lysine for enhanced adherence of cells; hydrocarbon chains for hydrophobicizing the surface in preparation for lipid monolayer adsorption; and antibodies, antigens, or lectins for producing specific reactivities.

Covalent derivatization generally involves pretreatment by an organosilane (see the catalog of Petrarch Systems). The protocol for covalent poly-L-lysine attachment to planar glass slides is similar to that described for the treatment of spherical glass beads.[50] The protocol for preparing lipid monolayers on hydrophobic glass is given by VonTscharner and McConnell.[51] Methods for preparing model membranes on planar surfaces suitable for TIR have been reviewed.[52–54]

Aluminum coating (for surface fluorescence quenching; see Section 7.3.5.5) can be accomplished in a standard vacuum evaporator; the amount of deposition can be made reproducible by completely evaporating a premeasured constant amount of aluminum. After deposition, the upper surface of the aluminum film spontaneously oxidizes in air very rapidly. This aluminum oxide layer appears to have some similar chemical properties to the silicon dioxide of a glass surface; it can be derivatized by organosilanes in much the same manner.

7.5. Applications of TIRF

7.5.1. Binding of Proteins and Probes to Artificial Surfaces

TIRF has been used to study equilibrium adsorption of proteins to artificial surfaces both to learn about the surface properties of various biomaterials that have medical applications and also to test the TIRF technique itself.

Several studies of the binding equilibria, kinetics, and conformational changes of proteins upon adsorption have employed extrinsic fluorophores attached to protein.[55–61] It is possible in principle that such extrinsic groups might themselves affect the adsorption process being investigated. To avoid this possibility, the intrinsic fluorescence of tryptophan or tyrosine residues in the protein can be monitored upon excitation by a $\lambda_0 = 280$ nm evanescent wave.[62–67] In certain cases, however, the greater susceptibility of proteins to photodegradation under UV illumination may outweigh the natural advantage of intrinsic fluorescence excitation.

Calibration of a TIRF intensity to derive an absolute concentration of adsorbate is a nontrivial problem, mainly because fluorescence quantum efficiencies are apt to change upon adsorption to a surface. One route around this problem is to measure the depletion of bulk solute (in epi-illumination mode or in a standard spectrofluorimeter) when it is allowed to adsorb onto a known surface area.[58] Hlady *et al.*[68] proposed another method for the approximate calibration of TIR fluorescence. This method, involving use of nonadsorbing species as standards and ^{125}I-labeled protein with a γ-detection system, introduces a small correction for incident light scattered beyond the evanescent wave volume and for changes in the fluorescence emission quantum efficiency of proteins upon adsorption. A third method[57] is self-contained, in the sense that no other measuring instruments outside the TIRF system are needed. The total fluorescence is due to a bulk dissolved contribution plus a surface contribution. Since the bulk concentration is usually known, we need only measure its fractional contribution to the total fluorescence to calculate the surface concentration (approximating that quantum and collection efficiencies are unchanged upon adsorption). The fractional contribution can be deduced simply either by abolishing the fluorescence from the surface (with a strong photobleaching pulse of incident light) or by replacing the solution with a fluorescence-free rinse. In either case, the method works only if the time scale of the reversible adsorption kinetics is much longer than the time of the bleaching pulse or the change of solution.

TIRF for the sensing of protein adsorption can be transformed into a practical medical procedure.[69–76] Designed to serve as a continuous sensor element in a remote sample, a single multimode optical fiber or planar waveguide both supports the evanescent wave on its surface used for excitation and also guides the captured near-field fluorescence which propagates into the fiber at greater than the critical angle.[77] To make the fiber biochemically specific, it is covalently coated with either an antibody or its complementary protein antigen. When introduced into the target liquid, the antigen, or its complementary antibody, specifically adsorbs. In some of these cases, the intrinsic fluorescence of tryptophan is used to assay specific adsorption onto the optical fiber. Further experimentation will show whether the increase of fluorescence resulting from specific binding will be evident above a large background of nonspecifically adsorbed protein fluorescence likely to be encountered in biological fluids.

If the soluble protein that specifically adsorbs to the fiber can be extrinsically labeled, the background problem can be avoided. Of course, *in vivo* proteins cannot be labeled. However, it is conceivable that a protein labeled with a bulky extrinsic group (e.g., fluorescent dextrans) could be confined by a molecular sieve membrane (e.g., a dialysis membrane) within a closed volume surrounding the specifically derivatized optical fiber. When exposed to the (unlabeled) protein in the biological fluid under investigation, the membrane-clad fiber would allow some unlabeled protein to permeate in and thereby compete with the confined fluorescent protein for surface binding sites.

Binding competition between a fluorescent and nonfluorescent species is the basis of a fluorosensor designed for detecting the presence of acetylcholine receptor (AChR) agonists and antagonists in solution.[78] Acetylcholine receptors are noncovalently but firmly attached to the optical fiber, and the target solution is spiked with a low concentration of the specific blocker α-bungarotoxin, labeled with fluorescein. It is found that the binding of the fluorescein bungarotoxin, as measured by TIRF through the fiber, is inhibited by the presence of known AChR agonists and antagonists.

By treating a glass surface with immobilized anti-human serum albumin (HSA) immunoglobulin in distinct spots, a spatially resolved TIRF pattern due to fluorescein-HSA binding from solution could be focused onto a CCD camera.[79] This spatially resolved TIRF technique offers the possibility of an internal control against background binding (giving rise to fluorescence between spots) and detecting the presence of several solution components simultaneously.

Some studies have used dye molecules themselves, rather than dye–protein conjugates, to investigate the surface charge properties of a solid/liquid interface. The emission spectra contain information on the hydrophobicity of the fluorophore environment. By studying the detailed structure of the vibronic bands of adsorbed pyrene at both a solid/liquid interface (using TIRF) and the corresponding solid/vapor interface, Hartner *et al.*[80] concluded that TIRF did successfully report the microenvironment of the adsorbed (rather than bulk) pyrene. The adsorption of a dye molecule at an SnO_2 thin film as a function of pH was measured.[81] Investigation of TIR-excited emission shifts of an adsorbed dye has successfully detected hydrophobicity gradients on silica surfaces which appear correlated with the ability of each local region to adsorb blood albumin protein.[82]

By preparing planar lipid monolayers or bilayers on hydrophobically derivatized or native hydrophilic glass, respectively, the adsorption equilibrium constants of a blood coagulation cascade protein, prothrombin, have been examined by TIRF on a surface that more closely models actual cell surfaces and is amenable to alterations of surface charge. It was found that membranes of phosphatidylcholine (PC) that contain some phosphatidylserine (PS) bind prothrombin more strongly than pure PC membranes.[83]

Most of the early TIRF work with proteins at surfaces involved nonspecific adsorption. Recently, Poglitsch and Thompson[84, 85] have shown that TIRF can also detect specific but still reversible ligand–receptor binding at a planar lipid membrane on a solid glass support. Macrophage Fc receptors were successfully reconstituted into supported lipid monolayers, and the equilibrium binding constant of fluorescent-labeled monoclonal Fab fragments of immunoglobulin G (IgG) was measured by TIRF. By using the competition with a labeled species (rhodamine-Fab) to measure binding strength of an unlabeled species (Fc-containing IgG), these authors were able to report an association constant between unlabeled polyclonal IgG and the reconstituted Fc receptors. Equilibrium binding constants of the antigen-binding site of both divalent monoclonal antibody and its Fab fragment on supported planar lipid monolayers doped with hapten-derivatized lipid were also obtained with TIRF, and the divalent antibody results were compared with theoretical models of two-step binding.[86]

7.5.2. Concentration of Molecules near Surfaces

The concentration of a solute or adsorbate may be a nontrivial function of the distance to the surface, a function which contains information about the thermodynamics of the surface interaction. To explore the fluorophore concentration $C(z)$ as a function of distance z from the surface, one can record the observed fluorescent intensity F as the characteristic depth d of the evanescent wave is varied. The mathematics of this is discussed immediately following Eqs. (7.44) and (7.45) above.

To vary d [or the related parameter β in Eqs. (7.44) and (7.45)], the angle of incidence θ can be varied. Experimentally, this is not trivial, because d is a very strong function of θ within only a few degrees greater than θ_c, and therefore θ must be measured to fractions of a degree. In addition, the presence of a solute (or the cytoplasm of a biological cell) alters the refractive index n_1 from its pure water value, and this altered value must also be known accurately. Rondelez *et al.*[87] have measured $F(d)$ versus θ to obtain information on the z-dependent concentration profile of artificial polymers adsorbed to glass or silica. Reichert *et al.*[88] have tested this approach, first with a fluorescein solution which was presumed to have a constant C independent of z, and second with a layer of fluorescein-labeled immunoglobulin adsorbed to quartz which was presumed to have a step-function $C(z)$. The theoretical expectations here are an approximation since their theory omits the normalization step discussed in Section 7.3. This omission, although not exactly correct, simplifies the calculation of $C(z)$ by converting it to an inverse Laplace transform of the observed fluorescence. The ability of this general approach to correctly report concentration profiles was checked on planes of fluorophores deposited in steps between layers of Langmuir–Blodgett films.[89]

A similar simplifying assumption has been used by Allain *et al.*[90] in analyzing their experiments on a flexible fluorescent anthracene–polystyrene copolymer coil in the vicinity of a nonadsorbing wall. The analysis appears to confirm a local decrease in $C(z)$ for small z at the solid/solution interface. Such a depletion layer is interpreted in terms of an "entropic repulsion" model, whereby certain conformations of polymer are sterically prohibited near the surface.

Another method of obtaining $C(z)$ involves varying the angle at which the emission is observed. As discussed in Section 7.3, the near field of the emitting dipole can produce propagating light at supercritical angles in the glass; the smaller the dipole distance z to the surface is, the greater the intensity and the wider the range of angles that are cast into the supercritical zone. Each supercritical observation angle represents a different (but, for a nonmetallic surface, always monotonically decaying) dependence of collection efficiency versus z. The method of varying observation angle has been used by assuming that the monotonic decay is exponential (an approximation) to measure the concentration of a stiff, high-molecular-weight polysaccharide near a solid surface.[91] This application, using TIRF (but not requiring it) in a nonmicroscopic configuration, also indicated the presence of a surface depletion zone.

An unusual application of TIRF for measuring dye concentrations on a woven fabric of silk versus distance into the surface of the silk was reported by Kurahashi *et al.*[92] By comparing the relative intensities of silk's own fluorescence emission peak around 340 nm with that of the dye at longer wavelengths under both normal and TIR illumination, they concluded that the dye tends to concentrate in the interior bulk of the silk rather than on the surface. Although qualitatively clear, the spectra would have to be interpreted for the wavelength-dependent intensity of the evanescent wave and the significant light scattering to yield a more quantitative result.

7.5.3. Orientation, Rotation, and Fluorescence Lifetime of Molecules near Surfaces

The polarization properties of the evanescent wave[93] can be used to excite selected orientations of fluorophores, for example, fluorescent-labeled phosphatidylethanolamine embedded in lecithin monolayers on hydrophobic glass.[21] When interpreted according to an approximate theory, the total fluorescence gathered by a high-aperture objective for different evanescent polarizations gives a measure of the probe's orientational order. The polarization properties of the emission field itself, expressed in a properly normalized theory,[94] can also be used to determine features of the orientational distribution of fluorophores near a surface.

Both the physics and the chemistry of proximity to a surface can alter the excited-state lifetime and rotational motion of a fluorescent molecule. An extrinsic label attached to BSA has been found to reduce its fluorescence lifetime upon BSA adsorption to fused silica.[95] The decrease is too large to arise from the physical near-field proximity effects discussed in Section 7.3;

some sort of chemical conformational change or quenching due to high local concentrations upon adsorption might explain the effect. TIRF has been combined with time-resolved polarized anisotropy decay to measure molecular rotation rates in fluorescence-doped polystyrene double-layered films, mainly as a test of the selectivity of the results for the layer nearest the TIR sapphire prism.(96) Similar experiments on pyrene-containing poly(methyl methacrylate) films(97) showed that rotational rates of the probe were restricted near the surface relative to the bulk. In both cases, the "bulk" was not a low-viscosity liquid but a rigid polymer matrix. By a similar non-microscopic time-resolved TIRF technique, the rotational mobility of pyrene-labeled serum albumin adsorbed to artificial polymer films has been measured.(98)

Itaya *et al.*(99) have described a TIR system for obtaining time-resolved fluorescence decay curves induced by laser flash illumination of polymer films in a microscope configuration. Presumably, use of this configuration can be extended to studies on biological cells.

In these time-resolved studies, a simplified, non-normalized theory [i.e., effectively lacking the division by P_T in Eq. (7.34)] was used for comparison with the experimental results, so that the observed fluorescence from any region was assumed to be proportional to the local evanescent intensity in that region. A more precise analysis must take into account that distance from the interface affects the angular distribution of emission and that fluorescence lifetimes are necessarily affected by the proximity of the dielectric interface.

Many substances preferentially concentrate at interfaces, including liquid/liquid ones. Although TIRF is most easily adaptable to solid/liquid interfaces, Morrison and Weber(100) succeeded in observing the preferential adsorption of certain amphiphilic dyes at the interface between two immiscible and optically dissimilar liquids. Steady-state TIR fluorescence polarization in that system showed that the rotational diffusion of the interfacially adsorbed dye was restricted.

Fraaije *et al.*(101) have investigated the orientation of reversibly adsorbed cytochrome *c* as a function of experimentally controlled electrical surface potential. This project contains a number of distinctive experimental features. The intrinsic fluorescence of the cytochrome's porphyrin ring, rather than an extrinsic probe, was used. Orientational order was deduced from steady-state fluorescence excited by varying the incident evanescent polarization, as discussed in the theory of Thompson and Burghardt.(16) The TIR surface was quartz, coated with a thin film of the (semi)conductor SnO_2 connected via conducting glue and a wire to a variable-potential source, thereby forming an optically transparent electrode. The results indicate that the adsorbate's orientation can be affected by the imposed interfacial potential during the adsorption process, but once the adsorption has occurred, the orientations appear to become "locked in."

In a microscope, standard polarized epi-illumination cannot distinguish order from disorder in the polar direction (defined as the optical axis) because epi-illumination is polarized transverse to the optical axis and observation is along the optical axis at 180°. However, microscope TIR illumination can be partially polarized in the optical axis direction (the *z*-direction of Section 7.2) and can thereby detect order in the polar angle direction. Timbs and Thompson[102] used this feature to confirm that the popular lipid probe 3,3′-dioctadecylindocarbocyanine (diI) resides in a supported lipid monolayer with its dipoles parallel to the membrane surface, but labeled antibodies bound to the membrane exhibit totally random orientations.

7.5.4. Qualitative Observation of Labeled Cells

The most straightforward application of TIRF is to observe the location and (with time-lapse video) the motion of cell/substrate contacts. For this purpose, cells may be labeled by a membrane lipid fluorescent analogue such as diI (see Figure 7.9 and Refs. 5, 7, 41, and 103 for more photographic examples). For qualitative viewing, the TIRF contrast of the cell/substrate contacts over the background and cell autofluorescence is excellent in comparison to the contrast obtained with nonfluorescent techniques such as interference reflection contrast.[104, 105] Of course, labeling of cells can be cytotoxic, particularly under illumination. However, TIRF seems particularly advantageous for long-term viewing of cells compared to other fluorescence techniques, since the thin evanescent wave minimizes exposure of the cells' organelles to excitation light.

Quantitative determination of the absolute distance from the surface to a labeled cell membrane at a cell/substrate contact region can be based on the variation of $F(d)$ with θ.[106] This effort is challenging because corrections have to be made for θ-dependent reflection and transmission through four stratified layers (glass, culture medium, membrane, and cytoplasm), all with different refractive indices. For 3T3 cells, Lanni *et al.*[106] derived a plasma membrane/substrate spacing of 49 nm for focal contacts and 69 nm for "close" contacts elsewhere. They were also able to calculate an approximate refractive index for the cytoplasm of 1.358 to 1.374.

Another complication in the quantitation of TIRF on cells is the effect of the membrane thickness itself on the profile of the evanescent wave. Reichert and Truskey[105] have calculated that, in theory, the thickness of the membrane should have a negligible effect on the fluorescence and that a simplified theory of three stratified layers (glass/water/cytoplasm) should be adequate. The theory approximates for simplicity that scattering plays a negligible role and that fluorescence intensity versus angle of observation and fluorescence lifetime are not functions of distance to the interface z. Experiments that determine actual cell/substrate contact distances by an independent technique are now needed to confirm the validity of these convenient assumptions.

Evanescent light scattered by cells can be viewed directly, simply by removing the barrier filter. The contrast between cells and the background is rather low, and the scattered intensity is many orders of magnitude less than the evanescent wave intensity. On this qualitative basis, one might tentatively assume that scattering is not a significant factor. Nevertheless, at incidence angles very near the critical angle, the cells do cast a noticeable "shadow" along the surface.

In many cases of membrane labeling, some probe becomes internalized in living cells. Epi-illumination excites this internalized fluorescence from out-of focus planes and leads to a diffuse fluorescence that obscures detail. However, TIRF "optically sections" the sample, allowing observation of a distinct surface pattern even in the presence of a large amount of internalized label. The optical sectioning is particularly useful in viewing submembrane cytoplasmic filaments on thick cells. Although TIRF cannot view deeply into the cell as can confocal microscopy, it can display the submembrane filament structure with high contrast and sensitivity in the regions of cell/substrate contact.

A possible spatial correlation between submembrane filaments and surface acetylcholine receptors (AChR) on developing muscle cells in culture was investigated by Bloch *et al.*[107] Double labeling was used: a rhodamine-labeled second antibody for the cytoplasmic filaments, and fluorescein-labeled α-bungarotoxin for the AChR. Somewhat fortuitously for the use of TIR, the receptor clusters in this biological system happen to be found predominately in the general regions where the myotube plasma membrane is near the glass substrate, which allows TIR to effectively excite their fluorescence. Figure 7.13 shows double-labeled TIRF views of the relative distribution of AChR (labeled by fluorescein-labeled α-bungarotoxin) clustered on the surface of cultured rat myotubes and of certain specific non-AChR proteins (labeled by rhodamine-labeled antibodies) in or immediately under the membrane. The figure shows that AChR codistributes with 43K protein but interdigitates with vinculin. With standard epi-illumination on these intact thick cells, the cytoplasmic filament images would have been obscured by out-of-focus light.

One TIRF study found that some membrane proteins behave just oppositely to AChR: they avoid the cell/substrate contact regions.[108] When endothelial cells are grown on a bare glass surface or are brought into suspension, a specific membrane protein marked with antibodies appears all over the cell surface, as evidenced by epi-illumination and TIRF. However, when the cells are grown on (or returned to) a surface coated with their own extracellular matrix material, the protein disappears from the basal (cell/substrate-contacting) side of the cells.

The interaction between immune system cells and their targets often involves a specific and as yet incompletely understood surface reaction. This

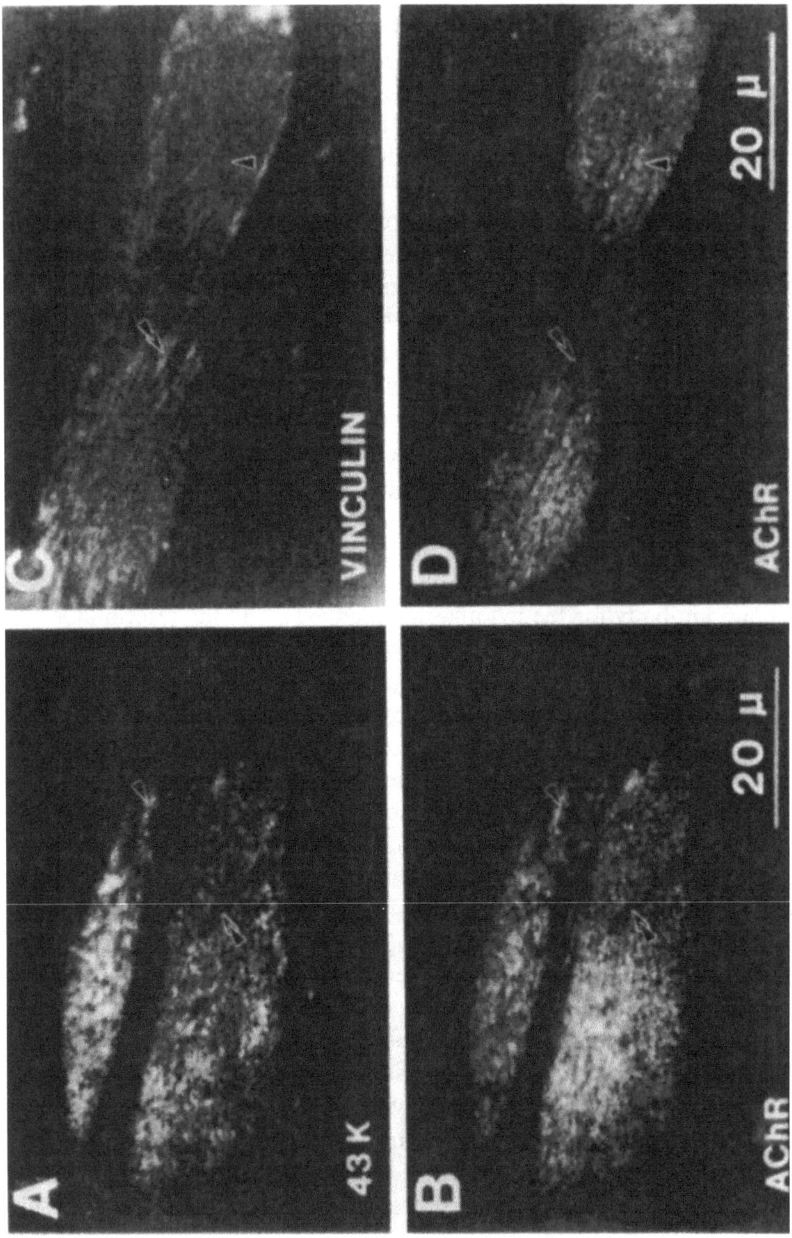

Figure 7.13. AChR clusters on a rat myotube in culture, visualized with (A) rhodamine anti-43K protein; (B) fluorescein-labeled α-bungarotoxin (corresponding field); (C) rhodamine-labeled anti-vinculin; and (D) fluorescein-labeled α-bungarotoxin (corresponding field). The antibodies were gifts from R. Bloch, University of Maryland Medical School. Note that AChR and 43K tend to aggregate in the same regions, whereas AChR and vinculin appear to exclude each other. Additional photographs are in Ref. 107.

interaction can be made optically accessible by modeling the target as a lipid monolayer or bilayer supported on glass.[53, 54] For example, Weis *et al.*[42] could visualize the contact region between basophils (which bear surface Fc receptors) and hapten-containing target model membranes by illuminating with TIR in the presence of fluoresceinated IgE antibodies. The contact region, where the Fc receptors are indirectly connected to the haptens through the IgE, appeared rather variegated and punctate, perhaps due to filopodia-like structures in the contact zone. This pattern could not be observed with conventional epi-fluorescence.

A variation of TIRF to observe cellular morphology, introduced by Gingell *et al.*,[109] produces essentially a negative of the standard fluorescence view of labeled cells. The solution surrounding the cells is doped with a non-adsorbing and nonpermeable fluorescent volume marker, fluorescein-labeled dextran. Focal contacts then appear as dark areas, and other areas appear brighter, depending on the depth of solution illuminated by the evanescent wave in the cell/substrate gap. A quantitative theory for converting fluorescence intensities into cell/substrate contact distances has been developed.[110] By using a high-refractive index glass ($n = 1.83$) as the TIR and cell substrate surface, a very shallow evanescent wave ($1/e$ decay distance ≈ 37 nm) can be produced[111] which minimizes the contribution from the cytoplasm and probes small undulations in the cell/substrate contact region.

7.5.5. Fluorescence Energy Transfer and TIRF

TIRF can be combined with fluorescence energy transfer to measure distances between fluorophores on a surface in the presence of a large background of bulk-dissolved fluorophores.

Burghardt and Axelrod[59] detected TIRF/energy transfer evidence of a change in the conformation of donor/acceptor-labeled bovine serum albumin upon the protein's adsorption to glass. In a TIRF/energy transfer study of relevance to cellular immunology, Watts *et al.*[112] explored whether helper T cells could force two nonidentical antigens in a target membrane into closer proximity with each other. These two antigens, one (a synthetic peptide) labeled with a fluorescence energy transfer donor and the other (a major histocompatibility complex) with an acceptor, were incorporated into a planar lipid bilayer on a TIR hydrophilic glass surface. Significant amount of the synthetic peptide remained in solution, so microscopic TIRF was needed to limit excitation to the region near the glass and overlaying lipid bilayer. TIRF also served to reduce the autofluorescence normally observed from the T cells that were allowed to settle on the lipid bilayer. It was found that fluorescence energy transfer occurred only in those microscopic lipid bilayer regions where the T-cell surface came into close apposition with the bilayer. The conclusion was that the T-cell surface forces the two membrane antigens to which it binds to within a distance of 4 nm of each other.

7.5.6. Reaction Rates at Biosurfaces

Consider a labeled molecule in equilibrium between a surface-bound state and a free solute state:

$$\text{free solute} + \text{vacant surface site} \underset{k_2}{\overset{k_1}{\rightleftharpoons}} \text{surface-bound solute}$$

If the solute is fluorescent, the TIRF intensity (which is proportional to the concentration of surface-bound solute) can be monitored as a function of time to measure the binding kinetic rates k_1 and k_2. What is required is some sort of perturbation to disturb the equilibrium of the fluorescent species.

Using a concentration jump as the perturbation, Sutherland *et al.*[113] measured the kinetics of binding of fluorescein-labeled human IgG (present as an antigen in solution) to surface-immobilized sheep anti-human IgG. Two TIRF surfaces were used: a planar slide and a fiber-optic cylinder. Also using a TIRF recovery after a concentration jump, Kalb *et al.*[114] measured the slow ($\sim 10^{-4}\ s^{-1}$) unbinding kinetics of anti-trinitrophenol (TNP) antibodies in solution and a TNP-derivatized lipid in a planar bilayer.

To increase the speed of the TIRF-based kinetic techniques, the perturbation can be optical rather than chemical. If the evanescent wave intensity is briefly flashed brightly, then some of the fluorophores associated with the surface will be photobleached. Subsequent exchange with unbleached dissolved fluorophores in equilibrium with the surface will lead to a recovery of fluorescence, excited by a continuous but much attenuated evanescent wave. The time course of this recovery is a measure of the desorption kinetic rate k_2. This technique[115] is called TIR/FRAP (or TIR/FPR) in reference to fluorescence recovery after photobleaching (or fluorescence photobleaching recovery).

Adsorption kinetics are especially interesting when compared with surface diffusion rates of the adsorbate. This is because of the theoretical possibility that nonspecific and reversible adsorption of a ligand (say, a hormone), followed by two-dimensional diffusion on the membrane, may enhance the reaction rate with a specific binding patch (say, a hormone receptor).[116, 117] A similar effect might enhance the reaction rates between a surface-immobilized enzyme and bulk-dissolved substrate, thereby speeding some reactions in industrial chemistry.

TIR/FRAP can be used to measure both surface diffusion coefficients and on/off kinetic rates, if the evanescent wave intensity is variegated over a distance on the surface that is short compared to the characteristic distance covered by surface diffusion within the time available before desorption.[115] Several studies have utilized TIR/FRAP in this manner. The adsorption-/desorption kinetics and surface diffusion of rhodamine-labeled bovine serum albumin (BSA) at a glass surface have been examined using a TIR illumination area focused into a thin line.[58] BSA was found to adsorb with a wide range of reversible kinetic rates, with more than half of the adsorption being reversible at higher bulk concentrations ($> \sim 1$ mg/ml). About 20% of the adsorbed BSA could surface diffuse, with a coefficient of about 5×10^{-9} cm^2/sec. This is fast enough to carry a BSA molecule at least ~ 1 μm on the average before it desorbs within 4 sec.

These results were extended by Tilton *et al.*[118] to adsorption of eosin-labeled BSA on polymer surfaces. They also found a component that surface diffuses, with coefficients ranging from 1.2×10^{-9} to 2.6×10^{-9} cm^2/s, depending on surface type. In this study, intersecting TIR laser beams rather than a focused stripe were used to define the spatial intensity variation. Surface diffusion was even noted for the most irreversibly adsorbed eosin-labeled BSA components; this was evident on samples rinsed for long periods with unlabeled BSA after exposure to eosin-labeled BSA. The surface diffusion coefficient of the irreversibly bound BSA was found to be a strong function of adsorbed concentration.[119]

A wide range of reversible adsorption kinetic rates was also found by TIR/FRAP for another protein, lysozyme, on a substrate with a different surface charge, alkylated silicon oxide.[61] It is possible that the wide range of rates results from a spectrum of surface binding site types and/or formation of multilayers of adsorbed protein.

A preliminary TIR/FRAP report[120] gives the desorption rate (k_2) for binding of prothrombin, a key protein in surface-activated thrombogenesis, with acidic (phosphatidylserine-containing) supported phospholipid bilayers. Another[121] gives desorption rates for specific binding between Fab antibody fragments and a lipid–hapten in a planar membrane. Knowledge of the desorption rates for such specific binding not only tells us how much time a bound pair have available to engage in more surface reactions together, but also allows us to calculate (with quantitative knowledge of the equilibrium binding constant) the specific adsorption rate (k_1). If (as in this Fab/lipid–hapten case) k_1 is calculated to be less than its theoretical diffusion-limited value, then one can conclude that the reaction is not diffusion-limited; that is, not every encounter leads to a successful binding event.

For TIR/FRAP to be useful for chemical kinetics studies on intact biological membranes as opposed to reconstituted or artificial surfaces, two problems must be confronted: (1) how to position the membrane in a TIR system; and (2) how to overcome background binding to the substrate to which the membrane is attached.

Proper positioning requires that at least part of the membrane be in the evanescent wave and that the surface under study (external or cytoplasmic) be accessible to chemical exchange with the bulk. This positioning has been successfully accomplished with erythrocyte ghosts.[122] After the glass substrate is covalently coated with poly-L-lysine, erythrocytes are allowed to adhere, followed by hyposmotic shock. Rather than floating away or crumpling up on the surface, the membrane ghosts flatten into circular disks on the glass with a characteristic tear that exposes the outer surface and the cytoplasmic surface to the solution in their own distinct regions (Figure 7.14). This technique or a modification of it may also work for other cell types.

This flattened erythrocyte preparation has been used to study reversible nonspecific adsorption kinetics and surface diffusion of insulin on the external surface of erythrocytes.[123] The nonspecific adsorption of insulin to the polylysine-coated substrate is very large compared to the adsorption to the flattened membrane adhered to the substrate. Fortunately, this nonspecific background fluorescence can be very successfully quenched simply by preparing the polylysine coating on an aluminum-film-coated glass surface, rather than on bare glass. As discussed in Section 7.3, the aluminum abolishes the fluorescence of fluorophores adsorbed directly onto the polylysine substrate, but the fluorophores adsorbed to the erythrocyte surface are not substantially quenched, because they are spaced at least two membrane thicknesses away.

The results of this TIR/FRAP study are that ~80% of the nonspecific binding of fluorescein-labeled insulin to the external face of red cell membranes is reversible within <100 s, and the mean residency time of the reversibly adsorbed insulin ranges from ~0.4 s to ~20 s. Surface diffusion of nonspecifically adsorbed insulin (as investigated by an TIR intersecting beam interference fringe pattern; see Figure 7.5) was immeasurably small: it is insufficient to carry a typical insulin more than ~0.3 μm before desorption.

At some point, these kinetic results for insulin on a biological membrane should be compared to kinetic results for insulin on an artificial lipid membrane, when such results become available. This comparison should be especially interesting in view of the suggestion by Sui *et al.*[124] that nonspecific equilibrium binding of insulin to planar membranes is a function not only of membrane charge but also of some sort of nonelectrostatic mechanism, based on their TIRF experiments with a chamber adapted to a standard spectrofluorimeter chamber.

Another TIR/FRAP study on biological cell membranes has examined the reversible but specific binding kinetics of fluorescence-labeled epidermal growth factor to the surface of cells.[125] The background problem here was solved simply by choosing cells with a very large concentration of epidermal

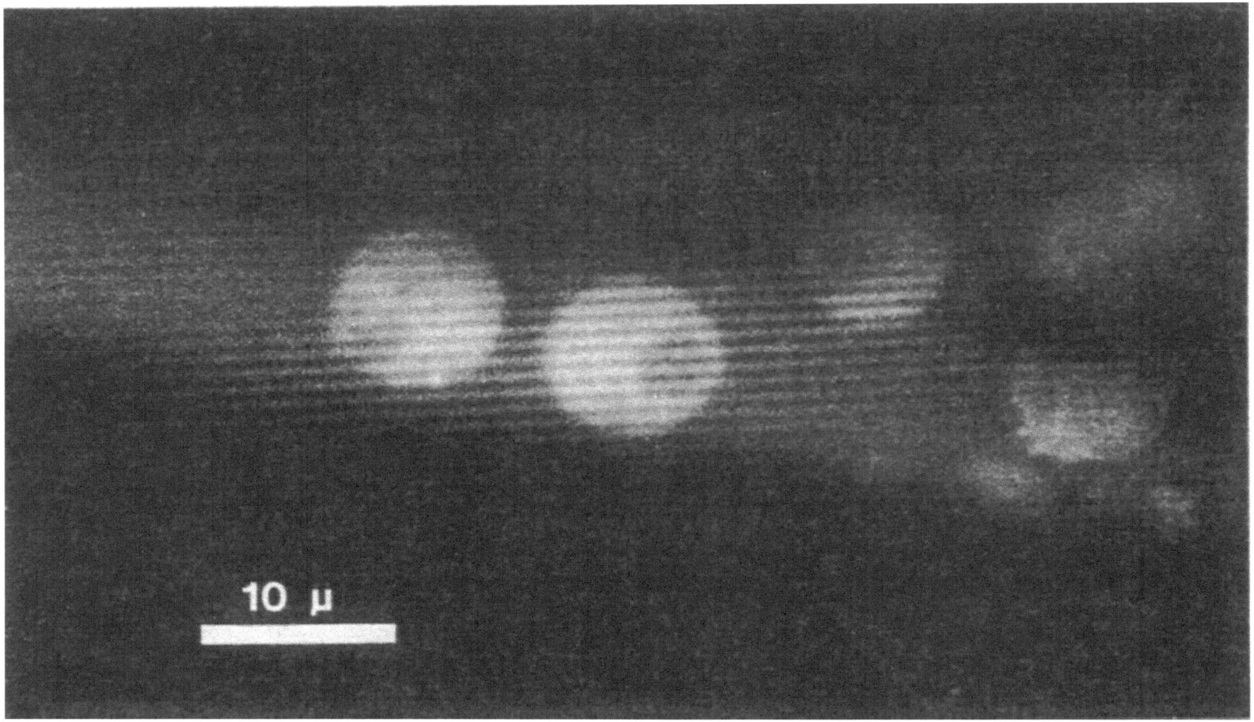

Figure 7.14. Erythrocyte ghosts adhered to aluminum film-coated glass covalently coated with polylysine. The TIR fluorescence is due to adsorbed fluorescein-labeled insulin; the background (off-cells) is dark only because of metal surface quenching. The small, darker circular regions in each ghost are open tears in the apical membrane which expose the cytoplasmic face of the basal membrane; this can be shown in other experiments by use of lipid, lectin, and cytoplasmic filament-specific labels. The stripes are due to interfering incident TIR laser beams, produced by the optical configuration shown in Figure 7.15.

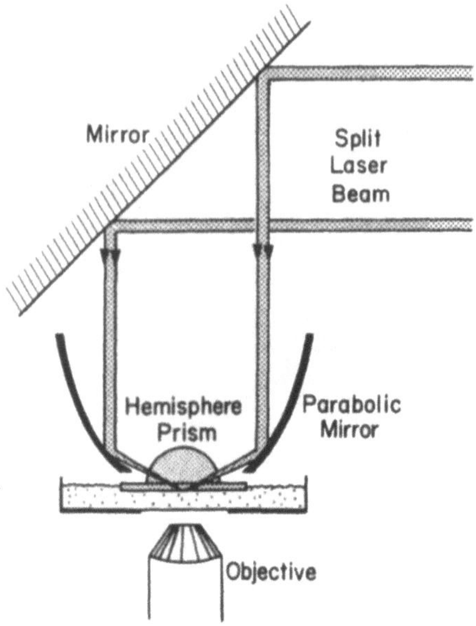

Figure 7.15. Intersecting TIR beams split from the same laser to produce interference fringes for viewing in an inverted microscope, as used for Figure 7.14. The beams reflect off the parabolic mirror and then propagate radially into a hemispherical prism. The drawing shows the beams intersecting at 180° for pictorial clarity only; for the photograph of Figure 7.14, the beams strike the parabolic mirror at adjustable nearby points and thereby intersect at an adjustable relative angle ϕ of $\sim 20°$. The TIR surface, at a glass slide surface just below the prism, is positioned at the focus of the parabola.

growth factor (EGF) receptors: the A431 human epidermoid cell line. No special procedures were used to flatten the cells; cell/substrate contact regions that were accessible enough to be labeled by EGF were observed. More than 85% of the EGF binds reversibly, with a range of characteristic times from ~ 20 s to ~ 250 s. Various control experiments and theoretical arguments show that the fluorescence recovery after photobleaching was indeed due to on/off kinetics of EGF binding to its receptors, and not to diffusion of the EGF receptors themselves, nor to restricted-access bulk diffusion of EGF to the membrane regions in the evanescent wave.

Given the recent successes in using TIRF to detect weak but specific equilibrium binding to surfaces, we can expect more results on the kinetics of such binding in the near future. Because of the close connection between chemical kinetics and dynamical processes in biology, TIR/FRAP measurements undoubtedly will be expanded to study reversible specific binding kinetic rates between a variety of soluble ligands and their cell surface receptors in natural or reconstituted biological membranes; the nonspecific but biologically important binding between cytoplasmic filaments and lipids in supported bilayer systems; and the attachment/detachment rates of cytoplasmic filaments with protein anchors in biological membranes.

7.5.7. TIRF Combined with Fluorescence Correlation Spectroscopy (FCS)

The volume defined by the depth of the evanescent wave in the area defined by the image plane diaphragm of a microscope can be extremely small, down to about 0.01 μm^3. Within this volume, the entrance or exit of a single fluorophore can cause a significant change in the fluorescence intensity. In fact, these TIRF fluctuations are clearly visible to the "naked eye" through the microscope. By autocorrelating (on-line) the random noise arising from such statistical fluctuations (a technique called fluorescence correlation spectroscopy, or FCS), one can obtain information about three parameters: the mean time of surface binding ($=1/k_2$), the surface diffusion coefficient, and the absolute mean number of fluorescent molecules bound per surface area (without requiring any information about quantum efficiencies or light collection efficiencies).

Two investigations have combined TIR with FCS thus far. The first[126] adapted TIR/FCS to measure the absolute concentration of virions in solution. The other[127] measured the adsorption/desorption kinetics of immunoglobulin on a protein-coated surface on the millisecond time scale.

Although TIR/FCS and TIR/FRAP both give similar information about kinetic rates and surface diffusion, and the mathematics of the two is similar,[115] there is an interesting and perhaps useful difference. Thompson[128] has shown theoretically that with TIR/FCS, but not with TIR/FRAP, one can infer kinetic rates of a nonfluorescent species as it competes with fluorescent species for the same nearly saturated surface sites.

7.6. Summary and Comparisons

TIRF is an experimentally simple technique for selective excitation of fluorophores on or near a surface. It can be set up on a standard upright or inverted microscope, preferably but not necessarily with a laser source, or in a nonmicroscopic custom setup or commercial spectrofluorimeter. In a microscope, the TIRF setup is compatible and rapidly interchangeable with bright-field, dark-field, phase contrast, and epi-illumination and accommodates a wide variety of common microscope objectives without alteration.

Confocal microscopy (CM) is another microscope technique for apparent optical sectioning, achieved by exclusion of out-of-focus emitted light with a set of image plane pinholes. CM has the clear advantage in versatility; its method of optical sectioning works at any plane of the sample, not just at an interface between substances having dissimilar refractive indices. However, other differences exist which, in some special applications, can favor the use of TIRF:

(a) The depth of the optical section in TIRF is ~0.1 μm whereas in CM it is relatively thick, ~0.6 μm.

(b) In some applications (e.g., FRAP, FCS, or on cells whose viability is damaged by light), illumination, not just detected emission, is best restricted to a thin section; this is possible only with TIRF.

(c) Since the TIRF setup can be adapted to and made interchangeable with existing standard microscope optics, even with "homemade" components, it is much less expensive than CM.
(d) TIRF has much better light throughput than currently available confocal microscopes.

Cell/substrate contacts can be located by a nonfluorescence technique completely distinct from TIRF, known as "internal reflection microscopy" (IRM).[129] Using conventional illumination sources, IRM visualizes cell/substrate contacts as dark regions. IRM has the advantage that it does not require the cells to be labeled, but the disadvantages that it contains no information about biochemical specificities in the contact regions and that it is less sensitive to changes in contact distance (relative to TIRF) within the critical first 100 nm from the surface.

Applications of TIRF in cell biology and surface chemistry include:

1. Localization of cell/substrate contact regions in cell culture.
2. High-contrast visualization of submembrane cytoskeletal structure on thick cells.
3. Measurement of the kinetic rates and surface diffusion of reversibly bound biomolecules at flattened biological and model membrane surfaces and at specifically derivatized glass surfaces (e.g., with immobilized enzymes).
4. Measurement of the concentration and orientational distributions of fluorescent molecules as a function of distance from the surface.
5. Measurement of intermolecular distances between fluorescent surface-bound molecules in the presence of a large excess of fluorophore or background fluorescence in the bulk.
6. Reduction of cell autofluorescence relative to fluorescence excited at cell/substrate contacts.
7. Construction of waveguide or optical-fiber fluorosensors usable for medical diagnoses.

Acknowledgments

We thank Dr. Nancy L. Thompson for helpful discussions and Dr. Marisela Velez, Ariane McKiernan, Andrea Stout, and Dong Wang of our lab for their contributions to various aspects of TIRF discussed here. This work was supported by a USPHS NIH grant NS 14565 and NSF grant DMB 8805296.

References

1. T. Hirschfeld, Total reflection fluorescence, *Can. Spectrosc. 10*, 128 (1965).
2. N. J. Harrick and G. I. Loeb, Multiple internal reflection spectrometry, *Anal. Chem. 45*, 687–691 (1973).
3. M. N. Kronick and W. A. Little, A new immunoassay based on fluorescence excitation by internal reflection spectroscopy, *J. Immunol. Methods 8*, 235–240 (1975).
4. R. W. Watkins and C. R. Robertson, A total internal reflection technique for the examination of protein adsorption, *J. Biomed. Mater. Res. 11*, 915–938 (1977).
5. D. Axelrod, T. P. Burghardt, and N. L. Thompson, Total internal reflection fluorescence, *Annu. Rev. Biophys. Bioeng. 13*, 247–268 (1984).
6. E. H. Hellen, R. M. Fulbright, and D. Axelrod, Total internal reflection fluorescence: Theory and applications at biosurfaces, in: *Spectroscopic Membrane Probes* (L. M. Loew, ed.), Vol. II, pp. 47–49, CRC Press, Boca Raton, Florida (1988).
7. D. Axelrod, Total internal reflection fluorescence at biological surfaces, in: *Noninvasive Techniques in Cell Biology* (J. K. Foskett and S. Grinstein, eds.), pp. 93–127, Wiley-Liss, New York (1990).
8. T. P. Burghardt and N. L. Thompson, Evanescent intensity of a focused Gaussian light beam undergoing total internal reflection in a prism, *Opt. Eng. 23*, 62–67 (1984).
9. H. Chew, D. Wang, and M. Kerker, Elastic scattering of evanescent electromagnetic waves, *Appl. Opt. 18*, 2679–2687 (1979).
10. M. Born and E. Wolf, *Principles of Optics*, 5th ed., Pergamon Press, Oxford (1975).
11. H. J. Simon, D. E. Mitchell, and J. G. Watson, Surface plasmons in silver film. A novel undergraduate experiment, *Am. J. Phys. 43*, 630–636 (1975).
12. G. W. Ford and W. H. Weber, Electromagnetic interactions of molecules with metal surfaces, *Phys. Rep. 113*, 195–287 (1984).
13. W. H. Weber and C. F. Eagen, Energy transfer from an excited dye molecule to the surface plasmons of an adjacent metal, *Opt. Lett. 4*, 236–238 (1979).
14. W. Lukosz and R. E. Kunz, Fluorescence lifetime of magnetic and electric dipoles near a dielectric surface, *Opt. Commun. 20*, 195–199 (1977).
15. W. Lukosz and R. E. Kunz, Light emission by magnetic and electric dipoles close to a plane interface. I. Total radiated power, *J. Opt. Soc. Am. 67*, 1607–1614 (1977).
16. N. L. Thompson and T. P. Burghardt, Total internal reflection fluorescence: Measurement of spatial and orientation distributions of fluorophores near planar dielectric interfaces, *Biophys. Chem. 25*, 91–97 (1986).
17. W. Lukosz and R. E. Kunz, Light emission by magnetic and electric dipoles close to a plane interface. II. Radiation patterns of perpendicularly oriented dipoles, *J. Opt. Soc. Am. 67*, 1615–1619 (1977).
18. C. K. Carniglia, L. Mandel, and K. H. Drexhage, Absorption and emission of evanescent photons, *J. Opt. Soc. Am. 62*, 479–486 (1972).
19. E.-H. Lee, R. E. Benner, J. B. Fen, and R. K. Chang, Angular distribution of fluorescence from liquids and monodispersed spheres by evanescent wave excitation, *Appl. Opt. 18*, 862–868 (1979).
20. T. P. Burghardt and N. L. Thompson, Effect of planar dielectric interfaces on fluorescence emission and detection: Evanescent excitation with high aperture collection, *Biophys. J. 46*, 729–737 (1984).
21. N. L. Thompson, H. M. McConnell, and T. P. Burghardt, Order in supported phospholipid monolayers detected by dichroism of fluorescence excited with polarized evanescent illumination, *Biophys. J. 46*, 739–747 (1984).
22. R. R. Chance, A. Prock, and R. Silbey, Molecular fluorescence and energy transfer near interfaces, *Adv. Chem. Phys. 37*, 1–65 (1978).

23. R. R. Chance, A. Prock, and R. Silbey, Comments on the classical theory of energy transfer, *J. Chem. Phys. 62*, 2245–2253 (1975).
24. K. H. Drexhage, Interaction of light with monomolecular dye lasers, *Prog. Opt. 12*, 163–232 (1974).
25. K. H. Drexhage, Influence of a dielectric interface on fluorescence decay time, *J. Lumin. 12*, 693–701 (1970).
26. C. F. Eagen, W. H. Weber, S. L. McCarthy, and R. W. Terhune, Time dependent decay of surface plasmon-coupled molecular fluorescence, *Chem. Phys. Lett. 75*, 274–277 (1980).
27. H. Kuhn, Classical aspects of energy transfer in molecular systems, *J. Chem. Phys. 53*, 101–108 (1970).
28. M. R. Philpott, Effect of surface plasmons on transitions in molecules, *J. Chem. Phys. 62*, 1812–1817 (1975).
29. T. Tamir, J. J. Burke, and G. I. Stegeman, Surface polariton-like waves guided in thin, lossy metal films, *Phys. Rev. B 33*, 5186–5201 (1986).
30. B. N. J. Persson, Theory of the damping of excited molecules located above a metal surface, *J. Phys. C 11*, 4251–4269 (1978).
31. W. H. Weber and G. W. Ford, Enhanced Raman scattering by adsorbates including the nonlocal response of the metal and excitation of nonradiative modes, *Phys. Rev. Lett. 44*, 1774–1777 (1980).
32. P. Ye and Y. R. Shen, Local field effect on linear and nonlinear optical properties of adsorbed molecules, *Phys. Rev. B 28*, 4288–4294 (1983).
33. E. H. Hellen and D. Axelrod, Fluorescence emission at dielectric and metal-film interfaces, *J. Opt. Soc. Am. B 4*, 337–350 (1987).
34. V. Hlady, R. A. VanWagenen, and J. D. Andrade, Total internal reflection intrinsic fluorescence (TIRIF) spectroscopy applied to protein adsorption, in: *Protein Adsorption* (J. D. Andrade, ed.), *Interfacial Aspects of Biomedical Polymers*, Vol. 2, Chapter 2, Plenum Press, New York (1985).
35. V. Hlady, D. R. Reinecke, and J. D. Andrade, Fluorescence of adsorbed protein layers: I. Quantitation of total internal reflection fluorescence, *J. Colloid Interface Sci. 111*, 555–569 (1986).
36. S. A. Rockhold, R. D. Quinn, R. A. VanWegenen, J. D. Andrade, and M. Reichert, Total internal reflection fluorescence (TIRF) as a quantitative probe of protein adsorption, *J. Electroanal. Chem. 150*, 261–275 (1983).
37. C. Allain, D. Ausserre, and F. Rondelez, Direct optical observation of interfacial depletion layers in polymer solutions, *Phys. Rev. Lett. 49*, 1694–1697 (1982).
38. D. Ausserre, H. Hervet, and F. Rondelez, Concentration profile of polymer solutions near a solid wall, *Phys. Rev. Lett. 54*, 1948–1951 (1985).
39. F. Rondelez, D. Ausserre, and H. Hervet, Experimental studies of polymer concentration profiles at solid–liquid and liquid–gas interfaces by optical and X-ray evanescent wave techniques, *Annu. Rev. Phys. Chem. 38*, 317–347 (1987).
40. M. Nakache, A. B. Schreiber, H. Gaub, and H. M. McConnell, Heterogeneity of membrane phospholipi in endothelial cells depends on cell substrate, *Nature 317*, 75–77 (1985).
41. D. Axelrod, Total internal reflection fluorescence microscopy, in: *Fluorescence Microscopy of Living Cells in Culture B* (D. L. Taylor and Y.-L. Wang, eds.), *Methods in Cell Biology*, Vol. 30, pp. 245–270, Academic Press, San Diego, California (1989).
42. R. M. Weis, K. Balakrishnan, B. A. Smith, and H. M. McConnell, Stimulation of fluorescence in a small contact region between rat basophil leukemia cells and planar lipid membrane targets by coherent evanescent radiation. *J. Biol. Chem. 257*, 6440–6445 (1982).
43. A. L. Stout and D. Axelrod, Evanescent field excitation of fluorescence by epi-illumination microscopy, *Appl. Opt. 28*, 5237–5242 (1989).
44. J. A. Bloom and W. W. Webb, Photodamage to intact erythrocyte membranes at high laser intensities; methods of assay and suppression, *J. Histochem. Cytochem. 32*, 608–616 (1984).

45. G. Bock, M. Hilchenbach, K. Schaenstein, and G. Wick, Photometric analysis of antifading reagents for immunofluorescence with laser and conventional illumination sources, *J. Histochem. Cytochem.* *33*, 699–705 (1985).
46. G. D. Johnson and G. M. deC. Noguerira Araujo, A simple method of reducing the fading of immunofluorescence during microscopy, *J. Immunol. Methods* *43*, 349–350 (1981).
47. G. L. Picciolo and D. S. Kaplan, Reduction of fading of fluorescent reaction product for microphotometric quantitation, *Adv. Appl. Microbiol.* *30*, 197–324 (1984).
48. H. Storz, Investigations of fading of immunofluorescence objects, *Acta Histochem. (Jena)* *71*, 2–9 (1982).
49. J. C. White and L. Stryer, Photostability studies of phycobiliprotein fluorescent labels, *Anal. Biochem.* *161*, 442–452 (1987).
50. B. S. Jacobson, J. Cronin, and D. Branton, Coupling polylysine to glass beads for plasma membrane isolation, *Biochim. Biophys. Acta* *506*, 81–96 (1978).
51. V. VonTscharner and H. M. McConnell, Physical properties of lipid monolayers on alkylated planar glass surfaces, *Biophys. J.* *36*, 421–427 (1981).
52. H. M. McConnell, T. H. Watts, R. M. Weis, and A. A. Brian, Supported planar membranes in studies of cell–cell recognition in the immune system, *Biochim. Biophys. Acta* *864*, 95–106 (1986).
53. N. L. Thompson and A. G. Palmer, Model cell membranes on planar substrates, *Commun. Mol. Cell. Biophys.* *5*, 39–56 (1988).
54. N. L. Thompson, A. G. Palmer, L. L. Wright, and P. E. Scarborough, Fluorescence techniques for supported planar model membranes, *Commun. Mol. Cell. Biophys.* *5*, 109–113 (1988).
55. V. Hlady and J. D. Andrade, Fluorescence emission from adsorbed bovine albumin and albumin-bound 1-anilinonaphthalene-8-sulfonate studied by TIRF, *Colloids Surf.* *32*, 359–368 (1988).
56. C.-G. Gölander, V. Hlady, K. Caldwell, and J. D. Andrade, Absorption of human lysozyme and adsorbate enzyme activity as quantified by means of total internal reflection fluorescence, ^{125}I labeling and ESCA, *Colloids Surf.* *50*, 113–130 (1990).
57. R. M. Zimmermann, C. F. Schmidt, and H. E. Gaub, Absolute quantities and equilibrium kinetics of macromolecular adsorption measured by fluorescence photobleaching in total internal reflection, *J. Colloid Interface Sci.* *139*, 268–280 (1990).
58. T. P. Burghardt and D. Axelrod, Total internal reflection/fluorescence photobleaching recovery study of serum albumin adsorption dynamics, *Biophys. J.* *33*, 455–468 (1981).
59. T. P. Burghardt and D. Axelrod, Total internal reflection fluorescence study of energy transfer in surface-adsorbed and dissolved bovine serum albumin, *Biochemistry* *22*, 979–985 (1983).
60. V. Hlady and J. D. Andrade, A TIRF titration study of 1-anilinonaphthalene-8-sulfonate binding to silica-adsorbed bovine serum albumin, *Colloids Surf.* *42*, 85–96 (1989).
61. C. F. Schmidt, R. M. Zimmermann, and H. E. Gaub, Multilayer adsorption of lysozyme on a hydrophobic substrate, *Biophys. J.* *57*, 577–588 (1990).
62. M. R. Rainbow, S. Arterton, and R. C. Eberhardt, Fluorescence lifetime measurements using total internal reflection fluorimetry: Evidence for a conformational change adsorbed to quartz, *J. Biomed. Mater. Res.* *21*, 539–555 (1987).
63. H. Bader, R. VanWagenen, J. D. Andrade, and H. Ringsdorf, Interactions of concanavalin A with polymerized monolayers, *J. Colloid Interface Sci.* *101*, 246–249 (1984).
64. G. K. Iwamoto, L. C. Winterton, R. S. Soker, R. A. VanWagenen, J. D. Andrade, and D. F. Mosher, Fibronectin adsorption detected by interfacial fluorescence, *J. Colloid Interface Sci.* *106*, 459–463 (1985).
65. R. Lowe, V. Hlady, J. D. Andrade, and R. A. VanWagenen, Human haptoglobin adsorption by a total internal reflection fluorescence method, *Biomaterials* *7*, 41–44 (1986).
66. V. Hlady, J. Rickel, and J. D. Andrade, Fluorescence of adsorbed protein layers. II. Adsorp-

tion of human lipoproteins studies by total internal reflection intrinsic fluorescence, *Colloids Surf. 34*, 171–183 (1988).
67. D. Horsley, J. Herron, V. Hlady, and J. D. Andrade, Human and hen lysozyme adsorption: A comparative study using total internal reflection fluorescence spectroscopy and molecular graphics, in: *Proteins at Interfaces: Physicochemical and Biochemical Studies* (J. L. Brash and T. A. Horbett, eds.), ACS Symposium Series No. 343, pp. 290–305, American Chemical Society, Washington, D.C. (1987).
68. V. Hlady, D. R. Reinecke, and J. D. Andrade, Fluorescence of adsorbed protein layers: Quantitation of total internal reflection fluorescence, *J. Colloid Interface Sci. 111*, 555–569 (1986).
69. K. Newby, W. M. Reichert, J. D. Andrade, and R. E. Benner, Remote spectroscopic sensing of chemical adsorption using a single multimode optical fiber, *Appl. Opt. 23*, 1812–1814 (1984).
70. J. D. Andrade, W. M. Reichert, D. E. Gregonis, and R. A. VanWagenen, Remote fiber-optic biosensors based on evanescent-excited fluoro-immunoassay: Concept and progress, *IEEE Trans. Electron Devices ED-32*, 1175–1179 (1985).
71. K. Newby, J. D. Andrade, R. E. Benner, and W. M. Reichert, Remote sensing of protein adsorption using a single optical fiber, *J. Colloid Interface Sci. 111*, 280–282 (1986).
72. C. Dahne, R. M. Sutherland, J. F. Place, and A. S. Ringrose, Detection of antibody–antigen reactions at a glass–liquid interface: A novel fibre-optic sensor concept, *Conf. Proc. OFS '84, 2nd International Conference on Optical Fiber Sensors*, pp. 75–79 (1984).
73. I. J. Higgins, W. G. Potter, and A. P. F. Turner, Opto-electronic immunosensors: A review of optical immunoassay at continuous surfaces, *Biosensors 1*, 321–353 (1985).
74. W. M. Reichert, J. T. Ives, P. A. Suci, and V. Hlady, Excitation of fluorescent emission from solutions at the surface of polymer thin-film waveguides: An integrated optics technique for the sensing of fluorescence at the polymer/solution interface, *Appl. Spectrosc. 41*, 636–639 (1987).
75. S. Zhao and W. M. Reichert, Protein adsorption using an evanescent chemical sensor with a fused optical fiber coupler, *J. Colloid Interface Sci. 140*, 294–297 (1990).
76. J. T. Ives and W. M. Reichert, Protein adsorption on the surface of a thin-film polymer integrated optical waveguide, *Appl. Spectrosc. 42*, 68–72 (1988).
77. T. R. Glass, S. Lackie, and T. Hirschfeld, Effect of numerical aperture on signal level in cylindrical waveguide evanescent fluorosensors, *Appl. Opt. 26*, 1218–1287 (1987).
78. K. R. Rogers, J. J. Valdes, and E. Eldefrawi, Acetylcholine receptor fiber-optic evanescent fluorosensor, *Anal. Biochem. 182*, 353–359 (1989).
79. V. Hlady, J. N. Lin, and J. D. Andrade, Spatially resolved detection of antibody–antigen reaction on solid/liquid interface using total internal reflection excited antigen fluorescence and charge-coupled device detection, *Biosensors Bioelectronics 5*, 291–301 (1990).
80. K. C. Hartner, J. W. Carr, and J. M. Harris, Total internal reflection fluorescence for adsorbed probe molecule studies of liquid/solid interfacial environments, *Appl. Spectrosc. 43*, 81–86 (1989).
81. T. Nakashima and A. Fujishima, Highly sensitive analysis of SnO_2/solution interface by internal reflection-fluorescence spectroscopy, *Chem. Lett. 1990* (11), 1995–1998.
82. V. Hlady, C. Golander, and J. D. Andrade, Hydrophobicity gradient on silica surfaces: A study using total internal reflection fluorescence spectroscopy, *Colloids Surf. 33*, 185–190 (1988).
83. S. W. Tendian, N. L. Thompson, and B. R. Lentz, Calcium-independent binding of prothrombin to negatively charged membranes, *Biophys. J. 57*, 72a (1990).
84. C. L. Poglitsch and N. L. Thompson, Interaction of antibodies with Fc receptors in substrate-supported planar membranes measured by total internal reflection fluorescence microscopy, *Biochemistry 29*, 248–254 (1990).

85. C. L. Poglitsch and N. L. Thompson, Substrate-supported planar membranes containing murine antibody Fc receptors: A total internal reflection fluorescence microscopy study, in: *Biosensor Technology, Fundamentals and Applications* (R. P. Buck, W. E. Hatfield, M. Umañia, and E. F. Bowden, eds.), pp. 375–382, Marcel Dekker, New York (1990).
86. M. L. Pisarchick and N. L. Thompson, Binding of a monoclonal antibody and its Fab fragment to supported phospholipid monolayers measured by total internal reflection fluorescence microscopy,. *Biophys. J. 58*, 1235–1239 (1990).
87. F. Rondelez, D. Ausserre, and H. Hervet, Experimental studies of polymer concentration profiles at solid–liquid and liquid–gas interfaces by optical and X-ray evanescent wave techniques, *Annu. Rev. Phys. Chem. 38*, 317–347 (1987).
88. W. M. Reichert, J. T. Suci, J. T. Ives, and J. D. Andrade, Evanescent detection of adsorbed protein concentration–distance profiles: Fit of simple models to variable-angle total internal reflection fluorescence data, *Appl. Spectrosc. 41*, 503–507 (1987).
89. P. A. Suci and W. M. Reichert, Determination of fluorescence density profiles of Langmuir–Blodgett deposited films using standing light waves, *Langmuir 4*, 1131–1141 (1988).
90. C. Allain, D. Ausserre, and F. Rondelez, Direct optical observation of interfacial depletion layers in polymer solutions, *Phys. Rev. Lett. 49*, 1694–1697 (1982).
91. D. Ausserre, H. Hervet, and F. Rondelez, Concentration profile of polymer solutions near a solid wall, *Phys. Rev. Lett. 54*, 1948–1951 (1985).
92. A. Kurahashi, A. Itaya, H. Masuhara, M. Sato, T. Yamada, and C. Koto, Depth distribution of fluorescent species in silk fabrics as revealed by total internal reflection fluorescence microscopy, *Chem. Lett. 1986*, 1413–1416.
93. A. I. Mahan and C. V. Bitterli, Total internal reflection: A deeper look, *Appl. Opt. 17*, 509–519 (1978).
94. T. P. Burghardt, Polarized fluorescent emission from probes near dielectric surfaces, *Chem. Phys. Lipids 50*, 271–287 (1989).
95. P. Suci and V. Hlady, Fluorescence lifetime components of Texas Red-labeled bovine serum albumin: Comparison of bulk and adsorbed states, *Colloids Surf. 51*, 89–104 (1990).
96. M. Masuhara, S. Tazuke, N. Tamai, and I. Yamazaki, Time-resolved total internal reflection fluorescence spectroscopy for surface photophysics studies, *J. Phys. Chem. 90*, 5830–5835 (1986).
97. A. Itaya, T. Yamada, K. Tokuda, and H. Masuhara, Interfacial characteristics of poly(methyl methacrylate) film: Aggregation of pyrene and micropolarity revealed by time-resolved total internal reflection fluorescence spectroscopy, *Polym. J. 22*, 697–704 (1990).
98. H. Fukumura and K. Hayashi, Time-resolved fluorescence anisotropy of labeled plasma proteins adsorbed to polymer surfaces, *J. Colloid Interface Sci. 135*, 435–442 (1990).
99. A. Itaya, A. Kurahashi, H. Masuhara, N. Tamai, and I. Yamazaki, Dynamic fluorescence microprobe method utilizing total internal reflection phenomena, *Chem. Lett. 1987*, 1079–1082.
100. L. E. Morrison and G. Weber, Biological membrane modeling with a liquid/liquid interface. Probing mobility and environment with total internal reflection excited fluorescence, *Biophys. J. 52*, 367–379 (1987).
101. J. G. E. M. Fraaije, J. M. Kleijn, M. van der Graaf, and J. C. Dijt, Orientation of adsorbed cytochrome c as a function of the electrical potential of the interface studied by total internal reflection fluorescence, *Biophys. J. 57*, 965–975 (1990).
102. M. M. Timbs and N. L. Thompson, Slow rotational mobilities of antibodies and lipids associated with substrate-supported phospholipid monolayers as measured by polarized fluorescence photobleaching recovery, *Biophys. J. 58*, 413–428 (1990).
103. D. Axelrod, Cell–substrate contacts illuminated by total internal reflection fluorescence, *J. Cell Biol. 89*, 141–145 (1981).

104. J. Bailey and D. Gingell, Contacts of chick fibroblasts on glass: Results and limitations of quantitative interferometry, *J. Cell Sci. 90*, 215–224 (1988).
105. W. M. Reichert and G. A. Truskey, Total internal reflection fluorescence (TIRF) microscopy. I. Modeling cell contact region fluorescence, *J. Cell Sci. 96*, 219–230 (1990).
106. F. Lanni, A. S. Waggoner, and D. L. Taylor, Structural organization of interphase 3T3 fibroblasts studied by total internal reflection fluorescence microscopy, *J. Cell Biol. 100*, 1091–1102 (1985).
107. R. J. Bloch, M. Velez, J. Krikorian, and D. Axelrod, Microfilaments and actin-associated proteins at sites of membrane–substrate attachment within acetylcholine receptor clusters, *Exp. Cell Res. 182*, 583–596 (1989).
108. M. Nakache, H. E. Gaub, A. B. Screiber, and H. M. McConnell, Topological and modulated distribution of surface markers on endothelial cells, *Proc. Natl. Acad. Sci. U.S.A. 83*, 2874–2878 (1986).
109. D. Gingell, I. Todd, and J. Bailey, Topography of cell-glass apposition revealed by total internal reflection fluorescence of volume markers, *J. Cell Biol. 100*, 1334–1338 (1985).
110. D. Gingell, O. S. Heavens, and J. S. Mellor, General electromagnetic theory of internal reflection fluorescence: The quantitative basis for mapping cell-substratum topography, *J. Cell Sci. 87*, 677–693 (1987).
111. I. Todd, J. S. Mellor, and D. Gingell, Mapping cell–glass contacts of *Dictyostelium amoebae* by total internal reflection aqueous fluorescence overcomes a basic ambiguity of interference reflection microscopy, *J. Cell Sci. 89*, 107–114 (1988).
112. T. H. Watts, H. E. Gaub, and H. M. McConnell, T-cell-mediated association of peptide antigen and major histocompatibility complex protein detected by energy transfer in an evanescent wave-field, *Nature 320*, 176–179 (1986).
113. R. M. Sutherland, C. Dahne, J. F. Place, and A. S. Ringrose, Optical detection of antibody–antigen reactions at a glass–liquid interface, *Clin. Chem. 30*, 1533–1538 (1984).
114. E. Kalb, J. Engel, and L. K. Tamm, Binding proteins to specific target sites in membranes measured by total internal reflection fluorescence microscopy, *Biochemistry 29*, 1607–1613 (1990).
115. N. L. Thompson, T. P. Burghardt, and D. Axelrod, Measuring surface dynamics of biomolecules by total internal reflection with photobleaching recovery or correlation spectroscopy, *Biophys. J. 33*, 435–454 (1981).
116. G. Adam and M. Delbruck, Reduction of dimensionality in biological diffusion processes, in: *Structural Chemistry and Molecular Biology* (A. Rich and N. Davidson, eds.), pp. 198–215, W. H. Freeman, San Francisco (1968).
117. H. Berg and E. M. Purcell, Physics of chemoreception, *Biophys. J. 20*, 193–219 (1977).
118. R. D. Tilton, C. R. Robertson, and A. P. Gast, Lateral diffusion of bovine serum albumin adsorbed at the solid–liquid interface, *J. Colloid Interface Sci. 137*, 192–203 (1990).
119. R. D. Tilton, A. P. Gast, and C. R. Robertson, Surface diffusion of interacting proteins. Effect of concentration on the lateral mobility of adsorbed bovine serum albumin, *Biophys. J. 58*, 1321–1326 (1990).
120. K. H. Pearce, R. G. Hiskey, and N. L. Thompson, Binding kinetics of fluorescently labeled bovine prothrombin fragment 1 at planar model membranes measured by total internal reflection fluorescence microscopy, *Biophys. J. 59*, 622a (1991).
121. M. L. Pisarchik and N. L. Thompson, Surface binding kinetics of a monoclonal Fab fragment on supported phospholipid monolayers measured by total internal reflection/fluorescence photobleaching recovery, *Biophys. J. 59*, 350a (1991).
122. D. Axelrod, R. M. Fulbright, and E. H. Hellen, Adsorption kinetics on biological membranes: Measurement by total internal reflection fluorescence, in: *Applications of Fluorescence in the Biomedical Sciences* (D. L. Taylor, A. S. Waggoner, F. Lanni, R. F. Murphy, and R. Birge, eds.), pp. 461–467, Alan R. Liss, New York (1986).

123. R. M. Fulbright, Adsorption kinetics of insulin at erythrocyte membranes, Ph.D. thesis, University of Michigan (1991), manuscript submitted.
124. S.-F. Sui, T. Urumow, and E. Sackmann, Interaction of insulin receptors with lipid bilayers and specific and nonspecific binding of insulin to supported membranes, *Biochemistry 27*, 7463–7469 (1988).
125. E. H. Hellen and D. Axelrod, Kinetics of epidermal growth factor/receptor binding on cells measured by total internal reflection/fluorescence recovery after photobleaching, *J. Fluor. 1*, 113–128 (1991).
126. T. Hirschfeld, M. J. Block, and W. Mueller, Virometer: An optical instrument for visual observation, measurement and classification of free viruses, *J. Histochem. Cytochem. 25*, 719–723 (1977).
127. N. L. Thompson and D. Axelrod, Immunoglobulin surface-binding kinetics studied by total internal reflection with fluorescence correlation spectroscopy, *Biophys. J. 43*, 103–114 (1983).
128. N. L. Thompson, Surface binding rates of nonfluorescent molecules may be obtained by total internal reflection with fluorescence correlation spectroscopy, *Biophys. J. 38*, 327–329 (1982).
129. D. Gingell and I. Todd, Interference reflection microscopy. A quantitative theory for image interpretation and its application to cell-substratum separation measurement, *Biophys. J. 26*, 507–526 (1979).

8

Microparticle Fluorescence and Energy Transfer

L. M. Folan and S. Arnold

8.1. Introduction

8.1.1. Fluorescence from a Microparticle

The invention of intense light sources (lasers) and real-time fluorometers (diode array devices) has allowed one, as never before, to perform fluorescence spectroscopy on extremely small samples. This enhanced experimental ability, in turn, has led to the discovery of previously unanticipated effects. Perhaps the most interesting of the recent discoveries is that the spectrum of fluorescence from a spherical hydrosol microparticle contains a pronounced structure which has little to do with molecular environment or other local effects, but which is the result of particle morphology.[1] This effect brings our general knowledge of photophysics to bear on a system of cellular size. As a result of this inquiry, it has been found that particle morphology can affect intermolecular energy transfer.[2,3] Both of these effects result from the stimulation of morphological resonances of the particle by an internally excited electronic state. Such resonances may also be stimulated by external narrow-band laser radiation. In this case the excitation spectrum can be utilized in obtaining information associated with the radial position of the fluorescent species and the orientation of fluorescent molecules at the particle surface.[4] Our focus in this review will be fluorescence from spherical microparticles whose characteristic dimensions are greater than a wavelength of visible light and limited to 30 μm in diameter. It is our intention in what follows to introduce the reader to this relatively new field and to provide a coherent framework for understanding the above effects.

L. M. Folan and S. Arnold • *M*icro*p*article *P*hoto*p*hysics *L*aboratory (MP³L), Department of Physics, Polytechnic University, Brooklyn, New York 11201.

Topics in Fluorescence Spectroscopy, Volume 3: Biochemical Applications, edited by Joseph R. Lakowicz. Plenum Press, New York, 1992.

8.1.2. Nature of the Effects

This book attests to the varied uses of fluorescence from biological systems. A review of the literature on such studies would therefore be superfluous. However, since we are particularly interested in luminescence from systems of high symmetry, certain experiments are worthy of note. One in particular is the observation of fluorescence from 3,3′-dioctadecylindocarbocyanine (diI) on the surface of an erythrocyte ghost.[5] In this case Axelrod has shown that one can use polarized excitation for the determination of the orientation of the diI chromophore relative to the surface of a cell. It was also noted that not all cells appeared with equal emission intensities. The analysis of this problem excluded from the local field the contribution due to scattering from the boundary of the ghosts. This should be a reasonable approach since only a small refractive index mismatch exists at the surface. However, more recent experiments have shown that amount of luminescence cannot be predicted without including these effects, even when the relative refractive index is 1.19.[1] The importance of including the scattered field can be best appreciated by observing the spectrum of emitted fluorescence from a dye-impregnated polystyrene microsphere in aqueous suspension. Figure 8.1 shows a set of fluorescence emission spectra taken by Benner *et al.*[1] As one can see, the spectra are laced with peaks which do not appear in the solution spectrum

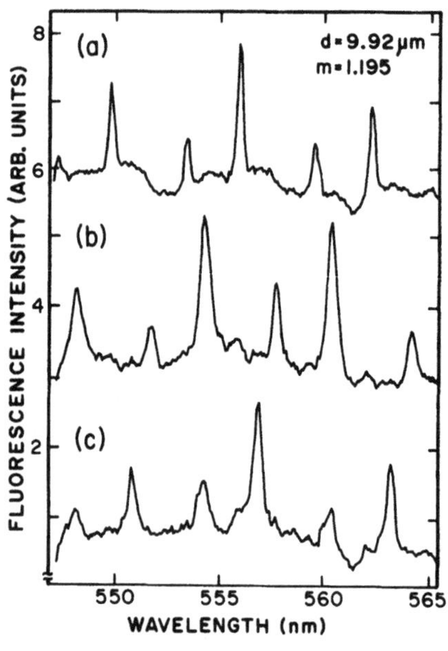

Figure 8.1. Fluorescence spectra for three individual dye-impregnated polystyrene spheres in a nominally monodispersed hydrosol (from Ref. 1).

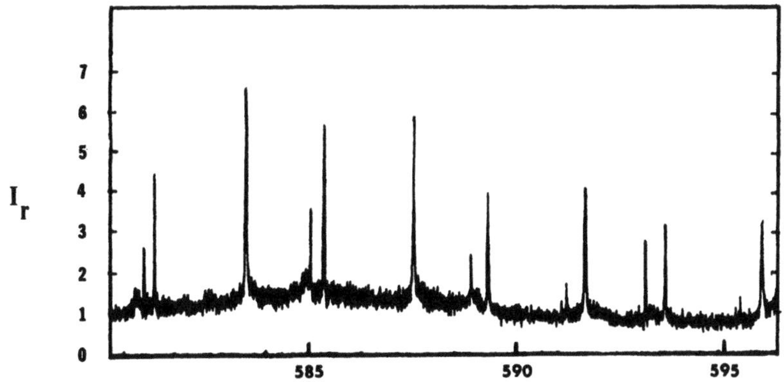

Figure 8.2. The excitation spectrum of a glycerol particle, $\sim 10\ \mu m$ in radius, containing $3 \times 10^{-6}\ M$ Sulforhodamine 101.

of the same dye. It is interesting to note that these peaks are associated with the natural electromagnetic resonances of the spherical particle. In fact, Conwell *et al.*[6] have shown that one can obtain a precise size for the particle from the optical frequencies of these peaks. Of course, one may argue that the appearance of these peaks does not necessarily change the amount of integrated luminescence from a particle. In fact, a considerable difference in integrated luminescence can occur. A good example of this phenomenon is demonstrated in Figure 8.2. Here we show the excitation spectrum of a single levitated glycerol particle, $10\ \mu m$ in radius and containing $3 \times 10^{-6}\ M$ Sulforhodamine 101. Here the dielectric mismatch is somewhat greater since the particle is levitated in air; however, the effect of changing wavelength can be quite large: in the spectrum shown, the ratio between the greatest and least luminescence is greater than 6. The particle size is of course constant; however, the effect of changing wavelength by a certain fraction at a fixed size is the equivalent of changing size at constant wavelength by the same fraction. Having demonstrated the importance of these effects, we will present a theoretical framework for understanding the experimental results in Figure 8.2 and elaborate further on its consequences.

8.2. Excitation Spectroscopy

8.2.1. Interaction of a Plane Wave with a Sphere

In the presence of a continuous plane (harmonic in time) wave of amplitude

$$\underline{\mathbf{E}}_i = E_0 e^{ikr\cos\theta} \underline{\mathbf{e}}_x \tag{8.1}$$

as defined by the coordinates shown in Figure 8.3, the internal field at the surface and throughout the interior of a particle may be derived by solving the vector wave equation for the electric and magnetic fields both inside and outside the particle and satisfying appropriate boundary conditions. Until recently, the local field was of much less interest than the far field scattering, for which the formalism is similar. The first far field scattering calculations were carried out as early as 1908 by Gustav Mie.[7] The solution for the internal field may be written in the form

$$\underline{\mathbf{E}} = \sum_{n=1}^{\infty} E_n(c_n \underline{\mathbf{M}}_{on} - i d_n \underline{\mathbf{N}}_{en}) \tag{8.2}$$

where $\underline{\mathbf{M}}_{on}$ and $\underline{\mathbf{N}}_{en}$ are vector spherical harmonics which are functions of r, θ, and the azimuthal angle ϕ (the first subscript, o or e, designates whether the function is odd or even with respect to ϕ, while the second indicates the mode number associated with the polar angle θ); E_n is a complex constant which is proportional to the incident field, and c_n and d_n are complex coefficients which depend on particle radius a, complex refractive index m, and incident wave vector k (see, e.g., Ref. 8). Equation (8.2) is particularly interesting because the $\underline{\mathbf{M}}_{on}$ and $\underline{\mathbf{N}}_{en}$ type vector spherical harmonics have considerably different physical characteristics. For example, $\underline{\mathbf{N}}_{en}$ spherical harmonics have radial, polar, and azimuthal components and are known as transverse magnetic or TM modes. On the other hand, the $\underline{\mathbf{M}}_{on}$ spherical harmonics

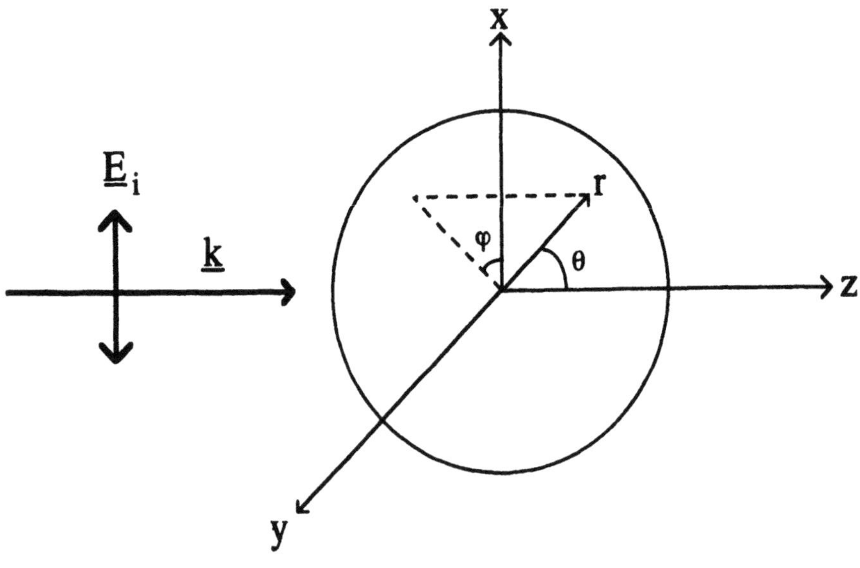

Figure 8.3. Spherical polar coordinate systems used to describe the fluorescence excitation problem.

have only polar and azimuthal components and are known as transverse electric or TE modes. This difference between the transverse electric and transverse magnetic modes may be exploited in determining the orientation of molecules at the surface of a particle since the absorption for a given molecule is proportional to the square of the projection of the local field along the direction of the transition moment. For now, let us deal with a particle which is homogeneously filled with scalar absorbers. In this case the orientation of the molecules is unimportant, and the overall absorption will be proportional to the volume average of the square modulus of the local electric field, $\langle \underline{\mathbf{E}}^* \cdot \underline{\mathbf{E}} \rangle_v$. Using Equation (8.2) and the fact that the vector spherical harmonics obey the orthogonality relations

$$\begin{aligned} \langle \underline{\mathbf{M}}^*_{pn} \cdot \underline{\mathbf{M}}_{p'm} \rangle_\Omega &= \delta_{nm}\, \delta_{pp'} \langle \underline{\mathbf{M}}^*_{pn} \cdot \underline{\mathbf{M}}_{pn} \rangle_\Omega \\ \langle \underline{\mathbf{M}}^*_{pn} \cdot \underline{\mathbf{N}}_{p'm} \rangle_\Omega &= 0 \end{aligned} \tag{8.3}$$

with the integrals taken over the solid angle, the volume average of the square modulus of the local field is given by

$$\langle \underline{\mathbf{E}}^* \cdot \underline{\mathbf{E}} \rangle_v = \sum_{n=1}^{\infty} |E_n|^2 \left(|c_n|^2 \langle \underline{\mathbf{M}}^*_{on} \cdot \underline{\mathbf{M}}_{on} \rangle_v + |d_n|^2 \langle \underline{\mathbf{N}}^*_{en} \cdot \underline{\mathbf{N}}_{en} \rangle_v \right) \tag{8.4}$$

Now we will assume in addition that the total measured fluorescence is proportional to the above volume average. This can be accomplished experimentally by suspending the fluorescent particle in an integrating enclosure and monitoring the fluorescence with an optical fiber which is pushed through a small hole in the side of the enclosure. Our interest in what follows is to use Eq. (8.4) to simulate a fluorescence excitation spectrum.

The principal frequency dependence in Eq. (8.4) comes through the c_n and d_n coefficients. These coefficients demonstrate resonant poles at which the fields at specific locations within the particle can rise by orders of magnitude over the incident field. The wave vector values at which this occurs can be easily found by examining the form of the coefficients:

$$\begin{aligned} d_n &= im/\{m\psi_n(mx)[\xi_n(x)]' - \xi_n(x)[\psi_n(mx)]'\} \\ c_n &= im/\{\psi_n(mx)[\xi_n(x)]' - m\xi_n(x)[\psi_n(mx)]'\} \end{aligned} \tag{8.5}$$

where ψ_n and ξ_n are Riccati–Bessel functions, m is the complex refractive index, and x is the optical size of the particle, ka. The resonant poles occur where the imaginary parts for each of the expressions within the denominators in Eqs. (8.5) reverse sign. For a given particle size a and mode number n, resonances can occur at a number of different wavelengths. These separate resonances are specified by an order number q. The lowest order resonance is the one with the longest wavelength. Unlike resonances for an undamped harmonic oscillator, the frequency at which a given resonance occurs in Eqs. (8.5) has a small imaginary part. This is a consequence of the scattering loss out of the resonant mode. For this reason, the resonances are termed virtual modes.

The effect which these resonances have on the ratio of the volume-averaged local field $\langle \underline{E}^* \cdot \underline{E} \rangle_v$ to the square of the incident field E_0^2 is best demonstrated from the graph in Figure 8.4. Here we see the ratio plotted versus excitation wavelength for a particle 5 μm in diameter and having a refractive index of $1.47 + 10^{-6} i$. One way to appreciate the effect which resonant excitation can have on the fluorescence excitation spectrum is to divide the ratio depicted on the left-hand ordinate by its average value "off resonance." This ratio I_r is depicted on the right-hand ordinate in Figure 8.4. We see that for the resonances calculated in Figure 8.4 the intensity at resonance can rise by a factor of 47 above the "off resonance" luminescence! Other resonances with even greater enhancements were also calculated; however, the widths were considerably narrower than the typical dye laser linewidth (0.2 Å) and were consequently rejected by the convolution used in the calculation. The actual enhancement depends on refractive index, particle size, and the linewidth of the excitation source. The local field enhancement varies considerably with position within the particle as we will see in what follows.

Figure 8.5 shows a plot of the angle-averaged field intensity as a function of position for a particle of refractive index 1.40 and optical size

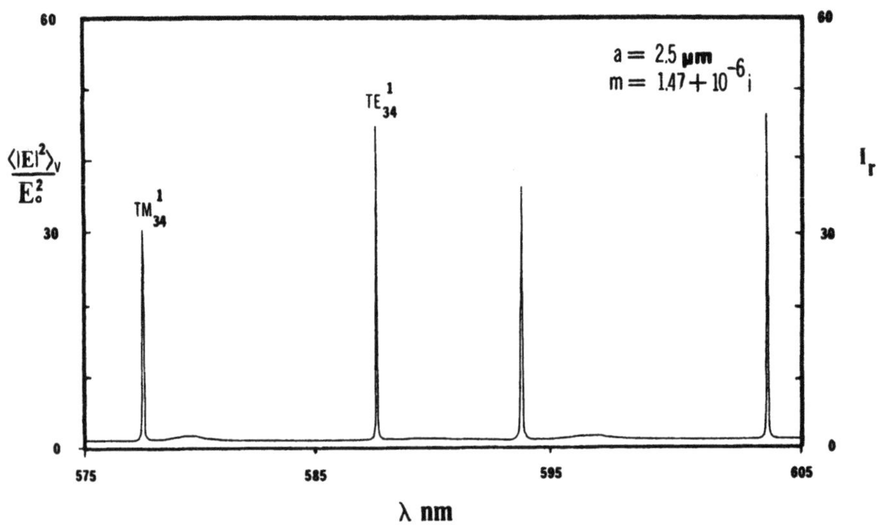

Figure 8.4. The ratio of the volume-averaged local field $\langle \underline{E}^* \cdot \underline{E} \rangle_v$ to the square of the incident field E_0^2, for a particle 5 μm in diameter and having a refractive index $m = 1.47 + 10^{-6} i$.

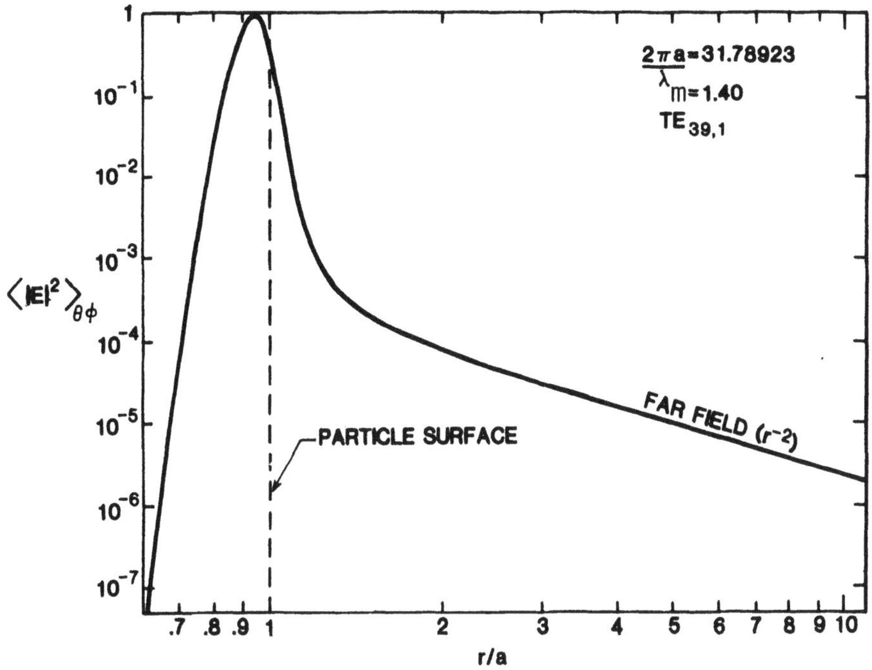

Figure 8.5. Normalized radial dependence of the angularly averaged internal intensity, $\langle \underline{E}^* \cdot \underline{E} \rangle_\Omega$ for a particle with $ka = 31.78923$ and $m = 1.40 + 0i$.

$ka = 31.78923$. This size corresponds to the $TE_{39,1}$ resonance of the particle. The angle-averaged intensity rises rapidly from the particle center and reaches a maximum at a radial position $r = 0.95a$. The intensity then decreases with increasing radial position and at large distances falls off as $1/r^2$ as would be expected for an outwardly propagating spherical wave. The major features to note are that most of the energy is localized within 10% of the particle radius from the surface, the field intensity at the surface is very large compared to that at positions near the particle center, and the evanescent field has a substantial range outside the actual boundary of the particle. The single peak near the surface is indicative of the order of the mode. Second-order modes would have two peaks, third-order three, and so on.

One can calculate the ratio of the surface-averaged intensity to the incident intensity. This simply involves changing the volume averages in Eq. (8.4) to surface averages. Such a procedure has been used by Messinger *et al.*[9] in treating surface-enhanced Raman scattering (SERS). Figure 8.6 shows the results of having done this for a particle having the same size and refractive index as the particle in Figure 8.4. We now see that the resonant enhan-

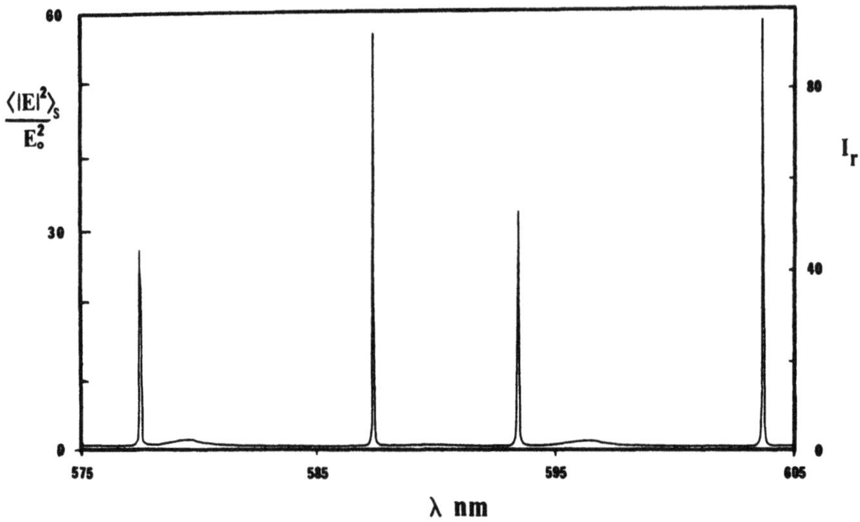

Figure 8.6. The ratio of the surface-averaged intensity to the incident intensity for the same particle as in Figure 8.4.

cements over the background are considerably larger than before. Thus, one has the ability through examination of the excitation spectrum to know whether the molecules are on the surface or throughout the bulk.

8.2.2. Excitation of a Dipole and Photoselection

The excitation spectrum proves even more useful near the surface. Since anisotropic molecules at the surface of a liquid tend to orient relative to the surface tangent, one might expect the excitation spectrum to be sensitive to such orientation. For example, suppose we take the extreme case in which molecules at the surface are oriented with their transition moments perpendicular to the surface tangent. Then the only field component which can excite these molecules is the radial field at the surface. When one recalls that only the $\underline{N}_{en}$ type vector field has radial components, one expects that a calculation of the excitation spectrum of such a molecular layer will yield half as many resonant features as shown in Figure 8.4. Indeed this is the case. Figure 8.7 shows the calculated surface average of the square modulus of the radial component of the local electric field, $\langle |\underline{E} \cdot \underline{e}_r|^2 \rangle_s$, where $\underline{e}_r$ is the radial unit vector.

A more realistic statement concerning the orientation of a molecule at the surface is that the transition moment establishes an angle to the normal θ_m but is random with respect to its projected orientation in the tangent plane. Figure 8.8 shows the associated coordinate system. The absorption by such a

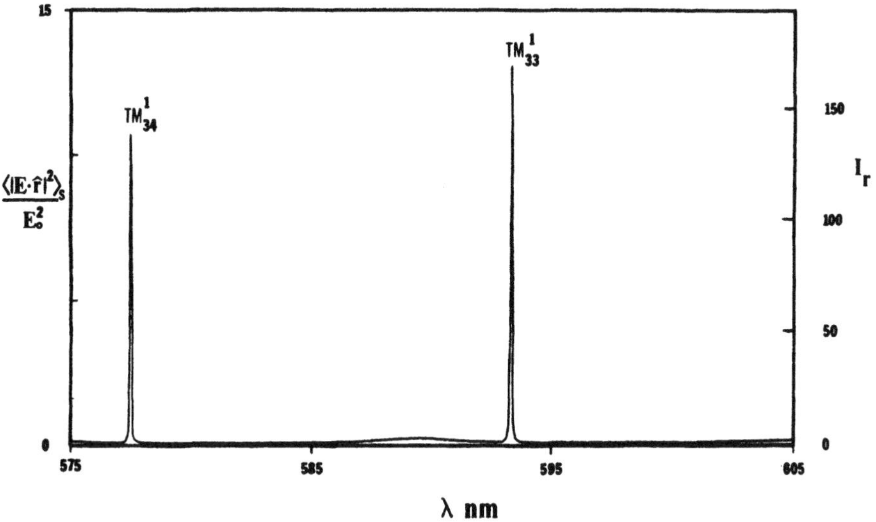

Figure 8.7. The ratio of the surface average radial intensity to the incident intensity for the same particle as in Figure 8.4.

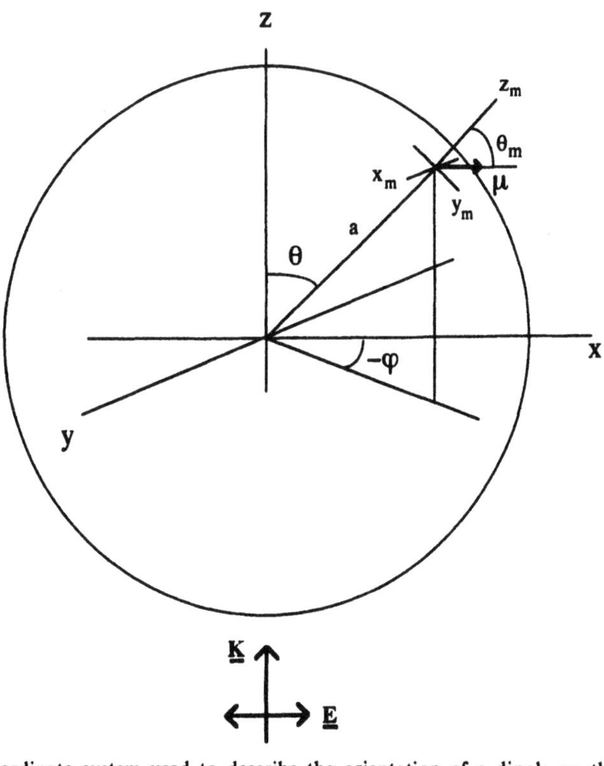

Figure 8.8. Coordinate system used to describe the orientation of a dipole on the surface of a spherical particle.

dipole is proportional to $|\underline{\mu}\cdot\underline{E}|^2$, where $\underline{\mu}$ is the vector transition moment of the molecule. This expression is easily evaluated for the case of a randomly oriented in-plane component of $\underline{\mu}$ with the result

$$\langle\langle|\underline{\mu}\cdot\underline{E}|^2\rangle_o\rangle_s=\mu^2[\langle|E_r|^2\rangle_s\cos^2\theta_m+\langle|E_t|^2\rangle_s(\sin^2\theta_m)/2] \tag{8.6}$$

where E_r and E_t are the radial and tangential components, respectively, of the local field, and $\langle\langle|\underline{\mu}\cdot\underline{E}|^2\rangle_o\rangle_s$ indicates that the quantity $|\underline{\mu}\cdot\underline{E}|^2$ is averaged first over the in-plane orientation of the dipole moment and then over the surface. Since the strength of the transition moment is arbitrary, we have plotted the normalized quantity $[\langle|E_r|^2\rangle_s\cos^2\theta_m+\langle|E_t|^2\rangle_s(\sin^2\theta_m)/2]/E_0^2$ in Figure 8.9 in representing the shape of the expected fluorescence excitation spectra. The spectrum in the foreground for θ_m equal to zero is identical with Figure 8.7 as expected. However, as θ_m is increased, other modes begin to grow into the spectra. The implication is that an experimental excitation spectrum taken on a single particle may be used in determining the angle θ_m and thus the orientation of the molecular species at the surface. The maximum luminescence at the peak of each transverse magnetic or electric resonance, $L^M_{n,q}$ or $L^E_{n,q}$, will be proportional to the value of $\langle\langle|\underline{\mu}\cdot\underline{E}|^2\rangle_o\rangle_s$ at the associated wavelength. By formally including the field components E_r and E_t into Eq. (8.6) from Eq. (8.2) and performing the averages indicated in Eq. (8.6), we arrive at the following equation for the ratio of the maximum luminescence at the $TE_{n,q}$ resonance to that at the $TM_{n,q}$ resonance:

$$\frac{L^E_{n,q}}{L^M_{n,q}}=\frac{|c_{n,q}|^2\{[\langle|M^\theta_{on}|^2\rangle_s+\langle|M^\phi_{on}|^2\rangle_s]_{n,q}(\sin^2\theta_m)/2\}}{\left(\begin{array}{l}|d_{n,q}|^2\{[\langle|N^r_{en}|^2\rangle_s]_{n,q}\cos^2\theta_m+[\langle|N^\theta_{en}|^2\rangle_s\\ \quad+\langle|N^\phi_{en}|^2\rangle_s]_{n,q}(\sin^2\theta_m)/2\}\end{array}\right)} \tag{8.7}$$

where the superscrits r, θ, and ϕ represent the radial, polar, and azimuthal components of the associated vector spherical harmonics. As one can see, the ratio vanishes at $\theta_m=0$; when the transition moment is radial, all TE modes are unable to excite the molecules.

In the preceding, we have assumed that the molecules are all oriented at a fixed angle θ_m relative to the surface normal. Thompson *et al.*[10] have utilized a distribution function in angle in place of our more restricted assumption. Inasmuch as our major interest is in showing the manner in which the particle resonances affect the excitation spectroscopy, we will continue to use the more restrictive assumption.

Equation (8.7) may be written more explicitly in terms of spherical Bessel functions:

$$\frac{L^E_{n,q}}{L^M_{n,q}}=\frac{|c_{n,q}|^2(mX^M_{n,q})^2|j_n(mX^E_{n,q})|^2(\sin^2\theta_m)/2}{\left(\begin{array}{l}|d_{n,q}|^2\{n(n+1)|j_n(mX^M_{n,q})|^2\cos^2\theta_m\\ \quad+|[mXj_n(mX)]'|^2_{n,q}(\sin^2\theta_m)/2\}\end{array}\right)}, \tag{8.8}$$

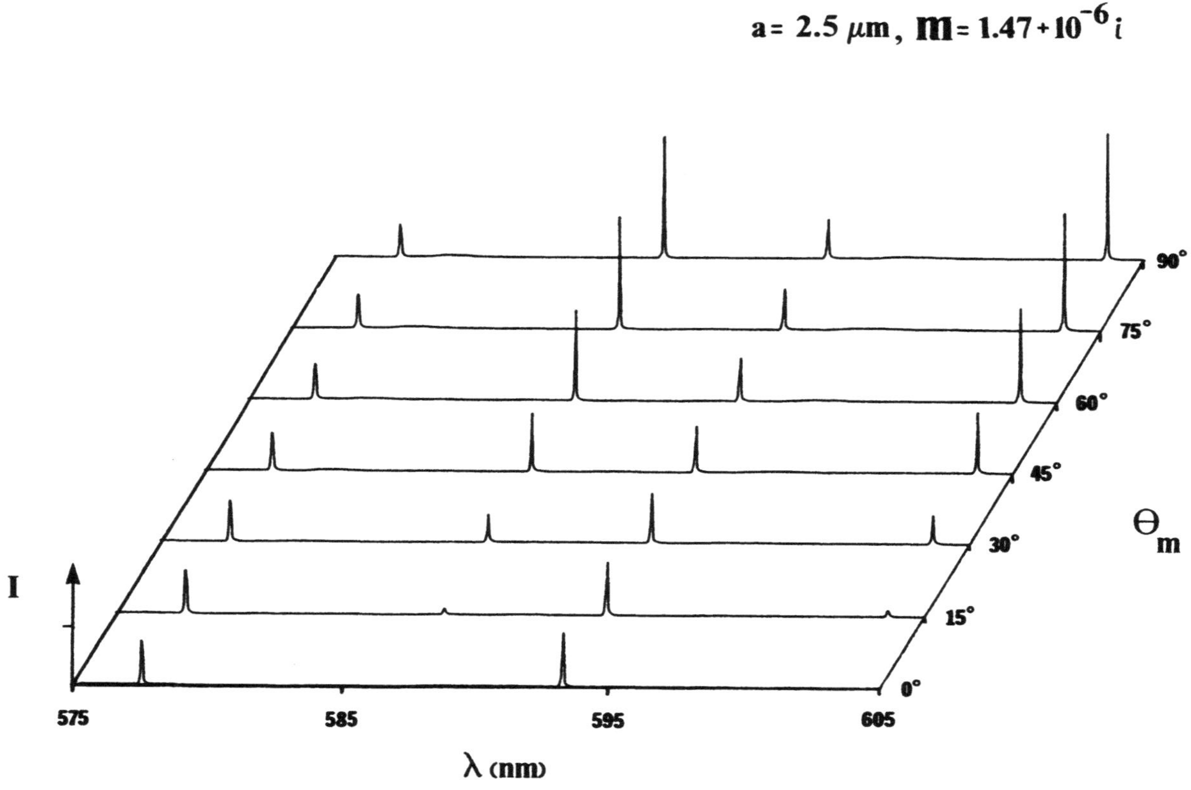

Figure 8.9. Calculated excitation spectra for different orientations of the absorption transition moment θ_m for the same particle as in Figure 8.4.

where $j_n(m\mathrm{X}^E_{n,q})$ is the nth spherical Bessel function evaluated at $m\mathrm{X}^E_{n,q}$, and $\mathrm{X}^E_{n,q}$ is the value of ka corresponding to a TE resonance of order q and mode number n. The expression $|[m\mathrm{X}j_n(m\mathrm{X})]'|^2_{n,q}$ indicates that the quantity $m\mathrm{X}j_n(m\mathrm{X})$ is to be differentiated with respect to the argument $m\mathrm{X}$ before being squared and evaluated at $m\mathrm{X}^M_{n,q}$. The sensitivity of $L^M_{n,q}/L^E_{n,q}$ to molecular orientation is best demonstrated by a simple example. If we take as an example the particle for which results have been presented in Figure 8.4 and evaluate Eq. (8.8) for the first-order resonances corresponding to a mode number of 34, we find

$$\frac{L^E_{34,1}}{L^M_{34,1}} = \frac{3.72(\sin^2\theta_m)/2}{0.66\cos^2\theta_m + 1.02(\sin^2\theta_m)/2}. \tag{8.9}$$

This result is plotted in Figure 8.10. As one can see, the ratio $L^M_{34,1}/L^E_{34,1}$ is extremely sensitive to angle, ranging from zero to 3.6 in a monotonic fashion with increasing angle. The flattening in the sensitivity of the ratio to angle near 90° is due to the in-plane components of the TM mode. An analysis of the fields on the outside when applied to the problem of determining the corresponding ratio gives similar sensitivity to that shown in Figure 8.10. In addition, other modes provide characteristics of an identical form although the maximum ratio is somewhat different.

8.2.3. Experiments

The preceding analysis yields a general formalism with which fluorescence excitation spectra of molecules in small particles can be theoreti-

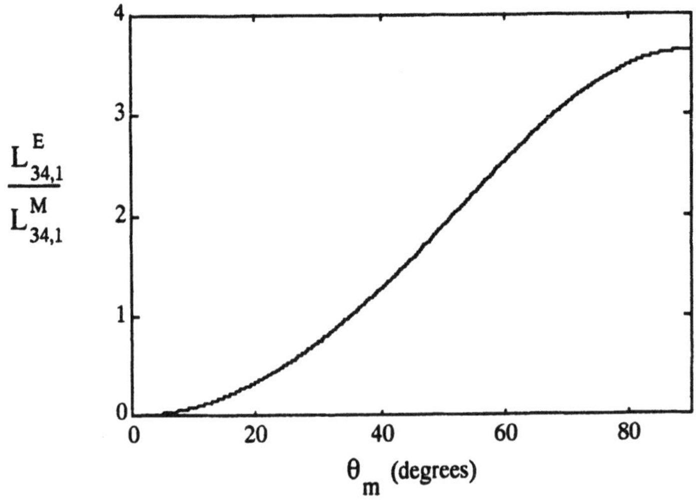

Figure 8.10. The luminescence ratio $L^E_{34,1}/L^M_{34,1}$ as a function of θ_m.

cally modeled. The observation of such spectra can best be achieved using an apparatus which allows both the isolation of individual particles and the angular integration of the emitted fluorescence. In the following section, some of the methods used to trap single particles will be discussed, specifically those which may find some use in biological experiments. Then the particular apparatus used to study fluorescence excitation spectra will be described, and some typical results presented.

8.2.3.1. Trapping Techniques

In order to stably levitate an object, the net force on it must be zero, and the forces on the body, if it is perturbed, must act to return it to its original position. The object must be at a local potential minimum; that is, the second derivatives with respect to all spatial coordinates of the potential must be positive. This may seem, at first sight, to be trivial to arrange. However, any system whose potential is a solution to Laplace's equation is automatically unstable! A statement in words of Laplace's equation is that the sum of the second partial derivatives of the potential is zero, and so not all can be simultaneously positive. This has long been known for electrostatic potentials, having been stated by Earnshaw[11]; Millikan's scheme for suspending charged particles is thus only neutrally stable, since the fields within a Millikan capacitor provide no lateral constraint.

A number of schemes have been developed which circumvent this restriction. Some require the particle to be charged, while others will work with neutral objects. Arnold[12] has recently reviewed the history, design, and operating principles of charged-particle levitators, but a brief description is provided here for completeness.

An electrodynamic levitator makes use of an ac potential to stably trap charged particles. This type of electrodynamic levitator has been used to confine objects as small as a single atomic ion[13] and is capable of trapping charged particles indefinitely. The trapping force is dynamic, with the time-averaged levitation force resulting from a spatial gradient in the electric field along the vertical symmetry axis and the phase difference between the particle motion and the applied field.[14] In vacuum the phase difference is 180°, and viscous drag introduced by a liquid or gas in the levitator acts to reduce the phase difference. With moderate applied ac voltage [about 300 V (rms)], a particle 2–30 μm in diameter, suspended in air, oscillates vertically along the axis of the levitator below the geometric center of the device. Introduction of a dc potential allows the particle to be brought to the levitator center, where the ac field is zero and the particle comes to rest. Any perturbation which displaces the particle from the center is countered by the ac force, which is always directed radially inward. The particles are confined to within a small fraction of their diameter and can be held indefinitely.

Neutral dielectric particles such as polystyrene spheres or cells provide more of a challenge. Recently, it has been demonstrated that such objects can be trapped using optical forces, even when dispersed in a liquid medium. Two forces act on an object placed near the focus of a laser beam. The first is the well-known force of photon pressure, which is directed away from the source. The second force is due to the interaction of the gradient of the fields with the induced dipole moment of the object. This force, which is in the direction of the gradient, is directed toward the beam focus. Thus, if the gradient force can be made to exceed the photon pressure force, a particle can be trapped. Ashkin *et al.*(15) have demonstrated that this is possible for a range of particle sizes and dielectric constants including for single cells.(16) The technique has recently been refined by Buican *et al.*(17) to allow automated cell manipulation and sorting.

8.2.3.2. Fluorescence Excitation

A schematic diagram of an electrodynamic levitator as used in elastic scattering and fluorescence measurements is shown in Figure 8.11. For

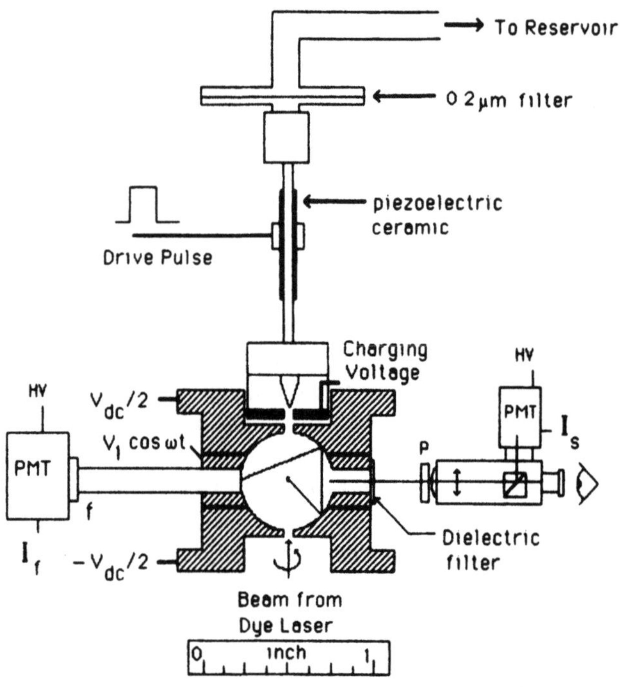

Figure 8.11. A schematic diagram of the excitation spectrometer. I_s and I_f are the elastic and fluorescence signals, respectively.

fluorescence excitation experiments, a special electrode configuration was chosen. The spherical void electrodynamic levitator (SVEL[18]) was constructed from three metal plates machined so the interior of the levitator is a perfect spherical void. This electrode configuration was chosen so that once the interior surface is coated with a diffuse reflector the SVEL become an integrating sphere. The SVEL used in the experiments described below was 0.5 in. in diameter, and the four holes constituted about 3% of the sphere surface area.

The liquid particles used were produced in a picopipette, a device akin to the printing industry's ink jet.[19] The body of the device is filled with an appropriate filtered solvent, and a small amount of the sample solution is drawn up into the glass tip by producing a small pressure differential in the stainless steel body. The orifice diameter in the glass tip ranges from 10 to 80 μm.

A voltage pulse is applied to the piezoelectric strips (PZT), and they constrict the stainless steel body and force a single droplet out through the orifice. The particle is charged by induction (by applying a dc potential to the charging electrode), and its momentum carries it into the SVEL, where it is trapped and levitated. The individual particles leaving the jet are typically the size of the orifice; however, smaller particles may be obtained by supplying a larger impulse to the PZT strips. Under such an impulse, the primary drop is left with enough energy in capillary modes at its surface that it splits apart into a number of smaller satellites.

Excitation was provided by a circularly polarized continuous-wave (cw) dye laser. Circular polarization was used to eliminate any possible azimuthal bias in the angular integration of emitted fluorescence. The elastically scattered light I_s was collected through a short-pass dielectric filter [λ(pass) < 600 nm] using a telescope with $f/6$ optics. A polarizer was used to select elastically scattered radiation polarized in the scattering plane. For the small acceptance angle used in the experiments, the elastic scattering near 90° is dominated by TM resonances.

The integrated fluorescence signal I_f was collected with a $\frac{1}{8}$-in. glass light pipe and detected through a combination of dielectric and colored glass filters with a photomultiplier tube. Fluorescence excitation and elastic scattering spectra were recorded simultaneously, in order to identify the type (TM or TE) of resonance responsible for the peaks seen in the excitation spectrum.

Figure 8.12 shows a typical pair of spectra obtained with the spectrometer. The particle was a droplet of glycerol which contained the dye Sulforhodamine 101 (SR101) at a concentration of 10^{-5} *M*. The upper curve is the excitation spectrum, and the lower one the parallel-polarized elastic scattering. The elastic scattering spectrum was used to accurately size the particle through a procedure similar to that of Chylek *et al.*[20]. This procedure involves matching the detailed shape of the scattering spectrum

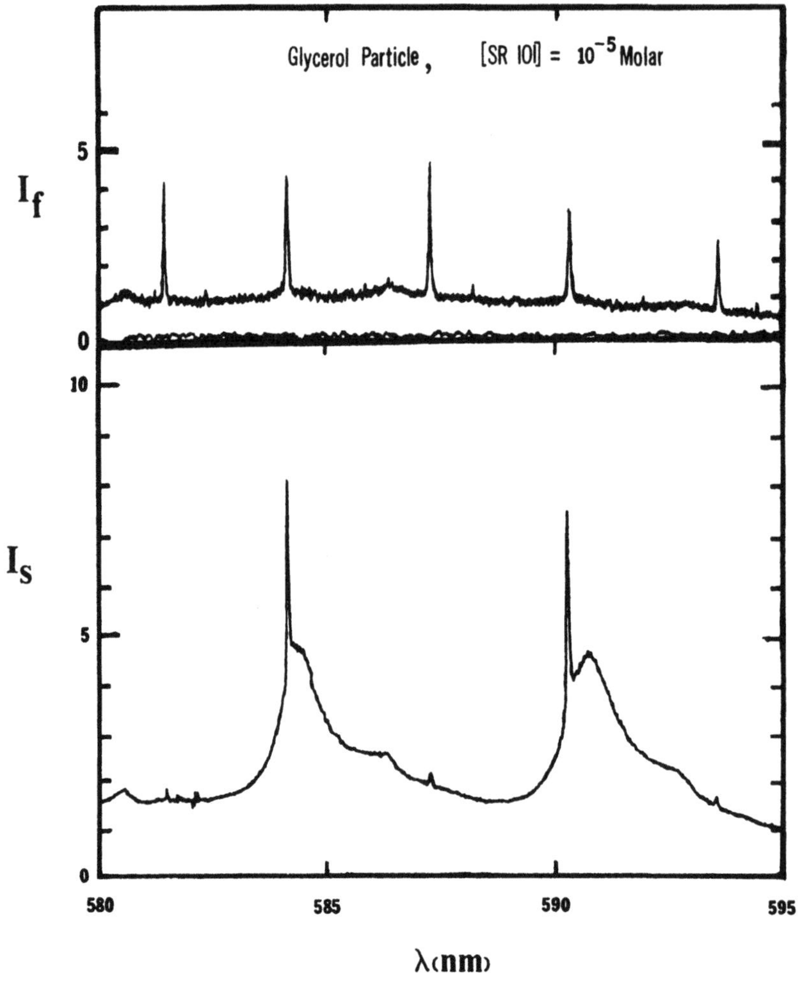

Figure 8.12. Fluorescence excitation (I_f) and elastic scattering (I_s) spectra of a glycerol particle containing 10^{-5} *M* SR101.

with calculations based on the well-known theory of Mie[7] for the elastic scattering by a homogeneous sphere.

Each of the resonances appearing in the spectra are identified and characterized by the type (TE or TM), mode number *n*, and mode order *s* (i.e., $TE_{n,s}$). Allowances were made in the fit for a small amount of scattered light polarized perpendicular to the scattering plane (due to imperfect alignment of the polarizer) and a small change in the particle radius due to evaporation during the experiment. Once the resonances are identified there are no adjustable parameters in the simulation of an excitation spectrum of a homogeneously distributed fluorescent dye. Each resonance position is a function of the incident wavelength, particle radius, and complex refractive index. The radius, wavelength, and real part of the refractive index are known, and the imaginary part of the refractive index, κ, is obtained from the concentration of the dye and its absorption spectrum.

Figure 8.13 shows the results of a simulation for the particle data obtained from Figure 8.12. The average radius was found to be 5.14 μm, and the refractive index used was $1.47 + 10^{-5}\,i$. The dimensionless optical size $x = 2\pi a/\lambda$ is displayed so that the simulation and the experimental data can be

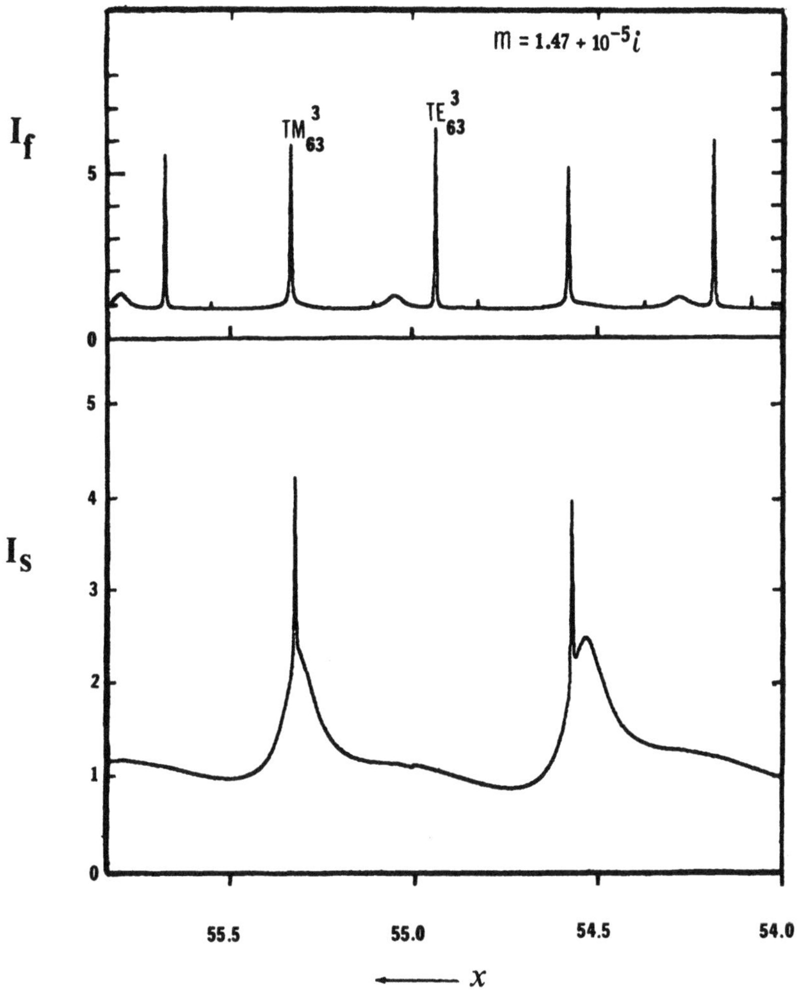

Figure 8.13. Model calculation of the data presented in Figure 8.12. The refractive index used was $m = 1.47 + 10^{-5}\,i$.

directly compared. The imaginary part of the index used corresponds to the peak molar extinction of SR101 of 10^5 liter/mol cm and the concentration of 10^{-5} mol/liter. The finite linewidth of the laser used in the experiment (0.025 nm) was accounted for in the simulation by making calculations in 0.0025 nm intervals and averaging every 10 points.

The spectra compare very well even though the variations in real and imaginary parts of the refractive index over the range considered were neglected. The five prominent peaks in the excitation spectrum are reproduced, as well as indications of both higher (broad) and lower (narrow) order resonances. The peak heights of the three peaks nearest to 585 nm (absorption maximum of SR101 in glycerol) are qualitatively reproduced. The excitation spectrum of a homogeneous particle can thus be modeled using the formalism developed above.

An inhomogeneous system of interest is one in which a monolayer or less of a material segregates to the surface of a particle. The fluorescent molecule dil(5) (Figure 8.14) is an example of a material which is expected to be surface active on polar liquids because of its hydrophilic head group and hydrophobic side chains. In fact, dil(5) has been used to prepare Langmuir–Blodgett films on water[21] and would be expected to be surface active on glycerol.

A glycerol particle with a submonolayer coating of dil(5) can be prepared in the excitation spectrometer as follows. A pure glycerol particle is produced and levitated as before, and then a second particle pipette containing 10^{-5} M dil(5) in chloroform is positioned above the levitator. Chloroform particles are injected into the SVEL and made to collide with the levitated glycerol particle. The collisions are assisted by making the charge on the chloroform particle opposite to and smaller than the charge on the glycerol particle. When a collision occurs, the levitated particle is seen to recoil and the dil(5) fluorescence is detected. The chloroform rapidly evaporates, leaving a composite dil(5)–glycerol particle. The spectra are then obtained in the usual way. Figure 8.15 shows scattering and excitation spectra for a composite particle. The resonances in the excitation spectrum are identified as before, and calculations were performed to simulate the spectrum. In the case of an inhomogeneous particle, the imaginary part of the index is a variable because

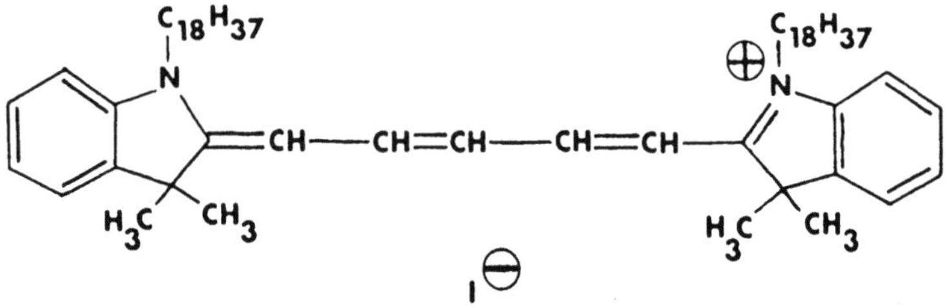

Figure 8.14. The chemical structure of dil(5).

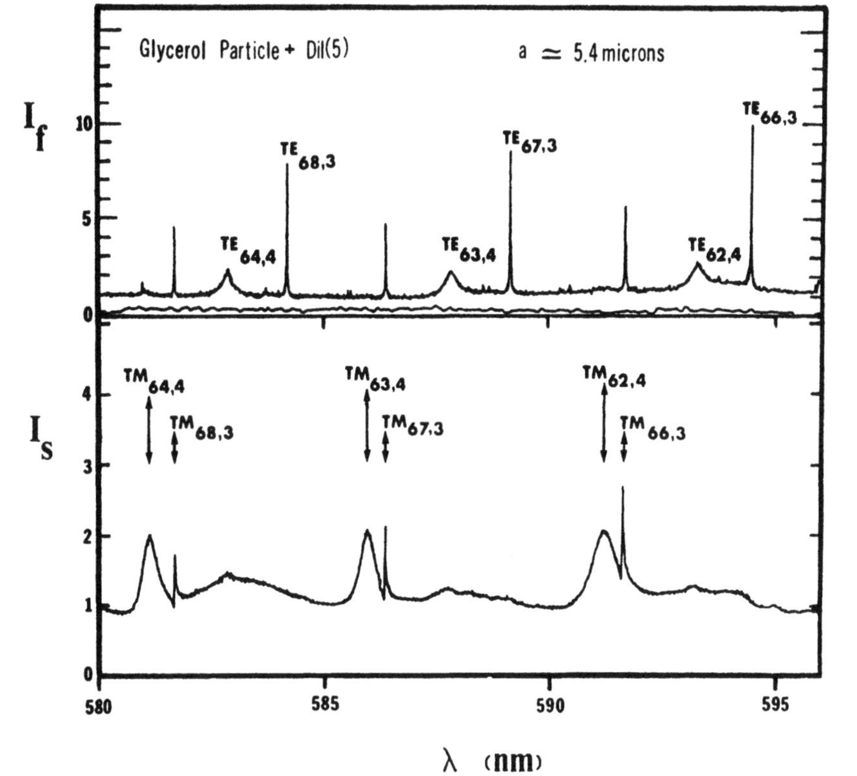

Figure 8.15. Spectra of a glycerol coated with a fraction of a monolayer of dil(5). The upper curve is the fluorescence excitation (I_f) spectrum, and the lower curve is the 90° elastic scattering (I_s) spectrum.

in the collision of the two particles it is difficult to accurately size the chloroform particle. The resonance peak height (peak height above the non-resonant background) was calculated for the resonances $TE_{68,3}$, $TM_{68,3}$, $TE_{64,4}$, and $TM_{64,4}$, using wavelength intervals of 0.0025 nm and averaging 10 points around the peak of interest. The effective imaginary part to the index κ_e was determined by forcing the peak height ratio $TE_{68,3}/TE_{64,4}$ to match the data. The ratios $TE_{68,3}/TM_{68,3}$ and $TE_{64,4}/TM_{64,4}$ were then used to decide whether the model calculations agreed with experiment.

The calculated ratios for a homogeneous distribution of absorbers gave peak ratios $TE_{68,3}/TM_{68,3}$ which were close to unity (similiarly to the SR101 simulation above), in sharp disagreement with the data for dil(5). The $TE_{68,3}/TE_{64,4}$ ratio was found to be very sensitive to the choice of κ_e while the $TE_{68,3}/TM_{68,3}$ and $TE_{64,4}/TM_{64,4}$ ratios were quite insensitive. In other words, the homogeneous model could not be made to fit the data for the dil(5)-glycerol system. Calculations for the case in which the fluorescent molecules are assumed to be at the surface of the particle and randomly oriented were also found to be incompatible with the experimental peak ratios.

The simplest nonrandom distribution is that in which the molecules are allowed to adopt an average angle with the surface normal but have random projected directions in the surface plane. Expressions for the intensity ratios for a particular mode pair are easily obtained from Eq. (8.9), and the results of such calculations are shown in Figure 8.16 for the $TE_{68,3}/TM_{68,3}$ and $TE_{64,4}/TM_{64,4}$ ratios as a function of the angle θ_m. The two data points with error bars indicate the measured ratios from Figure 8.15. The $TE_{68,3}/TM_{68,3}$ data point yields an angle $\theta_m = 66° \pm 2°$. The $TE_{64,4}/TM_{64,4}$ data point yields a less precise value for θ_m, but the two determinations agree within experimental uncertainty. The experimental results for the dil(5)-glycerol system are consistent with a model system where the absorption moments make an angle of 66° with the surface normal. A simulated spectrum for the composite particle assuming θ_m of 66° is shown in Figure 8.17. The simulation gives an accurate representation of the observed experimental spectrum.

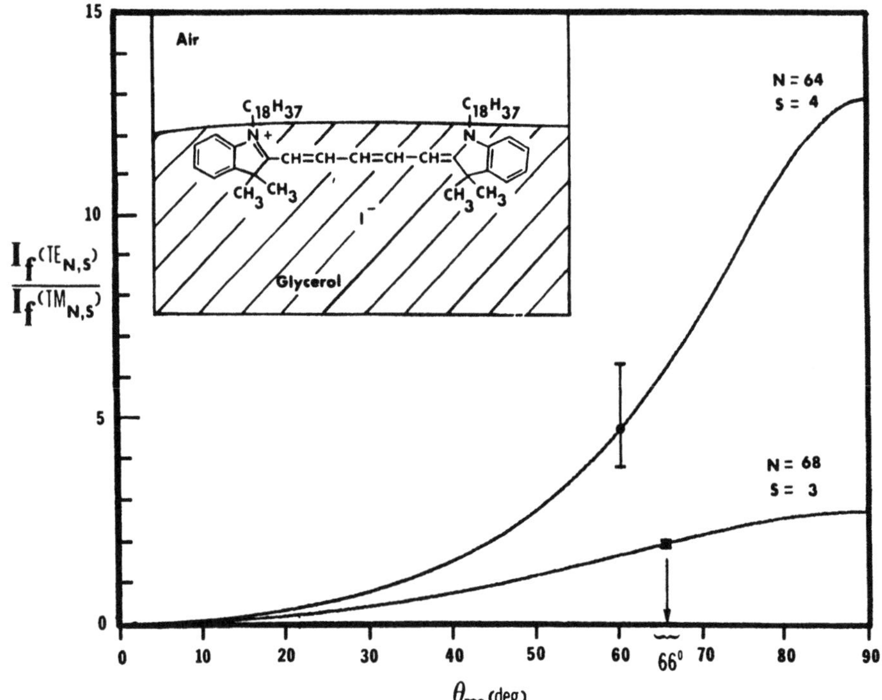

Figure 8.16. The luminescence ratio as a function of the angle θ_m for the mode pairs $TE_{64,4}/TM_{64,4}$ and $TE_{68,3}/TM_{68,3}$. The two points show the measured peak height ratios from the data in Figure 8.15.

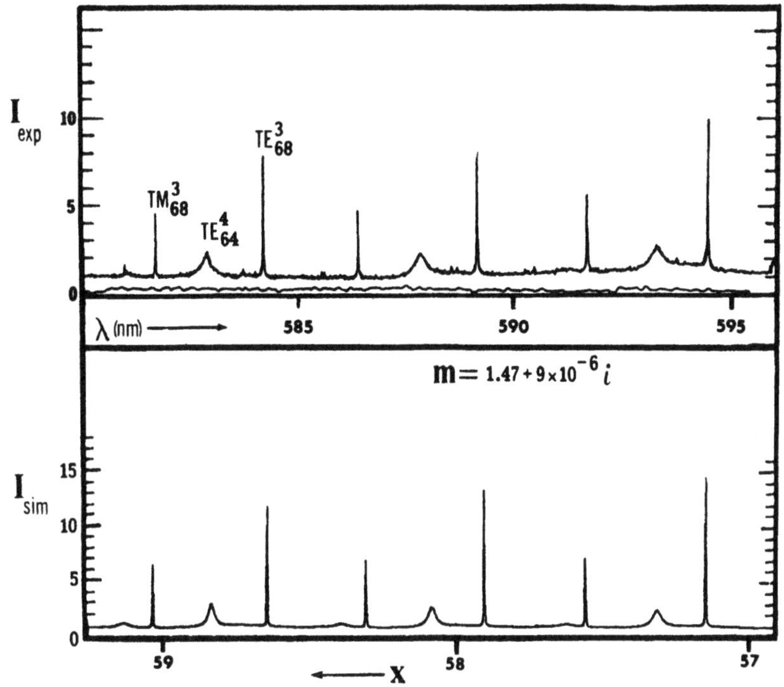

Figure 8.17. The lower curve (I_{sim}) shows the results of a model calculation for the dil(5) system, using $\theta_m = 66°$. The experimental excitation spectrum is shown in the upper curve for comparison (I_{exp}).

Studies with the homologue dil(3) have shown that its emission transition moment is parallel to the conjugated bridge of the molecule and that the absorption moment makes an angle of approximately 28° with the emission moment.[5] Direct comparison of fluorescence polarization of dil(5) and dil(3) indicates that the angle between the emission and absorption moments in the two homologues is the same to within a few degrees. Therefore, the dil(5) molecule sits at the surface of glycerol with the conjugated bridge approximately parallel to the surface plane (Figure 8.16), as one might expect from the structure of the molecule.

Fluorescence excitation spectroscopy is thus a powerful technique for obtaining molecular information about systems of cellular size. At present, the technique is restricted to single small objects because of the requirement of angular integration of the emitted fluorescence. As work progresses, similiar information will be obtainable from spectra taken at a particular angle with respect to the exciting beam. This will allow extension of the photoselection concept to suspensions of particles and perhaps to individual cells.

8.3. Emission Spectroscopy

Complementary information about molecular species can be obtained by examination of emission spectra and careful consideration of the perturbations introduced by the presence of a dielectric interface.

As we have seen, an external plane wave can excite resonances of a particle, which leads to significant variation in fluorescence intensity. A fluorescent molecule located in or near a particle can also excite the resonances of the particle. This can be modeled by again considering the molecule as a classical point dipole and obtaining the fields due to the dipole from the solution to the boundary value problem.

8.3.1. Interaction between an Excited Electronic State and a Microsphere: Radiative and Nonradiative Decay Rates

Several authors[22–30] have contributed to developing the formalism with which the effects of an interface on a dipole inside or near a particle can be treated. In the Rayleigh regime ($\lambda \gg a$), Gersten and Nitzan have made several contributions to the theory of molecular decay rates and energy transfer.[22–24] Kerker *et al.*[25] solved the boundary value problem for a dipole and a spherical particle of arbitrary size, and NcNulty *et al.*,[26] Ruppin,[27] Chew,[28] and Druger and co-workers[29,30] have used the solution to solve some of the problems of interest.

One can gain an understanding of the effect of a scattering boundary on an excited atom in a straightforward way by considering the excited state to be an oscillating dipole.[31] For simplicity, we will consider a one-dimensional case in which the scattered field induced at the position of the dipole is in the direction of the dipole. Since this scattered field is coherent with the motion of the dipole, it offers the possibility of strong feedback effects. If the feedback is positive, the oscillator will attempt to sustain itself against intrinsic losses (i.e., live for a relatively long time). However, if the feedback is negative, the oscillator will lose energy at a faster rate than its basic damping rate (i.e., "free-space" excited-state decay rate). The equation of motion of the dipole looks like the equation of motion of a driven pendulum:

$$\ddot{\mu} + \gamma_0 \dot{\mu} + \omega_0^2 \mu = \frac{e^2}{m} E_s \tag{8.10}$$

To express the fact that the scattered field is induced, we let $E_s = T\mu$, where T is in general complex, $T = T_0 e^{i\phi}$. Now we allow the dipole to oscillate as

$\mu = \mu_0 e^{-i\omega t}$ and solve for the rate of decay of the oscillation $\gamma/2$ [i.e., $\gamma/2 = \mathrm{Im}(\omega)$]. We find that

$$\gamma \approx \gamma_0 + \frac{e^2}{m} \frac{\mathrm{Im}(T)}{\omega_0} = \gamma_0 + \frac{e^2}{m} \frac{T_0 \sin \phi}{\omega_0} \tag{8.11}$$

where the approximation involves limiting the feedback so that $\mathrm{Re}(\omega) \approx \omega_0$. Thus, the rate of decay of the oscillator can be modified by the presence of the boundary. The rate can be most effectively enhanced if T is large and if ϕ is 90°. The 90° phase shift is reminiscent of the response which occurs at resonance for a mechanical system. For example, a driven pendulum has its displacement 90° out of phase with the driving force at resonance. Therefore, if an excited state stimulates a dynamic resonance in a small structure, it can be expected to alter its decay rate (note: at the moment we are referring to the total decay rate, i.e., the experimentally measured fluorescence decay rate).

To test the above ideas, Weitz *et al.*[32] performed experiments on the fluorescence decay from a thin layer of europium(III) thenoyltrifluoracetonate (ETA) deposited on a glass slide covered with Ag particles approximately 200 Å in diameter. The fluorescence decay rate was found to increase by three orders of magnitude in comparison with that of ETA in solid form. In addition to the large increase in decay rate, there was also evidence for an increase in overall fluorescence quantum efficiency. It is not possible from Eq. (8.11) to say anything about the manner in which γ is partitioned between radiative and nonradiative processes. γ should be written in terms of a radiative part γ_r and a nonradiative part γ_{nr}; $\gamma = \gamma_r + \gamma_{nr}$. Since the radiative rate for dipole emission is given by

$$\gamma_r = \frac{\omega^3}{12\pi\varepsilon_0 \hbar c^3} |\underline{\mu}|^2 \tag{8.12}$$

the increase in radiative rate can be explained by an enhancement in the dipole moment $\underline{\mu}$. Gersten and Nitzan[22] have provided a model for such an effect in which the dipole moment results from the collective moment of the particle and the excited atom. Since a metal particle has a dipolar surface plasmon mode at which it resonates, stimulation of this mode by the excited atom causes the collective dipole moment to be enhanced by orders of magnitude. On this basis, Eq. (8.12) would predict a much larger fluorescence decay rate. The nonradiative decay rate is also enhanced by orders of magnitude but not by as much as the radiative decay rate. The origin of the enhancement in the nonradiative decay rate is ohmic heating within the particle.

Since metals are highly polarizable in comparison to insulators, one might expect that effects such as those described above would not exist in the case of insulators. This is indeed the case for Rayleigh sized particles in the visible; however, for micron-sized particles an excited state can stimulate cavitylike resonances such as those mentioned in Section 8.2. Figure 8.18 shows a calculation of the phase ϕ of the scattered field, at the position of the emitting dipole, as a function of free-space wavenumber. The dipole is on the surface of the sphere with its moment radially directed. We see that the phase spectrum is similar to that of a simple mechanical oscillator; the phase shift is 90° at resonance. In addition to the 90° phase shift, the scattered field on resonance for this particular case is found to increase a millionfold in comparison with the scattered field off resonance. Thus, the interaction between an excited atom and a spherical dielectric particle can be expected, in accordance with our discussion of Eq. (8.11), to produce a considerable enhancement in the fluorescence decay rate. Indeed, Chew[28] has recently made a detailed calculation of the decay rate of an atom in the presence of a spherical microparticle and found an enhancement of several hundredfold in decay rate for one of the cases studied. In fact, in Chew's case the atom is supposed to be within the microparticle. In his calculations the atomic transition was directly at resonance with the particle. Aside from questions of the

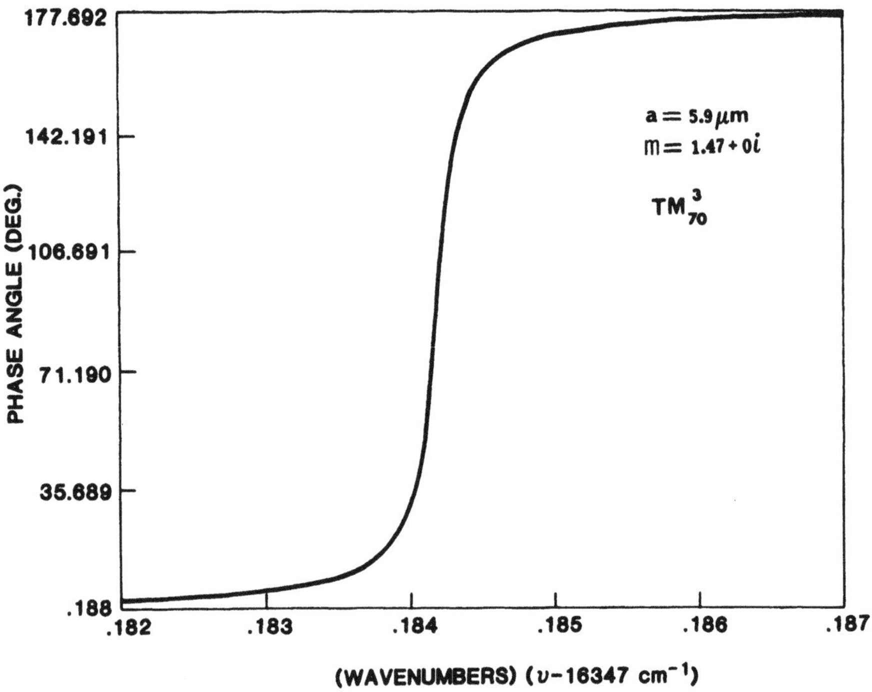

Figure 8.18. The phase of the scattered field, evaluated at the position of the source dipole, as a function of free-space wavenumber (cm^{-1}).

linewidth[33] of the atom in comparison with the particle resonance, such a coincidence is extremely difficult to manage in practice. In addition, from a biophysical standpoint, emitting species are usually molecules which are continually agitated by the local environment (e.g., a solvent) and which have a host of vibrational modes. Such molecules emit a distribution of energies characterized by the normalized distribution function $f(E)$; the integral of $f(E)$ over the entire emission band is one. The excited molecule interacts strongly with the particle only when its energy is on or near a particle resonance. As a consequence, the resulting decay rate is

$$\gamma_{tot} = \int \gamma(E)\, f(E)\, dE \tag{8.13}$$

where $\gamma(E)$ is determined from Eq. (8.11), with the field amplitude determined from

$$\underline{E}(\underline{r}) = \overleftrightarrow{T}(\underline{r}', \underline{r}, \omega) \cdot \underline{\mu} \tag{8.14}$$

where $\underline{r}'$ is the position of the dipole. The dyadic in Eq. (8.14) is obtained by solving the electrodynamic boundary value problem. To calculate decay rates, we are interested in the case for which $\underline{r}'$ is equal to $\underline{r}$; the scattered field is then the field reflected back onto the dipole. Only the component of that field in the direction of the dipole can alter its rate of decay. Thus, the field of interest is

$$E_n = (\hat{n} \cdot \overleftrightarrow{T} \cdot \hat{n})\,\mu = T_{nn}\,\mu \tag{8.15}$$

where T_{nn} is the projection of the dyadic onto unit vectors in the direction of the transition moment. The new object T_{nn} now replaces our old T in Eq. (8.11). One must calculate T_{nn} over the band of frequencies associated with the molecular emission and use the imaginary part of T_{nn} to determine the transition rate at a given frequency (or energy) from Eq. (8.11). The spectrum of transition rates is then used in Eq. (8.13) to calculate the overall rate of decay. Since the microparticle resonances are considerably narrower than the width of the molecular emission band, the overall rate of decay is not enhanced nearly as much as calculated by Chew[28] for the atomic case. Druger *et al.*[30] have anticipated this problem and have computed that for a molecule such as coumarin the overall decay rate is expected to be increased by no more than a factor of ~ 2. It should be emphasized that the contribution of the second term in Eq. (8.11) has a strong spatial dependence. The most loss-free modes (i.e., the modes having the highest quality factor) of a spherical particle have their intensities peaked near the surface of the particle as illustrated in Figure 8.5. Thus, one expects the effect on decay rate to be inhomogeneous, with molecules at the center of the particle being essentially uninfluenced.[30]

8.3.2. Angular Intensity Distribution

McNulty *et al.*[26] have investigated the effect of various dipole locations on the far-field angular intensity distribution. The calculations are interesting because they illustrate the point that the emission by a molecule in a dielectric particle cannot be treated as if the molecule were in free space. They chose an optical size of 5 and investigated the inelastic scattering intensity at selected angles as a function of the position of the dipole in the particle. The largest variation in scattered intensity reported was for the backscattering direction. The intensity varied by more than three orders of magnitude as the dipole was moved around inside the particle.[26]

Most situations of practical interest require treating a large number of radiating dipoles. The dipole distributions we shall consider radiate incoherently, and so stimulated processes will not be discussed. In passing, however, it is appropriate to mention the recent experimental observation of a number of coherent effects. R. K. Chang's group and others have observed stimulated Raman scattering, lasing, and other nonlinear effects in small dye-impregnated liquid droplets.[34–36] One of the striking things about the observations was the low threshold intensities required, in some cases as much as a factor of 10^3 lower than for comparable macroscopic samples.[34] The nonlinear optics of small particles is a rapidly developing field, and discoveries of phenomena such as optical bistability[37] will lead to expanded interest in the future.

The computation of far-field radiation from a collection of incoherently radiating dipoles is in general quite a complicated problem. To calculate the angular dependence of the far-field intensity, the volume distribution of excited states must first be obtained, which, as we have seen, depends on the volume distribution of the absorbers and the electromagnetic field which stimulates them. The fields in turn depend on the frequency and linewidth of the exciting light source. Then the emission problem for the excited-state distribution (both spatial and frequency) must be solved including reorientation and depolarization effects.

McNulty *et al.*[26] have investigated this problem for $ka \leqslant 5$ using some simplifing assumptions about the molecular properties and spatial distribution. They assumed "scalar molecules" (i.e., isotropically polarizable) uniformly distributed throughout the particle. Their comparison with experimental data on quite small particles ($ka = 1$) was encouraging but not particularly good. Druger and McNulty[29] were able to fit the data convincingly by allowing the molecules to have a preferred polarization direction and accounting for depolarization using a reorientation angle between the directions of the absorption moment and the emission moment. The fit to the polarized fluorescence data was improved and yielded a value of the reorientation angle of 29° for the dansyl chromophore excited at 366 nm. Independent measurements of fluorescence depolarization from dansyl amide in an organic glass at $-65\,^{\circ}\mathrm{C}$ yield a value of 15° for the intrinsic molecular reorientation angle.[38] The discrepancy is perhaps due to some preferential orientation of the chromophores near the particle surface as a result of the method of encapsulation of the dansyl amide in the polystyrene particles.

Theoretical work on inhomogeneous molecular distributions is as yet incomplete, but McNulty *et al.*[26] have made some calculations on adsorbed molecules on the surface of particles. Their conclusion was that the angular distributions have structure and that the inelastic scattering intensity in the forward and backward directions is particularly sensitive to particle size.

Chew and Wang[39] have pointed out the possibility of double resonance, that is, that the frequencies of both the excitation and inelastically scattered radiation are resonant. They presented the results of calculations which indicate that double resonance can have a significant effect on the angular intensity distribution of inelastically scattered radiation. This case is of some practical interest, particularly in Raman studies, where coincidence may lead to anomalous Raman band intensities, if both the excitation and the shifted frequency are resonant.

8.3.3. Energy Transfer

Less direct but convincing evidence for the effects of an interface on emission can be obtained by studying the rates of competing processes. Intermolecular energy transfer provides a probe of the environment of molecules confined to a small particle.

Energy transfer in solution occurs through a dipole–dipole interaction of the emission dipole of an excited molecule (donor) and the absorptive moment of a unexcited molecule (acceptor). Förster[40] treated the interaction quantum mechanically and derived and expression for the rate of transfer between isolated, stationary, homogeneously broadened donors and acceptors. Dexter[41] formulated the transfer rate using the Fermi golden rule and extended it to include quadrupole and higher transition moments in either the donor or the acceptor. Following the scheme of Dexter, the transfer rate for a specific transition is

$$W_t = \frac{2\pi}{\hbar} |\underline{\mu}_a \cdot \overleftrightarrow{T}(\underline{r}_d, \underline{r}_a, \omega) \cdot \underline{\mu}_d|^2 \, \delta(E_f - E_i) \tag{8.16}$$

where $\underline{\mu}_d$ and $\underline{\mu}_a$ are the transition moments for emission and absorption for the donor and acceptor molecules ptioned at $\underline{r}_d$ and $\underline{r}_a$, respectively, $\overleftrightarrow{T}$ is the dipole–dipole interaction dyadic, which in general is a function of frequency, E_i and E_f are the initial and final energies of the donor–acceptor pair, and the delta function guarantees that a transition will only take place if energy is conserved. The near-field interaction dyadic is

$$\overleftrightarrow{T}^{dd} = \frac{3\underline{R}\underline{R} - R^2\overleftrightarrow{1}}{R^5} \tag{8.17}$$

where $\underline{R}$ is the separation vector between the donor and the acceptor. Förster used this interaction in understanding the long-range energy transfer originally investigated by Perrin.(42) An explicit expression for the transfer rate is obtained by summing over all possible transitions in the molecules and expressing the rate as a function of the donor and acceptor emission and absorption spectra. This yields

$$W_t = \frac{9000 \ln(10)\, \chi^2}{128\pi^5 n^4 N_a \tau_0 R^6} \int \frac{f(\nu)\, \varepsilon(\nu)}{\nu^4}\, d\nu \tag{8.18}$$

where χ^2 is an orientational factor which depends on molecular rotation rates, n is the refractive index of the solution, N_a is Avogadro's number, τ_0 is the natural lifetime of the donor, R is the distance between the molecules, $f(\nu)$ is the normalized emission spectrum of the donor, and $\varepsilon(\nu)$ is the molar extinction of the acceptor.

The rate expression can be simplified by making use of the relationship between the fluorescence lifetime and the spectrum of the donor molecule and lumping together all the constants in one characteristic range R_0. The rate of transfer is then

$$W_t = \frac{1}{\tau}\left(\frac{R_0}{R}\right)^6 \tag{8.19}$$

R_0 characterizes a donor/acceptor pair of molecules and typically has a value between 10 and 100 Å.

The rate of transfer for a homogeneous system of donors and acceptors has been shown to be linear with acceptor concentration in dilute systems.(43, 44) This can be understood simply by presuming that the donor has a sphere of influence, the radius of which is equal to the Förster range R_0. If an acceptor molecule lies inside this sphere, the excitation is transferred; otherwise the donor deexcites by fluorescence. The probability that an acceptor will lie within the sphere of influence of an excited donor is directly proportional to the acceptor concentration, and so the transfer is linear with acceptor concentration in dilute systems.

Energy transfer has been used extensively in biological work as a "spectroscopic ruler"(45) and in numerous other studies,(46) the underlying assumption being that the Förster expressions are valid in all situations.

Gersten and co-workers(23, 24) investigated theoretically the effect of a small metal particle on the rate of energy transfer between two diametrically opposed external point dipoles. The emission dipole of the donor was polarized perpendicular to the particle surface. They used an electrostatic solution for the fields due to the donor dipole, and since the particle is assumed to be much smaller than the wavelength, this should accurately represent the fields. They took the ratio of the dipole–dipole interaction with the particle present to that without the particle. This yielded an enhancement factor for the interaction, which for a sphere and collinearly placed dipoles is

$$|A|^2 = \left\{1 + \frac{1}{2}\sum_N (-1)^N (N+1)^N \left(\frac{a}{r_a r_b}\right)^{N+2} \left[\frac{\varepsilon - 1}{\varepsilon + ((N+1)/N)}\right]\right\}^2 \quad (8.20)$$

where a is the metal particle radius, r_d and r_a are the distances from the center of the sphere to the donor and acceptor, respectively, and ε is the dielectric function for the metal. Resonances occur whenever ε is $-[(N+1)/N]$, and the rate of energy transfer is expected to be significantly enhanced over the free-space value. As we have noted previously, similar enhancements also occur for the rate of energy loss to the metal and far-field radiation by the donor. Thus, it may be difficult to observe enhanced energy transfer due to small metal particles.

Folan *et al.*[2] recently investigated energy transfer in micron-sized dielectric particles. They found that the transfer was enhanced by as much as a factor of 100 in the particles compared to the transfer observed in bulk quantities of the same material. An interesting aspect of this work is that the concentrations used in the experiment place the average separation between molecules at over 1000 Å. Thus, no nearfield model such as the dyadic used by Förster (Eq. 8.17) or the electrostatic approach used by Gersten and Nitzan is applicable. In what follows, we will attempt in a simple manner to show how such an effect might take place. Following this, we will review the experiments and present a more comprehensive model for the effect.

For the present, let us suppose that a donor molecule is sitting on the surface of a particle with its emission moment ($\underline{\mu}_d$) perpendicular to the surface, as shown in Figure 8.19. Only one acceptor molecule is available, and it is also on the surface with its absorption moment ($\underline{\mu}_a$) perpendicular to the particle surface. For a typical dielectric particle 5 μm in radius, the maximum distance between the donor and acceptor would be 100,000 Å. We would now like to calculate the effect which this sphere has on the energy transfer rate. In other words, we would like to calculate the ratio of the transfer rate with the sphere present to that without the sphere present. This ratio $|A(\omega)|^2$ is

$$|A(\omega)|^2 = \frac{|\underline{\mu}_a \cdot \overleftrightarrow{T}^{\text{sp}}(\underline{r}_d, \underline{r}_a, \omega) \cdot \underline{\mu}_d|^2}{|\underline{\mu}_a \cdot \overleftrightarrow{T}^{\text{dip}}(\underline{r}_d, \underline{r}_a, \omega) \cdot \underline{\mu}_d|^2} \quad (8.21)$$

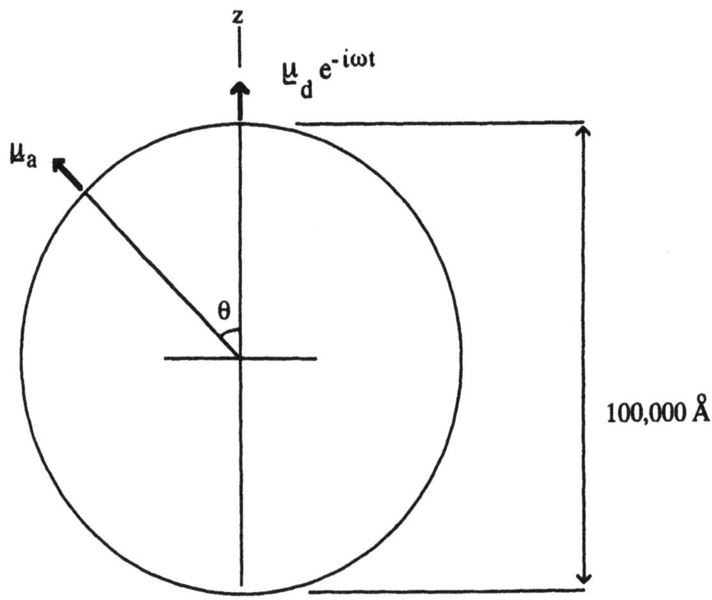

Figure 8.19. Coordinate system used to describe the positions and orientations of the donor and acceptor dipoles on the surface of a sphere.

Unlike the near-field dyadic of Förster, which has no frequency dependence, the dyadics appearing in the above expression are explicitly frequency-dependent due to the range of the interaction. In particular, $\overleftrightarrow{T}^{\mathrm{sp}}$ is the appropriate dyadic with the sphere in place, and $\overleftrightarrow{T}^{\mathrm{dip}}$ is the dyadic in the absence of the sphere. Although $\overleftrightarrow{T}^{\mathrm{dip}}$ is easily obtained from dipole radiation theory, $\overleftrightarrow{T}^{\mathrm{sp}}$ must be obtained by solving the appropriate boundary value problem. When one considers that $\overleftrightarrow{T}(\underline{r}_d, \underline{r}_a, \omega) \cdot \underline{\mu}_d$ is the electric field at the acceptor [see Eq. (8.14)], it becomes apparent that $|A(\omega)|^2$ is simply a ratio of intensities. For the case of transition moments which are normal to the surface as depicted in Figure 8.19, the numerator of Eq. (8.21) reduces to

$$|\underline{\mu}_a \cdot \overleftrightarrow{T}^{\mathrm{sp}}(\underline{r}_d, \underline{r}_a, \omega) \cdot \underline{\mu}_d|^2 = |\underline{\mu}_d|^2\, |\underline{\mu}_a|^2\, |\hat{r} \cdot \overleftrightarrow{T}^{\mathrm{sp}}(R, 0; R, \theta_a; \omega) \cdot \hat{r}|^2 \tag{8.22}$$

The projection of $\overleftrightarrow{T}^{\mathrm{sp}}$ on each of the radial unit vectors can be evaluated in terms of the basic angular functions which make up the vector spherical harmonics.[27] Although these functions are associated Legendre polynomials for an arbitrarily oriented donor dipole, for the case of full azimuthal symmetry shown in Figure 8.19 the angular functions are ordinary Legendre functions, P_n (i.e., $m = 0$). Under these circumstances,

$$|\underline{\mu}_d|^2\, |\underline{\mu}_a|^2\, |\hat{r} \cdot \overleftrightarrow{T}(R, 0; R, \theta_a; \omega) \cdot \hat{r}|^2 \propto |a_n|^2\, [P_n(\cos\theta_a)]^2 \tag{8.23}$$

where a_n is a frequency-dependent coefficient associate with far-field scattering by TM modes. The Legendre function is plotted in Figure 8.20 for $n = 10$. The most striking feature about Figure 8.20 is that the maximum transfer (for angles larger than zero degrees) occurs when the acceptor dipole is farthest from the donor! The frequency-dependent coefficient a_n has its own peculiar features. Just as the interal TM modes resonate at particular frequencies [see Eqs. (8.5)], the associated external field contains resonant poles at these same frequencies. Figure 8.21 shows $|A(\omega)|^2$ as a function of wavelength over a limited spectral region for $\theta_a = 180°$. Resonant transfer is apparent from this figure with an enhancement of 10^{15} over the case with no sphere present. It should not be surprising that the mode associated with this enhanced intensity is of TM character. In fact, no TE modes are stimulated by our donor at any frequency. This effect is the reciprocal to the property found in the photoselection case; a TE resonance cannot stimulate a radially oriented dipole. The 10^9-fold enhancement seen off resonance is a consequence of geometrical focusing of the dipole radiation by the sphere. Figure 8.21 clearly shows that the same resonances which are excited by external laser radiation can also be excited by a dipole in close proximity to the sphere. A simple calculation shows that the maximum enhancement in Figure 8.21 would lead to a transfer rate equivalent to that predicted by the Förster theory if the acceptor were moved to a separation of ~350 Å. However, the entire particle is now accessible, and therefore one can anticipate a large overall enhancement over Förster transfer. This enhancement is expected to be greatest at dilute concentrations since absorptive losses are known to damp the resonant modes.

Before leaving this section, it is important to note that the huge enhance-

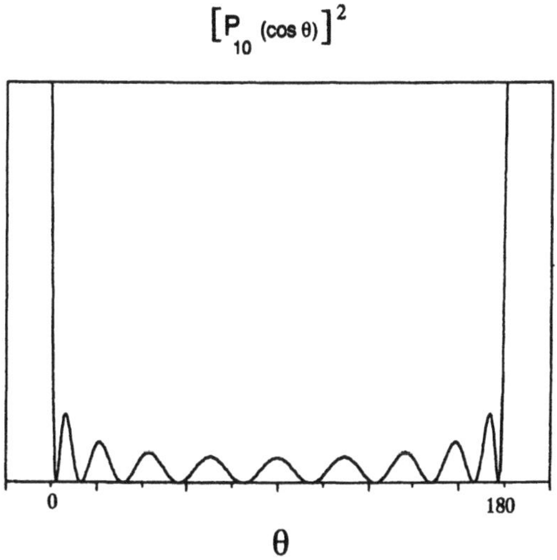

Figure 8.20. A plot of $[P_{10}(\cos\theta)]^2$ as a function of θ.

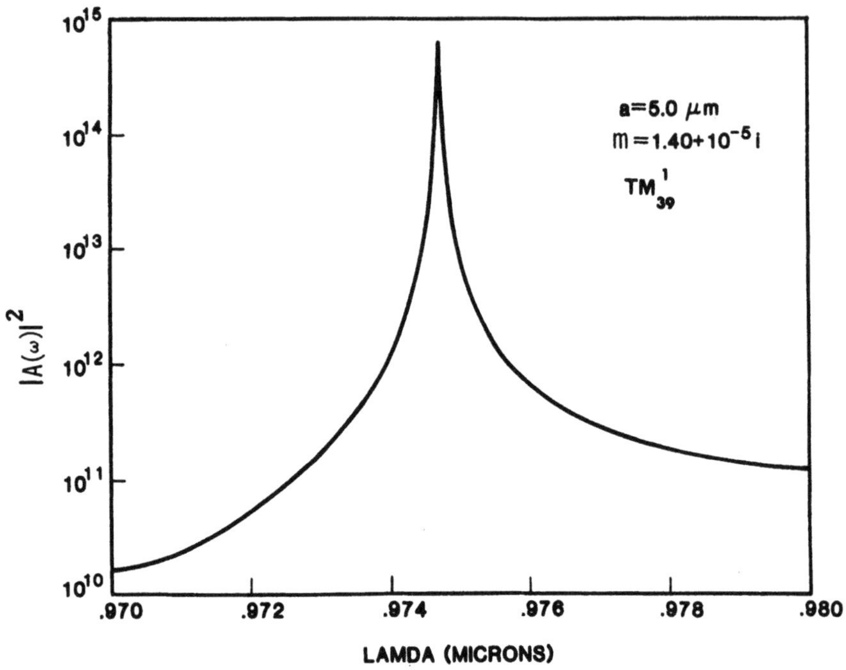

Figure 8.21. Results of a calculation of $|A(\omega)|^2$ for a sphere of radius 5 μm plotted as a function of the dipole free-space wavelength.

ment exhibited in the energy transfer rate does not guarantee that the probability for transfer to the acceptor will be enhanced. Up till now, we have only considered the enhancement in transfer rate between the donor and acceptor. We should also consider the rate at which the donor-particle system radiates into the far field. If this radiative rate is more greatly enhanced than the transfer rate, the probability for electronic energy transfer will actually be reduced. So, on to the experiments.

8.3.4. Experiments

A schematic diagram of the apparatus used in the energy transfer experiments[3] is shown in Figure 8.22. The particles are produced and levitated in an electrodynamic levitator as described previously. Excitation is provided by the filtered output of either a Xe or Hg–Xe high-pressure arc. The intensity produced at the particle was found to be 10–50 mW/cm^2. The fluorescence emitted from each of the levitated particles was monitored at 90° to the exciting beam using $f/3$ optics, dispersed with a $\frac{1}{4}$-m monochromator, and detected with an optical multichannel analyzer. The levitator could be

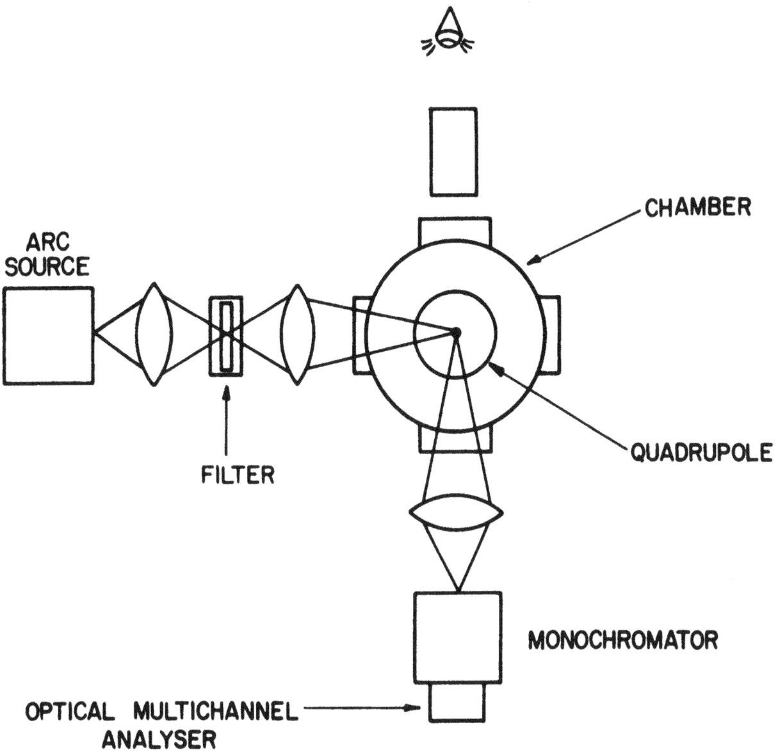

Figure 8.22. Schematic diagram of the apparatus used in the energy transfer experiments.

cooled below room temperature using a combination of a thermoelectric device and circulating ice water. Particles containing 10^{-5} mol/liter of an organic laser dye produced sufficient intensity that 0.5-s detector exposure time gave an adequate signal-to-noise ratio for real-time studies. Longer exposure times were used to enhance the signals at low dye concentration.

Two donor–acceptor systems were examined, with Coumarin 1 (C1) and 9-aminoacridine (9AA) as donors and Rhodamine 6G (R6G) as the acceptor. Initial experiments were performed to compare the amount of transfer observed in bulk solution and in particles made of the same material. Glycerol was chosen as the solvent, mainly because of its low vapor pressure and high viscosity. The low vapor pressure was necessary so that particles would be relatively stable in size, and the high viscosity ensures that the excited donor is essentially stationary for the lifetime of the excited state. The concentrations used were chosen to minimize donor reabsorption and to make the extinction of the donor considerably larger than the extinction of the acceptor at the excitation wavelength. Excitation wavelengths of either 365 or 387 nm were used in the experiments. Concentration ratios, donor to acceptor, of 10:1 and

100:1 were used with the C1/R6G system, and ratios of 50:1 and 500:1 were used with the 9AA/R6G system.

Figure 8.23 shows emission spectra characteristic of the energy transfer systems studied in Ref. 2. In each case the principal excitation in the ultraviolet is of the donor, C1. The lowest curve (a) is for a neat C1 solution in glycerol at 3×10^{-4} M. The second curve (b) is the spectrum of a bulk sample containing the donor and an acceptor (R6G at a concentration of 3×10^{-5} M). The upper curve (c) shows the spectrum of a 10-μm-diameter spherical particle of the same material used to obtain curve b. The emission intensity, normalized to the donor peak, is considerably enhanced at the acceptor peak, indicative of extra transfer in the particle compared to the corresponding bulk sample.

The spectra are very smooth when compared to the emission spectra shown for polystyrene particles in Figure 8.1. The anticipated resonances are not observed in Figure 8.23. The evaporation of the particle at room temperature is slow, but is still rapid enough that over the integration time of the detection system the resonance structure is completely washed out. Figure 8.24 shows the effect of cooling the levitating chamber to 13 °C. The upper curve shows the emission spectrum of a cooled particle. The next lower curve shows a room temperature spectrum of a similar particle. The lowest

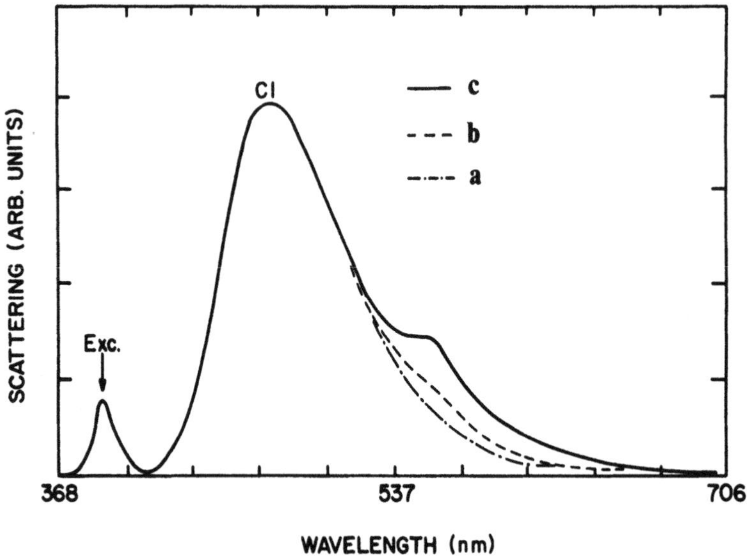

Figure 8.23. Emission spectra characteristic of the C1/R6G energy transfer system. The lowest curve (a) is the spectrum of the donor (C1) in bulk solution. The center curve (b) is the spectrum of a 10:1 mixture of the donor and acceptor (R6G) in bulk solution. The upper curve (c) is the spectrum of a particle made of the same mixed solution as used for curve (b).

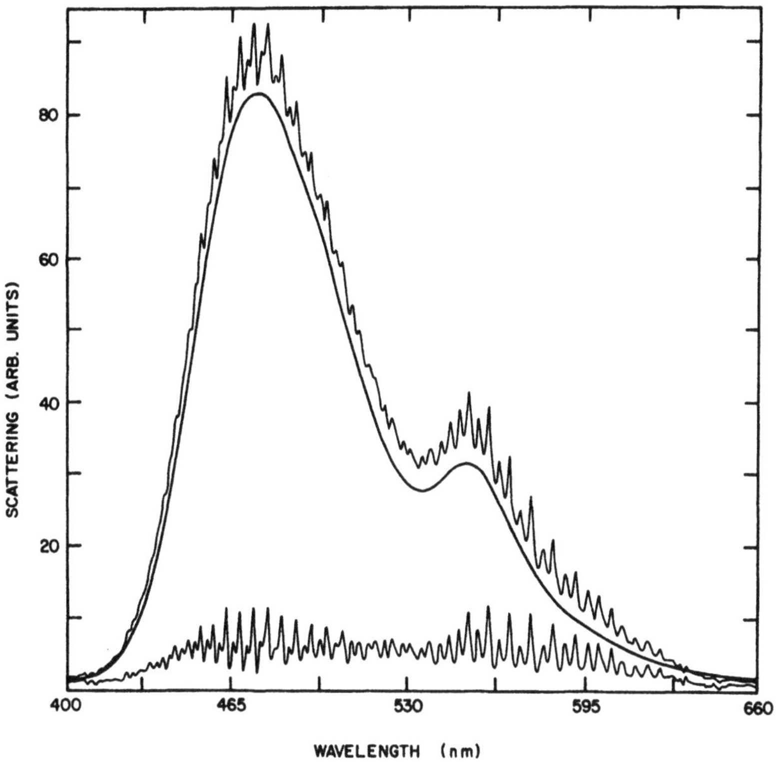

Figure 8.24. The effect of cooling the levitator is to reveal the resonances in the upper curve. The curve just below it is a room temperature spectrum of a similar particle. The difference is displayed below.

curve is the difference between the low-temperature curve and the room temperature curve. One clearly sees the spectrum of the resonances. The pronounced minimum in the amplitude near 530 nm is indicative of a particle-assisted mechanism, since the position of this minimum corresponds to the position of the maximum overlap between the emission of the donor and the absorption of the acceptor.

The resonance structure in Figure 8.24 is resolution limited; however, a higher resolution spectrum taken on a similar particle is shown in Figure 8.25. The spectrum is centered near the peak of the acceptor emission band. The smooth curve is the extinction spectrum of the acceptor R6G. The modes at the right are identified by mode number, and two orders (3 and 4) are present. The third-order resonances are narrower, and as the progression to higher mode number is followed (71, 72, ...), the amplitudes of the peaks rapidly decrease, even though the underlying fluorescence intensity is

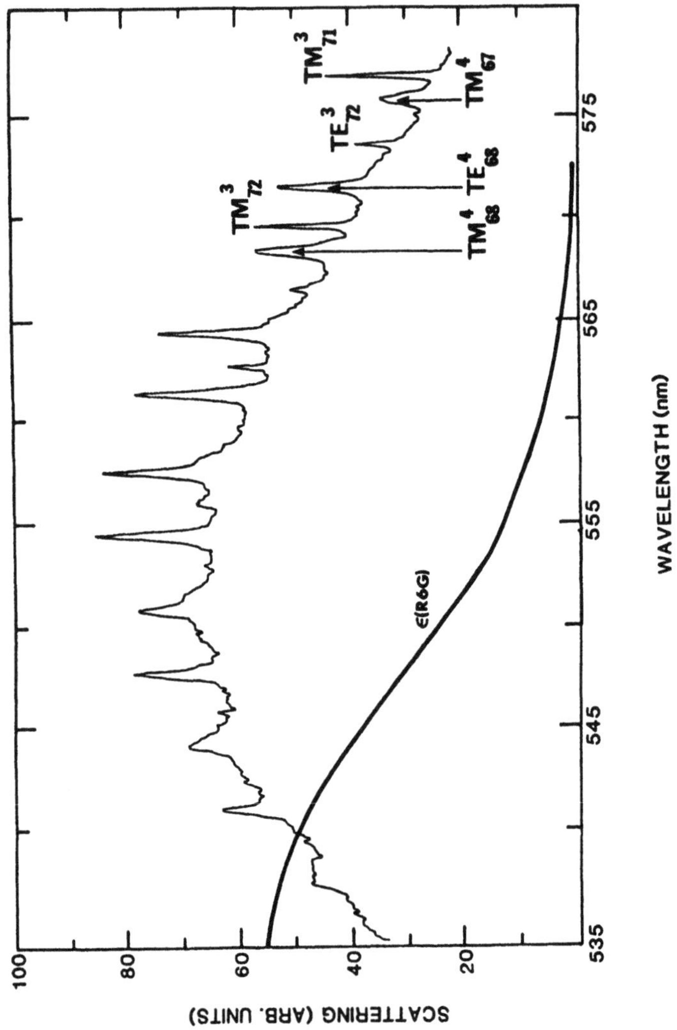

Figure 8.25. A higher resolution spectrum than the upper curve in Figure 8.24, showing the region near the acceptor peak. The smooth curve is the extinction spectrum of R6G.

increasing. The fourth-order resonances show similar behavior, but these resonances have lower intrinsic Q factors and thus are damped at higher extinction.

More convincing proof for a particle-enhanced energy transfer mechanism comes from a study of the concentration dependence of the transfer. Bulk Förster transfer leads to a linear dependence on acceptor concentration with constant donor-to-acceptor ratio. The resonance mechanism would be expected to saturate at (relatively) high concentrations and fall off linearly at very low concentrations.

To get a measure of the transfer, the ratio of the acceptor to donor peak heights was measured for several particles at each of several acceptor concentrations. Figure 8.26 shows the experimental results for the two donor–acceptor pairs, C1/R6G and 9AA/R6G. The data were corrected for a small

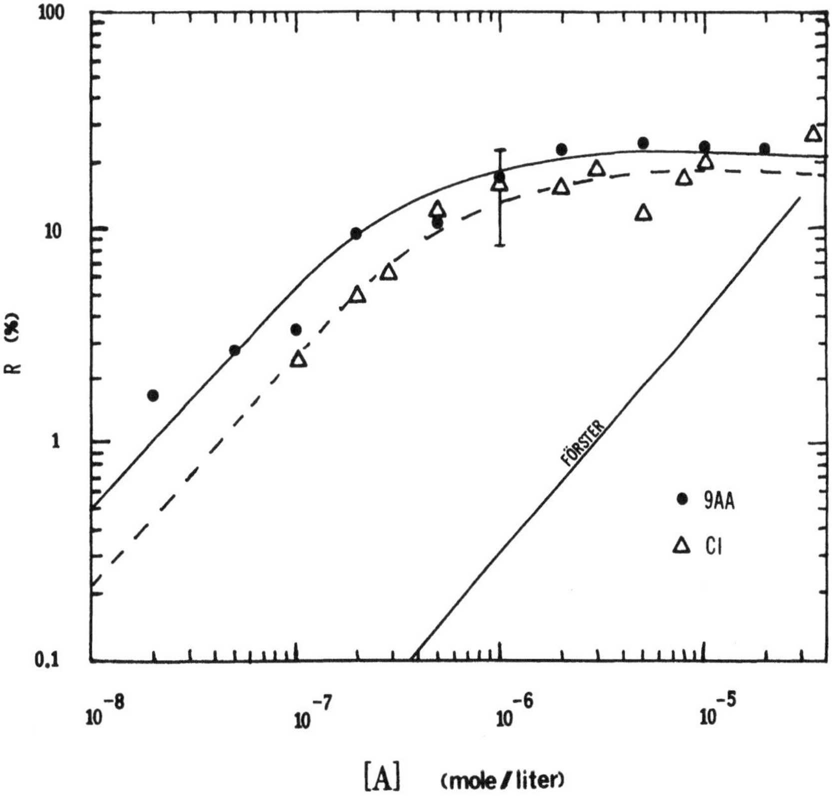

Figure 8.26. Plot of the acceptor/donor peak height ratio as a function of the acceptor concentration, for the two systems C1/R6G and 9AA/R6G. The straight line is an extrapolation to low concentrations of bulk data for the C1/R6G system.

amount of direct excitation of the acceptor by the UV excitation. The ratio of the acceptor peak height to the donor peak height, R (%), is plotted against acceptor concentration. The straight line is an extrapolation to low concentration of bulk measurements on the C1/R6G system. The particle transfer is enhanced by as much as a factor of 100 over Förster transfer.

Figure 8.26 may be modeled by considering the detailed interaction between pairs of molecules as suggested by the theory in Section 8.3.4. This involves calculating the rate of transfer between a given donor–acceptor pair at particular locations within the particle using Eq. (8.16). The interaction must then be averaged over random orientations for both the emission moment of the donor and the absorption moment of the acceptor. Finally, the overall rate for a given donor is determined by the addition of all pair interactions with acceptors at random locations within the particle. This rate is then compared with the radiative decay rate which is determined by calculating the rate of radiative loss into the far field. The efficiency for energy transfer within the particle is then determined by dividing the overall rate of transfer by the sum of the rate of radiative decay and the rate of transfer. Druger *et al.*[30] have made such a calculation. This calculation clearly demonstrates that energy transfer in a particle is assisted by the particle. In fact, Druger *et al.* found that an enhancement over conventional Förster transfer of 100–1000 was reasonable. It would be wrong to give the reader the impression that the calculation of Druger *et al.* is *ab initio*. A key piece of information is the photon lifetime τ_r of the longest lived mode. Fortunately, this lifetime can be estimated from the data in Figure 8.26.

The photon lifetime τ_r is related to the so-called quality factor Q for the mode. Let us suppose that there is no absorption in the particle. Under these circumstances, the photon lifetime

$$\tau_r = 1/\Gamma_r = Q_u/\omega_r \tag{8.24}$$

where Γ_r is the radiative decay rate, ω_r is the resonant frequency, and Q_u is the unloaded quality factor (i.e., the quality factor with no absorption). If absorption is added, we expect the photon lifetime to decrease due to processes which do not give rise to far-field radiation. We will call this process a "nonradiative" loss, with the associated rate Γ_{nr}. If Γ_{nr} can be estimated, then the probability for absorption is $P = \Gamma_{nr}/(\Gamma_{nr} + \Gamma_r)$. This is not the energy transfer probability, since one expects that the amount of energy coupled into a mode will be dependent on the concentration of acceptors. However, at low concentrations where $\Gamma_{nr}/\Gamma_r \ll 1$, one expects the coupling process to be virtually independent of absorption; the concentration dependence of P should mimic the concentration dependence of the energy transfer at low concentrations. Now, on to the task of estimating Γ_{nr}. We will take a simple classical point of view in which the loss per unit volume is given by $\underline{E} \cdot (\partial \underline{P}/\partial t)$, where P is the polarization density. Thus, the rate of loss of energy $d\varepsilon/dt$ is given by

$$\frac{d\varepsilon}{dt} = -\varepsilon\Gamma_{nr} = -\int_{sp} \left\langle \underline{E} \cdot \frac{\partial \underline{P}}{\partial t} \right\rangle_t dV$$

$$= -\frac{\omega\chi_i}{2} \int_{sp} \underline{E}^* \cdot \underline{E}\, dV \tag{8.25}$$

where the integrals are taken over the entire sphere, the bracket within the first integrand indicates that the dot product is to be time averaged, and χ_i is the imaginary part of the susceptibility and may be estimated from the molar extinction represented by the acceptor molecules. In order to calculate Γ_{nr}, we must estimate the energy in a given mode in terms of the internal field. Fortunately, for such high-Q resonances the time-averaged energy is equally distributed between magnetic and electric forms so that

$$\varepsilon \approx \int_{as} \langle \underline{D} \cdot \underline{E} \rangle_t\, dV = \frac{f\varepsilon_r}{2} \int_{sp} \underline{E}^* \cdot \underline{E}\, dV \tag{8.26}$$

where the first integral is taken over all of space and the second is taken only over the sphere, ε_r is the permittivity of the particle, and f is a fill factor. This fill factor f is the fraction of resonant energy contained within the sphere. For high-Q modes this factor is very nearly one (i.e., little of the total energy of such a resonance is contained within the evanescent field on the outside of the particle). Combining Eqs. (8.25) and (8.26), we find that $\Gamma_{nr} = f\chi_i\omega/\varepsilon_r$, and consequently

$$P = \frac{\Gamma_{nr}}{\Gamma_r + \Gamma_{nr}} = \frac{1}{1 + (\varepsilon_r/Q_u\chi_i f)} \tag{8.27}$$

P has a very suggestive form in relation to Figure 8.26. For a large concentration of acceptors, the second term in the denominator can be made considerably smaller than 1 (i.e., χ_i is proportional to acceptor concentration [A]), and P will be independent of concentration. On the other hand, for a small concentration of acceptors, the second term in the denominator can be made considerably larger than 1, and P will fall off linearly as the concentration is reduced. The scale factor in all of this is Q. With Q large, the transition from concentration independence to linear concentration dependence will be at low acceptor concentrations. P falls to $\frac{1}{2}$ when the second term in the denominator of Eq. (8.27) is equal to 1, and so a critical concentration of acceptors $[\mathrm{A}]_{1/2}$ can be defined to characterize the falloff. Expressing χ_i in terms of molecular parameters ($\chi_i = n\varepsilon_m[\mathrm{A}]\ln(10)/k$, where n is the particle refractive index, ε_m is the molar decadic extinction coefficient, [A] is the concentration of acceptors, and k is $2\pi/\lambda$) yields

$$[\mathrm{A}]_{1/2} = nk/f\varepsilon_m Q_u \ln(10) \tag{8.28}$$

The preceding analysis is valid in the region of acceptor concentration below that where bulk Förster transfer occurs.

The experimental concentration dependence shown in Figure 8.26 can thus be used to estimate the highest Q resonances participating in the microparticle energy transfer. The value obtained is 6×10^6, and this value compares favorably with the value obtained recently by Zhang *et al.*[47] who have estimated Q from time-resolved measurements. In these experiments the decay of both elastic and stimulated Raman scattering signals was measured following excitation with a 100-ps pulse of radiation.[47]

8.4. Conclusions

Morphology-dependent resonances are found in other structures of high symmetry,[48] and so their possible existence in biological systems should not be ignored. The presence of such resonances should alter the rates of decay of excited species. A considerable amount of work remains to be done on investigating the optical properties of small, regularly shaped objects. Continuing studies of fluorescence from such systems may lead to improved probes of the microscopic environment of molecules, remote sensing techniques, and surface probes to investigate both solid and liquid surfaces. The study of interactions higher intensity is almost sure to produce a host of interesting effects. The work described in this chapter will hopefully introduce a reader to the field, and we have endeavored to make the citations as current as possible in order to allow the interested reader to follow subsequent developments.

Acknowledgments

We would like to acknowledge the support of the National Science Foundation and the Chemical Research and Engineering Development Center of the Army (ATM-89-05871).

References

1. R. E. Benner, P. W. Barber, J. F. Owen, and R. K. Chang, Observation of structure resonances in the fluorescence of microspheres, *Phys. Rev. Lett. 44*, 475–478 (1980).
2. L. M. Folan, S. Arnold, and S. D. Druger, Enhanced energy transfer within a microparticle, *Chem. Phys. Lett. 118*, 322–327 (1985).
3. S. Arnold and L. M. Folan, Fluorescence spectrometer for a single electrodynamically levitated microparticle, *Rev. Sci. Instrum 57*, 2250–2253 (1986).
4. L. M. Folan and S. Arnold, Determination of molecular orientation at the surface of an aerosol particle by morphology-dependent photoselection, *Opt. Lett. 13*, 1–3 (1988).

5. D. Axelrod, Carbocyanine dye orientation in red cell membrane studied by microscopic fluorescence polarization, *Biophys. J. 26*, 557–573 (1979).
6. P. R. Conwell, C. K. Rushforth, R. E. Benner, and S. C. Hill, Efficient automated algorithm for the sizing of dielectric microspheres using the resonance spectrum, *J. Opt. Soc. Am. A 1*, 1181–1187 (1984).
7. G. Mie, Contributions to the optics of turbid media, especially colloidal suspensions of metals, *Ann. Physik 25*, 377–445 (1908).
8. C. F. Bohren and D. R. Huffman, *Absorption and Scattering of Light by Small Particles*, Chapter 4, Wiley Interscience, New York (1983).
9. B. J. Messinger, K. Ulrich von Raben, R. K. Chang, and P. W. Barber, Local fields at the surface of noble-metal microspheres, *Phys. Rev. B 24*, 649–657 (1981).
10. N. L. Thompson, H. M. McConnell, and Thomas P. Burghardt, Order in supported phospholipid monolayers detected by dichroism or fluorescence excited with polarized evanescent illumination, *Biophys. J. 46*, 739–747 (1984).
11. S. Earnshaw, On the nature of the molecular forces which regulate the constitution of the luminiferious ether, *Trans. Cambridge Phil. Soc. 7*, 97–112 (1842).
12. S. Arnold, Spectroscopy of single levitated micron sized particles, in: *Optical Effects Associated with Small Particles* (P. W. Barber and R. K. Chang, eds.), World Scientific, New York (1988).
13. W. Nauhauser, M. Hohenstatt, P. Toschek, and H. Dehmelt, Localized visible Ba^+ mono-ion oscillator, *Phys. Rev. A 22*, 1137 (1980).
14. S. Arnold and N. Hessel, Photoemission from single electrodynamically levitated microparticles, *Rev. Sci. Instrum. 56*, 2066–2069 (1985).
15. A. Ashkin, J. M. Dziedzic, J. E. Bjorkholm, and S. Chu, Observation of a single-beam gradient force optical trap for dielectric particles, *Opt. Lett. 11*, 288–290 (1986).
16. A. Ashkin and J. M. Dziedzic, Optical trapping and manipulation of viruses and bacteria, *Science 235*, 1517–1520 (1987).
17. T. N. Buican, M. J. Smyth, H. A. Crissman, G. C. Salzman, C. C. Stewart, and J. C. Martin, Automated single-cell manipulation and sorting by light trapping, *Appl. Opt. 26*, 5311–5316 (1987).
18. S. Arnold and L. M. Folan, Spherical void electrodynamical levitator, *Rev. Sci. Instrum. 58*, 1732–1735 (1987).
19. E. L. Kyser, L. F. Collins, and N. Herbert, Design of an impulse ink jet, *J. Appl. Photogr. Eng. 7*, 73–79 (1981).
20. P. Chylek, V. Ramaswamy, A. Ashkin, and J. M. Dziedzic, Simultaneous determination of refractive index and size of spherical dielectric particles from light scattering data, *Appl. Opt. 22*, 2302–2307 (1983).
21. A. Ruaudel-Teixier and M. Vandevyver, Energy transfer in dye monomolecular layers, *Thin Solid Films 68*, 129–133 (1980).
22. J. I. Gersten and A. Nitzan. Spectroscopic properties of molecules interacting with small dielectric particles, *J. Chem. Phys. 75*, 1139–1152 (1981).
23. J. I. Gersten and A. Nitzan, Accelerated energy transfer between molecules near a solid particle, *Chem. Phys. Lett. 104*, 31–37 (1984).
24. X. M. Hua, J. I. Gersten, and A. Nitzan, Theory of energy transfer between molecules near solid state particles, *J. Chem. Phys. 83*, 3650–3659 (1985).
25. M. Kerker, D.-S. Wang, and H. Chew, Surface enhanced Raman scattering (SERS) by molecules absorbed at spherical particles: errata, *Appl. Opt. 19*, 4159–4174 (1980).
26. P. J. McNulty, H. Chew, and M. Kerker, in: *Aerosol Microphysics I* (W. H. Marlow, ed.), Chapter 4, Springer-Verlag, New York (1980).
27. R. Ruppin, Decay of an excited molecule near a small metal sphere, *J. Chem. Phys. 76*, 1681–1684 (1982).

28. H. Chew, Transition rates of atoms near spherical surfaces, *J. Chem. Phys. 87*, 1355–1360 (1987).
29. S. D. Druger and P. J. McNulty, Radiation pattern of fluorescence from molecules embedded in small particles: General case, *Appl. Opt. 22*, 75–82 (1983).
30. S. D. Druger, S. Arnold, and L. M. Folan, Theory of enhanced energy transfer between molecules embedded in spherical dielectric particles, *J. Chem. Phys. 87*, 2649–2659 (1987).
31. R. R. Chance, A. Prock, and R. Silby, Molecular fluorescence and energy transfer near surfaces, in: *Advances in Chemical Physics*, Vol. XXXVII (I. Prigogine and S. A. Rice, eds.), pp. 1–65, Wiley, New York (1978).
32. D. A. Weitz, S. Garoff, C. D. Hanson, T. J. Gramila, and J. I. Gersten, Fluorescent lifetimes of molecules on silver island films, *Opt. Lett. 7*, 89 (1982).
33. H. M. Lai, P. T. Leung, and K. Young, Electromagnetic decay rates into narrow resonances in an optical cavity, *Phys. Rev. A 37*, 1597 (1988).
34. H.-M. Tzeng, K. F. Wall, M. B. Long, and R. K. Chang, Laser emission from individual droplets at wavelengths corresponding to morphology-dependent resonances, *Opt. Lett. 9*, 499–501 (1984).
35. J. B. Snow, S.-X. Qian, and R. K. Chang, Stimulated Raman scattering from individual water and ethanol droplets at morphology-dependent resonances, *Opt. Lett. 10*, 37–39 (1985).
36. S.-X. Qian, J. B. Snow, and R. K. Chang, Coherent Raman mixing and coherent anti-Stokes Raman scattering from individual micrometer-sized droplets, *Opt. Lett. 10*, 499–501 (1985).
37. S. Arnold, K. M. Leung, and A. B. Pluchino, Optical bistability of an aerosol particle, *Opt. Lett. 11*, 800–802 (1986).
38. J. R. Lakowicz, private communication.
39. H. Chew and D.-S. Wang, Double resonance in fluorescent and Raman scattering by molecules in small particles, *Phys. Rev. Lett. 49*, 490–492 (1982).
40. T. Förster, Intermolecular energy transfer and fluorescence, *Ann. Physik. 2*, 55–75 (1948).
41. D. L. Dexter, A theory of sensitized luminescence in solids, *J. Chem. Phys. 21*, 836–850 (1953).
42. J. Perrin, Fluorescence and molecular induction by resonance, *C. R. Acad. Sci. 184*, 1097–1100 (1927).
43. V. M. Agranovich and M. D. Galanin, *Electronic Excitation Energy Transfer in Condensed Matter*, Chapter 2, North-Holland, New York (1982).
44. J. B. Birks, *Photophysics of Aromatic Molecules*, pp. 567–576, Wiley, London (1970).
45. L. Stryer and R. P. Haugland, Energy transfer: a spectroscopic ruler, *Proc. Natl. Acad. Sci. U.S.A. 58*, 719–726 (1967).
46. J. R. Lakowicz, *Principles of Fluorescence Spectroscopy*, Chapter 10, Plenum, New York (1983).
47. J.-Z. Zhang, D. H. Leach, and R. K. Chang, Photon lifetime within a droplet: Temporal determination of elastic and stimulated Raman scattering, *Opt. Lett. 13*, 270–272 (1988).
48. P. W. Barber, J. F. Owen, and R. K. Chang, Resonant scattering for characterization of axisymmetric dielectric objects, *IEEE Trans. Antennas Propagation, AP-30*, 168–172 (1982).

Index

Italic numbers indicate a detailed description of the topic.

MIX
Papier aus verantwortungsvollen Quellen
Paper from responsible sources
FSC® C105338

If you have any concerns about our products,
you can contact us on
ProductSafety@springernature.com

In case Publisher is established outside the EU,
the EU authorized representative is:
Springer Nature Customer Service Center GmbH
Europaplatz 3, 69115 Heidelberg, Germany

Printed by Libri Plureos GmbH
in Hamburg, Germany